P9-DBQ-576

A
Comprehensive Introduction
to
DIFFERENTIAL GEOMETRY

VOLUME FIVE

A Comprehensive Introduction to DIFFERENTIAL GEOMETRY

VOLUME FIVE
Second Edition

MICHAEL SPIVAK

Publish or Perish, Inc.
Houston, Texas

ISBN 0-914098-88-8
Cloth Set, Vols. 3,4,5 ISBN 0-914098-82-9
Cloth Set, Vols. 1-5 ISBN 0-914098-83-7
Library of Congress Card Catalog Number 78-71771

PUBLISH OR PERISH, INC.
HOUSTON, TEXAS (U.S.A.)

In Japan distributed exclusively by
KINOKUNIYA COMPANY LTD
TOKYO, JAPAN

Printed in the United States of America

CONTENTS

[Although the Chapters are not divided into sections, except for the major subdivisions of Chapters 10 and 13, the list for each Chapter gives some indication which topics are treated, and on which pages.]

Contents of Volume III

Contents of Volume IV

A
Comprehensive Introduction *to* DIFFERENTIAL GEOMETRY

VOLUME FIVE

Chapter 10. And Now a Brief Message From our Sponsor

Partial differential equations have played a decisive role in our investigations ever since they were first introduced in Chapter 6 of Volume I. To be sure, at times we have suppressed the equations themselves in favor of a more geometric conception involving k-dimensional distributions, and on other occasions we have instead expressed things in terms of differential forms. But, in one form or another, the Frobenius Theorem (which represents everything we know about partial differential equations) was used in discussing Lie groups, ordinary and affine theory of curves and surfaces in space (where Lie group methods were used), in all our proofs of the Test Case, in the proof of the Fundamental Theorem of Surface Theory, and in the generalizations of this theorem which were given in Chapter 7. The partial differential equations involved are of the form

$$\frac{\partial\alpha^i}{\partial x_j}(x_1,\dots,x_m) = f^i_j(x_1,\dots,x_m,\ \alpha^1(x_1,\dots,x_m),\dots,\alpha^n(x_1,\dots,x_m))$$

$$i = 1,\dots,n\ ; \qquad j = 1,\dots,m\ .$$

Now it's really rather laughable to call these things partial differential equations at all. True, we are considering functions α^i defined on $\mathbb{R}^n$, and therefore partial derivatives are involved, but the equations do not posit any relationship between <u>different</u> partial derivatives; this comes out quite clearly in the proof, where the equations are reduced to ordinary differential equations. The only reason we get anything interesting at all in this situation is because we are dealing with a <u>system</u> of equations, and this system is "overdetermined": there are more equations (namely mn) than there are unknown functions (namely n). Our particular overdetermined system happens to be one where it is not too hard to determine the additional "integrability conditions" which must hold for the functions f^i_j if the strain of satisfying so many equations is not to hopelessly overburden the poor functions α^i.

With only this superficial knowledge of partial differential equations, one can make his way through a good part of differential geometry ("the good part", you may be inclined to say after looking at this chapter). But there are some topics in differential geometry, to be covered in the next two chapters, where a more intimate acquaintanceship with partial differential equations is required. I should say right away, that even in the next two chapters there are only a few occasions where this knowledge is necessary, and one could easily decide to take on faith any theorems from this chapter which happen to be quoted later. On the other hand, many theorems cannot even be stated without some definitions that arise in the first attempts to understand partial differential equations; these definitions involve basic facts about the behavior of partial differential equations, and this behavior is often reflected in geometric phenomena in a surprisingly nice way.

This chapter is not meant to be a substitute for a course in partial differential equations; we will try to reach in as short a space as possible those particular properties of partial differential equations which will be of importance to us in the next two chapters, even if they are of only secondary importance to analysis. Consequently, we will omit much material that is contained in elementary courses, and at the same time prove special cases of results which are usually found only in more advanced treatments, where they are proved in much greater generality, and with much more effort. (Just to keep the presentation from being too one-sided, I have sometimes given passing mention to matters which are of great importance to analysts, but of no importance to us). Since we are going to be totally immersed in the study of partial differential equations for quite a while, we might as well admit it, and henceforth resort to the standard abbreviation PDE.

A few general considerations might be made before we begin in earnest. When we consider an ordinary differential equation

$$u'(x) = f(x, u(x)) ,$$

we find that there are solutions u with any desired value for $u(x_0)$. This dependence on the "initial condition" $u(x_0)$ usually manifests itself, if we explicitly solve the equation, by the presence of an arbitrary constant of integration. For example, the equation

$$\frac{du}{dx} = -u^2 , \qquad \left(\Rightarrow \frac{du}{-u^2} = dx \Rightarrow \frac{1}{u} = x + C\right)$$

has the "general" solution

$$u(x) = \frac{1}{x+C} ,$$

which gives all desired initial conditions $u(x_0)$ except $u(x_0) = 0$; for this one needs the "singular" solution $u(x) = 0$. Equations of order n, on the other hand, will involve n constants of integration.

When we solve a PDE, we usually obtain arbitrary <u>functions</u> in the answer. For example, to be as simple-minded about the thing as we can, we note that the equation

$$\frac{\partial u}{\partial y}(x,y) = 0$$

has the solutions $u(x,y) = A(x)$; the only restrictions on A are ones which follow from restrictions we might choose to place on u (e.g., that u be differentiable with respect to x). The equally stupid looking, but actually quite important, second order equation

$$\frac{\partial^2 u}{\partial x \partial y}(x,y) = 0$$

leads to

$$\frac{\partial u}{\partial x}(x,y) = \alpha(x) ,$$

and hence to

$$u(x,y) = A(x) + B(y) , \qquad A'(x) = \alpha(x) .$$

Without belaboring the point any further, we simply note that when we look for precise theorems, we should expect the hypotheses to reflect the presence of these "arbitrary functions" in the same way that the precise theorem for ordinary differential equations reflects the presence of arbitrary constants.

1. First Order PDEs

In this section we will consider those equations which involve a function u on $\mathbb{R}^n$ and only its first partial derivatives u_{x_i}. For simplicity of writing, and convenience of visualization, we will first deal exclusively with the case of $\mathbb{R}^2$, denoting a typical point of $\mathbb{R}^2$ by (x,y) and adopting the standard notation

$$u_x = p , \qquad u_y = q .$$

By a <u>first order PDE</u> we then mean an equation of the form

$$F(x,\ y,\ u(x,y),\ u_x(x,y),\ u_y(x,y)) = 0\ ,$$

or, to use the standard abbreviated form,

$$F(x,y,u,p,q) = 0\ .$$

It will be convenient to denote the various partial derivatives of F by F_x, F_y, F_u, F_p, and F_q. Naturally, the function $F\colon \mathbb{R}^5 \to \mathbb{R}$ shouldn't be too badly behaved; for example, it wouldn't be very interesting if F were never 0. Just what hypotheses we really need will come out soon enough. To begin with, we might imagine that F is differentiable and satisfies $F_p \neq 0$ or $F_q \neq 0$, so that by the implicit function theorem we can solve for p in terms of q, or <u>vice versa</u>. Our main result is, that we can always completely reduce any first order PDE to a system of ordinary differential equations. This holds both in a "practical" and in a theoretical sense: We can actually write down a system of ordinary differential equations whose solutions, if we can find them, will give us the solution of our original problem; and the method by which this is done enables us to state and prove exact theorems. We will not deal at the very outset with the most general first order PDE, but will approach it in stages.

We consider first the most general <u>linear first order PDE</u>

$$(1)\qquad A(x,y)u_x(x,y)\ +\ B(x,y)u_y(x,y) = C(x,y)u(x,y)\ +\ D(x,y)\ .$$

Usually this is simply written

$$A(x,y)u_x + B(x,y)u_y = C(x,y)u + D(x,y)\ ,$$

with the arguments (x,y) appearing in A, B, C, and D just to emphasize that we are not considering an equation like $A(x,y,u(x,y))u_x + \cdots$.

Consider the vector field X on $\mathbb{R}^2$ defined by

$$X = A\frac{\partial}{\partial x} + B\frac{\partial}{\partial y} . \tag{2}$$

The value of X at (x_0,y_0) is

$$A(x_0,y_0)\frac{\partial}{\partial x}\bigg|_{(x_0,y_0)} + B(x_0,y_0)\frac{\partial}{\partial y}\bigg|_{(x_0,y_0)} ;$$

using the standard identification of the tangent space $\mathbb{R}^2_{(x_0,y_0)}$ with $\mathbb{R}^2$, we can also write

$$X(x_0,y_0) = (A(x_0,y_0),\ B(x_0,y_0)) .$$

We will call X the <u>characteristic vector field</u> of equation (1); the integral curves of this vector field are called the <u>characteristic curves</u> of equation (1). Thus $c = (c_1,c_2)$ is a characteristic curve if and only if

$$\frac{dc_1(t)}{dt} = A(c(t)) , \qquad \frac{dc_2(t)}{dt} = B(c(t)) . \tag{3}$$

We then have, for any C^1 function $u\colon \mathbb{R}^2 \to \mathbb{R}$,

$$\begin{aligned}\frac{du(c(t))}{dt} &= u_x(c(t))\frac{dc_1(t)}{dt} + u_y(c(t))\frac{dc_2(t)}{dt}\\ &= A(c(t))\cdot u_x(c(t)) + B(c(t))\cdot u_y(c(t)) .\end{aligned}$$

So any solution u of equation (1) satisfies

$$\text{(4)} \qquad \frac{du(c(t))}{dt} = C(c(t))\cdot u(c(t)) + D(c(t)) \qquad \text{for any characteristic curve } c.$$

For any fixed characteristic curve $t \mapsto c(t)$, equation (4) is an ordinary differential equation for the function $u \circ c$. Consequently, $u \circ c$ is uniquely

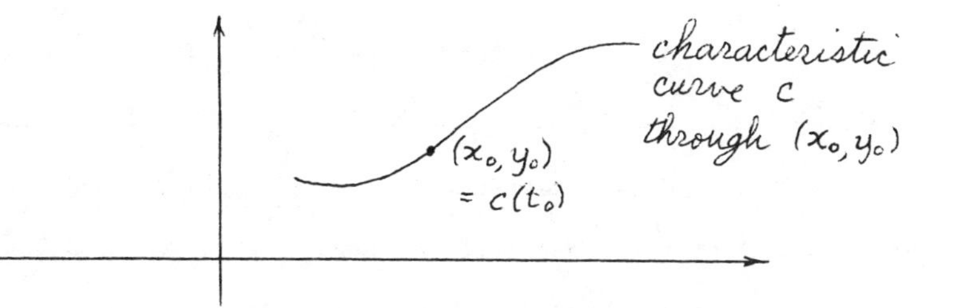

determined once $u(c(t_0))$ is specified. In other words, once we prescribe a value $u(x_0, y_0)$ for a solution u of equation (1), the solution u will then be completely determined along the characteristic curve c through (x_0, y_0). Now suppose we have any curve σ which cuts a family of characteristic curves.

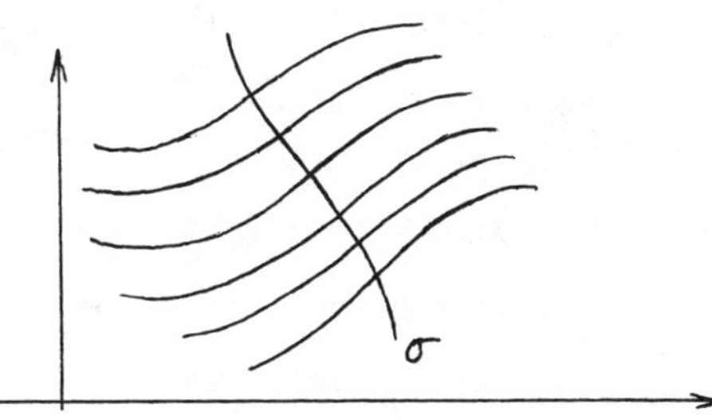

If we arbitrarily specify the values of u at each point of σ, then the solution u will be determined in a neighborhood of σ. Moreover, we ought to be able to produce this solution u simply by solving equation (4) for each of the characteristic curves through each point of σ. Of course, we clearly have to rule out the possibility that a portion of σ itself is a characteristic curve, for then we could not arbitrarily specify the values of u along σ. We even have to rule out the possibility that σ is tangent to some integral curve c

at some point $(x_0,y_0) = c(t_0)$; for in this case, the directional derivative

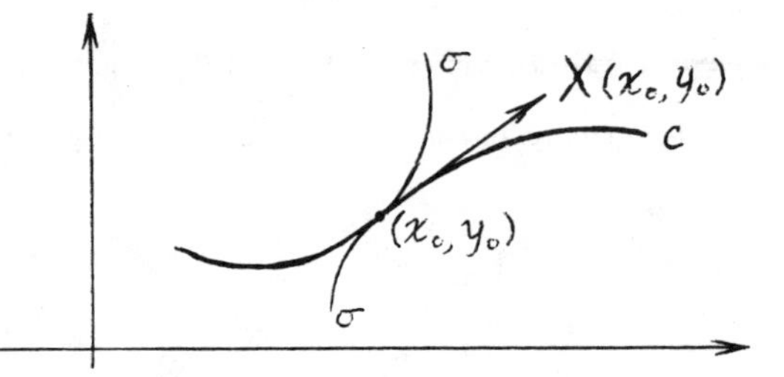

$X(x_0,y_0)(u)$ would be determined both by equation (4) and (in a possibly conflicting way) by the arbitrarily assigned values of u along σ. We must thus assume that the vectors

$$\sigma'(s) = (\sigma_1'(s),\sigma_2'(s) \qquad \text{and} \qquad (A(\sigma(s)),B(\sigma(s)))$$

are always linearly independent. Equivalently, we must require that

$$0 \neq \det\begin{pmatrix} \sigma_1'(s) & A(\sigma(s)) \\ \sigma_2'(s) & B(\sigma(s)) \end{pmatrix} = \sigma_1'(s)B(\sigma(s)) - \sigma_2'(s)A(\sigma(s))$$

for all s. In particular, $\sigma'(s) \neq (0,0)$, so σ is an imbedding. Although we will later have a much more general result, we summarize this information in a theorem, in order to get all the details cleaned up before we carry the discussion any further.

1. THEOREM. Let A, B, C, and D be C^k functions defined in an open set $U \subset \mathbb{R}^2$, and let $\sigma\colon [a,b] \to U$ be a one-one C^k curve such that

$$\sigma_1'(s)B(\sigma(s)) \neq \sigma_2'(s)A(\sigma(s)) \qquad \text{for all } s \in [a,b] .$$

Let $\mathring{u}\colon [a,b] \to \mathbb{R}$ be a C^k function. Then there is a C^k function u, defined in a neighborhood V of $\sigma([a,b])$, such that u satisfies

$$(1) \qquad A\cdot u_x + B\cdot u_y = C\cdot u + D \qquad \text{on } V ,$$

with the initial condition

$$u(\sigma(s)) = \mathring{u}(s) \qquad \text{for all } s \in [a,b] .$$

Moreover, any two functions u with this property agree on a neighborhood of $\sigma([a,b])$.

Proof. There is a C^k map

$$\gamma\colon [a,b] \times (-\varepsilon,\varepsilon) \longrightarrow U$$

such that each curve

$$t \mapsto \gamma(s,t)$$

is a characteristic curve with

$$\gamma(s,0) = \sigma(s) .$$

Clearly

$$\frac{\partial\gamma}{\partial s}(s,0) = \sigma'(s) = (\sigma_1'(s),\sigma_2'(s))$$

$$\frac{\partial\gamma}{\partial t}(s,0) = (A(\sigma(s)),B(\sigma(s))) .$$

So, by the hypothesis on σ, the Jacobian of γ at $(s,0)$ is always non-singular; consequently, if ε is sufficiently small, then γ is a C^k diffeomorphism onto a neighborhood V of $\sigma([a,b])$.

By choosing ε still smaller, if necessary, we can insure that for each $s \in [a,b]$ there is a C^k function $\beta_s\colon (-\varepsilon,\varepsilon) \to \mathbb{R}$ satisfying

$$\begin{cases} \dfrac{d\beta_s(t)}{dt} = C(\gamma(s,t))\cdot\beta_s(t) + D(\gamma(s,t)) \\ \beta_s(0) = \mathring{u}(s) \end{cases}$$

[this is just the equation (4) which should be satisfied by $u \circ c$ along the integral curve $t \mapsto \gamma(s,t)$]. We would actually like to know that $\beta_s(t)$ is C^k as a function of s and t; in other words, if we define $\beta\colon [a,b] \times (-\varepsilon,\varepsilon) \to \mathbb{R}$ by

$$\beta(s,t) = \beta_s(t) ,$$

then we would like to know that β is C^k. To prove this, we must consider the equation "depending on parameters"

$$\begin{cases} \alpha(0,s,r) = r \qquad \text{for } r \in \mathbb{R} \\ \dfrac{\partial}{\partial t}\alpha(t,s,r) = C(\gamma(s,t))\cdot\alpha(t,s,r) + D(\gamma(s,t)) . \end{cases}$$

Problem I.5-5 shows that α is C^k; consequently

$$\beta(s,t) = \alpha(t,s,\mathring{u}(s))$$

is also C^k.

Now the solution u, if it exists, clearly must be the C^k function

$$u(x,y) = \beta(\gamma^{-1}(x,y)) \qquad \text{or equivalently} \qquad u(\gamma(s,t)) = \beta(s,t) .$$

To prove that u really is a solution, we note that through any point

$(x,y) \in V$ there is a characteristic curve $t \mapsto \gamma(s,t)$, and that

$$\frac{du(\gamma(s,t))}{dt} = \frac{d\beta(s,t)}{dt} = C(\gamma(s,t))\cdot\beta(s,t) + D(\gamma(s,t))$$

$$= C(\gamma(s,t))\cdot u(\gamma(s,t)) + D(\gamma(s,t)) ,$$

while we also have

$$\frac{du(\gamma(s,t))}{dt} = u_x(\gamma(s,t))\cdot\frac{\partial\gamma_1}{\partial t}(s,t) + u_y(\gamma(s,t))\cdot\frac{\partial\gamma_2}{\partial t}(s,t)$$

$$= u_x(\gamma(s,t))\cdot A(\gamma(s,t)) + u_y(\gamma(s,t))\cdot B(\gamma(s,t)) ,$$

since $t \mapsto \gamma(s,t)$ is a characteristic curve. ■

Notice that Theorem 1 involves exactly the sort of "arbitrary function" that our general considerations would lead us to expect: in a neighborhood of the "initial curve" σ, the solution u is uniquely determined by the "initial condition" $u(\sigma(s)) = \mathring{u}(s)$. The only requirement is that σ be nowhere tangent to a characteristic curve; we will express this by saying that σ is **free** (sometimes the term "non characteristic" is used, but this seems a little misleading). In general, the problem of finding a solution of a PDE with an appropriate initial condition is called the "Cauchy problem" for this equation. Thus we have solved the Cauchy problem for the linear PDE (1) for any initial condition along any free curve. In particular, we can solve the Cauchy problem along the x-axis $\sigma(s) = (s,0)$ if the x-axis is free, which is equivalent to the condition that $B \neq 0$ along the x-axis. In this case we can use the given equation (1) to solve for u_y in terms of u_x along the x-axis:

$$u_y = -\frac{A}{B}u_x + \frac{C}{B}u + \frac{D}{B}.$$

If we were interested in the Cauchy problem only along the x-axis, then we could simply demand this very natural condition in our hypotheses, and not mention the characteristic curves at all; but the characteristic curves are still the most important ingredient in the proof, and their generalizations will play decisive roles in all other equations we discuss.

If our initial curve σ actually happens to be a characteristic curve (thus failing in the worst possible way to be free), then we will be unable to solve the Cauchy problem, and this inability will be manifested in the worst possible way: the possible initial condition along σ is almost uniquely determined -- it is determined by the value at only one point, by the equation (4). On the other hand, if we are given an initial condition $\overset{\circ}{u}$ along σ which does satisfy (4), then there will be infinitely many solutions u with this initial condition; for we can consider any free curve ρ with $\rho(0) = \sigma(0)$, and choose any initial data ϕ along ρ with $\phi(0) = \overset{\circ}{u}(0)$.

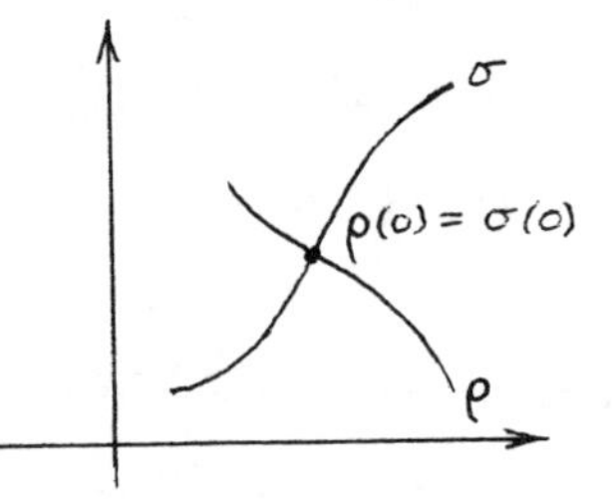

Thus, the characteristic curves are the places where different solutions agree.

From Theorem 1 we can see immediately that an arbitrary linear first order PDE has, in common with the simple-minded equation $\partial u/\partial y = 0$, a property which sharply distinguishes it from an ordinary differential equation

$$u'(x) = f(x,u(x)) .$$

For the ordinary differential equation, any solution u will clearly be at least one time more differentiable than f is, and if f is analytic, the solution will also be analytic (Problem I.6-9). But there are solutions of the equation in Theorem 1 which are only C^ℓ $(1 \le \ell \le \infty)$ even when A, B, C, D are C^k $(\ell < k \le \omega)$. For we may choose σ to be a C^k curve and $\overset{\circ}{u}$ to be a function which is C^ℓ, but not $C^{\ell+1}$; then the solution u cannot be $C^{\ell+1}$, since its restriction to the C^k curve σ is not $C^{\ell+1}$.

We next consider the most general quasi-linear first order PDE

$$(1) \quad A(x,y,u(x,y))u_x(x,y) + B(x,y,u(x,y))u_y(x,y) = C(x,y,u(x,y)) ,$$

or, more briefly,

$$A(x,y,u)u_x + B(x,y,u)u_y = C(x,y,u) .$$

The functions A, B, and C are now defined on $\mathbb{R}^3$, and we consider the vector field X in $\mathbb{R}^3$ defined by

$$(2) \qquad X = A\frac{\partial}{\partial x} + B\frac{\partial}{\partial y} + C\frac{\partial}{\partial z} .$$

This vector field will be called the characteristic vector field of equation (1); the integral curves of X are called the characteristic curves of equation (1). Thus $c = (c_1,c_2,c_3)$ is a characteristic curve if and only if

$$(3) \quad \frac{dc_1(t)}{dt} = A(c(t)) , \qquad \frac{dc_2(t)}{dt} = B(c(t)) , \qquad \frac{dc_3(t)}{dt} = C(c(t)) .$$

The slight discrepancy between this terminology and that adopted in the linear case is easily explained. Notice that if A and B depend only on x and y, then all characteristic vectors $X(x_0,y_0,z_0)$ have the same projection on the (x,y)-plane, namely $(A(x_0,y_0),B(x_0,y_0))$. So the characteristic curves of a linear equation are really the projections on the (x,y)-plane of the characteristic curves in $\mathbb{R}^3$.

For the quasi-linear PDE (1), the characteristic curves in $\mathbb{R}^3$ have the following significance. Any C^1 function $u:\mathbb{R}^2 \to \mathbb{R}$ determines a surface $M_u = \{(x,y,u(x,y))\} \subset \mathbb{R}^3$, and the vector

$$(u_x(x,y),\ u_y(x,y),\ -1)$$

is normal to M_u at $(x,y,u(x,y))$. Equation (1) is therefore equivalent to saying that $X(x,y,u(x,y))$ lies in the tangent space of M_u at $(x,y,u(x,y))$.

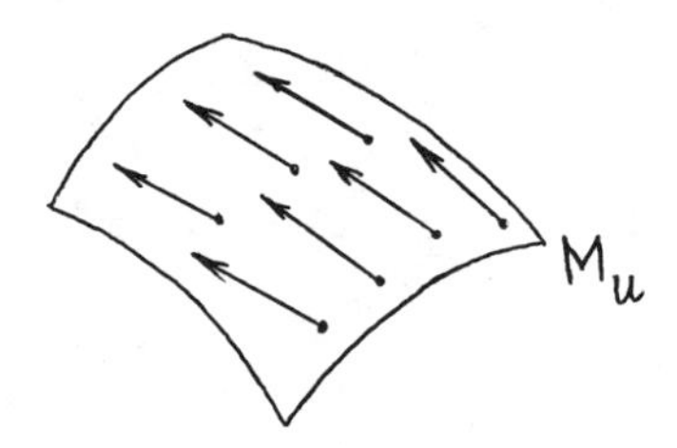

So the characteristic vectors at the various points of M_u give a vector field on M_u. Thus M_u is the union of integral curves of this vector field; that is, M_u is the union of characteristic curves. If we are given an arbitrary initial condition $\mathring{u}$ along an initial curve σ in $\mathbb{R}^2$, then we ought to be able to construct a solution u through the curve $s \mapsto (\sigma_1(s),\sigma_2(s),\mathring{u}(s))$ in $\mathbb{R}^3$ simply by taking the union of the characteristic curves through all the points of this curve. We will clearly have to require that the vectors

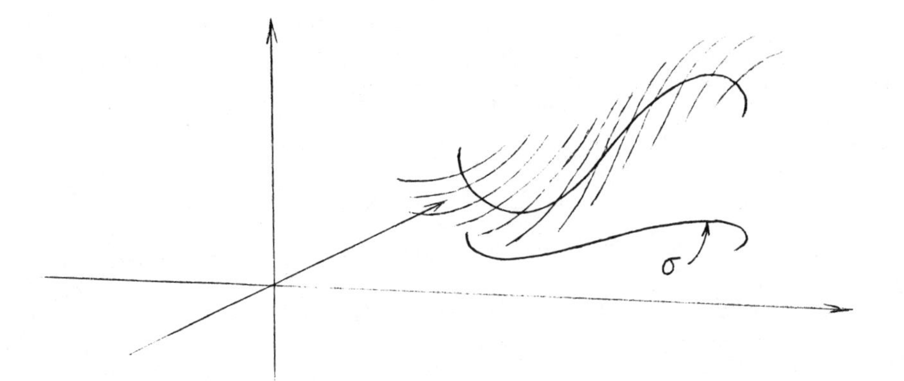

$(\sigma_1'(s),\sigma_2'(s))$ and $(A(\sigma_1(s),\sigma_2(s),\mathring{u}(s)),\ B(\sigma_1(s),\sigma_2(s),\mathring{u}(s)))$ are linearly independent for all s.

2. THEOREM. Let A, B, C be C^k functions defined in an open set $U \subset \mathbb{R}^3$. Let $\sigma\colon [a,b] \to \mathbb{R}^2$ be a one-one C^k function, and $\mathring{u}\colon [a,b] \to \mathbb{R}$ a C^k function such that $(\sigma_1(s),\sigma_2(s),\mathring{u}(s)) \in U$ for all $s \in [a,b]$. Suppose moreover that

$$\sigma_1'(s)\cdot B(\sigma_1(s),\sigma_2(s),\mathring{u}(s)) \neq \sigma_2'(s)\cdot A(\sigma_1(s),\sigma_2(s),\mathring{u}(s)) \qquad \text{for all } s \in [a,b] .$$

Then there is a C^k function u, defined in a neighborhood V of $\sigma([a,b])$, which satisfies the equation

$$(1) \qquad A(x,y,u)u_x + B(x,y,u)u_y = C(x,y,u) \qquad \text{on } V$$

with the initial condition

$$u(\sigma(s)) = \mathring{u}(s) \qquad \text{for all } s \in [a,b] .$$

Moreover, any two functions u with this property agree on a neighborhood of $\sigma([a,b])$.

<u>Proof</u>. By Problem I.5-5 there is a C^k function $\alpha = (\alpha_1,\alpha_2,\alpha_3)$ with

$$(*)\qquad \begin{cases} \alpha(0,s,r) = r & \text{for } r \in \mathbb{R}^3 \\ \dfrac{\partial}{\partial t}\alpha_1(t,s,r) = A(\alpha(t,s,r)) \\ \dfrac{\partial}{\partial t}\alpha_2(t,s,r) = B(\alpha(t,s,r)) \\ \dfrac{\partial}{\partial t}\alpha_3(t,s,r) = C(\alpha(t,s,r)) . \end{cases}$$

Let

$$\beta(s,t) = \alpha(t,s,\sigma_1(s),\sigma_2(s),\mathring{u}(s)) ,$$

so that β is also C^k. In particular,

$$\begin{aligned} \beta(s,0) &= (\sigma_1(s),\sigma_2(s),\mathring{u}(s)) \\ &= \bullet, \quad \text{for short} \end{aligned}$$

[so for each s, the curve $t \mapsto \beta(s,t)$ is a characteristic curve through $\bullet$]. If we define

$$\gamma(s,t) = (\beta_1(s,t),\beta_2(s,t)) \in \mathbb{R}^2 ,$$

then the Jacobian of γ at $(s,0)$ is

$$\begin{aligned} \begin{pmatrix} \dfrac{\partial\beta_1}{\partial s}(s,0) , & \dfrac{\partial\beta_1}{\partial t}(s,0) \\ \dfrac{\partial\beta_2}{\partial s}(s,0) , & \dfrac{\partial\beta_2}{\partial t}(s,0) \end{pmatrix} &= \begin{pmatrix} \sigma_1'(s) , & \dfrac{\partial\alpha_1}{\partial t}(0,s,\bullet) \\ \sigma_2'(s) , & \dfrac{\partial\alpha_2}{\partial t}(0,s,\bullet) \end{pmatrix} \\ &= \begin{pmatrix} \sigma_1'(s) , & A(\bullet) \\ \sigma_2'(s) , & B(\bullet) \end{pmatrix} \qquad \text{by } (*), \end{aligned}$$

and this is non-singular, by hypothesis. So if ε is sufficiently small, then

$\gamma\colon [a,b]\times(-\varepsilon,\varepsilon)\to\mathbb{R}^2$ is a C^k diffeomorphism onto a neighborhood V of $\sigma([a,b])$.

The solution u, if it exists, clearly must be the C^k function

$$u(x,y) = \beta_3(\gamma^{-1}(x,y)) \qquad \text{or equivalently} \qquad u(\gamma(s,t)) = \beta_3(s,t) .$$

To prove that u is a solution, we note that for any point $(x,y)\in V$, there is a characteristic curve $t\mapsto\beta(s,t)$ through $(x,y,u(x,y))$, and that

$$\frac{du(\gamma(s,t))}{dt} = \frac{d\beta_3(s,t)}{dt} = C(\beta(s,t)) \qquad \text{by } (*) ,$$

while we also have

$$\begin{aligned}
\frac{du(\gamma(s,t))}{dt} &= u_x(\gamma(s,t))\cdot\frac{\partial\gamma_1}{\partial t}(s,t) + u_y(\gamma(s,t))\cdot\frac{\partial\gamma_2}{\partial t}(s,t) \\
&= u_x(\gamma(s,t))\cdot\frac{\partial\beta_1}{\partial t}(s,t) + u_y(\gamma(s,t))\cdot\frac{\partial\beta_2}{\partial t}(s,t) \qquad \text{by definition of } \gamma \\
&= u_x(\gamma(s,t))\cdot A(\beta(s,t)) + u_y(\gamma(s,t))\cdot B(\beta(s,t)) \qquad \text{by } (*). \ \blacksquare
\end{aligned}$$

We will say that the initial curve σ is <u>free for the initial condition</u> $\mathring{u}$ when it satisfies

$$\sigma_1'(s)\cdot B(\sigma_1(s),\sigma_2(s),\mathring{u}(s)) \neq \sigma_2'(s)\cdot A(\sigma_1(s),\sigma_2(s),\mathring{u}(s)) .$$

Thus we can solve the Cauchy problem for a quasi-linear PDE (1) for any initial condition along any curve which is free for this initial condition. (In the linear case things are simpler, since the condition that σ be free doesn't depend on the initial condition $\mathring{u}$.)

The worst way in which the initial curve $\sigma\colon [a,b] \to \mathbb{R}^2$ can fail to be free for the initial condition $\overset{\circ}{u}$ is when $\sigma'(s) = (\sigma_1'(s), \sigma_2'(s))$ and $(A(\sigma_1(s), \sigma_2(s), \overset{\circ}{u}(s)),\ B(\sigma_1(s), \sigma_2(s), \overset{\circ}{u}(s))) = (A(\bullet), B(\bullet))$ are everywhere linearly dependent. In this case, it is customary to say that σ is *characteristic* for $\overset{\circ}{u}$; this does *not* mean that σ is a characteristic curve (indeed, σ isn't even a curve in $\mathbb{R}^3$). If we assume that σ is an imbedding, then σ is characteristic if and only if $(A(\bullet), B(\bullet))$ is always a multiple of the tangent vector $\sigma'(s)$; by reparameterizing σ we can then arrange that

$$(A(\bullet), B(\bullet)) = \sigma'(s) .$$

Then if $\overset{\circ}{u}$ is to be the initial condition for a solution u of (1) we must have

$$\begin{aligned} C(\bullet) &= \sigma_1'(s)\cdot u_x(\sigma(s)) + \sigma_2'(s)\cdot u_y(\sigma(s)) \\ &= \frac{d}{ds}u(\sigma(s) = \frac{d}{ds}\overset{\circ}{u}(s) . \end{aligned}$$

These equations show that the reparameterized curve $s \mapsto (\sigma_1(s), \sigma_2(s), \overset{\circ}{u}(s))$ must be a characteristic curve; equivalently, the original curve $s \mapsto (\sigma_1(s), \sigma_2(s), \overset{\circ}{u}(s))$ must be a characteristic curve up to reparameterization in order for the Cauchy problem to be solvable when σ is characteristic for $\overset{\circ}{u}$. If our initial condition $\overset{\circ}{u}$ does have this property, then there will be infinitely many solutions u with this initial condition along σ. The characteristic curves in $\mathbb{R}^3$ are the places where the graphs of different solutions intersect; the projections of the characteristic curves onto $\mathbb{R}^2$ are the places where different solutions agree.

It should be clear once again that a quasi-linear first order PDE has solutions which are less differentiable than its coefficients.

We are now ready to consider the most general first order PDE

$$(1) \qquad F(x,y,u,p,q) = F(x,y,u(x,y),u_x(x,y),u_y(x,y)) = 0 .$$

This equation can also be reduced to a system of ordinary differential equations, but in this case the system will involve five functions; the geometric analysis will be correspondingly more complicated.

At each point $(x_0,y_0,z_0) \in \mathbb{R}^3$, we can consider the set of all vectors $(a,b,-1)$ with

$$F(x_0,y_0,z_0,a,b) = 0 ,$$

and the corresponding family $\mathcal{F}(x_0,y_0,z_0)$ of planes perpendicular to such vectors. If u is a solution of (1), and M_u is the surface

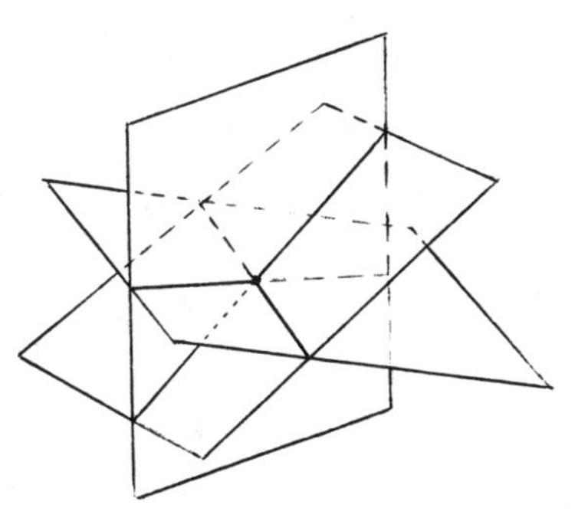

$M_u = \{(x,y,u(x,y))\}$, then the tangent space of M_u at $(x_0,y_0,u(x_0,y_0))$ is a member of the family $\mathcal{F}(x_0,y_0,u(x_0,y_0))$. In order to describe this situation more geometrically, we would like to have a more geometric way of describing the families $\mathcal{F}(x_0,y_0,z_0)$. Now the relation

$$F(x_0,y_0,z_0,a,b) = 0$$

is one equation in the two unknowns, a and b, so $\mathcal{F}(x_0,y_0,z_0)$ ought to be a one-parameter family of planes; this suggests that there is a cone $K(x_0,y_0,z_0)$, having its vertex at (x_0,y_0,z_0), with the property that a plane P is in $\mathcal{F}(x_0,y_0,z_0)$ if and only if P is tangent to $K(x_0,y_0,z_0)$ along a generator of this cone. If we consider a quasi-linear equation

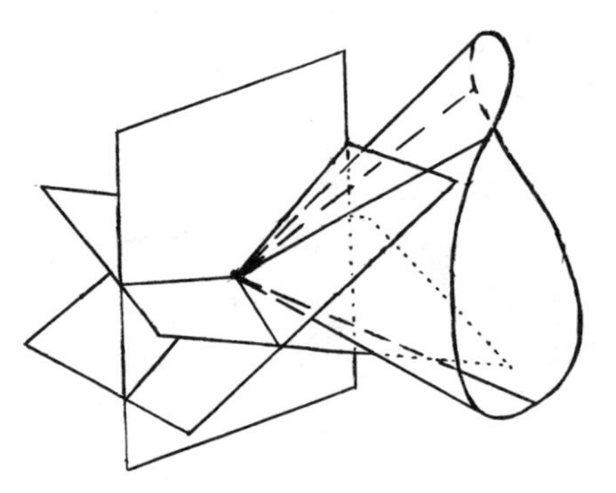

$$F(x,y,u,p,q) = A(x,y,u)\cdot p + B(x,y,u)\cdot q - C(x,y,u) = 0 ,$$

we immediately see that this is not always so. For in this case, the family $\mathcal{F}(x_0,y_0,z_0)$ consists of planes perpendicular to vectors $(a,b,-1)$ with

$$a\cdot A(x_0,y_0,z_0) + b\cdot B(x_0,y_0,z_0) = C(x_0,y_0,z_0) .$$

These planes all contain the characteristic vector

$$(A(x_0,y_0,z_0),\ B(x_0,y_0,z_0),\ C(x_0,y_0,z_0)) .$$

Thus our "cone" degenerates into a straight line through (x_0,y_0,z_0), pointing in the direction of the characteristic vector at that point. Clearly things

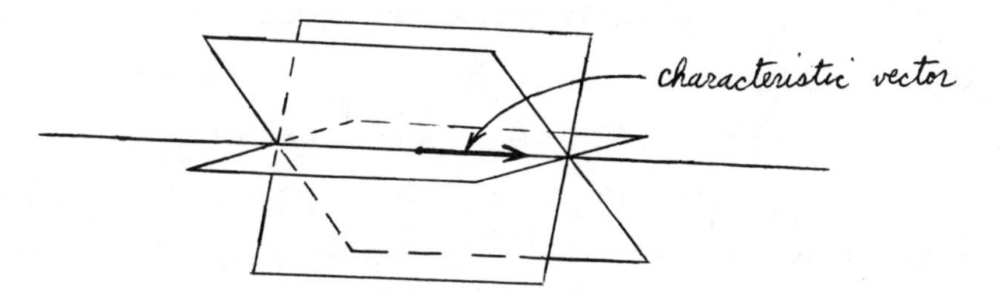

might be even messier if the analytic properties of the function F are sufficiently nasty.

Despite these difficulties, we can obtain a great deal of geometric motivation by temporarily pretending that each family $\mathcal{F}(x_0,y_0,z_0)$ _is_ determined by a cone $K(x_0,y_0,z_0)$, which happens to degenerate to a straight line in the case of a quasi-linear equation. This semi-mythical cone is called the **Monge cone** at (x_0,y_0,z_0). Having accepted this fiction, we can now imagine a field of cones in $\mathbb{R}^3$; a C^1 function $u\colon \mathbb{R}^2 \to \mathbb{R}$ is a solution of equation (1) if

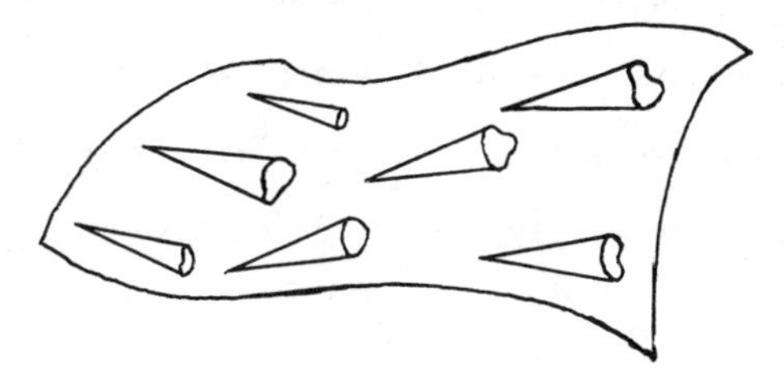

and only if the corresponding surface $M_u = \{(x,y,u(x,y))\}$ is tangent to the Monge cone $K(x_0,y_0,u(x_0,y_0))$ at each point $(x_0,y_0,u(x_0,y_0))$. This gives us a field of directions at each point of M_u, namely the direction which lies along a generator of the Monge cone at that point. The integral manifolds of this field of directions could be called the "characteristic curves of the solution u." This definition is easily seen to be compatible with the one already given in the quasi-linear case, where the Monge cones degenerate to straight lines. For these straight lines must be the field of directions for any solution u, and the "characteristic curves of the solution u" are simply those characteristic curves of the quasi-linear equation which happen to lie on M_u. But in the general case, we cannot write $\mathbb{R}^3$ as a disjoint union of curves in such a way that each M_u is the union of a certain subset of these curves; we cannot describe the "characteristic curves of a solution u" at all until we already know u. This might make the concept seem rather useless, but the requisite supplementary considerations will appear quite naturally when we seek an analytic description of these geometric pictures.

How would we go about finding an analytic description of the Monge cone? The Addendum to Chapter 3 suggests that the Monge cone $K(x_0,y_0,z_0)$ should be the "envelope" of the family of planes $\mathcal{F}(x_0,y_0,z_0)$; geometrically, the generators of $K(x_0,y_0,z_0)$ should be the limits of the intersections of two planes of the family $\mathcal{F}(x_0,y_0,z_0)$, the limit being formed as the two planes approach each other. Until we explicitly say the opposite, everything we now do will be based on the assumption that these limits really exist; the ensuing discussion is consequently merely a route to discovery, and does not purport to prove anything.

Let us assume for the moment that the equation

$$F(x_0,y_0,z_0,a,b) = 0$$

can be solved for b in terms of a. In other words, assume there is a function ϕ with

$$\text{(i)} \qquad F(x_0,y_0,z_0,a,\phi(a)) = 0 .$$

One plane of the family $\mathcal{F}(x_0,y_0,z_0)$ may be described by the equation

$$z - z_0 = a(x - x_0) + \phi(a)(y - y_0) .$$

A nearby plane may be described by the equation

$$z - z_0 = (a+h)(x - x_0) + \phi(a+h)(y - y_0) .$$

The points (x,y,z) in the intersection then satisfy

$$0 = h(x - x_0) + [\phi(a+h) - \phi(a)](y - y_0) ,$$

and hence

$$0 = (x - x_0) + \left[\frac{\phi(a+h) - \phi(a)}{h}\right](y - y_0) .$$

Therefore points in the limiting intersection ought to satisfy

$$\text{(ii)} \qquad \begin{cases} z - z_0 = a(x - x_0) + \phi(a)(y - y_0) \\ 0 = (x - x_0) + \phi'(a)(y - y_0) . \end{cases}$$

On the other hand, equation (i) shows that

$$\begin{aligned} 0 &= \frac{d}{da}F(x_0,y_0,z_0,a,\phi(a)) \\ &= F_p(x_0,y_0,z_0,a,\phi(a)) + \phi'(a)\cdot F_q(x_0,y_0,z_0,a,\phi(a)) , \end{aligned}$$

and hence

$$\text{(iii)} \qquad \phi'(a) = -\frac{F_p(x_0,y_0,z_0,a,\phi(a))}{F_q(x_0,y_0,z_0,a,\phi(a))} .$$

From (ii) and (iii) we find that the points (x,y,z) on the Monge cone $K(x_0,y_0,z_0)$ should satisfy

$$\text{(iv)} \quad \begin{cases} z - z_0 = a(x-x_0) + b(y-y_0), & \text{where } a \text{ and } b \text{ are numbers such that:} \\ F(x_0,y_0,z_0,a,b) = 0 & \\ \dfrac{x-x_0}{F_p} = \dfrac{y-y_0}{F_q} & [F_p \text{ and } F_q \text{ evaluated at } (x_0,y_0,z_0,a,b)] . \end{cases}$$

Now consider a solution u of (1), and let

$$z_0 = u(x_0,y_0), \qquad p_0 = u_x(x_0,y_0), \qquad q_0 = u_y(x_0,y_0) .$$

The tangent plane of M_u at (x_0,y_0,z_0) consists of points (x,y,z) satisfying

$$z - z_0 = p_0(x-x_0) + q_0(y-y_0) .$$

Equations (iv) show that points (x,y,z) which are on both this tangent plane and the Monge cone $K(x_0,y_0,z_0)$ ought to satisfy

$$\text{(v)} \qquad \frac{x-x_0}{F_p} = \frac{y-y_0}{F_q} = \frac{z-z_0}{p_0F_p + q_0F_q}$$

$$[F_p \text{ and } F_q \text{ evaluated at } (x_0,y_0,z_0,p_0,q_0)] .$$

Therefore, these points ought to lie along the line through (x_0,y_0,z_0) with direction

$$(F_p,\ F_q,\ p_0F_p + q_0F_q) \qquad [F_p \text{ and } F_q \text{ evaluated at } (x_0,y_0,z_0,p_0,q_0)].$$

We have finally reached the stage where we can make a perfectly sensible definition, involving no assumptions at all. Let u be a solution of (1), and for a point (x_0,y_0), define z_0, p_0, and q_0 as before. We then define the characteristic vector of u at (x_0,y_0) to be the vector

$$(2) \qquad X(u;x_0,y_0) = (F_p,\ F_q,\ p_0F_p + q_0F_q) ,$$

where F_p and F_q are to be evaluated at (x_0,y_0,z_0,p_0,q_0); this vector is to be considered as an element of $\mathbb{R}^3_{(x_0,y_0,z_0)}$. If $M_u = \{(x,y,u(x,y))\}$, then the tangent plane of M_u at (x_0,y_0,z_0) is perpendicular to the vector $(p_0,q_0,-1)$. The vector $X(u;x_0,y_0)$ clearly has this property, so every characteristic vector of u is tangent to M_u, and the set of characteristic vectors of u forms a vector field on M_u. The integral curves of this vector field are called the characteristic curves of the solution u, and they are clearly curves on M_u.

A characteristic curve c of u is thus a curve in $\mathbb{R}^3$ satisfying the equations

$$(3) \quad \begin{cases} \dfrac{dc_1(t)}{dt} = F_p(\bullet) \\[2ex] \dfrac{dc_2(t)}{dt} = F_q(\bullet) \\[2ex] \dfrac{dc_3(t)}{dt} = u_x(c_1(t),c_2(t))\cdot F_p(\bullet) + u_y(c_1(t),c_2(t))\cdot F_q(\bullet) \\[2ex] \quad \text{where} \quad \bullet = (c_1(t),c_2(t),c_3(t),u_x(c_1(t),c_2(t)),u_y(c_1(t),c_2(t))) . \end{cases}$$

Now if we assume that u is C^2, then we can also obtain equations for $u_x(c_1(t),c_2(t))$ and $u_y(c_1(t),c_2(t))$. For equations (3) allow us to write

$$(4)\quad \begin{cases} \dfrac{du_x(c_1(t),c_2(t))}{dt} = u_{xx}(c_1(t),c_2(t))\dfrac{dc_1(t)}{dt} + u_{xy}(c_1(t),c_2(t))\dfrac{dc_2(t)}{dt} \\ \qquad = u_{xx}(c_1(t),c_2(t))F_p(\bullet) + u_{xy}(c_1(t),c_2(t))F_q(\bullet) \\ \dfrac{du_y(c_1(t),c_2(t))}{dt} = u_{yx}(c_1(t),c_2(t))F_p(\bullet) + u_{yy}(c_1(t),c_2(t))F_q(\bullet) . \end{cases}$$

On the other hand, since u satisfies

$$F(x,y,u(x,y),u_x(x,y),u_y(x,y)) = 0 ,$$

we also have

$$(5)\quad \begin{cases} F_x + u_xF_u + u_{xx}F_p + u_{yx}F_q = 0 \\ F_y + u_yF_u + u_{xy}F_p + u_{yy}F_q = 0 , \end{cases}$$

where all partials of F are evaluated at $(x,y,u(x,y),u_x(x,y),u_y(x,y))$. Thus equations (4) become

$$(6)\quad \begin{cases} \dfrac{du_x(c_1(t),c_2(t))}{dt} = -F_x(\bullet) - u_x(c_1(t),c_2(t))\cdot F_u(\bullet) \\ \dfrac{du_y(c_1(t),c_2(t))}{dt} = -F_y(\bullet) - u_y(c_1(t),c_2(t))\cdot F_u(\bullet) . \end{cases}$$

Let us now define a curve Γ in $\mathbb{R}^5$ by

$$(7)\quad \Gamma(t) = (c_1(t),c_2(t),c_3(t),u_x(c_1(t),c_2(t)),u_y(c_1(t),c_2(t))) .$$

Then equations (3) and (6) may be written

$$(8)\quad \begin{cases} \dfrac{d\Gamma_1(t)}{dt} = F_p(\Gamma(t)) \\[2ex] \dfrac{d\Gamma_2(t)}{dt} = F_q(\Gamma(t)) \\[2ex] \dfrac{d\Gamma_3(t)}{dt} = \Gamma_4(t)\cdot F_p(\Gamma(t)) + \Gamma_5(t)\cdot F_q(\Gamma(t)) \\[2ex] \dfrac{d\Gamma_4(t)}{dt} = -F_x(\Gamma(t)) - \Gamma_4(t)F_u(\Gamma(t)) \\[2ex] \dfrac{d\Gamma_5(t)}{dt} = -F_y(\Gamma(t)) - \Gamma_5(t)F_u(\Gamma(t)) . \end{cases}$$

Now although the curve Γ was defined in terms of a solution u, the final equations (8) involve only the original equation (1). This will allow us to define geometrically meaningful objects which do not depend on knowing a solution u. We may regard a point $(x_0,y_0,z_0,a,b) \in \mathbb{R}^5$ as a plane in the tangent space $\mathbb{R}^3_{(x_0,y_0,z_0)}$, namely, as the plane perpendicular to the vector $(a,b,-1)$. A curve Γ in $\mathbb{R}^5$ may then be regarded as a family of planes, with $\Gamma(t)$ in the tangent space of $\mathbb{R}^3$ at $c(t) = (\Gamma_1(t),\Gamma_2(t),\Gamma_3(t))$; it will be convenient to refer to this curve c as the base curve of Γ. An arbitrary curve Γ is called a strip if the tangent vector $c'(t)$ of the base

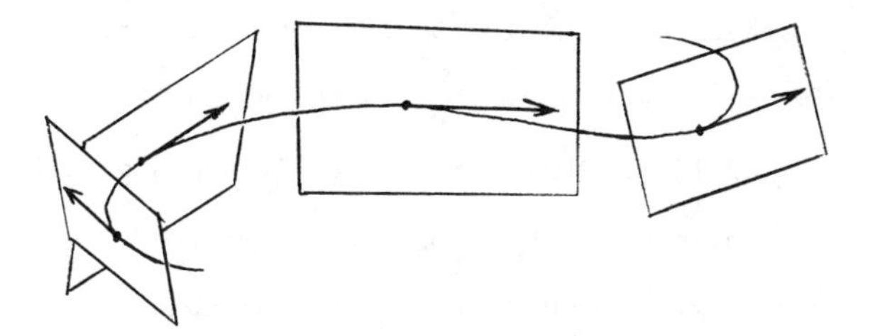

curve c always lies in the plane determined by Γ at time t. This means that

$$c'(t) = (\Gamma_1'(t),\Gamma_2'(t),\Gamma_3'(t)) \quad \text{is perpendicular to} \quad (\Gamma_4(t),\Gamma_5(t),-1) .$$

So Γ is a strip if and only if it satisfies the strip condition:

$$\frac{d\Gamma_3(t)}{dt} = \Gamma_4(t)\frac{d\Gamma_1(t)}{dt} + \Gamma_5(t)\frac{d\Gamma_2(t)}{dt} . \tag{9}$$

Notice that any solution of (8) is automatically a strip. A curve Γ will be called a characteristic strip of the PDE (1) if Γ satisfies (8) and also

$$F(\Gamma(t)) = F(\Gamma_1(t),\Gamma_2(t),\Gamma_3(t),\Gamma_4(t),\Gamma_5(t)) = 0 . \tag{10}$$

This last restriction is not as stringent as it might first seem, for if Γ satisfies (8), then

$$\begin{aligned} \frac{d}{dt}F(\Gamma(t)) &= F_x\frac{d\Gamma_1(t)}{dt} + \cdots + F_q\frac{d\Gamma_5(t)}{dt} \\ &\qquad [\text{all partials of } F \text{ evaluated at } \Gamma(t)] \\ &= F_xF_p + F_yF_q + F_z\cdot(\Gamma_4(t)F_p + \Gamma_5(t)F_q) \\ &\qquad + F_p\cdot(-F_x - \Gamma_4(t)F_z) + F_q\cdot(-F_y - \Gamma_5(t)F_z) \\ &= 0 . \end{aligned} \tag{11}$$

So if Γ satisfies (8) and also satisfies (10) for one t, then it satisfies (10) for all t, and is consequently a characteristic step.

Now how are characteristic strips related to solutions? We have seen that if u is a solution of (1), then M_u is the union of certain characteristic curves [solutions of (3)]. Moreover, if c is a characteristic curve, then the set of tangent planes of M_u along c gives the curve Γ of equation (7), which is a characteristic strip. So M_u is the union of base curves of characteristic strips.

Now suppose that we have an arbitrary curve Σ in $\mathbb{R}^5$, with base curve σ, and that $F(\Sigma(s)) = 0$ for all s. There is a unique solution of (8) through each point $\Sigma(s)$, and by the remark after equation (11), this solution is a characteristic strip. We thus obtain a family of characteristic strips Γ.

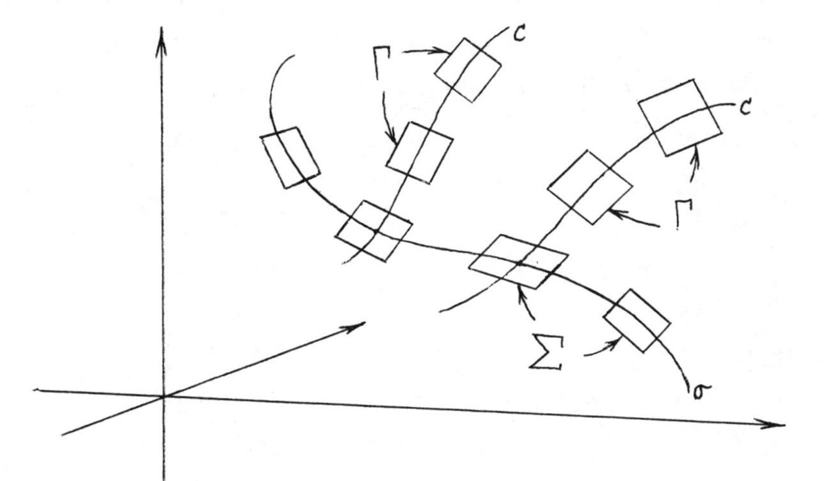

The union of the corresponding base curves c is a surface M_u, containing the base curve σ. Is it reasonable to suppose now that u is a solution of (1)? The answer is no, for there is clearly no hope unless Σ is also a strip. When this condition is satisfied, then everything works out. We will prove that if $\sigma\colon [a,b] \to \mathbb{R}^2$ is a given curve, $\mathring{u}\colon [a,b] \to \mathbb{R}$ is a given function, and $\mathring{p}, \mathring{q}\colon [a,b] \to \mathbb{R}$ are two functions satisfying

(a) $$F(\Sigma(s)) = F(\sigma_1(s),\sigma_2(s),\mathring{u}(s),\mathring{p}(s),\mathring{q}(s)) = 0\ ,$$

and the strip condition

(b) $$\frac{d\mathring{u}(s)}{ds} = \mathring{p}(s)\frac{d\sigma_1(s)}{ds} + \mathring{q}(s)\frac{d\sigma_2(s)}{ds}\ ,$$

then there is a unique solution u of (1) satisfying

$$u(\sigma(s) = \overset{\circ}{u}(s) , \quad u_x(\sigma(s)) = \overset{\circ}{p}(s) , \quad u_y(\sigma(s)) = \overset{\circ}{q}(s)$$

[naturally, (b) is a necessary consequence of these equations]. We will clearly have to assume that $\sigma'(s)$ is linearly independent of the vector obtained by projecting the characteristic vector (2) on the (x,y)-plane. In other words, we will have to require that $\sigma'(s)$ and $(F_p(\Sigma(s)), F_q(\Sigma(s)))$ are linearly independent, or that

$$\text{(c)} \qquad \sigma_1'(s)\cdot F_q(\Sigma(s)) \neq \sigma_2'(s)\cdot F_p(\Sigma(s)) .$$

Before we proceed to prove the theorem, we should insert a remark about the hypotheses, which will involve σ, $\overset{\circ}{u}$, $\overset{\circ}{p}$, and $\overset{\circ}{q}$ satisfying (a)-(c). At first sight, we seem to be contradicting our basic philosophy about first order equations, for we seem to be saying that we can arbitrarily specify not only the values $\overset{\circ}{u}$ of u along σ, but <u>also</u> the values $\overset{\circ}{p}$ and $\overset{\circ}{q}$ of u_x and u_y along σ. This is not really the case, for $\overset{\circ}{p}$ and $\overset{\circ}{q}$ are practically determined by the equations (a) and (b) which they must satisfy. This is most apparent when our initial curve σ is the x-axis, $\sigma(s) = (s,0)$. Then equation (b) already determines $\overset{\circ}{p}$. Moreover, condition (c) says that $F_q \neq 0$ along $\{(s,0,\overset{\circ}{u}(s),\overset{\circ}{p}(s),\overset{\circ}{q}(s))\}$, so the implicit function theorem shows that equation (a) can be solved for $\overset{\circ}{q}(s)$ in terms of $\overset{\circ}{p}(s)$ -- there is a function ϕ with

$$F(s,0,\overset{\circ}{u}(s),\overset{\circ}{p}(s),\phi(\overset{\circ}{p}(s))) = 0 .$$

Of course, there may be several possible ϕ, but once $\overset{\circ}{q}(0)$ is determined, there will be only one continuous choice of $\mathring{q}$ satisfying (a). [In the quasi-linear case, $\mathring{q}(s)$ will actually be uniquely determined.] It is not hard to see that a similar situation prevails when σ is any curve satisfying (c): we are essentially specifying only the values $\mathring{u}$ of u along σ, and then making certain that we have a continuous choice of the limited possibilities for $\mathring{p}$ and $\mathring{q}$. In order to emphasize this point we will refer to $(\mathring{u},\mathring{p},\mathring{q})$ as "initial data," rather than as initial conditions.

3. THEOREM. Let F be a function of class C^k, $k \geq 3$, defined in an open set $U \subset \mathbb{R}^5$. Let $\sigma\colon [a,b] \to \mathbb{R}^2$ be a one-one C^{k-1} function, and let $\mathring{u}, \mathring{p}, \mathring{q}\colon [a,b] \to \mathbb{R}$ be C^{k-1} functions such that for all $s \in [a,b]$ we have

(a) $$\Sigma(s) = (\sigma_1(s),\sigma_2(s),\mathring{u}(s),\mathring{p}(s),\mathring{q}(s)) \in U \quad \text{and} \quad F(\Sigma(s)) = 0 ,$$

(b) $$\frac{d\mathring{u}(s)}{ds} = \mathring{p}(s)\frac{d\sigma_1(s)}{ds} + \mathring{q}(s)\frac{d\sigma_2(s)}{ds} ,$$

(c) $$\sigma_1'(s)\cdot F_q(\Sigma(s)) \neq \sigma_2'(s)\cdot F_p(\Sigma(s)) .$$

Then there is a C^{k-1} function u, defined in a neighborhood V of $\sigma([a,b])$, which satisfies the equation

$$F(x,y,u(x,y),u_x(x,y),u_y(x,y)) = 0 \quad \text{on } V$$

and also

$$u(\sigma(s)) = \mathring{u}(s) , \quad u_x(\sigma(s)) = \mathring{p}(s) , \quad u_y(\sigma(s)) = \mathring{q}(s) , \quad \text{for } s \in [a,b] .$$

Moreover, any two functions u with this property agree on a neighborhood of $\sigma([a,b])$.

<u>Proof</u>. As in the proof of Theorems 1 and 2, we use Problem I.5-5 to conclude that there is a C^{k-1} function $\alpha = (\alpha_1,\dots,\alpha_5)$ with

$$(*)\quad \begin{cases} \alpha(0,s,r) = r \qquad \text{for } r \in \mathbb{R}^5 \\ \dfrac{\partial}{\partial t}\alpha_1(t,s,r) = F_p(\alpha(t,s,r)) \\ \dfrac{\partial}{\partial t}\alpha_2(t,s,r) = F_q(\alpha(t,s,r)) \\ \dfrac{\partial}{\partial t}\alpha_3(t,s,r) = \alpha_4(t,s,r)\cdot F_p(\alpha(t,s,r)) + \alpha_5(t,s,r)\cdot F_q(\alpha(t,s,r)) \\ \dfrac{\partial}{\partial t}\alpha_4(t,s,r) = -F_x(\alpha(t,s,r)) - \alpha_4(t,s,r)\cdot F_u(\alpha(t,s,r)) \\ \dfrac{\partial}{\partial t}\alpha_5(t,s,r) = -F_y(\alpha(t,s,r)) - \alpha_5(t,s,r)\cdot F_u(\alpha(t,s,r)) . \end{cases}$$

Let

$$\beta(s,t) = \alpha(t,s,\sigma_1(s),\sigma_2(s),\mathring{u}(s),\mathring{p}(s),\mathring{q}(s)) ,$$

so that β is also C^{k-1}. In particular,

$$\beta(s,0) = (\sigma_1(s),\sigma_2(s),\mathring{u}(s),\mathring{p}(s),\mathring{q}(s)) = \Sigma(s) .$$

If we define

$$\gamma(s,t) = (\beta_1(s,t),\beta_2(s,t)) \in \mathbb{R}^2 ,$$

then the Jacobian of γ at $(s,0)$ is

$$\begin{pmatrix} \dfrac{\partial\beta_1}{\partial s}(s,0) , & \dfrac{\partial\beta_1}{\partial t}(s,0) \\ \dfrac{\partial\beta_2}{\partial s}(s,0) , & \dfrac{\partial\beta_1}{\partial t}(s,0) \end{pmatrix} = \begin{pmatrix} \sigma_1'(s) , & \dfrac{\partial\alpha_1}{\partial t}(0,s,\Sigma(s)) \\ \sigma_2'(s) , & \dfrac{\partial\alpha_2}{\partial t}(0,s,\Sigma(s)) \end{pmatrix}$$

$$= \begin{pmatrix} \sigma_1'(s) & , F_p(\Sigma(s)) \\ \sigma_2'(s) & , F_q(\Sigma(s)) \end{pmatrix} \qquad \text{by } (*) ,$$

and this is non-singular by hypothesis. So if ε is sufficiently small, then $\gamma\colon [a,b] \times (-\varepsilon,\varepsilon) \to \mathbb{R}^2$ is a C^{k-1} diffeomorphism onto a neighborhood V of $\sigma([a,b])$.

The solution u, if it exists, clearly must be the C^{k-1} function

$$u(x,y) = \beta_3(\gamma^{-1}(x,y)) \qquad \text{or equivalently} \qquad u(\gamma(s,t)) = \beta_3(s,t) .$$

We claim that

$$u_x(\gamma(s,t)) = \beta_4(s,t) \qquad \text{and} \qquad u_y(\gamma(s,t)) = \beta_5(s,t) .$$

This will prove that

$$F(x,y,u(x,y),u_x(x,y),u_y(x,y)) = 0 ;$$

for we have already seen (equation 11) that $F(\alpha(t,s,r))$ is constant for fixed s and r, while $F(\alpha(0,s,\Sigma(s))) = 0$ by (a), so that we will have

$$\begin{aligned} 0 = F(\alpha(t,s,\Sigma(s))) &= F(\beta(s,t)) = F(\beta_1(s,t),\beta_2(s,t),\beta_3(s,t),\beta_4(s,t),\beta_5(s,t)) \\ &= F(\gamma(s,t),u(\gamma(s,t)),u_x(\gamma(s,t)),u_y(\gamma(s,t))) . \end{aligned}$$

To prove the claim, we consider the function

$$\Delta = \frac{\partial\beta_3}{\partial s} - \beta_4 \cdot \frac{\partial\beta_1}{\partial s} - \beta_5 \cdot \frac{\partial\beta_2}{\partial s} .$$

We have

$$\begin{aligned} \Delta(s,0) &= \frac{d\mathring{u}(s)}{ds} - \mathring{p}(s)\cdot\frac{d\sigma_1(s)}{ds} - \mathring{q}(s)\cdot\frac{d\sigma_2(s)}{ds} \\ &= 0 \qquad \text{by (b)}. \end{aligned}$$

Moreover,

$$\frac{\partial\Delta}{\partial t} = \frac{\partial^2\beta_3}{\partial s\partial t} - \frac{\partial\beta_4}{\partial t}\frac{\partial\beta_1}{\partial s} - \frac{\partial\beta_5}{\partial t}\frac{\partial\beta_2}{\partial s} - \beta_4\cdot\frac{\partial^2\beta_1}{\partial s\partial t} - \beta_5\cdot\frac{\partial^2\beta_2}{\partial s\partial t}$$

$$= \frac{\partial}{\partial s}\left(\frac{\partial\beta_3}{\partial t} - \beta_4\cdot\frac{\partial\beta_1}{\partial t} - \beta_5\cdot\frac{\partial\beta_2}{\partial t}\right) + \frac{\partial\beta_4}{\partial s}\frac{\partial\beta_1}{\partial t} + \frac{\partial\beta_5}{\partial s}\frac{\partial\beta_2}{\partial t} - \frac{\partial\beta_4}{\partial t}\frac{\partial\beta_1}{\partial s} - \frac{\partial\beta_5}{\partial t}\frac{\partial\beta_2}{\partial s}$$

$$= 0 + F_p\cdot\frac{\partial\beta_4}{\partial s} + F_q\cdot\frac{\partial\beta_5}{\partial s} + (F_x + F_u\beta_4)\frac{\partial\beta_1}{\partial s} + (F_y + F_u\beta_5)\frac{\partial\beta_2}{\partial s}$$

by (*) [where all partials of F are evaluated at $\beta(s,t)$]

$$= F_x\cdot\frac{\partial\beta_1}{\partial s} + F_y\cdot\frac{\partial\beta_2}{\partial s} + F_u\cdot\frac{\partial\beta_3}{\partial s} + F_p\cdot\frac{\partial\beta_4}{\partial s} + F_q\cdot\frac{\partial\beta_5}{\partial s}$$

$$- F_u\cdot\left(\frac{\partial\beta_3}{\partial s} - \beta_4\cdot\frac{\partial\beta_1}{\partial s} - \beta_5\cdot\frac{\partial\beta_2}{\partial s}\right)$$

$$= \frac{\partial}{\partial s}(F(\beta(s,t))) - F_u\cdot\Delta$$

$$= - F_u\cdot\Delta ,$$

since we have already seen that $F(\beta(s,t)) = 0$. Now for each fixed s, we have an ordinary differential equation

$$\frac{\partial\Delta}{\partial t} = - F_u\cdot\Delta ,$$

with the initial condition

$$\Delta(s,0) = 0 ,$$

so the unique solution is $\Delta(s,t) = 0$. In other words, we have shown that

$$\frac{\partial\beta_3}{\partial s} = \beta_4\cdot\frac{\partial\beta_1}{\partial s} + \beta_5\cdot\frac{\partial\beta_2}{\partial s} .$$

Also

$$\frac{\partial\beta_3}{\partial t} = \beta_4\cdot\frac{\partial\beta_1}{\partial t} + \beta_5\cdot\frac{\partial\beta_2}{\partial t} \qquad \text{by } (*) .$$

On the other hand, differentiating the definition $u(\gamma(s,t)) = \beta_3(s,t)$ gives

$$\frac{\partial\beta_3}{\partial s} = u_x(\gamma(s,t))\cdot\frac{\partial\beta_1}{\partial s} + u_y(\gamma(s,t))\cdot\frac{\partial\beta_2}{\partial s}$$

$$\frac{\partial\beta_3}{\partial t} = u_y(\gamma(s,t))\cdot\frac{\partial\beta_1}{\partial t} + u_y(\gamma(s,t))\cdot\frac{\partial\beta_2}{\partial t} .$$

These last 4 equations give two solutions for two linear equations in two unknowns, whose determinant

$$\det\begin{pmatrix} \dfrac{\partial\beta_1}{\partial s} & \dfrac{\partial\beta_2}{\partial s} \\[2ex] \dfrac{\partial\beta_1}{\partial t} & \dfrac{\partial\beta_2}{\partial t} \end{pmatrix}$$

is $\neq 0$ for $(s,t) \in [a,b] \times (-\varepsilon,\varepsilon)$. So the two solutions must be the same, i.e.

$$u_x(\gamma(s,t)) = \beta_4(s,t) \qquad \text{and} \qquad u_y(\gamma(s,t)) = \beta_5(s,t) ,$$

as desired. ■

We will say that the initial curve σ is *free for the initial data* $\mathring{u}$, $\mathring{p}$, $\mathring{q}$, when condition (c) in Theorem (3) is satisfied. Thus we can solve the Cauchy problem for a first order PDE (1) for any initial strip $\Sigma = (\sigma_1,\sigma_2,\mathring{u},\mathring{p},\mathring{q})$ for which the initial curve σ is free for the initial data $\mathring{u}$, $\mathring{p}$, $\mathring{q}$.

Again we consider the case where our initial curve σ fails to be free for the initial data $\mathring{u}$, $\mathring{p}$, $\mathring{q}$ in the worst possible way, namely when $\sigma'(s)$ and $(F_p(\Sigma(s)), F_q(\Sigma(s)))$ are everywhere linearly dependent. Once again we say that σ is *characteristic* for $\mathring{u}$, $\mathring{p}$, $\mathring{q}$. Assuming that σ is an imbedding, we can reparameterize σ so that $\sigma'(s) = (F_p(\Sigma(s)), F_q(\Sigma(s)))$. This gives us the first two equations in (8) for the curve $(\sigma_1,\sigma_2,\mathring{u},\mathring{p},\mathring{q})$. The third equation of (8) is just the strip condition (b). The argument on pp. 25-26 shows that these three equations imply the last two if there is a solution u of (1) with

$$u(\sigma(s)) = \mathring{u}(s) , \qquad u_x(\sigma(s)) = \mathring{p}(s) , \qquad u_y(\sigma(s)) = \mathring{q}(s) .$$

So when σ is characteristic, the Cauchy problem is solvable for the initial data $\mathring{u}$, $\mathring{p}$, $\mathring{q}$ along σ only if $(\sigma_1,\sigma_2,\mathring{u},\mathring{p},\mathring{q})$ is a characteristic strip. When this is the case, there will be infinitely many solutions with this initial data along σ. The base curves of characteristic strips are the intersection curves of the graphs of different solutions meeting tangentially.

We can now describe the situation for first order PDEs in n variables very easily, without bothering to write down all the results as formal theorems. Consider first the quasi-linear PDE

$$\sum_{i=1}^{n} A_i(x_1,\ldots,x_n,u)\cdot u_{x_i} = C(x_1,\ldots,x_n,u) .$$

The characteristic vector field of this equation is the vector field X in $\mathbb{R}^{n+1}$ defined by

$$X = \sum_{i=1}^{n} A_i \frac{\partial}{\partial x_i} + C \frac{\partial}{\partial z} ;$$

the integral curves of X are the characteristic curves of the equation. As in the case $n = 2$, it is clear that if u is a solution of (1), then the hypersurface

$$M_u = \{(x_1,\dots,x_n,u(x_1,\dots,x_n))\} \subset \mathbb{R}^{n+1}$$

is a union of characteristic curves. Now suppose we are given a one-one map

$$\sigma\colon \mathcal{D} \to \mathbb{R}^n ,$$

where $\mathcal{D} \subset \mathbb{R}^{n-1}$ is a compact $(n-1)$-dimensional manifold-with-boundary, and a function $\mathring{u}\colon \mathcal{D} \to \mathbb{R}$. We can produce a solution u of (1) with

$$u(\sigma(s)) = \mathring{u}(s) \qquad \text{for all} \quad s \in \mathcal{D}$$

by taking the union of the characteristic curves through all points $(\sigma(s),\mathring{u}(s)) \in \mathbb{R}^{n+1}$. The proof is exactly analogous to the proof of Theorem 2, except that we will now require that the matrix

$$\begin{pmatrix} D_1\sigma_1(s) & \cdots & D_{n-1}\sigma_1(s) & A_1(\sigma(s),\mathring{u}(s)) \\ \vdots & & \vdots & \vdots \\ D_1\sigma_n(s) & \cdots & D_{n-1}\sigma_n(s) & A_n(\sigma(s),\mathring{u}(s)) \end{pmatrix}$$

be non-singular for all $s \in \mathcal{D}$. This means, first of all, that the matrix $(D_j\sigma_i(s))$ must have rank $n-1$, so that σ is an imbedding and $\sigma(\mathcal{D}) \subset \mathbb{R}^n$ is a hypersurface. In addition, the vector $(A_1(\sigma(s),\mathring{u}(s)),\dots,A_n(\sigma(s),\mathring{u}(s)))$ must not lie in the tangent space of $\sigma(\mathcal{D})$; we express this by saying that the "initial manifold" $\sigma(\mathcal{D})$ is **free for the initial condition** $\mathring{u}$ (for linear equations the initial condition $\mathring{u}$ is irrelevant). Thus we can solve the Cauchy problem for any initial condition along an initial $(n-1)$-manifold which is free for the initial condition.

Now we consider the general first order PDE

$$F(x_1,\dots,x_n,u(x_1,\dots,x_n),u_{x_1}(x_1,\dots,x_n),\dots,u_{x_n}(x_1,\dots,x_n)) = 0 .$$

We denote the partials of F by

$$F_{x_i}\ ,\quad F_u\ ,\quad F_{p_i}\ .$$

Consider curves Γ in $\mathbb{R}^{2n+1}$ satisfying

$$\begin{cases} \dfrac{d\Gamma_i(t)}{dt} = F_{p_i}(\Gamma(t)) & i = 1,\dots,n \\ \dfrac{d\Gamma_{n+1}(t)}{dt} = \displaystyle\sum_{i=1}^{n} \Gamma_{n+1+i}(t)\cdot F_{p_i}(\Gamma(t)) & \\ \dfrac{d\Gamma_{n+1+i}(t)}{dt} = -F_{x_i}(\Gamma(t)) - \Gamma_{n+1+i}(t)F_u(\Gamma(t)) & i = 1,\dots,n\ . \end{cases}$$

As before, we easily check that if Γ satisfies these equations, then $F(\Gamma(t))$ is constant in t. A solution Γ with $F(\Gamma(t)) = 0$ for all t is called a <u>characteristic strip</u>. Now suppose we have a one-one map

$$\sigma\colon \mathscr{D} \longrightarrow \mathbb{R}^n$$

with $\mathscr{D} \subset \mathbb{R}^{n-1}$ as before, and functions

$$\mathring{u},\ \mathring{p}_1,\dots,\mathring{p}_n\colon \mathscr{D} \longrightarrow \mathbb{R}$$

with

$$F(\Sigma(s)) = F(\sigma_1(s),\dots,\sigma_n(s),\mathring{u}(s),\mathring{p}_1(s),\dots,\mathring{p}_n(s)) = 0 \qquad \text{for all } s \in \mathscr{D}\ .$$

Then there is a unique characteristic strip Γ through each point $\Sigma(s)$, and the union of the corresponding base curves is a hypersurface M_u. In order for the function u to be a solution to our PDE we will need two conditions, which allow us to extend the proof of Theorem 3 essentially without change. First, the matrix

$$\begin{pmatrix} D_1\sigma_1(s) & \cdots & D_{n-1}\sigma_1(s) & F_{p_1}(\Sigma(s)) \\ \vdots & & \vdots & \vdots \\ D_1\sigma_n(s) & \cdots & D_{n-1}\sigma_n(s) & F_{p_n}(\Sigma(s)) \end{pmatrix}$$

must be non-singular. Thus $\sigma(\mathcal{D}) \subset \mathbb{R}^n$ must be an $(n-1)$-manifold, and $(F_{p_1}(\Sigma(s)),\ldots,F_{p_n}(\Sigma(s)))$ must not lie in its tangent space -- once again, we express this by saying that the initial manifold $\sigma(\mathcal{D})$ is free for the initial data $\mathring{u}_0, \mathring{p}_1,\ldots,\mathring{p}_n$. Second, we must have

$$\frac{\partial \mathring{u}}{\partial s_j} = \sum_{i=1}^{n} \mathring{p}_i \cdot \frac{\partial \sigma_i}{\partial s_j} .$$

In terms of Σ, this condition reads

$$\frac{\partial \Sigma_{n+1}}{\partial s_j} = \sum_{i=1}^{n} \Sigma_{n+1+i} \cdot \frac{\partial \Sigma_i}{\partial s_j} ,$$

and is called the strip manifold condition. If we think of a point $(x_1,\ldots,x_n,z,p_1,\ldots,p_n)$ in $\mathbb{R}^{2n+1}$ as a hyperplane in $\mathbb{R}^{n+1}_{(x_1,\ldots,x_n,z)}$, namely as the hyperplane perpendicular to the vector $(p_1,\ldots,p_n, -1)$, then $\Sigma\colon \mathcal{D} \to \mathbb{R}^{2n+1}$ may be regarded as a family of hyperplanes along the $(n-1)$-dimensional submanifold $\sigma(\mathcal{D})$. It is easy to see that Σ satisfies the strip manifold condition if and only if the tangent space of $\sigma(\mathcal{D})$ at any point $\sigma(s)$ always lies in the hyperplane determined by Σ at s. We may summarize by saying that we can solve the Cauchy problem for any strip manifold

$(\sigma_1,\ldots,\sigma_n,\overset{\circ}{u},\overset{\circ}{p}_1,\ldots,\overset{\circ}{p}_n)$ for which the initial $(n-1)$-dimensional submanifold $\sigma(\mathcal{D})$ is free for the initial data $\overset{\circ}{u},\overset{\circ}{p}_1,\ldots,\overset{\circ}{p}_n$.

2. Free Initial Manifolds for Higher Order Equations

In the previous section we found that the characteristic curves or characteristic strips for first order equations were the clue to solving them, while the free hypersurfaces were the appropriate initial manifolds for which we could solve the Cauchy problem. For higher order equations things are not nearly so simple, but we can at least decide at the outset what the free initial manifolds ought to be. To do this, we will first consider the special case where the initial manifold is $M = \{x \in \mathbb{R}^n\colon x_n = 0\} \subset \mathbb{R}^n$.

First a review of the situation for first order equations. For the quasi-linear equation

$$\sum_{i=1}^{n} A_i(x_1,\ldots,x_n,u)\cdot u_{x_i} = C(x_1,\ldots,x_n,u) ,$$

the manifold $M = \{x \in \mathbb{R}^n\colon x_n = 0\}$ is free for the initial condition $\overset{\circ}{u}$ on M if and only if

$$A_n(x_1,\ldots,x_{n-1},0,\overset{\circ}{u}(x_1,\ldots,x_{n-1})) \neq 0 \qquad \text{on } M .$$

If this condition holds, then in a neighborhood of M we can write our equation as an equation for u_{x_n} in terms of u and the other u_{x_i} :

$$u_{x_n} = -\sum_{i=1}^{n-1} \frac{A_i(x_1,\ldots,x_n,u)}{A_n(x_1,\ldots,x_n,u)} u_{x_i} + \frac{C(x_1,\ldots,x_n,u)}{A_n(x_1,\ldots,x_n,u)}$$

$$= f(x_1,\ldots,x_n,u,u_{x_1},\ldots,u_{x_{n-1}}) ,$$

where the function f is defined in a neighborhood of all points

$$(x_1,\ldots,x_{n-1},0,\mathring{u}(x_1,\ldots,x_{n-1}),p_1,\ldots,p_{n-1}) .$$

On the other hand, if M fails to be free at some point, then our original equation gives us a relationship between the u_{x_i} for $i < n$, which means that there are additional conditions which $\mathring{u}$ would have to satisfy for a solution to exist.

For the general first order PDE

$$F(x,u(x),u_{x_1}(x),\ldots,u_{x_n}(x)) = 0 \qquad [x = (x_1,\ldots,x_n)] ,$$

the initial data $\mathring{u},\mathring{p}_1,\ldots,\mathring{p}_n$ must satisfy

$$\text{(a)} \quad 0 = F(x_1,\ldots,x_{n-1},0,\mathring{u}(x_1,\ldots,x_{n-1}),\mathring{p}_1(x_1,\ldots,x_{n-1}),\ldots,\mathring{p}_n(x_1,\ldots,x_{n-1}))$$

$$= F(\Sigma(s)) \qquad s = (x_1,\ldots,x_{n-1}) ,$$

as well as the obvious compatibility conditions

$$\text{(b)} \qquad \frac{\partial\mathring{u}}{\partial x_j} = \mathring{p}_j \qquad j = 1,\ldots,n-1 ,$$

which is what the strip manifold condition on p. 39 boils down to in this case; otherwise expressed, the initial value $\mathring{u}$ of u along M already determines the values of u_{x_i} along M for $i < n$, so the only other initial data that we need is a value $\overset{\circ}{p}_n$ of u_{x_n} along M satisfying (a), when the $\overset{\circ}{p}_j$ for $j < n$ are defined by (b). Now M is free for this initial data if and only if

$$F_{p_n}(\Sigma(s)) \neq 0 \qquad \text{on } M .$$

In this case, the implicit function theorem tells us that there is a unique function f defined in a neighborhood of any given

$$\bullet = (x_1,\ldots,x_{n-1},0,\overset{\circ}{u}(x_1,\ldots,x_{n-1}),\overset{\circ}{p}_1(x_1,\ldots,x_{n-1}),\ldots,\overset{\circ}{p}_{n-1}(x_1,\ldots,x_{n-1})$$

such that

$$\begin{cases} F(x,z,p_1,\ldots,p_{n-1},f(x,z,p_1,\ldots,p_{n-1})) = 0 & \text{in this neighborhood} \\ f(\bullet) = \overset{\circ}{p}_n(x_1,\ldots,x_{n-1}) . \end{cases}$$

So our PDE is equivalent to the equation

$$u_{x_n} = f(x_1,\ldots,x_n,u,u_{x_1},\ldots,u_{x_{n-1}})$$

expressing u_{x_n} in terms of u and the other u_{x_i}. On the other hand, if M is not free at some point, so that $F_{p_n}(\Sigma(s_0)) = 0$ for some s_0, then we generally cannot find any continuous initial data $\overset{\circ}{p}_n$ satisfying (a).

We will now generalize this discussion to decide when $M = \{x \in \mathbb{R}^n : x_n = 0\}$ should be called free for a second order equation. First consider the quasi-linear second order equation

$$(1) \qquad \sum_{i,j=1}^{n} A_{ij} u_{x_i x_j} = C ,$$

where the functions A_{ij} and C depend not only on $x_1,\ldots,x_n$, and u, but also on the u_{x_i}. It seems reasonable that the initial conditions for the Cauchy problem should be the values $\mathring{u},\mathring{p}_1,\ldots,\mathring{p}_n$ of u and its first derivatives on M. But, as we have already noted, $\mathring{u}$ already determines the $\mathring{p}_i$ for $i < n$. So the initial conditions for the Cauchy problem* should be the values $\mathring{u}$, $\mathring{p}_n$ of u and u_{x_n} along M. For the PDE (1) we will define M to be <u>free for the initial conditions</u> $\mathring{u}$, $\mathring{p}_n$ if

$$A_{nn}(\Sigma(s))$$

$$= A_{nn}(x_1,\ldots,x_{n-1},0,\mathring{u}(x_1,\ldots,x_{n-1}),\mathring{p}_1(x_1,\ldots,x_{n-1}),\ldots,\mathring{p}_n(x_1,\ldots,x_{n-1}))$$

$$\neq 0 \qquad \text{on } M ,$$

where $\mathring{p}_i = \partial\mathring{u}/\partial x_i$ for $i < n$. If this condition holds, then in a neighborhood of any point of M we can write our equation as an equation for $u_{x_n x_n}$ in terms of u, the first partials u_{x_i}, and the other second partials $u_{x_i x_j}$:

$$u_{x_n x_n} = - \sum_{\substack{i,j=1 \\ (i,j)\neq(n,n)}}^{n} \frac{A_{ij}}{A_{nn}} u_{x_i x_j} + \frac{C}{A_{nn}}$$

$$= f(x_1,\ldots,x_n,u,u_{x_1},\ldots,u_{x_n},\ldots.u_{x_i x_j}\ldots.) ,$$

$$[(i,j) \neq (n,n)]$$

where the function f is defined in a neighborhood of all points

$$(x_1,\ldots,x_{n-1},0,\mathring{u}(x_1,\ldots,x_{n-1}),\mathring{p}_1(x_1,\ldots,x_{n-1}),\ldots,\mathring{p}_n(x_1,\ldots,x_{n-1}),\ldots.p_{ij}\ldots.) .$$

---

*To avoid any confusion about the basic philosophy of the Cauchy problem, we emphasize that for a <u>quasi-linear second order</u> equation the "initial conditions" $\mathring{u}$, $\mathring{p}_n$ are completely arbitrary, while for the <u>general first order</u> equation the "initial data" $\mathring{p}_n$ must satisfy (a) on p. 41 , and are essentially uniquely determined by the value at a single point -- we have to include $\mathring{p}_n$ in the initial data just to show which of the possible solutions of (a) we are considering.

Now consider the general second order equation

$$(2) \qquad F(x,u(x),\ldots,u_{x_i}(x)\ldots,u_{x_ix_j}(x)\ldots) = 0 .$$

Appropriate initial data will be functions

$$\mathring{u}, \quad \mathring{p}_i, \quad \mathring{p}_{ij}$$

satisfying

(a) $0 =$

$$F(x_1,\ldots,x_{n-1},0,\mathring{u}(x_1,\ldots,x_{n-1})\ldots\mathring{p}_i(x_1,\ldots,x_{n-1})\ldots\mathring{p}_{ij}(x_1,\ldots,x_{n-1})\ldots)$$
$$= F(\Sigma(s)) ,$$

and

$$\text{(b)} \qquad \begin{cases} \dfrac{\partial \mathring{u}}{\partial x_j} = p_{j_0} & j < n \\[2ex] \dfrac{\partial \mathring{p}_i}{\partial x_j} = \mathring{p}_{ij} & i \leq n,\ j < n ; \end{cases}$$

in other words, we really need only $\mathring{u}$, $\mathring{p}_n$, $\mathring{p}_{nn}$ satisfying (a), when the other $\mathring{p}_j$ and $\mathring{p}_{ij}$ are defined by (b). For the PDE (2) we define M to be free for the initial data $\mathring{u}$, $\mathring{p}_n$, $\mathring{p}_{nn}$ if

$$F_{p_{nn}}(\Sigma(s)) \neq 0 \qquad \text{on } M .$$

In this case there is a unique function f defined in a neighborhood of any given point

$$\bullet = (x_1,\ldots,x_{n-1},0,\mathring{u}(x_1,\ldots,x_{n-1})\ldots\mathring{p}_i(x_1,\ldots,x_{n-1})\ldots\mathring{p}_{ij}(x_1,\ldots,x_{n-1})\ldots)$$
$$[(i,j) \neq (n,n)]$$

such that

$$\begin{cases} F(x,z,\dots p_i \dots p_{ij} \dots f(x,z,\dots p_i \dots p_{ij} \dots)) = 0 & \text{in this neighborhood} \\ f(\bullet) = \overset{\circ}{p}_{nn}(x_1,\dots,x_{n-1}) . \end{cases}$$

So our PDE is equivalent to the equation

$$u_{x_n x_n} = f(x_1,\dots,x_n,u,\dots u_{x_i} \dots u_{x_i x_j} \dots) \qquad [(i,j) \neq (n,n)]$$

expressing $u_{x_n x_n}$ in terms of u, the partials u_{x_i}, and the other second partials $u_{x_i x_j}$.

Now we are ready to decide when an arbitrary $(n-1)$-dimensional submanifold $M \subset \mathbb{R}^n$ should be called free for a second order PDE. Again we begin with the quasi-linear equation

$$(1) \qquad \sum_{i,j=1}^{n} A_{ij} u_{x_i x_j} = C .$$

It seems reasonable that the initial conditions for the Cauchy problem on M should be the value $\overset{\circ}{u}$ of u on M, together with the normal derivative u' of u on M,

$$u'(p) = \lim_{h \to 0} \frac{u(p + h \cdot \nu(p)) - u(p)}{h} ,$$

where $\nu\colon M \longrightarrow \mathbb{R}^n$ is the unit normal on M. This normal derivative is the same as

$$u'(p) = \nabla_{\nu(p)} u ,$$

where ∇ denotes the ordinary covariant derivative in $\mathbb{R}^n$. From the value $\overset{\circ}{u}$ of u on M we can calculate any directional derivative $\nabla_{X_p} u$ for which X_p is tangent to M at p. So from $\overset{\circ}{u}$ and u' we can calculate <u>all</u> directional derivatives $\nabla_{Y_p} u$, for $p \in M$ and $Y_p \in \mathbb{R}^n{}_p$.

Now choose a diffeomorphism $\phi\colon \mathbb{R}^n \to \mathbb{R}^n$ such that $\phi(M) \subset \{x \in \mathbb{R}^n\colon x^n = 0\}$. We look for a solution of (1) of the form $u = \tilde{u}\circ\phi$. Substituting the expressions

$$(*)\qquad \begin{cases} u_{x_i} = \sum_k \tilde{u}_{x_k} \cdot \phi^k_{x_i} \\ u_{x_i x_j} = \sum_{k,\ell} \tilde{u}_{x_k x_\ell} \phi^k_{x_i} \phi^\ell_{x_j} + \sum_k \tilde{u}_{x_k} \phi^k_{x_i x_j} \end{cases}$$

into (1), we obtain a quasi-linear equation for $\tilde{u}$,

$$(1')\qquad \sum_{i,j=1}^{n} \tilde{A}_{ij} \tilde{u}_{x_i x_j} = \tilde{C}\,.$$

Prescribing the value $\overset{\circ}{u}$ of u on M is equivalent to prescribing the value $\overset{\circ}{\tilde{u}}$ of $\tilde{u}$ on $\{x \in \mathbb{R}^n\colon x^n = 0\}$. If we also know u' on M, then we know all directional derivatives of u on M, and consequently all directional derivatives of $\tilde{u}$ on $\{x \in \mathbb{R}^n\colon x^n = 0\}$; in particular, we know $\partial\tilde{u}/\partial x_n$. So solving the Cauchy problem for (1) for the initial conditions $\overset{\circ}{u}$, u' on M is equivalent to solving the Cauchy problem for (1') for given initial conditions $\overset{\circ}{\tilde{u}}$, $\tilde{p}_n$ on $\{x \in \mathbb{R}^n\colon x^n = 0\}$. Now we ask: what conditions on $(M, \overset{\circ}{u}, u')$ will make $\{x \in \mathbb{R}^n\colon x^n = 0\}$ free for the initial conditions $\overset{\circ}{\tilde{u}}$, $\overset{\circ}{\tilde{p}}_n$ for equation (1')? In other words, when will the coefficient $\tilde{A}_{nn}$ of $\tilde{u}_{x_n x_n}$ in (1') be non-zero? From the derivation of equation (1') we see immediately that

$$\tilde{A}_{nn} = \sum_{i,j} A_{ij} \phi^n_{x_i} \phi^n_{x_j}\,.$$

Since $M = (\phi^n)^{-1}(0)$, the vector $(\phi^n_{x_1}, \ldots, \phi^n_{x_n})$ is a multiple of the normal ν of M (compare p.II.3B-2). So for the PDE (1) we define the $(n-1)$-dimensional submanifold $M \subset \mathbb{R}^n$ to be <u>free for the initial conditions</u> $\overset{\circ}{u}$, u' if

$$\sum_{i,j} A_{ij}\nu_i\nu_j \neq 0 \qquad \text{on } M$$

(in order to write this equation out for a point of M, we have to know the values of u and the u_{x_i} at this point, since these occur as arguments of A_{ij}; but we can compute these from the initial values $\overset{\circ}{u}$, u'). If M is free for the initial conditions $\overset{\circ}{u}$, u', then equation (1) with the initial conditions $\overset{\circ}{u}$, u' is equivalent to an equation for a function $\tilde{u}$ expressing $\tilde{u}_{x_n x_n}$ in terms of $\tilde{u}$, first partials of $\tilde{u}$, and the remaining second partials of $\tilde{u}$, with initial conditions giving the value of $\tilde{u}$ and $\tilde{u}_{x_n}$ at points $(x_1, \ldots, x_{n-1}, 0)$.

Now consider the general second order PDE

$$(2) \qquad F(x, u(x), \ldots . u_{x_i}(x) \ldots . u_{x_i x_j}(x) \ldots .) = 0 .$$

Appropriate initial data for the Cauchy problem on an $(n-1)$-dimensional submanifold $M \subset \mathbb{R}^n$ will be functions

$$\overset{\circ}{u}, \overset{\circ}{p}_i, \overset{\circ}{p}_{ij} ,$$

giving the values of u and its first and second partial derivatives on M. Of course, the $\overset{\circ}{p}_i$ can be determined by giving the normal derivative u' along M, which can be prescribed arbitrarily. But the $\overset{\circ}{p}_{ij}$ must satisfy

$$(a) \quad 0 = F(x, \overset{\circ}{u}(x), \ldots . \overset{\circ}{p}_i(x) \ldots . \overset{\circ}{p}_{ij}(x) \ldots .) = 0 \qquad \text{for } x \in M ,$$

as well as certain compatibility conditions; if M is the image of the map

$$(s_1,\dots,s_{n-1}) \mapsto (\sigma_1(s_1,\dots,s_{n-1}),\dots,\sigma_n(s_1,\dots,s_{n-1})) \in \mathbb{R}^n ,$$

and we regard $\mathring{u}$, $\mathring{p}_i$, $\mathring{p}_{ij}$ as functions of $(s_1,\dots,s_{n-1})$, then these conditions can be written as

$$\text{(b)} \qquad \begin{cases} \dfrac{\partial \mathring{u}}{\partial s_j} = \displaystyle\sum_{i=1}^{n} \mathring{p}_i \cdot \frac{\partial \sigma_i}{\partial s_j} \\[2ex] \dfrac{\partial \mathring{p}_i}{\partial s_j} = \displaystyle\sum_{i=1}^{n} \mathring{p}_{ij} \cdot \frac{\partial \sigma_i}{\partial s_j} . \end{cases}$$

Once again, choose a diffeomorphism $\phi\colon \mathbb{R}^n \to \mathbb{R}^n$ such that $\phi(M) \subset \{x \in \mathbb{R}^n : x^n = 0\}$, and consider a solution of (2) of the form $u = \tilde{u}\circ\phi$. Equation (2) is equivalent to a second order PDE for $\tilde{u}$

$$\text{(2')} \qquad \tilde{F}(x,\tilde{u}(x),\dots,\tilde{u}_{x_i}(x)\dots.\tilde{u}_{x_i x_j}(x)\dots.) = 0 ,$$

and prescribing the functions $\mathring{u}$, $\mathring{p}_i$, $\mathring{p}_{ij}$ on M is equivalent to prescribing functions $\tilde{\mathring{u}}$, $\tilde{\mathring{p}}_i$, $\tilde{\mathring{p}}_{ij}$ on $\{x \in \mathbb{R}^n : x_n = 0\}$ which satisfy the conditions (a) and (b) for the equation (2') on this initial manifold. We want to know when $\{x \in \mathbb{R}^n : x_n = 0\}$ will be free for this initial data; thus we want to know when $\tilde{F}_{\tilde{p}_{nn}}$, evaluated at suitable points, is non-zero. Since we get equation (2') by substituting (*) into (2), we easily see that

$$\tilde{F}_{\tilde{p}_{nn}} = \sum_{i,j=1}^{n} F_{p_{ij}} \phi^n_{x_i} \phi^n_{x_j} .$$

So for the PDE (2) we define M to be <u>free for the initial data</u> $\mathring{u}$, $\mathring{p}_i$, $\mathring{p}_{ij}$

if

$$\sum_{i,j} F_{p_{ij}} \nu_i \nu_j \neq 0 \qquad \text{on } M$$

(in order to write this equation out for a point of M we need to know the values of $u, u_{x_i}, u_{x_i x_j}$ on M, which are given to us). If M is free for the data $\overset{\circ}{u}, \overset{\circ}{p}_i, \overset{\circ}{p}_{ij}$, then the Cauchy problem for the general second order PDE (2) is equivalent to the Cauchy problem for an equation for a function $\tilde{u}$ expressing $\tilde{u}_{x_n x_n}$ in terms of $\tilde{u}$, first partials of $\tilde{u}$, and the remaining second partials of $\tilde{u}$.

As in the case of first order equations we define M to be characteristic for the initial data $\overset{\circ}{u}, \overset{\circ}{p}_i, \overset{\circ}{p}_{ij}$ if M fails to be free in the worst possible way, that is, if

$$\sum_{i,j} F_{p_{ij}} \nu_i \nu_j = 0 \qquad \text{on } M .$$

Since we will never consider PDEs of order higher than 2, we will not bother to carry out a similar discussion for these equations. We merely note that with the appropriate definitions, solving the Cauchy problem for a kth order PDE when the initial manifold is free for the initial data is always equivalent to solving an equation

$$\frac{\partial^k u}{\partial x_n{}^k}(x) = f\left(x, u(x), \ldots, \frac{\partial^\ell u}{\partial x_1{}^{i_1} \cdots \partial x_n{}^{i_n}}, \ldots\right),$$

in which the order i_n of any partial on the right with respect to x_n is $\leq k-1$, with initial conditions for

$$u(x_1,\ldots,x_{n-1},0),\ldots,\frac{\partial^{k-1}u}{\partial x_n{}^{k-1}}(x_1,\ldots,x_{n-1},0) .$$

3. Systems of First Order PDEs

For an ordinary differential equation

$$u'(x) = f(x,u(x)) , \tag{1}$$

we found that the existence of solutions was no harder to prove for the case of a function $u: \mathbb{R} \to \mathbb{R}^n$ than it was for the case of a function $u: \mathbb{R} \to \mathbb{R}$. So we could consider (1) to be a _system_ of equations

$$u_i'(x) = f_i(x,u_1(x),\ldots,u_n(x)) .$$

This enabled us to solve an nth order equation

$$u^{(n)}(x) = f(x,u(x),u'(x),\ldots,u^{(n-1)}(x)) , \tag{2}$$

for equation (2) is equivalent to the system of equations

$$\left\{\begin{aligned} u' &= u_1 \\ u_2' &= u_2 \\ &\vdots \\ u_{n-2}' &= u_{n-1} \\ u_{n-1}'(x) &= f(x,u(x),\ldots,u_{n-1}(x)) . \end{aligned}\right. \tag{3}$$

More precisely, if u satisfies (2), then $(u,u',\ldots,u^{(n-1)})$ satisfies (3); and conversely, if $(u,u_1,\ldots,u_{n-1})$ satisfies (3), then u satisfies (2) [and moreover $u_i = u^{(i)}$]. Since (3) can be solved with any initial conditions $(u(x_0),\dots,u_{n-1}(x_0))$, equation (2) can be solved with any intial conditions $u(x_0),u'(x_0),\dots,u^{(n-1)}(x_0)$.

There is no such general theorem about systems of first order PDEs. If there were, the study of PDEs would certainly be much simpler, because, as we will now point out, the Cauchy problem for any PDE can be reduced to the Cauchy problem for a system of first order PDEs*. Because of the considerations in the previous section, we will assume that the partials of u with respect to one of the variables, which we will call y, are explicitly expressed in terms of the partials with respect to the other variables, which we call $x_1,\dots,x_n$. Thus we consider the equation

$$\frac{\partial^k u}{\partial y^k}(x_1,\dots,x_n,y) = f\left(x_1,\dots,x_n,y,u(x_1,\dots,x_n,y),\dots,\frac{\partial^\ell u}{\partial x_1^{i_1}\cdots\partial x_n^{i_n}\partial y^j},\dots\right);$$

the partial derivatives appearing on the right are all of order $\ell \le k$, and the order j with respect to y is $\le k-1$.

In order to make the notation less unwieldy, let us take the simple, but completely representative, case of a second order equation in 2 variables

$$(1)\qquad \frac{\partial^2 u}{\partial y^2}(x,y) = f(x,y,u(x,y),u_x(x,y),u_y(x,y),u_{xx}(x,y),u_{xy}(x,y)),$$

with the initial conditions

$$(1_0)\qquad \begin{aligned} u(x,0) &= \xi(x)\\ \frac{\partial u}{\partial y}(x,0) &= \eta(x)\,. \end{aligned}$$

---

*See p. III.448, §5.

To reduce this to a system of first order equations, we introduce new unknown functions

$$p_1, p_2, p_{11}, p_{12}, p_{22} \qquad [\text{for } u_x, u_y, u_{xx}, u_{xy}, u_{yy}] .$$

(When one is dealing with a second order equation in 2 variables, the symbols

$$p, q, r, s, t \qquad \text{for } u_x, u_y, u_{xx}, u_{xy}, u_{yy}$$

are more standard.) We will denote the various partial derivatives of f in (1) by

$$f_x, f_y, f_u, f_{p_1}, f_{p_2}, f_{p_{11}}, f_{p_{12}} .$$

Consider the following system of equations for $(u,p_1,p_2,p_{11},p_{12},p_{22})$ [where the last equation is obtained by differentiating (1) with respect to y]:

$$(2)\quad \begin{cases} \text{(i)} & \dfrac{\partial u}{\partial y} = p_2 \\[2ex] \text{(ii)} & \dfrac{\partial p_1}{\partial y} = p_{12} \\[2ex] \text{(iii)} & \dfrac{\partial p_2}{\partial y} = p_{22} \\[2ex] \text{(iv)} & \dfrac{\partial p_{11}}{\partial y} = \dfrac{\partial p_{12}}{\partial x} \\[2ex] \text{(v)} & \dfrac{\partial p_{12}}{\partial y} = \dfrac{\partial p_{22}}{\partial x} \\[2ex] \text{(vi)} & \dfrac{\partial p_{22}}{\partial y}(x,y) = f_y + f_u\cdot p_2(x,y) + f_{u_x}\cdot p_{12}(x,y) + f_{u_y}\cdot p_{22}(x,y) \\ & \qquad + f_{u_{xx}}\cdot\dfrac{\partial p_{12}}{\partial x}(x,y) + f_{u_{xy}}\cdot\dfrac{\partial p_{22}}{\partial x} \end{cases}$$

[all partials of f evaluated at $(x,y,u(x,y),p_1(x,y),\ldots,p_{12}(x,y))$].

This system, like the original equation (1), expresses partials with respect to y in terms of partials with respect to x alone (it was precisely in order to achieve this that we had to introduce new unknowns for all the partials of u up to order 2, unlike the case of an ordinary second order differential equation, where we only introduce the first derivative as a new unknown). The system is first order, and also quasi-linear. Clearly, if u is a C^3 function satisfying (1), then $(u,u_x,u_y,u_{xx},u_{xy},u_{yy})$ satisfies (2). However, it is not true that every solution $(u,p_1,p_2,p_{11},p_{12},p_{22})$ of (2) has u satisfying (1). For one thing, the first 5 equations do not even allow us to identify p_1, p_2, p_{11}, p_{12}, p_{22} with u_x, u_y, u_{xx}, u_{xy}, u_{yy}. Equations (i) and (iii) give

$$p_2 = u_y \quad \text{and thus} \quad p_{22} = \frac{\partial p_2}{\partial y} = u_{yy} ,$$

but equation (v), for example, gives only

$$\frac{\partial p_{12}}{\partial y} = \frac{\partial p_{22}}{\partial x} = u_{yyx} = u_{xyy} ,$$

and hence

$$p_{12}(x,y) = u_{xy}(x,y) + A(x)$$

for an arbitrary function A. Moreover, even if we knew that the p's were the partial derivatives of u, equation (vi) would still not imply equation (1), for the two sides could differ by an arbitrary function of x.

On the other hand, there is a Cauchy problem for the system (2) which is equivalent to the Cauchy problem for (1): If $(u,p_1,p_2,p_{11},p_{12},p_{22})$ satisfies (2) with the initial conditions

$$(2_0)\quad \begin{cases} u(x,0) = \xi(x) \\ p_1(x,0) = \xi'(x) \\ p_2(x,0) = \eta(x) \\ p_{11}(x,0) = \xi''(x) \\ p_{12}(x,0) = \eta'(x) \\ p_{22}(x,0) = f(x,0,\xi(x),\xi'(x),\eta(x),\xi''(x),\eta'(x)) \ , \end{cases}$$

and u is C^3, then u satisfies (1) with the initial conditions (1_0) [and moreover, $u_x = p_1$, $u_y = p_2$, $u_{xx} = p_{11}$, $u_{xy} = p_{12}$, $u_{yy} = p_{22}$]. To prove this, we note first that equations (i) and (iii) of (2) give

$$\text{(A)}\qquad p_2 = u_y \ , \qquad p_{22} = u_{yy} \ .$$

Then equation (v) of (2) gives

$$\frac{\partial p_{12}}{\partial y} = \frac{\partial p_{22}}{\partial x} = \frac{\partial u_{yy}}{\partial x} = \frac{\partial u_{xy}}{\partial y} \ ,$$

while the initial conditions give

$$\begin{aligned} p_{12}(x,0) &= \eta'(x) = \frac{\partial p_2}{\partial x}(x,0) = \frac{\partial u_y}{\partial x}(x,0) \qquad \text{by (A)} \\ &= u_{xy}(x,0) \ , \end{aligned}$$

so that we must have

$$\text{(B)}\qquad p_{12} = u_{xy} \ .$$

Then equation (ii) of (2) gives

$$\frac{\partial p_1}{\partial y} = p_{12} = u_{xy} \qquad \text{by (B)} ,$$

while the initial conditions give

$$p_1(x,0) = \xi'(x) = u_x(x,0) ,$$

so that we must have

$$\text{(C)} \qquad p_1 = u_x .$$

Finally, equation (iv) of (2) gives

$$\frac{\partial p_{11}}{\partial y} = \frac{\partial p_{12}}{\partial x} = \frac{\partial u_{xy}}{\partial x} = \frac{\partial u_{xx}}{\partial y} ,$$

while the initial conditions give

$$p_{11}(x,0) = \xi''(x) = \frac{\partial p_1}{\partial x}(x,0) = u_{xx}(x,0) \qquad \text{by (C)} ,$$

so that we must have

$$\text{(D)} \qquad p_{11} = u_{xx} .$$

Equation (vi) of (2) now shows that the two sides of (1) have the same partial derivatives with respect to y. The initial conditions then imply that the two sides are equal.

Exactly the same procedure, but with considerably more complicated notation, proves that the equation

$$\frac{\partial^k u}{\partial y^k}(x_1,\ldots,x_n,y) = f\left(x_1,\ldots,x_n,y,u(x_1,\ldots,x_n,y)\ldots\frac{\partial^\ell u}{\partial x_1^{i_1}\ldots\partial x_n^{i_n}\partial y^j}\ldots\right)$$

with the initial conditions

$$\begin{cases} u(x_1,\ldots,x_n,0) = \xi_0(x) \\ \vdots \\ \dfrac{\partial^{k-1}u}{\partial y^{k-1}}(x_1,\ldots,x_n,0) = \xi_{k-1}(x) \end{cases}$$

is equivalent to a system of first order quasi-linear equations with initial conditions. The equations will all express the partial derivatives of the unknown functions with respect to y in terms of partial derivatives with respect to $x_1,\ldots,x_n$; the number of unknown functions will be the number of distinct derivatives

$$\frac{\partial^\ell}{\partial x_1^{i_1}\ldots\partial x_n^{i_n}\partial y^j}$$

with $0 \le \ell \le k$ and $j \le k-1$.

Similar procedures allow us to reduce a Cauchy problem for a (not necessarily quasi-linear) system of equations to the Cauchy problem for a quasi-linear system (in more unknowns). To take a simple example, which will come up later on, consider a first order system

$$u_y^i = F^i(x,y\ldots u^j\ldots u_x^j\ldots) \ ,$$

with initial conditions

$$u^i(x,0) = \xi^i(x) \ .$$

We simply construct the new system

$$u^i_y = F^i(x,y,\dots u^j \dots p^j \dots)$$

$$p^i_y = F^i_x + \sum_j F^i_{u^j}\cdot u^j_x + \sum_j F^i_{p^j}\cdot p^j_x$$

with initial conditions

$$u^i(x,0) = \xi^i(x)$$

$$p^i(x,0) = \xi^{i\prime}(x) \ .$$

This system is quasi-linear, and the solution $\{u^i,p^i\}$ clearly gives us a solution $\{u^i\}$ of the original system.

4. The Cauchy-Kowalewski Theorem

In this section we will consider the most general system of first order quasi-linear equations in the variables $x_1,\dots,x_n,y$, where the partials with respect to y are expressed in terms of the partials with respect to $x_1,\dots,x_n$. We thus have N equations for N unknown functions $u_1,\dots,u_N$:

$$\frac{\partial u_\alpha}{\partial y} = \sum_{\beta=1}^{N} \sum_{i=1}^{n} A^\beta_{\alpha i}(x_1,\dots,x_n,y,u_1,\dots,u_N)\frac{\partial u_\beta}{\partial x_i} + B_\alpha(x_1,\dots,x_n,y,u_1,\dots,u_N) \ .$$

[Notice that the symbol $A^\beta_{\alpha i}(x_1,\dots,x_n,y,u_1,\dots,u_N)$, for example, is really an abbreviation for

$$A^\beta_{\alpha i}(x_1,\dots,x_n,y,u_1(x_1,\dots,x_n,y),\dots,u_N(x_1,\dots,x_n,y)) \ .]$$

We will prove that this system of equations has a solution $u_1,\ldots,u_N$ with given initial conditions

$$u_\alpha(x_1,\ldots,x_n,0) = \xi_\alpha(x_1,\ldots,x_n) .$$

The hitch is that we will have to assume that both the coefficients $A^\beta_{\alpha i}$, B_α and the initial condition ξ_α are <u>real analytic</u>. Recall that a function $f\colon U \subset \mathbb{R}^m \to \mathbb{R}$ is real analytic if it can be expressed as a convergent sum

$$\sum_{\sigma_1,\ldots,\sigma_m=0}^{\infty} c_{\sigma_i,\ldots,\sigma_m}(x_1 - a_1)^{\sigma_1}\cdots(x_m - a_m)^{\sigma_m}$$

in a neighborhood of each point $(a_1,\ldots,a_m)$ in its domain. We will also write this in the abbreviated form

$$\sum_\sigma c_\sigma(x-a)^\sigma .$$

<u>4. THEOREM (THE CAUCHY-KOWALEWSKI THEOREM)</u>. Let ξ_α $(\alpha = 1,\ldots,N)$ be analytic functions in a neighborhood of $(a_1,\ldots,a_n)$ in $\mathbb{R}^n$, set $b_\alpha = \xi_\alpha(a_1,\ldots,a_n)$, and let $A^\beta_{\alpha i}$ and B_α $(\alpha, \beta = 1,\ldots,N;\ i = 1,\ldots,n)$ be analytic functions in a neighborhood of $(a_1,\ldots,a_n,0,b_1,\ldots,b_N)$ in $\mathbb{R}^{N+n+1}$. Then there are unique analytic functions $u_1,\ldots,u_N$ in a neighborhood of $(a_1,\ldots,a_n)$ in $\mathbb{R}^n$ satisfying

$$\frac{\partial u_\alpha}{\partial y} = \sum_{\beta=1}^{N}\sum_{i=1}^{n} A^\beta_{\alpha i}(x_1,\ldots,x_n,y,u_1,\ldots,u_N)\frac{\partial u_\beta}{\partial x_i} + B_\alpha(x_1,\ldots,x_n,y,u_1,\ldots,u_N)$$

with the initial conditions

$$u_\alpha(x_1,\ldots,x_n,0) = \xi_\alpha(x_1,\ldots,x_n) .$$

Proof. We first make three slight simplifications.

(1) We can assume that all $a_i = 0$. For if we define

$$v_\alpha(x_1,\ldots,x_n,y) = u_\alpha(x_1+a_1,\ldots,x_n+a_n,y) ,$$

then our equations and initial conditions are equivalent to equations of the same sort for v_α,

$$\frac{\partial v_\alpha}{\partial y} = \sum_{\beta=1}^{N} \sum_{i=1}^{n} A^\beta_{\alpha i}(x_1+a_1,\ldots,x_n+a_n,y,v_1,\ldots,v_N)\frac{\partial v_\beta}{\partial x_i}$$
$$+ B_\alpha(x_1+a_1,\ldots,x_n+a_n,y,v_1,\ldots,v_N) ,$$

with the initial conditions

$$v_\alpha(x_1,\ldots,x_n,0) = \xi_\alpha(x_1+a_1,\ldots,x_n+a_n) .$$

The functions $\bar{\xi}_\alpha(x_1,\ldots,x_n) = \xi_\alpha(x_1+a_1,\ldots,x_n+a_n)$ are analytic in a neighborhood of 0 in $\mathbb{R}^n$, and the coefficients of the new equation are analytic in a neighborhood of $(0,\ldots,0,0,\bar{\xi}_1(0,\ldots,0),\ldots,\bar{\xi}_N(0,\ldots,0))$ in $\mathbb{R}^{N+n+1}$.

(2) We can further assume that the ξ_α are all 0. For if we now define

$$v_\alpha(x_1,\ldots,x_n,y) = u_\alpha(x_1,\ldots,x_n,y) - \xi_\alpha(x_1,\ldots,x_n) ,$$

then our equations and initial conditions are equivalent to equations of the same sort for the v_α,

$$\frac{\partial v_\alpha}{\partial y} = \sum_{\beta=1}^{N} \sum_{i=1}^{n} A_{\alpha i}^{\beta}(x_1,\ldots,x_n,y,v_1+\xi_1(x_1,\ldots,x_N),\ldots,v_N+\xi_N(x_1,\ldots,x_N))\frac{\partial v_\beta}{\partial x_i}$$

$$+ \Big[\sum_{\beta=1}^{N} \sum_{i=1}^{n} A_{\alpha i}^{\beta}(x_1,\ldots,x_n,y,v_1+\xi_1(x_1,\ldots,x_n),\ldots,v_N+\xi_N(x_1,\ldots,x_n))$$

$$\cdot\frac{\partial \xi_\beta}{\partial x_i}(x_1,\ldots,x_n)$$

$$+ B_\alpha(x_1,\ldots,x_n,y,v_1+\xi_1(x_1,\ldots,x_n),\ldots,v_N+\xi_N(x_1,\ldots,x_n))\Big] ,$$

with the initial conditions

$$v_\alpha(x_1,\ldots,x_n,0) = 0 .$$

Notice that the coefficients of the new equation are analytic in a neighborhood of 0 in $\mathbb{R}^{N+n+1}$.

(3) We can assume finally that the $A_{\alpha i}^{\beta}$ and B_α do not depend on y. For we can consider the equations in $N+1$ unknowns $\eta,u_1,\ldots,u_N$,

$$\frac{\partial \eta}{\partial y} = 1$$

$$\frac{\partial u_\alpha}{\partial y} = \sum_{\beta=1}^{N} \sum_{i=1}^{n} A_{\alpha i}^{\beta}(x_1,\ldots,x_n,\eta,u_1,\ldots,u_N)\frac{\partial u_\beta}{\partial x_i} + B_\alpha(x_1,\ldots,x_n,\eta,u_1,\ldots,u_N) ,$$

with initial conditions

$$\eta(x_1,\ldots,x_n,0) = 0$$

$$u_\alpha(x_1,\ldots,x_n,0) = 0 .$$

To sum up, we can consider equations

$$(1)\qquad \frac{\partial u_\alpha}{\partial y} = \sum_{\beta=1}^{N} \sum_{i=1}^{n} A^\beta_{\alpha i}(x_1,\ldots,x_n,u_1,\ldots,u_N)\frac{\partial u_\beta}{\partial x_i} + B_\alpha(x_1,\ldots,x_n,u_1,\ldots,u_N)$$

with initial conditions

$$(2)\qquad u_\alpha(x_1,\ldots,x_n,0) = 0 \ ;$$

the functions $A^\beta_{\alpha i}$ and B_α are analytic in a neighborhood of 0 in $\mathbb{R}^{N+n}$, and we are looking for a solution u in a neighborhood of 0. We expand the analytic functions $A^\beta_{\alpha i}$ and B_α around 0 as

$$(3)\qquad A^\beta_{\alpha i}(x_1,\ldots,x_n,z_1,\ldots,z_N) = \Sigma\, a^\beta_{\alpha i;\sigma_1,\ldots,\sigma_n,\tau_1,\ldots,\tau_N}\, x_1^{\sigma_1}\cdots z_1^{\tau_1}\cdots$$
$$= \sum_{\sigma,\tau} a^\beta_{\alpha i;\sigma,\tau} x^\sigma z^\tau$$

$$(4)\qquad B_\alpha(x_1,\ldots,x_n,z_1,\ldots,z_N) = \Sigma\, b_{\alpha;\sigma_1,\ldots,\sigma_n,\tau_1,\ldots,\tau_N}\, x_1^{\sigma_1}\cdots z_1^{\tau_1}\cdots$$
$$= \sum_{\sigma,\tau} b_{\alpha;\sigma,\tau} x^\sigma z^\tau \ .$$

We claim, first of all, that there is at most one analytic solution

$$(5)\qquad u_\alpha(x_1,\ldots,x_n,y) = \Sigma\, c_{\alpha;\sigma_1,\ldots,\sigma_n,\rho}\, x_1^{\sigma_1}\cdots x_n^{\sigma_n} y^\rho$$
$$= \sum_{\sigma,\rho} c_{\alpha;\sigma,\rho} x^\sigma y^\rho$$

of (1) and (2). We just have to show that the coefficients $c_{\alpha;\sigma,\rho}$ are completely determined by those of $A^\beta_{\alpha i}$ and B_α. From the particular way they are determined, we will then be able to show that the resulting series (5) converges, thereby also proving existence.

For a given n-tuple $\sigma = (\sigma_1,\ldots,\sigma_n)$, let $\sigma+\delta_i$ be the n-tuple

$$\sigma + \delta_i = (\sigma_1,\ldots,\sigma_i+1,\ldots,\sigma_n) .$$

Then if the u_α are given by (5), we can write

$$(6)\quad \begin{cases} \dfrac{\partial u_\beta}{\partial x_i}(x_1,\ldots,x_n,y) = \sum\limits_{\sigma,\rho} (\sigma_i+1)c_{\beta;\sigma+\delta_i,\rho}x^\sigma y^\rho \\ \dfrac{\partial u_\alpha}{\partial y}(x_1,\ldots,x_n,y) = \sum\limits_{\sigma,\rho} (\rho+1)c_{\alpha;\sigma,\rho+1}x^\sigma y^\rho . \end{cases}$$

So if the u_α in (5) satisfy (1), then

$$(7)\quad \begin{cases} \sum\limits_{\sigma,\rho} (\rho+1)c_{\alpha;\sigma,\rho+1}x^\sigma y^\rho \\ \quad = \sum\limits_{\beta=1}^{N}\sum\limits_{i=1}^{n}\left\{\sum\limits_{\sigma,\tau} a^\beta_{\alpha i;\sigma,\tau}x^\sigma\left(\sum\limits_{\sigma,\rho} c_{1;\sigma,\rho}x^\sigma y^\rho\right)^{\tau_1}\cdots\left(\sum\limits_{\sigma,\rho} c_{N;\sigma,\rho}x^\sigma y^\rho\right)^{\tau_N}\right\}\cdot \\ \qquad\qquad \cdot \sum\limits_{\sigma,\rho}(\sigma_i+1)c_{\beta;\sigma+\delta_i,\rho}x^\sigma y^\rho \\ \quad + \sum\limits_{\sigma,\tau} b_{\alpha;\sigma,\tau}x^\sigma\left(\sum\limits_{\sigma,\rho} c_{1;\sigma,\rho}x^\sigma y^\rho\right)^{\tau_1}\cdots\left(\sum\limits_{\sigma,\rho} c_{N;\sigma,\rho}x^\sigma y^\rho\right)^{\tau_N} . \end{cases}$$

Now there is no need to become unduly frightened by this expression. After we expand everything out, we will have an expression of the form

$$(8)\quad \sum_{\sigma,\rho}(\rho+1)c_{\alpha;\sigma,\rho+1}x^\sigma y^\rho = \sum_{\alpha,\sigma,\rho} P_{\alpha,\sigma,\rho}\left(a^\eta_{\xi j;\mu,\nu},\ b_{\xi;\mu\nu},\ c_{\xi;\mu\nu}\right)$$

$$= \sum_{\alpha,\beta,\rho} P_{\alpha,\sigma,\rho}(a,b,c) \qquad \text{for short,}$$

where each $P_{\alpha,\sigma,\rho}$ is a polynomial in certain of the $a^{\eta}_{\xi j;\mu\nu}$, $b_{\xi;\mu\nu}$, and $c_{\xi;\mu\nu}$. Just which of these appear as arguments of $P_{\alpha,\sigma,\rho}$ depends on (σ,ρ); the only important thing for us to note is that

(A) $P_{\alpha,\sigma,\rho}$ depends only on those $c_{\xi;\mu,\nu}$ with $\nu \leq \rho$.

Notice that all the information in equation (1) enters into (8) as the <u>arguments</u> of the polynomials $P_{\alpha,\sigma,\rho}$. These polynomials themselves <u>do not depend</u> on the $A^{\beta}_{\alpha i}$ or B_{α}, or on the $c_{\xi;\mu\nu}$:

(B) The polynomials $P_{\alpha,\sigma,\rho}$ are "universal polynomials" depending only on N and n.

Finally, we note that

(C) The coefficients of $P_{\alpha,\sigma,\rho}$ are <u>non-negative</u> integers.

Now if (8) is to hold for all sufficiently small $(x_1,\ldots,x_n,y)$, then we must have

$$(\rho+1)c_{\alpha;\sigma,\rho+1} = P_{\alpha,\sigma,\rho}(a,b,c) \ . \tag{9}$$

Together with the initial condition (2), which gives us $c_{\alpha;\sigma,0} = 0$, we can now calculate all $c_{\alpha;\sigma,\rho}$ recursively from (9), since (A) shows us that the right side involves only $c_{\xi;\mu,\nu}$ with $\nu \leq \rho$. This proves uniqueness.

To prove existence, we must show that the series (5) converges, when the $c_{\alpha;\sigma,\rho}$ are computed from (10). We will show absolute convergence for sufficiently small $(x_1,\ldots,x_n,y)$. This is done by the following trick, called the <u>method of majorants</u>. Consider another set of equations

$$(1') \quad \frac{\partial u_\alpha}{\partial y} = \sum_{\beta=1}^{N} \sum_{i=1}^{n} \bar{A}^{\beta}_{\alpha i}(x_1,\ldots,x_n,u_1,\ldots,u_N)\frac{\partial u_\beta}{\partial x_i} + \bar{B}_\alpha(x_1,\ldots,x_n,u_1,\ldots,u_N)$$

which "majorizes" (1), that is, which satisfies

$$(10) \quad \begin{cases} |a^{\eta}_{\xi j;\mu,\nu}| \leq \bar{a}^{\eta}_{\xi j;\mu,\nu} \\ |b_{\xi;\mu,\nu}| \leq \bar{b}_{\xi;\mu,\nu} \end{cases} \qquad \text{for all } \xi,\ \eta,\ j,\ \mu,\ \nu\ .$$

Suppose that equation (1'), with the same initial condition (2), has an analytic solution

$$(5') \quad u_\alpha(x_1,\ldots,x_n,y) = \sum_{\sigma,\rho} \bar{c}_{\alpha;\sigma,\rho} x^\sigma y^\rho\ ,$$

Then we must have

$$(8') \quad (\rho+1)\bar{c}_{\alpha;\sigma,\rho+1} = P_{\alpha,\sigma,\rho}(\bar{a},\bar{b},\bar{c})\ ,$$

where, by (B), the $P_{\alpha,\sigma,\rho}$ are the <u>same</u> universal polynomials as in (8). We claim that we can then conclude that

$$(11) \quad |c_{\alpha;\sigma,\rho}| \leq \bar{c}_{\alpha;\sigma,\rho}\ .$$

The proof is by induction. It is clear for $\rho = 0$. Now assume it is true for ρ. Then

$$\begin{aligned} |c_{\alpha;\sigma,\rho+1}| &= \frac{1}{\rho+1}|P_{\alpha,\sigma,\rho}(a,b,c)| && \text{by (8)} \\ &\leq \frac{1}{\rho+1} P_{\alpha,\sigma,\rho}(|a|,|b|,|c|) && \text{by (C)} \\ &\leq \frac{1}{\rho+1} P_{\alpha,\sigma,\rho}(\bar{a},\bar{b},\bar{c}) && \text{by (C) and (10)} \\ &= \bar{c}_{\alpha;\sigma,\rho+1} && \text{by (8').} \end{aligned}$$

This completes the induction proof. But now the convergence of (5'), together with (11), proves the absolute convergence of (5). So the proof of the theorem will be complete once we show that some majorant of equation (1) has an analytic solution. This will be comparatively easy.

For some $r > 0$, the power series (3) converges for the point (x,z) with all $x_j = z_\xi = r$. Then the terms

$$a^{\beta}_{\alpha i;\sigma,\tau}\, r^{\sigma_1+\cdots+\tau_N}$$

in this infinite sum must approach 0, and consequently are surely bounded: there is some M with

$$|a^{\beta}_{\alpha i;\sigma,\tau}| \leq \frac{M}{r^{\sigma_1+\cdots+\tau_N}}$$

for all σ, τ. For r small enough and M large enough, this equation holds for each $\alpha, \beta \leq N$ and $i \leq n$. Similarly, we can assume that we have the same estimate for each $|b_{\alpha;\sigma,\tau}|$. We thus also have the weaker estimates

$$|a^{\beta}_{\alpha i;\sigma,\tau}|,\ |b_{\alpha;\sigma,\tau}| \leq \frac{M}{r^{\sigma_1+\cdots+\tau_N}} \cdot \frac{(\sigma_1+\cdots+\tau_N)!}{\sigma_1!\cdots\tau_N!} .$$

So if we take $\bar{a}^{\beta}_{\alpha i;\sigma,\tau}$ and $\bar{b}_{\alpha;\sigma,\tau}$ to be the expression on the right side of this inequality, then equation (1') majorizes equation (1), and we just have to show that the solution of (1') with the initial condition (2) is analytic. Since

$$\sum_{\sigma,\tau} \bar{a}^{\beta}_{\alpha i;\sigma,\tau} x^{\sigma} z^{\tau} = \sum_{\sigma,\tau} \bar{b}_{\alpha;\sigma,\tau} x^{\sigma} z^{\tau}$$

$$= M \sum_{\sigma,\tau} \left(\frac{x}{r}\right)^{\sigma} \left(\frac{z}{r}\right)^{\tau} \frac{(\sigma_1+\cdots+\tau_N)!}{\sigma_1!\cdots\tau_N!}$$

$$= \frac{M}{1 - \dfrac{(x_1+\cdots+z_N)}{r}} ,$$

we are dealing with the equations

$$\begin{cases} \dfrac{\partial u_\alpha}{\partial y} = \dfrac{M}{1 - \dfrac{(x_1+\cdots+u_N)}{r}} \left(\sum_{\beta=1}^{N} \sum_{i=1}^{n} \dfrac{\partial u_\beta}{\partial x_i} + 1 \right) \\ u_\alpha(x_1,\ldots,x_n,0) = 0 . \end{cases}$$

Since the equations for the different u_α are all the same, and since the x_i enter only in the combination $X = x_1 + \cdots + x_n$, it seems reasonable to look for a solution with all

$$u_\alpha(x_1,\ldots,x_n,y) = U(x_1 + \cdots + x_n,y) = U(X,y) .$$

This gives us the single equation

$$\begin{cases} \dfrac{\partial U}{\partial y} = \dfrac{M}{1 - \dfrac{X + NU}{r}} \left(Nn \dfrac{\partial U}{\partial X} + 1 \right) \\ U(X,0) = 0 . \end{cases}$$

This is a first order equation, and hence one which, in theory, we can deal with. The simplest thing, at the moment, is just to check that

$$U(X,y) = \frac{r - X}{(n+1)N} - \frac{\sqrt{(r-X)^2 - 2N(n+1)Mry}}{N(n+1)}$$

is the desired solution, and that it is analytic in a neighborhood of 0. ■

Although the Cauchy-Kowalewski Theorem is the most general result in the theory of PDEs, its usefulness is greatly restricted by the fact that both the coefficients and the initial conditions must be real analytic. We would like to know whether these hypotheses are somehow dictated by the very nature of the problem, or whether they represent merely a defect in our method of proof. [One source of difficulty may be the fact that in one respect the theorem proves too much, since it is formulated for an arbitrary system of first order quasi-linear equations. Although it would be nice to solve any such system, this problem doesn't bear directly on the problem of solving a single higher order PDE, because only very special sorts of systems of first order equations are derived from higher order equations; given an arbitrary system of first order equations with initial conditions, we generally cannot find a single higher order equation with initial conditions that is equivalent to it.]

The necessity of having analytic coefficients in the Cauchy-Kowalewski Theorem is demonstrated by the following simple example, due to Perron [1], of two linear first order equations for two unknown functions $u_1, u_2 \colon \mathbb{R}^2 \to \mathbb{R}$,

$$\begin{cases} \dfrac{\partial u_1}{\partial x} = \dfrac{\partial u_1}{\partial y} + \dfrac{\partial u_2}{\partial y} \\[2ex] \dfrac{\partial u_2}{\partial x} = a\,\dfrac{\partial u_1}{\partial y} + \dfrac{\partial u_2}{\partial y} + f(x+y) \;. \end{cases}$$

If the constant a is negative, and f is not analytic, then this system has no solution with initial conditions $u_i(x,0) = 0$. The famous example of Hans Lewy [1],

$$\begin{cases} \dfrac{\partial u_1}{\partial x_1} = \dfrac{\partial u_2}{\partial x_2} - 2x_2 \dfrac{\partial u_1}{\partial x_3} - 2x_1 \dfrac{\partial u_2}{\partial x_3} - f'(x_3) \\[2ex] \dfrac{\partial u_2}{\partial x_1} = -\dfrac{\partial u_1}{\partial x_2} + 2x_1 \dfrac{\partial u_1}{\partial x_3} - 2x_2 \dfrac{\partial u_2}{\partial x_3} , \end{cases}$$

has no solutions _whatsoever_ if f is C^∞ but not analytic! This system can be considered as the following single equation for the complex-valued function $u = u_1 + iu_2$:

$$-\frac{\partial u}{\partial x_1} - i\frac{\partial u}{\partial x_2} + 2i(x_1 + ix_2)\frac{\partial u}{\partial x_3} = f'(x_3) .$$

There are also cases where _analytic initial conditions_ are necessary, for we will soon see that there are simple PDEs, with analytic coefficients, that cannot have solutions with given initial conditions unless these conditions are analytic. So in a certain respect the Cauchy-Kowalewski theorem gives the best possible result in these cases. On the other hand, it turns out that the Cauchy problem isn't even the one which we want to pose for these equations. Moreover, there is a wide class of equations where the Cauchy problem is a natural one, but where analyticity is much too severe a restriction.

5. Classification of Second Order PDEs

In our first forage into the unchartered lands of higher order PDEs, it is natural that we first restrict our attention to those of second order:

$$F\left(x_1,\ldots,x_n,u(x_1,\ldots,x_n),\frac{\partial u}{\partial x_1}(x_1,\ldots,x_n),\ldots,\frac{\partial^2 u}{\partial x_n{}^2}(x_1,\ldots,x_n)\right) = 0 .$$

As a matter of fact, we will never get anywhere beyond this. Moreover, we will deal almost exclusively with equations in only two variables,

$$F\left(x,y,u(x,y),\frac{\partial u}{\partial x}(x,y),\frac{\partial u}{\partial y}(x,y),\frac{\partial^2 u}{\partial x^2}(x,y),\frac{\partial^2 u}{\partial x\partial y}(x,y),\frac{\partial^2 u}{\partial y^2}(x,y)\right) = 0 ,$$

or in abbreviated form

$$F(x,y,u,p,q,r,s,t) = 0 .$$

Even nowadays there are certain phenomena about second order PDEs which are much more completely understood in the two variable case than in higher dimensions, but the particular results that we are after can all be handled in a uniform way that works in all dimensions. However, a whole book would be required in order to reach them. So we will instead use quite classical methods to analyse second order PDEs in just 2 variables. Fortunately, the 2 variable case happens to be just the one we are interested in.

We begin by singling out the <u>semi-linear</u> equations

$$\begin{aligned}\text{(I)}\qquad 0 &= a(x,y)u_{xx} + 2b(x,y)u_{xy} + c(x,y)u_{yy} + f(x,y,u,u_x,u_y)\\ &= L(u) + f .\end{aligned}$$

For such an equation, $L(u)$ is called the "principal part," and we will often denote the remaining part, involving only lower order derivatives, by "...". There is a classification for these equations which is closely related to the classification of algebraic equations of the form

$$z = ax^2 + 2bxy + cy^2 = \left\langle (x,y),\ (x,y)\cdot\begin{pmatrix} a & b\\ b & c\end{pmatrix}\right\rangle ,$$

with a, b, and c not all 0. We briefly remind the reader how this classification goes (it is essentially the same as the classification of points on a surface in Chapter 2). We choose two orthonormal eigenvectors $X_1, X_2 \in \mathbb{R}^2$ for the symmetric matrix $\begin{pmatrix} a & b \\ b & c \end{pmatrix}$, with corresponding eigenvalues λ_1, λ_2. Then

$$\left\langle \phi X_1 + \psi X_2, (\phi X_1 + \psi X_2)\cdot\begin{pmatrix} a & b \\ b & c \end{pmatrix}\right\rangle = \langle \phi X_1 + \psi X_2, \phi\lambda_1 X_1 + \psi\lambda_2 X_2 \rangle$$
$$= \lambda_1\phi^2 + \lambda_2\psi^2 .$$

So if we use ϕ and ψ as new coordinates for $\mathbb{R}^2$, our equation becomes

$$z = \lambda_1\phi^2 + \lambda_2\psi^2 .$$

We can express this statement a little more precisely in terms of the linear transformation $S = (\phi,\psi): \mathbb{R}^2 \to \mathbb{R}^2$ defined by

$$S(X_1) = (1,0) \qquad S(X_2) = (0,1) .$$

Since

$$S(x,y) = (\phi(x,y),\psi(x,y)) = \phi(x,y)S(X_1) + \psi(x,y)S(X_2) ,$$

the functions ϕ and ψ are just the coordinates of (x,y) with respect to X_1 and X_2:

$$(x,y) = \phi(x,y)X_1 + \psi(x,y)X_2 .$$

So we obtain

$$\text{(II)}\qquad ax^2 + 2bxy + cy^2 = \left\langle (x,y),\ (x,y)\cdot\begin{pmatrix} a & b \\ b & c \end{pmatrix} \right\rangle$$

$$= \left\langle \phi(x,y)X_1 + \psi(x,y)X_2, (\phi(x,y)X_1 + \psi(x,y)X_2)\cdot\begin{pmatrix} a & b \\ b & c \end{pmatrix} \right\rangle$$

$$= \langle \phi(x,y)X_1 + \psi(x,y)X_2,\ \phi(x,y)\lambda_1 X_1 + \psi(x,y)\lambda_2 X_2 \rangle$$

$$= \lambda_1[\phi(x,y)]^2 + \lambda_2[\psi(x,y)]^2 .$$

These algebraic manipulations are often expressed slightly differently. We can write equation (II) as

$$(x,y)\begin{pmatrix} a & b \\ b & c \end{pmatrix}\begin{pmatrix} x \\ y \end{pmatrix} = (\phi(x,y),\psi(x,y))\begin{pmatrix} \lambda_1 & 0 \\ 0 & \lambda_2 \end{pmatrix}\begin{pmatrix} \phi(x,y) \\ \psi(x,y) \end{pmatrix}$$

$$= S(x,y)\begin{pmatrix} \lambda_1 & 0 \\ 0 & \lambda_2 \end{pmatrix}[S(x,y)]^t ,$$

where t denotes the transpose. If Q is the matrix of the linear transformation S, then we can write

$$(x,y)\begin{pmatrix} a & b \\ b & c \end{pmatrix}\begin{pmatrix} x \\ y \end{pmatrix} = [(x,y)\cdot Q]\begin{pmatrix} \lambda_1 & 0 \\ 0 & \lambda_2 \end{pmatrix}[(x,y)\cdot Q]^t$$

$$= (x,y)\left[Q\begin{pmatrix} \lambda_1 & 0 \\ 0 & \lambda_2 \end{pmatrix}Q^t\right]\begin{pmatrix} x \\ y \end{pmatrix} \qquad \text{for all } (x,y) ,$$

which implies that

$$\text{(III)}\qquad \begin{pmatrix} a & b \\ b & c \end{pmatrix} = Q\begin{pmatrix} \lambda_1 & 0 \\ 0 & \lambda_2 \end{pmatrix}Q^t .$$

It is not hard to see exactly what the matrix Q is. Since S^{-1} takes $(1,0)$ to X_1 and $(0,1)$ to X_2, its matrix has X_1 and X_2 as its two columns,

$$Q^{-1} = (X_1{}^t, X_2{}^t) .$$

Moreover, since X_1 and X_2 are orthonormal, we have $Q^{-1}(Q^{-1})^t = I$. So

$$Q = (Q^{-1})^t = \begin{pmatrix} X_1 \\ X_2 \end{pmatrix} .$$

Since $Q^{-1} = Q^t$, we can also write (III) is

$$\text{(IV)} \qquad \begin{pmatrix} \lambda_1 & 0 \\ 0 & \lambda_2 \end{pmatrix} = Q^t \begin{pmatrix} a & b \\ b & c \end{pmatrix} Q .$$

The reduction (II) [or its equivalent (IV)] shows that the equations $z = ax^2 + 2bxy + cy^2$ fall into three classes:

Elliptic Case: $ac - b^2 > 0$; equivalently, λ_1 and λ_2 have the same sign

Hyperbolic Case: $ac - b^2 < 0$; equivalently, λ_1 and λ_2 has opposite signs

Parabolic Case: $ac - b^2 = 0$; λ_1 or $\lambda_2 = 0$.

We introduce a similar classification for semi-linear PDEs

$$\text{(I)} \qquad a(x,y)u_{xx} + 2b(x,y)u_{xy} + c(x,y)u_{yy} + \cdots = 0 .$$

If a_0, b_0, and c_0 are the values of the functions a, b, and c at (x_0,y_0), then we say that equation (I) is

<u>elliptic at</u> (x_0,y_0) if $a_0c_0 - b_0^2 > 0$

<u>hyperbolic at</u> (x_0,y_0) if $a_0c_0 - b_0^2 < 0$

<u>parabolic at</u> (x_0,y_0) if $a_0c_0 - b_0^2 = 0$

(but not all of a_0, b_0, c_0 are 0) .

Naturally we say that equation (I) is <u>elliptic</u> in an open set U if it is elliptic at each point $(x,y) \in U$, etc. The simplest examples of equations of these three types are the "normal forms"

$$\text{(E)} \qquad u_{xx} + u_{yy} + \cdots = 0$$

$$\text{(H)} \qquad u_{xx} - u_{yy} + \cdots = 0$$

$$\text{(P)} \qquad u_{xx} \qquad + \cdots = 0 .$$

As always, "..." denotes terms which do not involve any second derivatives. There is in addition an alternative normal form in the hyperbolic case,

$$\text{(H')} \qquad u_{xy} + \cdots = 0$$

(corresponding to the possibility of writing the equation for a hyperbola in the form $xy = 1$).

We would like to see if equation (I) can be reduced to a normal form by writing it in terms of a function v defined by

$$\text{(V)} \qquad u(x,y) = v(\phi(x,y),\psi(x,y)) ;$$

here (ϕ,ψ) is supposed to be a differentiable map from $\mathbb{R}^2$ to $\mathbb{R}^2$ with differentiable inverse. Denoting a typical point in the domain of v by (ξ,η), and the partials of v by v_ξ and v_η, we compute that

$$u_x = v_\xi\phi_x + v_\eta\psi_x , \qquad u_y = v_\xi\phi_y + v_\eta\psi_y$$

and then that

$$\text{(VI)} \quad \begin{cases} u_{xx} = v_{\xi\xi}\phi_x{}^2 + 2v_{\xi\eta}\phi_x\psi_x + v_{\eta\eta}\psi_x{}^2 + \cdots \\ u_{xy} = v_{\xi\xi}\phi_x\phi_y + v_{\xi\eta}(\phi_x\psi_y + \phi_y\psi_x) + v_{\eta\eta}\psi_x\psi_y + \cdots \\ u_{yy} = v_{\xi\xi}\phi_y{}^2 + 2v_{\xi\eta}\phi_y\eta_y + v_{\eta\eta}\psi_y{}^2 + \cdots \end{cases}$$

where "..." again denotes terms which do not involve any second derivatives. [Naturally, if the derivatives of u are evaluated at (x,y), then the derivatives of ϕ and ψ are evaluated at (x,y), while those for v are evaluated at $(\phi(x,y),\psi(x,y))$.] From this we easily see that

$$\text{(VII)} \quad \begin{cases} au_{xx} + 2bu_{xy} + cu_{yy} = \alpha v_{\xi\xi} + 2\beta v_{\xi\eta} + \gamma v_{\eta\eta} + \cdots \\ \qquad\qquad \underline{\text{where}} \\ \alpha = a\phi_x{}^2 + 2b\phi_x\phi_y + c\phi_y{}^2 \\ \beta = a\phi_x\psi_x + b(\phi_x\psi_y + \phi_y\psi_x) + c\phi_y\psi_y \\ \gamma = a\psi_x{}^2 + 2b\psi_x\psi_y + c\psi_y{}^2 \\ \qquad\qquad \underline{\text{or equivalently}} \\ \begin{pmatrix}\alpha & \beta\\ \beta & \gamma\end{pmatrix} = \begin{pmatrix}\phi_x & \phi_y\\ \psi_x & \psi_y\end{pmatrix}\begin{pmatrix}a & b\\ b & c\end{pmatrix}\begin{pmatrix}\phi_x & \psi_x\\ \phi_y & \psi_y\end{pmatrix} . \end{cases}$$

Notice that the last part of (VII) shows that

$$\text{(VIII)} \qquad \alpha\gamma - \beta^2 = (ac - b^2)(\phi_x\psi_y - \phi_y\psi_x)^2 \; ;$$

therefore the type of the equation for v is always the same as the type for u.

In one case, purely algebraic manipulations will reduce our equation to normal form:

5. PROPOSITION. Suppose that the equation

$$\text{(I)} \qquad a u_{xx} + b u_{xy} + c u_{yy} + \cdots = 0$$

has constant coefficients a, b, and c in the principal part. Then there is a non-singular linear transformation $(\phi,\psi): \mathbb{R}^2 \to \mathbb{R}^2$ having the property that if v is defined by

$$\text{(V)} \qquad u(x,y) = v(\phi(x,y),\psi(x,y)) \ ,$$

then u satisfies (I) if and only if v satisfies a certain equation of the form (E), (H), or (P). In the hyperbolic case, we can also find (ϕ,ψ) so that u satisfies (I) if and only if v satisfies a certain equation of the form (H').

Proof. Equation (IV) shows that we can choose a constant matrix $Q = \begin{pmatrix} \phi_x & \phi_x \\ \phi_y & \psi_y \end{pmatrix}$ so that

$$\begin{pmatrix} \alpha & \beta \\ \beta & \gamma \end{pmatrix} = \begin{pmatrix} \lambda_1 & 0 \\ 0 & \lambda_2 \end{pmatrix} \qquad \text{in equation (VII)} \ .$$

Since $\det Q \neq 0$, the linear transformation (ϕ,ψ) is non-singular. If we define v by (V), then (VII) shows that equation (I) for u is equivalent to the equation

$$\text{(1)} \qquad \lambda_1 v_{\xi\xi} + \lambda_2 v_{\eta\eta} + \cdots = 0$$

for v. If we make a further change of coordinates by defining

$$\tilde{v}(\rho,\sigma) = v(r\rho, s\sigma) \qquad r,\ s \ \text{ constants} \ ,$$

then

$$\tilde{v}_{\rho\rho} = r^2 v_{\xi\xi} \quad \text{and} \quad \tilde{v}_{\sigma\sigma} = s^2 v_{\eta\eta} .$$

So we can also arrange for λ_1 and λ_2 to be ± 1 in (1). Then we have equations equivalent to (E), (H), or (P) [we may have to interchange the names of ξ and η].

The form (H') is obtained by analogy with the fact that the equation $x^2 - y^2 = 1$ becomes $4\bar{x}\bar{y} = 1$ when we perform the substitution $\bar{x} = \frac{1}{2}(x+y)$, $\bar{y} = \frac{1}{2}(x-y)$. We define

$$\text{(2)} \qquad w(\rho,\sigma) = v\left(\frac{\rho+\sigma}{2}, \frac{\rho-\sigma}{2}\right) .$$

Then

$$w_\rho = \frac{1}{2}v_\xi + \frac{1}{2}v_\eta$$
$$w_{\rho\sigma} = \frac{1}{2}\left(\frac{1}{2}v_{\xi\xi} - \frac{1}{2}v_{\xi\eta}\right) + \frac{1}{2}\left(\frac{1}{2}v_{\eta\xi} - \frac{1}{2}v_{\eta\eta}\right) = \frac{1}{4}\left(v_{\xi\xi} - v_{\eta\eta}\right) .$$

So an equation of the form (H) for v is equivalent to one of the form (H') for w. ■

The same method that was used in this proof will clearly enable us to reduce the general semi-linear equation (I) to an equation which has the normal form <u>at one point</u> (x_0,y_0). But to obtain the normal form in a whole neighborhood, we have to work much harder. We consider the elliptic case first.

<u>6. THEOREM.</u> Suppose that the equation

$$\text{(I)} \qquad a(x,y)u_{xx} + 2b(x,y)u_{xy} + c(x,y)u_{yy} + \cdots = 0$$

is elliptic in a neighborhood of (x_0,y_0). Then there is a differentiable map (ϕ,ψ) from a neighborhood of (x_0,y_0) into $\mathbb{R}^2$, with differentiable inverse, having the property that if v is defined by

$$\text{(V)} \qquad u(x,y) = v(\phi(x,y),\psi(x,y)) ,$$

then u satisfies (I) if and only if v satisfies a certain equation in the normal form (E).

Proof. It obviously suffices to find (ϕ,ψ) so that in equation (VII) we have $\alpha = \gamma$ and $\beta = 0$. So it suffices to find (ϕ,ψ) with

$$\text{(1)} \qquad a{\phi_x}^2 + 2b\phi_x\phi_y + c{\phi_y}^2 = a{\psi_x}^2 + 2b\psi_x\psi_y + c{\psi_y}^2$$

$$\text{(2)} \qquad a\phi_x\psi_x + b(\phi_x\psi_y + \phi_y\psi_x) + c\phi_y\psi_y = 0 ,$$

and $\phi_x\psi_y - \phi_y\psi_x \neq 0$ at (x_0,y_0). This is precisely the problem of introducing isothermal coordinates for the metric $a\, dx\otimes dx + b[dx\otimes dy + dy\otimes dx] + c\, dy\otimes dy$, which we solved in Addendum 1 to Chapter 9. ∎

In our proof of the existence of isothermal coordinates, we showed that (1) and (2) are equivalent to the "Beltrami equations"

$$\text{(a)} \qquad \phi_x = \frac{b\psi_x + c\psi_y}{\sqrt{ac - b^2}} , \qquad \phi_y = -\frac{a\psi_x + b\psi_y}{\sqrt{ac - b^2}} .$$

Note that if (a) is to hold, then we must have

$$\text{(b)} \qquad \frac{\partial}{\partial x}\left(\frac{a\psi_x + b\psi_y}{W}\right) + \frac{\partial}{\partial y}\left(\frac{b\psi_x + c\psi_y}{W}\right) = 0 , \qquad W = \sqrt{ac - b^2} .$$

Conversely, if (b) holds for some ϕ, then there is ψ satisfying (a); moreover, the Jacobian of (ϕ,ψ) is

$$\phi_x\psi_y - \phi_y\psi_x = -\frac{1}{W}(a\phi_x{}^2 + 2b\phi_x\phi_y + c\phi_y{}^2) ,$$

which is non-zero if $(\phi_x,\phi_y) \neq (0,0)$. So solving (a) is equivalent to solving (b), which is itself elliptic [with the very same principal part as (I)]. Had we not already solved equation (a), we would be in the embarrasing position of needing to know that elliptic equations have solutions before we could reduce them to normal form.

In the hyperbolic case, the same line of reasoning leads us into precisely this difficulty, and thus requires results from section 7. However, there is also an elementary argument.

7. THEOREM. Suppose that the equation

$$\text{(I)} \qquad a(x,y)u_{xx} + 2b(x,y)u_{xy} + c(x,y)u_{yy} + \cdots = 0$$

is hyperbolic at (x_0,y_0). Then there is a differentiable map (ϕ,ψ) from a neighborhood of (x_0,y_0) into $\mathbb{R}^2$, with differentiable inverse, having the property that if v is defined by

$$\text{(V)} \qquad u(x,y) = v(\phi(x,y),\psi(x,y)) ,$$

then u satisfies (I) if and only if v satisfies a certain equation in the normal form (H). The same result holds for a certain equation in the normal form (H').

<u>First Proof</u>. We claim, first of all, that we can assume that $c_0 = c(x_0,y_0) \neq 0$. For suppose that $c_0 = 0$. Choose (ϕ,ψ) to be a linear transformation with matrix $\begin{pmatrix}1&0\\ \lambda&1\end{pmatrix}$. Then at (x_0,y_0) the coefficients α, β, γ of the equation for v are, by (VII),

$$\begin{pmatrix}\alpha&\beta\\ \beta&\gamma\end{pmatrix} = \begin{pmatrix}1&0\\ \lambda&1\end{pmatrix}\begin{pmatrix}a_0&b_0\\ b_0&0\end{pmatrix}\begin{pmatrix}1&\lambda\\ 0&1\end{pmatrix} = \begin{pmatrix}a_0 & \lambda a_0+b_0\\ \lambda a_0+b_0 & \lambda^2 a_0+2\lambda b_0\end{pmatrix}.$$

Since we must have $a_0 \neq 0$ or $b_0 \neq 0$, we can certainly choose λ with $\lambda^2 a_0 + 2\lambda b_0 \neq 0$.

So we assume that $c(x_0,y_0) \neq 0$. To achieve the normal form (H), it now clearly suffices to choose (ϕ,ψ) so that in equation (VII) we have $\alpha = -\gamma$ and $\beta = 0$. Solving as before, we end up with the system of equations

$$\phi_x = \frac{b\psi_x + c\psi_y}{\sqrt{b^2 - ac}}$$

$$\phi_y = -\frac{a\psi_x + b\psi_y}{\sqrt{b^2 - ac}},$$

or equivalently

$$\begin{cases} \phi_y = -\dfrac{b}{c}\,\phi_x + \dfrac{\sqrt{b^2 - ac}}{c}\,\psi_x \\[2ex] \psi_y = \dfrac{\sqrt{b^2 - ac}}{c}\,\phi_x - \dfrac{b}{c}\,\psi_x . \end{cases}$$

Section 7 shows that we can solve this ("hyperbolic") system.

To obtain the normal form (H'), we start with v satisfying an equation of the normal form (H),

$$v_{\xi\xi} - v_{\eta\eta} + \cdots = 0 ,$$

and define

$$(2) \qquad w(\rho,\sigma) = v\left(\frac{\rho+\sigma}{2},\frac{\rho-\sigma}{2}\right) .$$

As in the proof of Proposition 5, we find that w satisfies an equation of the normal form (H').

<u>Second Proof</u>. We can instead try for the normal form (H') directly; the normal form (H) is then obtained by the same change of coordinates used in (2), which is equal to its own inverse, up to a factor of 2. So it suffices to find (ϕ,ψ) so that in equation (VII) we have $\alpha = \gamma = 0$; thus we need

$$(1) \qquad \begin{cases} a\phi_x^{\;2} + 2b\phi_x\phi_y + c\phi_y^{\;2} = 0 \\ a\psi_x^{\;2} + 2b\psi_x\psi_y + c\psi_y^{\;2} = 0 \end{cases}$$

and

$$(2) \qquad \phi_x\psi_y - \phi_y\psi_x \neq 0 .$$

As in the previous proof, we can assume that $c(x_0,y_0) \neq 0$. If there is any hope of solving (1) and (2), we clearly must have $\phi_x, \psi_x \neq 0$. This suggests that we look at the equations

$$(1') \qquad \begin{cases} a + 2b\left(\dfrac{\phi_y}{\phi_x}\right) + c\left(\dfrac{\phi_y}{\phi_x}\right)^2 = 0 \\[2ex] a + 2b\left(\dfrac{\psi_y}{\psi_x}\right) + c\left(\dfrac{\psi_y}{\psi_x}\right)^2 = 0 \end{cases}$$

(2')
$$\frac{\phi_y}{\phi_x} \neq \frac{\psi_y}{\psi_x} .$$

Clearly (1') and (2') imply (1) and (2). Now $ac - b^2 < 0$, so the equation

(3)
$$a(x,y) + 2b(x,y)\mu + c(x,y)\mu^2 = 0 \qquad c(x,y) \neq 0$$

always has two <u>distinct</u>, <u>real</u> roots $\mu_1(x,y)$ and $\mu_2(x,y)$, varying continuously with (x,y). So we just have to find ϕ and ψ satisfying

(4)
$$\begin{cases} \phi_y - \mu_1\phi_x = 0 & \phi_x \neq 0 \text{ at } (x_0,y_0) \\ \psi_y - \mu_2\psi_x = 0 & \psi_x \neq 0 \text{ at } (x_0,y_0) , \end{cases}$$

in order for (1') and (2') to hold. But the two equations in (4) are each linear first order PDEs, and the line $y = y_0$ is free, for any given initial conditions $\phi(x,y_0)$ and $\psi(x,y_0)$; in particular, we can assure that $\phi_y(x_0,y_0), \psi_y(x_0,y_0) \neq 0$. ■

Finally, there is no problem in the parabolic case.

<u>8. THEOREM</u>. Suppose that the equation

(I)
$$a(x,y)u_{xx} + 2b(x,y)u_{xy} + c(x,y)u_{yy} + \cdots = 0$$

is parabolic in a neighborhood of (x_0,y_0). Then there is a differentiable map (ϕ,ψ) from a neighborhood of (x_0,y_0) into $\mathbb{R}^2$, with differentiable inverse, having the property that if v is defined by

(V)
$$u(x,y) = v(\phi(x,y),\psi(x,y)) ,$$

then u satisfies (I) if and only if v satisfies a certain equation in the normal form (P).

Proof. It obviously suffices to find (ϕ,ψ) with $\phi_x\psi_y - \phi_y\psi_x \neq 0$ so that in equation (VII) we have $\gamma = 0$. For then β must be 0 by (VIII), while the last part of (VII) shows that α cannot be zero. We thus want

$$a\psi_x^{\ 2} + 2b\psi_x\psi_y + c\psi_y^{\ 2} = 0 \ . \tag{1}$$

We obviously must have $a(x_0,y_0) \neq 0$ or $c(x_0,y_0) \neq 0$, say the latter. It suffices to find ψ with $\psi_x(x_0,y_0) \neq 0$ and

$$a + 2b\left(\frac{\psi_y}{\psi_x}\right) + c\left(\frac{\psi_y}{\psi_x}\right)^2 = 0 \tag{1'}$$

in a neighborhood of (x_0,y_0); for we can then take $\phi(x,y) = x$, and $(\phi_x\psi_y - \phi_y\psi_x)(x_0,y_0) = \psi_y(x_0,y_0) \neq 0$. But equation (1') is simply equivalent to

$$\frac{\psi_y}{\psi_x} = -\frac{b}{c} \qquad \text{or} \qquad c\psi_x = -b\psi_y \ .$$

This is a first order linear PDE, and the line $y = y_0$ is free near (x_0,y_0), since $c(x_0,y_0) \neq 0$. So we can find a solution with arbitrary values of $\psi(x,y_0)$ for x near x_0, in particular with $\psi_x(x_0,y_0) \neq 0$. If $c(x_0,y_0) = 0$, we look at ψ_x/ψ_y instead. ∎

There is an especially important characterization of elliptic semi-linear PDEs

$$\text{(I)} \qquad a(x,y)u_{xx} + 2b(x,y)u_{xy} + c(x,y)u_{yy} + \cdots = 0 ,$$

which is the basis for extending the definition of ellipticity to more general equations. Consider an initial curve c in $\mathbb{R}^2$. According to the definitions of section 2, when we are considering the PDE (I), the curve c is free if and only if

$$a\nu_1^2 + 2b\nu_1\nu_2 + c\nu_2^2 \neq 0 \qquad \text{on } c$$

(the initial conditions are irrelevant in the semi-linear case). Now if $ac - b^2 > 0$, then the equation

$$0 = a\lambda^2 + 2b\lambda\mu + c\mu^2$$

has no real roots at all, except $\lambda = \mu = 0$. So if (I) is elliptic, then <u>any</u> initial curve c is free for any initial conditions. On the other hand, if (I) is hyperbolic, then there is a two parameter family of characteristic curves which fail to be free at all points.

For a general second order equation

$$\text{(1)} \qquad F(x,y,u,p,q,r,s,t) = 0 ,$$

we define a given solution u to be <u>elliptic</u> if

$$4F_rF_t - F_s^2 > 0$$

at all points

$$(x,y,u(x,y),u_x(x,y),u_y(x,y),u_{xx}(x,y),u_{xy}(x,y),u_{yy}(x,y)) .$$

In this case, any initial curve c is free for the initial data

$$u|c\ , \quad u_{x_i}|c\ , \quad u_{x_i x_j}|c\ .$$

We define a given solution u to be hyperbolic or parabolic in the obvious analogous way. Notice that for a given PDE which is not semi-linear, and even for a quasi-linear PDE, there may be solutions which are elliptic and also solutions which are hyperbolic or parabolic. The simplest example is the equation

$$u_{xx} + u \cdot u_{yy} = 0\ .$$

The solution $u = 1$ is elliptic, and the solution $u = -1$ is hyperbolic. This may seem like a ridiculous distinction, but, as we shall learn in sections 8 and 9, solutions near 1 will have entirely different properties from solutions near -1. A more natural example is the equation

$$(1-u_x^2)u_{xx} - 2u_x u_y u_{xy} + (1-u_y^2)u_{yy} = 0\ ,$$

which occurs in gas dynamics. The solution u is elliptic if and only if $u_x^2 + u_y^2 < 1$. Such solutions represent "subsonic" flow, while hyperbolic solutions represent "supersonic" flow. Thus we see that the terms elliptic, hyperbolic, and parabolic, do not make sense for the general second order equation; these terms apply to solutions of the equation, rather than to the equation itself. On the other hand, it clearly makes sense to apply the terms elliptic, hyperbolic, and parabolic to given initial data along a given initial curve.

It is important to observe that the type of a solution remains the same under a change of variable, just as in the semi-linear case. Suppose that u

is a solution of

$$(1) \qquad F(x,y,u,p,q,r,s,t) = 0 ,$$

and that we write

$$(V) \qquad u(x,y) = v(\phi(x,y),\psi(x,y))$$

for a diffeomorphism (ϕ,ψ). Using equations (VI), we see that v satisfies an equation

$$(2) \qquad G(x,y,v,v_\xi,v_\eta,v_{\xi\xi},v_{\xi\eta},v_{\eta\eta}) = 0 ,$$

where G has the form

$$G(\_\_,r,s,t) = F(\_\_,r',s',t')$$

for

$$r' = \phi_x{}^2 r + 2\phi_x\psi_x s + \psi_x{}^2 t$$

$$s' = \phi_x\phi_y r + (\phi_x\psi_y + \phi_y\psi_x)s + \psi_x\psi_y t$$

$$t' = \phi_y{}^2 r + 2\phi_y\eta_y s + \psi_y{}^2 t .$$

Hence

$$G_r = F_r\phi_x{}^2 + F_s\phi_x\phi_y + F_t\phi_y{}^2 , \qquad \text{etc.},$$

and we find that

$$\begin{pmatrix} G_r & \frac{1}{2}G_s \\ \frac{1}{2}G_s & G_t \end{pmatrix} = \begin{pmatrix} \phi_x & \phi_y \\ \psi_x & \psi_y \end{pmatrix} \begin{pmatrix} F_r & \frac{1}{2}F_s \\ \frac{1}{2}F_s & F_t \end{pmatrix} \begin{pmatrix} \phi_x & \psi_x \\ \phi_y & \psi_y \end{pmatrix} .$$

Therefore

$$G_r G_t - \frac{1}{4} G_s^2 = (F_r F_t - \frac{1}{4} F_s^2) \cdot (\phi_x \psi_y - \phi_y \psi_x)^2 .$$

Consequently, u is an elliptic or hyperbolic solution of (1) if and only if v is an elliptic or hyperbolic solution of (2).

A few remarks might be made concerning the definitions in higher dimensions (which we do not actually use). A semi-linear PDE

$$\sum_{i,j=1}^{n} a_{ij}(x_1,\ldots,x_n) u_{x_i x_j} + f(x_1,\ldots,x_n,u.\ldots u_{x_i}\ldots.) = 0$$

is called elliptic at a point $x = (x_1,\ldots,x_n)$ if the matrix $(a_{ij}(x))$ is definite, and elliptic in a region if it is elliptic at each point of the region. For an elliptic semi-linear PDE, any initial manifold will be free. We can define several different sorts of hyperbolicity and parabolicity, depending on the rank and signature of this matrix when it is not definite. Proposition 5 generalizes; if the a_{ij} are constants then our equation is equivalent to a normal form

$$u_{x_1 x_1} + \cdots + u_{x_r x_r} - u_{x_{r+1} x_{r+1}} - \cdots - u_{x_s x_s} + \cdots = 0 .$$

But Theorems 6-8 do not generalize; there are too many conditions to be satisfied by the diffeomorphism which changes u to v. We can also define when a given solution u of the general second order PDE is elliptic, in a fairly obvious way which is left to the reader.

The fact that any initial curve for an elliptic semi-linear 2nd order equation in 2 variables is free, while a hyperbolic semi-linear equation always has initial curves which are characteristic, certainly suggests that elliptic and hyperbolic equations might have quite different properties. But some of the most important reasons for this classification of 2nd order equations come from physics, which provides the motivation for many of the basic problems about them. So we will first make a brief excursion into this forbidding domain.

6. The Prototypical PDEs of Physics

We are going to begin by deriving certain classical PDEs which describe important (somewhat idealized) physical situations. The word "derive" had better be taken with a hefty grain of salt, however. What I have really tried to do is give plausible reasons why the physical situations should be governed by those PDEs which the physicists have agreed upon. I've never really been able to understand which parts of the standard derivations are supposed to be obvious, which are mathematically simplifying assumptions, which steps are supposed to correspond to empirically discovered physical laws, or even what all the words are supposed to mean.

The first idealized physical situation which we want to describe is a vibrating string which is not acted upon by any outside forces. We naturally consider this string to be one-dimensional, and we will assume that the motion actually takes place in a plane. We may regard our string as being either of finite length with fixed endpoints, or of infinite length. The first possibility corresponds to a string stretched between two prongs, while the second is a more idealized conception. We will let $u(x,t)$ denote the height of the string above $(x,0)$ at time t.

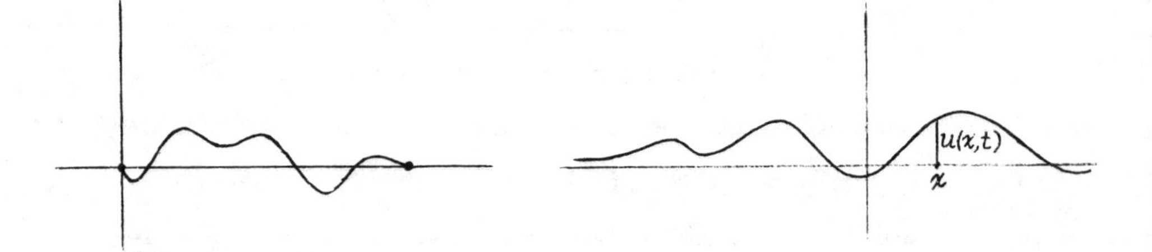

In order to apply the laws of mechanics to the string, we will first regard it as a discrete collection of point masses, whose x-coordinates are all some small distance h apart. In the course of time, the kth particle moves up

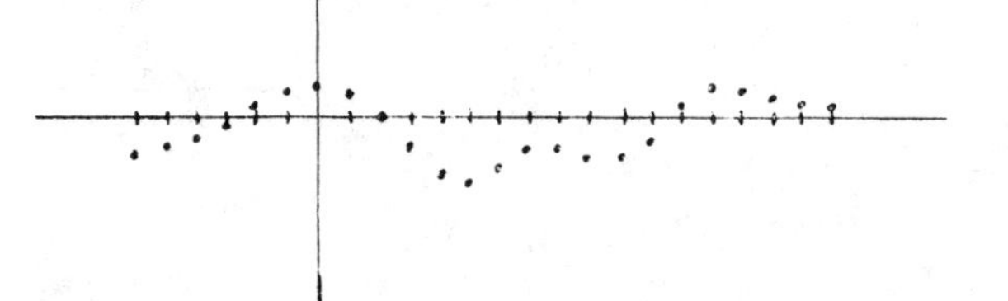

and down (but not sideways); its height above the x-axis at time t will be denoted by $u(x_k,t)$. The only forces acting are the "tension" forces between pairs of particles; physically these come about because the motion of the string involves slight changes in the distances between molecules, to which the intermolecular forces are extremely sensitive. We will assume that each particle

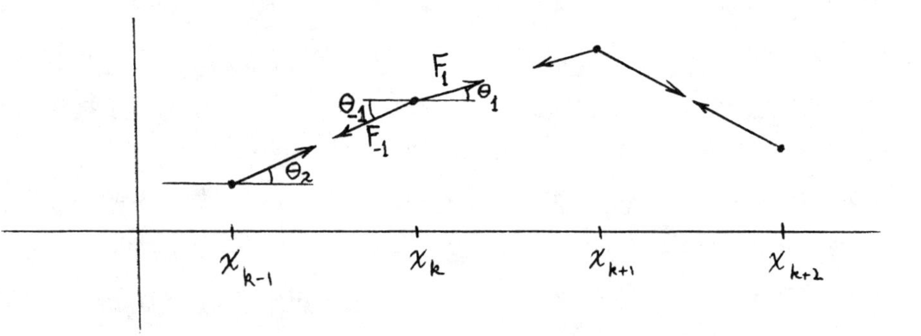

is acted upon only by the particle immediately to its left and right. It is thus influenced by two forces, $F_1(x_k,t)$ and $F_{-1}(x_k,t)$, which are vectors pointing along the line from it to its neighboring particles. These vectors make angles of $\theta_1(x_k,t)$ and $\theta_{-1}(x_k,t)$ with the horizontal rays pointing right and left from the position of the particle. Assuming that our particles all have mass m, we use the law $F = ma$ to obtain the following equation for the vertical motion $u(x_k,t)$ of our particle:

$$\text{(a)} \quad m\cdot u_{tt}(x_k,t) = |F_1(x_k,t)|\cdot \sin\theta_1(x_k,t) - |F_{-1}(x_k,t)|\cdot \sin\theta_{-1}(x_k,t) .$$

Since the particle does not move sideways, we also have

$$\text{(b)} \quad 0 = |F_1(x_k,t)|\cdot \cos\theta_1(x_k,t) - |F_{-1}(x_k,t)|\cdot \cos\theta_{-1}(x_k,t) .$$

We clearly have

$$\text{(c)} \quad \begin{cases} \cos\theta_1(x_k,t) = \dfrac{h}{\sqrt{h^2 + \{u(x_{k+1},t) - u(x_k,t)\}^2}} \\[2ex] \qquad = \dfrac{1}{\sqrt{1 + D_k^{\,2}}}, \qquad \text{where } D_k = \dfrac{u(x_{k+1},t) - u(x_k,t)}{h}, \\[2ex] \sin\theta_1(x_k,t) = \dfrac{u(x_{k+1},t) - u(x_k,t)}{h\sqrt{1 + D_k^{\,2}}} . \end{cases}$$

Noting that $\theta_{-1}(x_k,t) = \theta_1(x_{k-1},t)$, we find that (a)-(c) lead to

$$\text{(d)} \quad m\cdot u_{tt}(x_k,t) = |F_1(x_k,t)|\cdot\left[\sin\theta_1(x_k,t) - \sin\theta_{-1}(x_k,t)\cdot\frac{\cos\theta_1(x_k,t)}{\cos\theta_{-1}(x_k,t)}\right]$$

$$= |F_1(x_k,t)| \cdot \left[\frac{u(x_{k+1},t) - u(x_k,t)}{h\sqrt{1+D_k^2}} - \frac{u(x_k,t) - u(x_{k-1},t)}{h\sqrt{1+D_{k-1}^2}} \cdot \frac{\sqrt{1+D_{k-1}^2}}{\sqrt{1+D_k^2}} \right]$$

$$= \frac{|F_1(x_k,t)|}{\sqrt{1+D_k^2}} \cdot \left[\frac{u(x_{k+1},t) + u(x_{k-1},t) - 2u(x_k,t)}{h} \right] .$$

This is a system of differential equations for the (possibly infinitely many) functions $u(x_k,t)$. It depends, of course, on knowing F_1, which would depend on the particular molecular forces involved. Leaving side that objection for the moment, we now seek a PDE which will describe a uniform string, not a discrete collection. To obtain this, we want to let the number of particles increase, by decreasing h. Of course, we also want to change m in the process, so as not to have an infinitely heavy string at the end. On a piece of thread of length 1, there will be about $1/h$ particles, with a total mass of m/h. So we keep m/h equal to a constant ρ, the density. We will also assume that F_1 approaches a function T, the tension of the string (it measures with how much force the string will snap apart if it is cut at some point). Now it is well-known (use Taylor's theorem for a proof) that

$$\lim_{h\to 0} \frac{f(x+h) + f(x-h) - 2f(x)}{h^2} = f''(x) .$$

So if we divide equation (d) by h, and then take the limit as $h \to 0$, we find that $u(x,t)$ should satisfy

$$\rho u_{tt} = \frac{T}{\sqrt{1+u_x^2}} u_{xx} .$$

Quite apart from the fact that we don't know how to find T, this equation suffers the defect of being non-linear. We can simplify things by restricting ourselves to the case of small vibrations; then T is practically constant, and $\sqrt{1+u_x{}^2}$ is practically 1. We have thus completed our devious path to the 1-dimensional "wave equation"

$$u_{xx} = \rho u_{tt} .$$

Since $\rho > 0$, a simple change of coordinates always gives us the equation

$$u_{xx} = u_{tt} .$$

This equation also describes sound waves in a long thin pipe; in this case,

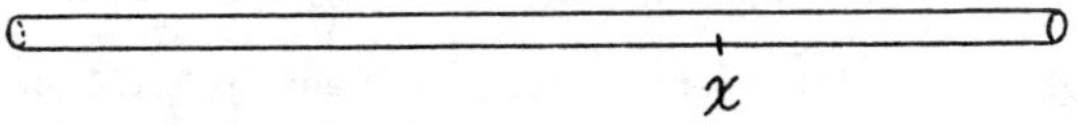

$u(x,t)$ represents the density of the air at distance x and time t. The 2-dimensional wave equation

$$u_{xx} + u_{yy} = u_{tt}$$

will describe the motion of a vibrating membrane, while the 3-dimensional wave equation

$$u_{xx} + u_{yy} + u_{zz} = u_{tt}$$

will describe sound waves, as well as certain phenomena involving electromagnetic waves. For our purposes, the 1-dimensional wave equation will be quite adequate.

The second idealized physical situation which we want to describe is the temperature distribution within a material body. It is important here to distinguish between the temperature and the heat energy of a body. The _tempera-ture_ of a body, which is operationally defined by putting it in contact with a thermometer, is the average kinetic energy of its molecules. We define the temperature $u(x,y,z)$ of a body B at the point (x,y,z) to be the limit of the temperatures of small parts of B which contain (x,y,z). Naturally this doesn't really make much sense for a physical body made of molecules, so we must deal with an idealized situation when we consider temperature to be a function u defined on a certain subset $B \subset \mathbb{R}^3$.

Heat energy is something different. It takes a certain amount of energy to produce a unit increase in temperature within a unit amount of matter. How much energy depends on the particular kind of matter we have. There are two reasons for this. First of all, the molecules in two different kinds of matter have different weights, so different amounts of energy will be required to increase the average kinetic energy of the same number of molecules by the same amount. In perfect gases, this is the only influencing factor. In other cases the strength of the intermolecular forces will also influence how much energy has to be put in to increase the average kinetic energy. The _specific heat_ or _heat capacitance_ C of a piece of matter is the amount of energy required to increase the temperature of a unit mass by a unit amount; we will consider only bodies with uniform specific heat. If $u(x,y,z)$ is the temperature at (x,y,z) of an object B with specific heat C and density ρ, then the total _heat energy_ of B is

$$\text{(a)} \qquad \text{heat energy} = C\rho \int_B u \, dV .$$

The basic experimental fact about heat is that when two bodies of different temperature are placed next to each other, the temperature of the hotter one decreases while the temperature of the cooler one increases, and the rate of change of temperature is proportional to the difference. So if we have two bodies which at each time t have uniform temperatures $T_1(t)$ and $T_2(t)$, then

$$\frac{d}{dt}T_1(t) = \text{constant} \cdot (T_1(t) - T_2(t)) .$$

This constant will depend on the amount of surface area which the two bodies have in common, as well as on the nature of the material of which they are made. The simplest case to consider is that of a single piece of matter B with two parts, B_1 and B_2, initially at different temperatures, T_1 and T_2. Of

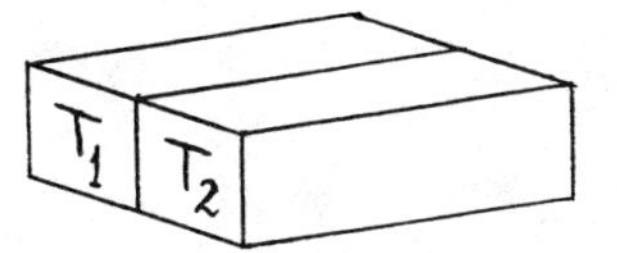

course, the two parts will not continue to have uniform temperatures, but at least we can say that

$$\left.\frac{d}{dt}\right|_{t=0} T_1(t) = \text{constant} \cdot A \cdot (T_1(t) - T_2(t)) ,$$

where A is the area between them. By changing the constant, we can also write this as

$$\text{(b)} \qquad \left.\frac{d}{dt}\right|_{t=0} C\rho T_1(t) = \kappa \cdot A \cdot (T_1(t) - T_2(t))$$

for some constant κ. This constant κ is called the <u>heat conductivity</u> of the matter in question, since it measures the rate at which heat energy is transferred. Roughly speaking, κ must depend on the way the molecules are arranged;

this arrangement will somehow determine to what extent faster moving molecules can influence slower moving ones.

Now let us consider a body $B \subset \mathbb{R}^3$ with uniform density ρ, specific heat C, and heat conductivity κ, but with temperature $u(x,y,z,t)$ varying both with position and time. What will be the analogue of equation (b)? The left side of (b) represents the rate of change of the heat energy of the part of B with temperature T_1. So for any subset $M \subset B$, equation (a) suggests that the analogue of the left side of (b) is

$$\text{(L)} \qquad \frac{d}{dt} C\rho \int_M u \, dV = C\rho \int_M u_t \, dV .$$

Let us suppose that $M \subset B$ is a 3-dimensional manifold-with-boundary, and that ν is the outward unit normal on ∂M. If X is the vector field

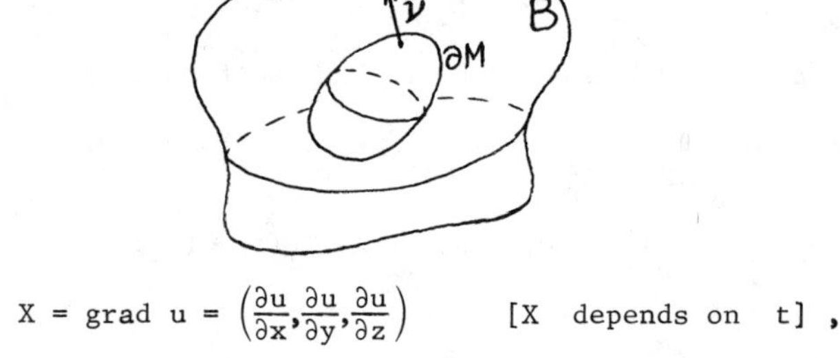

$$X = \text{grad } u = \left(\frac{\partial u}{\partial x}, \frac{\partial u}{\partial y}, \frac{\partial u}{\partial z}\right) \qquad [X \text{ depends on } t] ,$$

then for fixed t the function $-\langle X,\nu\rangle$ measures how fast the temperature is decreasing as we cross ∂M from the inside to the outside; roughly speaking it measures the difference $T_1 - T_2$ on the two sides of ∂M. So the analogue of the right side of (b) is

$$\text{(R)} \qquad (-\kappa) \int_{\partial M} -\langle X,\nu\rangle \, dA = \kappa \int_{\partial M} \langle X,\nu\rangle \, dA .$$

Setting (L) = (R), we are led to the equation

(c) $$C\rho \int_M u_t \, dV = \kappa \int_{\partial M} \langle X,n\rangle \, dA \, .$$

We could also obtain equation (c) by breaking M up into small cubes, each of which may be regarded as having constant temperature, and applying (b) to each cube; the term on the right of (b) is to be replaced by a sum over the faces of the cube.

Now applying the Divergence Theorem (Problem I.9-13), we are led from (c) to

(d) $$\begin{aligned} C\rho \int_M u_t \, dV &= \kappa \int_M \operatorname{div} X \, dV \\ &= \kappa \int_M u_{xx} + u_{yy} + u_{zz} \, dV \, . \end{aligned}$$

Since this is supposed to hold for all M, the integrands must be equal, and we obtain the 3-dimensional "heat equation"

$$\frac{\kappa}{C\rho} \cdot (u_{xx} + u_{yy} + u_{zz}) = u_t \, .$$

Of course, we usually replace the positive constant $\kappa/C\rho$ by 1. The 1-dimensional heat equation

$$u_{xx} = u_t$$

describes temperature distribution in a long thin rod, while the 2-dimensional heat equation

$$u_{xx} + u_{yy} = u_t$$

describes temperature distribution in a thin plate.

We obtain a very special equation when we seek the "steady state" temperature distribution of a body. This is the temperature distribution it has when the temperature is <u>not</u> varying with time. For example, if we keep both ends of a bar at fixed temperatures by attaching them to "heat resevoirs," mechanisms which maintain a fixed temperature at a point, then the temperature distribution will rapidly approach a linear function between these two values. To find the steady state temperature distribution, we just set $u_t = 0$ in the heat equation. Thus in the 1-dimensional case we obtain simply $u_{xx} = 0$, whose solutions are simply linear functions on $\mathbb{R}$. In the 2- and 3-dimensional cases, we obtain

$$u_{xx} + u_{yy} = 0 \qquad \text{2-dimensional Laplace equation}$$

$$u_{xx} + u_{yy} + u_{zz} = 0 \qquad \text{3-dimensional Laplace equation.}$$

Among these important equations of mathematical physics, we find representatives of each of the three types of second order PDEs. In particular, for 2 variables we have the following standard examples:

<u>Elliptic equations</u>

$$\boxed{u_{xx} + u_{yy} = 0 \qquad \text{the 2-dimensional Laplace equation}}$$

<u>Hyperbolic equations</u>

$$\boxed{u_{xx} - u_{yy} = 0 \qquad \text{the 1-dimensional wave equation}}$$

The equation $u_{xx}(x,y) = 0$ is a parabolic equation in 2 variables, but obviously a little too simple to be very representative. The standard representative is

Parabolic equations

$u_{xx} = u_y$	the 1-dimensional heat equation

Parabolic equations are often slighted in introductory treatments of PDEs, and they will suffer the same treatment in our hands -- we will never look at them again. We therefore say good-bye to the heat equation, and consider only the special case of Laplace's equation.

Now let us see what sort of mathematical questions these physical situations suggest. Consider first the 1-dimensional wave equation, which we will write as $u_{xx} - u_{tt} = 0$, to remind us that it describes a process involving time. In such processes, it is naturally of interest to predict what will happen later from a knowledge of what is happening now. It seems perfectly reasonable to hope that we can predict the motion of a string in terms of its initial position and initial velocity,

$$u(x,0) \ , \qquad u_t(x,0) \ .$$

Moreover, there seems to be no reason why we should have to limit ourselves to analytic initial conditions. For example, if we "pluck" a string, then the simplest description involves an initial condition $u(x,0)$ which is not even differentiable everywhere. This example suggests that for hyperbolic

equations the Cauchy problem is the right one to pose, and that we should not have to restrict ourselves to analytic initial data (which is all we can treat when we rely on the Cauchy-Kowalewski Theorem).

Another mathematical problem is suggested by the fact that in actuality vibrating strings are always secured at two ends. Given two functions $\phi, \psi\colon [0,L] \to \mathbb{R}$ with

$$\phi(0) = \phi(L) = 0$$
$$\psi(0) = \psi(L) = 0 ,$$

we can ask for a solution u of the 1-dimensional wave equation with

$$\begin{aligned} u(x,0) &= \phi(x) && 0 \le x \le L \\ u_t(x,0) &= \psi(x) && 0 \le x \le L \\ u(0,t) &= u(L,t) = 0 && \text{for all } t . \end{aligned}$$

This is an example of an "initial-boundary value" problem; although such problems are also quite important, we will not consider them at all.

Quite different questions are suggested by Laplace's equation

$$u_{xx} + u_{yy} + u_{zz} = 0 .$$

Here time is not involved at all; the equation describes the steady state heat distribution of some object $B \subset \mathbb{R}^3$. The Cauchy problem for this equation would correspond to the physical problems of predicting the temperature everywhere in B from a knowledge of its values along some plane in B, together with knowledge of its derivative in the perpendicular direction. Now this is

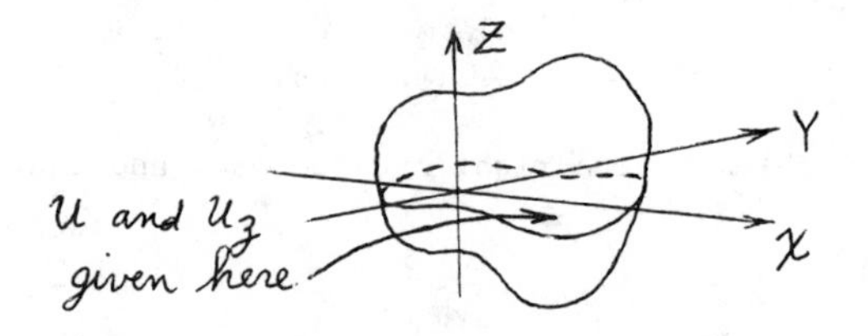

hardly a reasonable problem, since it isn't very easy to measure the temperature at various points inside a solid object. This is the sort of information we would like to *predict*. The sort of thing we can *measure* is the temperature along the boundary. Similarly, for a region B in $\mathbb{R}^2$, we would like to find the solutions u of the 2-dimensional Laplace equation

$$\frac{\partial^2 u}{\partial x^2} + \frac{\partial^2 u}{\partial y^2} = 0$$

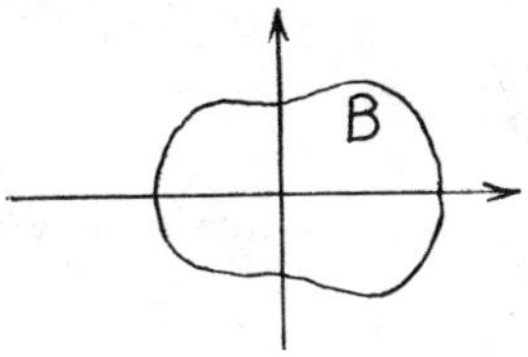

which has given values along the boundary of B. A problem of this sort is called a *Dirichlet problem*. The physics seems to suggest that for elliptic equations it is the Dirichlet problem rather than the Cauchy problem which should be of interest.

Now let us see how these physical speculations correspond to mathematical reality.

The 1-dimensional wave equation

$$(1) \qquad u_{xx} - u_{yy} = 0$$

is admirably suited to illustrate the general behavior of hyperbolic equations, because the most general solution of (1) can be written down completely. The trick for doing this is simply to use the alternative standard form for a hyperbolic equation. We define v by

$$(2) \qquad v(\xi,\eta) = u\left(\frac{\xi+\eta}{2}, \frac{\xi-\eta}{2}\right) \qquad u(x,y) = v(x+y,\, x-y) .$$

Then equation (1) for u gives

$$v_{\xi\eta} = 0 \,. \tag{3}$$

At the very beginning of this Chapter we mentioned that the general solution of equation (3) is

$$v(\xi,\eta) = f(\xi) + g(\eta) \,.$$

So the general solution of (1) is

$$\begin{aligned} u(x,y) &= v(x+y,\ x-y) \\ &= f(x+y) + g(x-y) \,. \end{aligned} \tag{4}$$

If we think of our equation in terms of position x and time t,

$$u_{xx} - u_{tt} = 0 \,,$$

then the solution

$$u(x,t) = f(x+t) + g(x-t)$$

represents the sum of two "waves," the first moving to the left as t increases, the second moving to the right.

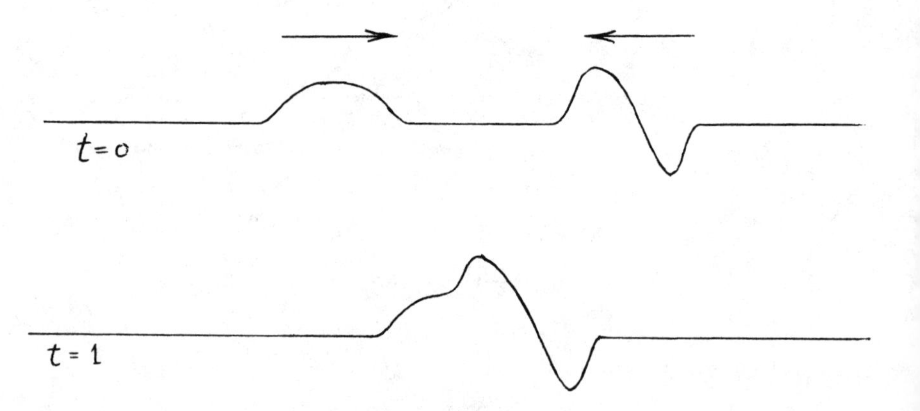

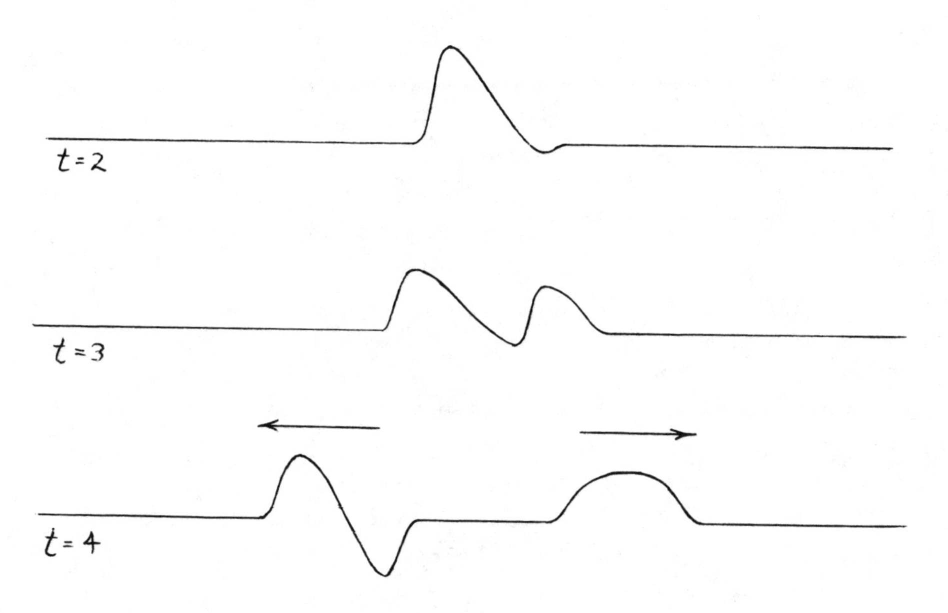

Using the representation (4) for the solution u of (1), it is easy to find solutions with given initial conditions

$$(5) \qquad \begin{cases} u(x,0) = \phi(x) \\ u_y(x,0) = \psi(x) . \end{cases}$$

Clearly we must have

$$(6) \qquad \begin{cases} f(x) + g(x) = \phi(x) \\ f'(x) - g'(x) = \psi(x) , \end{cases}$$

and therefore

$$\begin{cases} f'(x) + g'(x) = \phi'(x) \\ f'(x) - g'(x) = \psi(x) , \end{cases}$$

which implies

$$f'(x) = \frac{\phi'(x)+\psi(x)}{2} \qquad g'(x) = \frac{\phi'(x)-\psi(x)}{2} .$$

This means that we must have

$$(7) \qquad f(x) = \frac{\phi(x)}{2} + \frac{1}{2}\int_0^x \psi(s)ds + C_1 , \qquad g(x) = \frac{\phi(x)}{2} - \frac{1}{2}\int_0^x \psi(s)ds + C_2$$

for certain constants C_1, C_2; and to satisfy (6) we must have $C_1 = - C_2$. Using (4) we then find that u must be

$$(8) \qquad u(x,y) = \frac{\phi(x+y)+\phi(x-y)}{2} + \frac{1}{2}\int_{x-y}^{x+y} \psi(s)ds .$$

It is clear, moreover, that this u _is_ a solution of equation (1), with initial conditions (5).

Notice that the boundary values ϕ, ψ enter into the solution in quite a different way than for first order PDEs. If u is a solution of

$$F(x,y,u,p,q) = 0$$

with initial data $\mathring{u}$, $\mathring{p}$, $\mathring{q}$ along a free curve σ, then the value $u(x,y)$ of u at a particular point (x,y) depends on the value of the initial data at one point $(\bar{x},\bar{y})$ on σ, namely the intersection of σ with the base curve of the characteristic strip through $(x,y,u(x,y),u_x(x,y),u_y(x,y))$. Changing the initial data on an interval which does not contain $(\bar{x},\bar{y})$ will not change the value of the solution u at (x,y). But in the solution (8) of the 1-dimensional wave equation (1), we need to know the values of ϕ and ψ on the whole interval $[x-y, x+y]$ (or $[x+y, x-y]$ if $y < 0$). This interval is

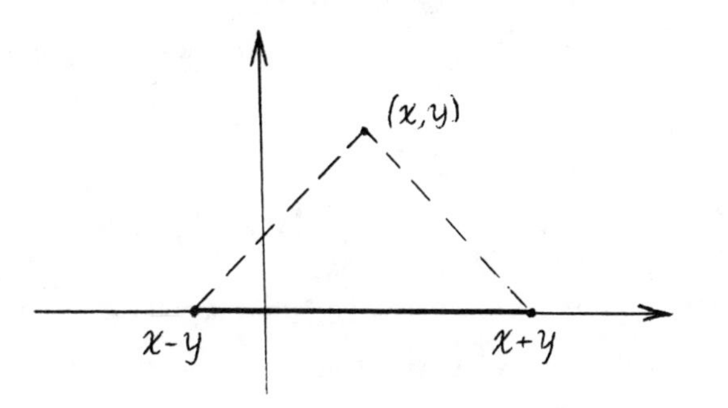

therefore called the "domain of dependence" of the point (x,y). Conversely, if we are given initial conditions ϕ, ψ defined only on an interval $[a,b]$, then equation (8) defines u only on the set A of all points (x,y) whose domain of dependence is contained in $[a,b]$. Notice that A is bounded by the

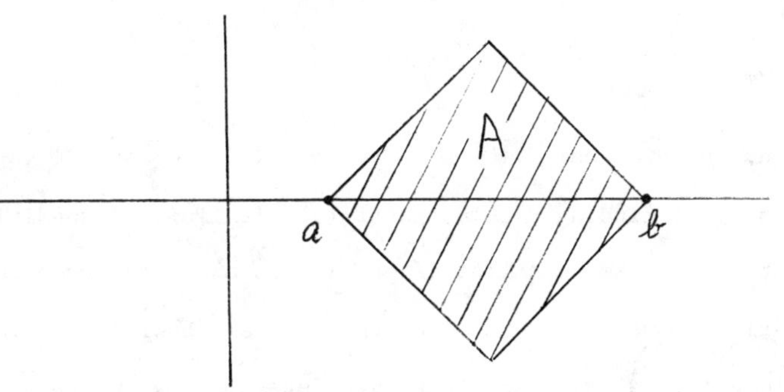

curves through a and b which are characteristic for the PDE (1).

Naturally, we might ask about solutions of (1) along free curves other than the x-axis. In section 8 we will consider this question for even more general hyperbolic equations. If the 1-dimensional wave equation is indeed representative of general hyperbolic equations, then we should be able to solve the Cauchy problem for any hyperbolic equation, along any free curve, and without any assumptions about analyticity of the initial conditions.

The situation is completely different for the 2-dimensional Laplace equation

$$(9) \qquad u_{xx} + u_{yy} = 0 .$$

Solutions of this equation are called harmonic functions (on $\mathbb{R}^2$), and, as we have pointed out in Chapter 9, their study is closely related to the theory of complex analytic functions. In a simply connected open subset of $\mathbb{R}^2$, every harmonic function u is the real part of a complex analytic function $u+iv$; and conversely, the real part of a complex analytic function is always harmonic. This means, in particular, that every solution of (9) is automatically real analytic on $\mathbb{R}^2$. So we cannot hope to solve equation (9) with initial conditions

$$(10) \qquad \begin{cases} u(x,0) = \phi(x) \\ u_y(x,0) = \psi(x) , \end{cases}$$

unless ϕ and ψ are real analytic.* Moreover, if ϕ and ψ are real analytic, then the problem of finding a solution of (9) with initial conditions (10) is essentially trivial. We note that if $u+iv$ is analytic, then the Cauchy-Riemann equations give $v_x = -u_y$. So the initial conditions allow us to determine v_x along the x-axis, and therefore determine v up to a constant along the x-axis. So the complex analytic function $u+iv$ is determined up to an imaginary constant on the x-axis, which means that $u+iv$ is determined up to an imaginary constant on the plane.

These remarks really amount to a restatement of the fact, already observed in the proof of the Cauchy-Kowalewski Theorem, that for analytic equations with analytic data, the coefficients of the presumptive analytic solution are easily

*See p. III.449, §6.

determined. It is perhaps of interest to note that we can formally solve (9) by analogy with the wave equation (1). If we formally define

$$(11) \qquad v(\xi,\eta) = u\left(\frac{\xi+\eta}{2}, \frac{\xi-\eta}{2i}\right) \qquad u(x,y) = v(x+iy,\ x-iy)\ ,$$

then equation (9) becomes $v_{\xi\eta} = 0$, which leads us to

$$v(\xi,\eta) = f(\xi) + g(\eta)$$

$$u(x,y) = f(x+iy) + g(x-iy)\ .$$

Remembering the initial conditions (10), we are led to the formal solution

$$(12) \qquad u(x,y) = \frac{\phi(x+iy)+\phi(x-iy)}{2} + \frac{1}{2}\int_{x-iy}^{x+iy} \psi(z)\,dz\ .$$

If ϕ and ψ are real analytic, and hence have complex analytic extensions, then this formula makes sense -- the integral may be taken along any path from $x-iy$ to $x+iy$. Because ϕ and ψ are real on the real axis, the function u is real-valued, and is easily seen to satisfy (9) and (10).

Our physical considerations suggest that we should be able to solve the Dirichlet problem for (9): given a function $f\colon \partial B \longrightarrow \mathbb{R}$ on the boundary of a region $B \subset \mathbb{R}^2$, we ought to be able to find a solution u of (9) with $u = f$ on ∂B. In complex analysis courses it is shown that this is indeed the case.

7. Hyperbolic Systems in Two Variables

In this section we will consider first order quasi-linear systems in two variables. Our initial manifolds for the Cauchy problem will therefore be curves in $\mathbb{R}^2$, and we are naturally only interested in initial data for which the initial curve is free. So without loss of generality, we assume that our initial curve is an interval $[a,b]$ of the x-axis, and that our quasi-linear system of n equations for n unknown functions $u^i\colon \mathbb{R}^2 \to \mathbb{R}$ is of the form

$$u^i_y(x,y) = \sum_{j=1}^{n} a_{ij}(x,y,u^1(x,y),\dots,u^n(x,y))\cdot u^j_x(x,y) + b^i(x,y,u_1(x,y),\dots,u_n(x,y)) .$$

We will often consider $u = (u^1,\dots,u^n)$ to be a column vector, just so that we can multiply on the left by a matrix. Then we can write our system as

$$u_y(x,y) = A(x,y,u(x,y))\cdot u_x(x,y) + b(x,y,u(x,y))$$

where A is an $n\times n$ matrix of functions, and b is a column vector. More briefly, we have the system

$$u_y = A\cdot u_x + b .$$

This system is called <u>hyperbolic for given initial conditions</u> $\mathring{u} = (\mathring{u}^1,\dots,\mathring{u}^n)$ on an interval $[a,b]$ of the x-axis if A is diagonalizable in a neighborhood of all points $(x,0,\mathring{u}(x))$ for $x \in [a,b]$; more precisely, it is <u>C^k hyperbolic</u> if there is a C^k matrix T such that TAT^{-1} is diagonalizable [this more precise formulation is necessary, because even if A is C^k and always diagonalizable, it may not be possible to choose the diagonalizing matrix T to be C^k]. The basic result is that a quasi-linear system with hyperbolic initial conditions has a solution with these initial conditions; in the next section we will apply this to a single second order equation.

In order to explain the main points of the argument, we will first sketch how the proof would go in the "semi-linear" case,

$$\text{(a)} \qquad u_y(x,y) = A(x,y)u_x(x,y) + b(x,y,u(x,y)) \;,$$

where A does not depend on u (although b might). Let T be the matrix which diagonalizes A, and define $v = (v^1,\ldots,v^n)$ by

$$u(x,y) = T(x,y)\cdot v(x,y) \;,$$

so that

$$u_x = T_x v + Tv_x \;, \qquad u_y = T_y v + Tv_y \;.$$

Substituting into (a), we obtain

$$T_y v + Tv_y = AT_x v + ATv_x + b \;,$$

so

$$\text{(b)} \qquad \begin{aligned} v_y &= (T^{-1}AT)v_x + (T^{-1}AT_x v + T^{-1}b - T^{-1}T_y v) \\ &= Cv_x + d \;. \end{aligned}$$

Clearly u satisfies (a) with initial conditions $\mathring{u}$ if and only if v satisfies (b) with initial conditions

$$\mathring{v}(x) = T^{-1}(x,0)\cdot\mathring{u}(x) \;.$$

So we might as well assume that we have the equation

$$\text{(1)} \qquad u_y = Cu_x + d \;,$$

where the matrix

$$C(x,y) = \begin{pmatrix} \lambda^1(x,y) & & 0 \\ & \ddots & \\ 0 & & \lambda^n(x,y) \end{pmatrix}$$

is a diagonal matrix. Thus our equation reads

$$(1') \qquad -\lambda^i(x,y)\cdot u_x^i + u_y^i = d^i(x,y,u^1(x,y),\ldots,u^n(x,y)) .$$

The vector field

$$X_i(x,y) = (-\lambda^i(x,y),\ 1)$$

is called the $i^{\underline{th}}$ characteristic vector field of equation (1'), and the integral curves of X_i are the $i^{\underline{th}}$ family of characteristic curves. We might as well consider only characteristic curves of the form $t \mapsto (c(t),t)$. If u is a solution of (1') and $t \mapsto (c(t),t)$ is a characteristic curve of the $i^{\underline{th}}$ family, then

$$\begin{aligned}(2) \qquad \frac{d}{dt}u^i(c(t),t) &= -\lambda^i(c(t),t)\cdot u_x^i(c(t),t) + u_y^i(c(t),t) \\ &= d^i(c(t),t,u(c(t),t)) .\end{aligned}$$

Consequently,

$$(3) \qquad u^i(c(\eta),\eta) = u^i(c(0),0) + \int_0^\eta d^i(c(t),t,u(c(t),t))\,dt .$$

In particular, let $t \mapsto (c^i(\xi,\eta),t)$ be the characteristic curve of the $i^{\underline{th}}$ family which satisfies

$$(4) \qquad c^i(\xi,\eta)(\eta) = \xi .$$

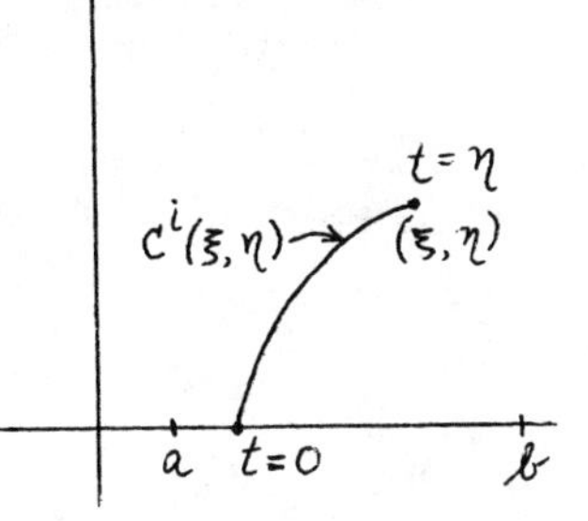

If $c^i(\xi,\eta)(0) \in [a,b]$, then equations (3) and (4) give

$$(5)\quad u^i(\xi,\eta) = \mathring{u}^i(c^i(\xi,\eta)(0)) + \int_0^\eta d^i(c^i(\xi,\eta)(t),t,u(c^i(\xi,\eta)(t),t))\,dt\,.$$

Conversely, if u satisfies (5) in a region obtained by following all characteristic curves from $[a,b]\times\{0\}$ for some time interval in either direction, then u will be a solution of (1') with initial conditions $\mathring{u}$ on $[a,b]\times\{0\}$. Equation (5) is something like the integral equation which we solved in Chapter I.5, when we proved that differential equations have solutions, but it is more complicated, because the different components of u are integrated over different curves. Nevertheless, it can be solved in essentially the same way. We define an operator S which takes u to the n-tuple of functions Su given by

$$(Su)^i(\xi,\eta) = \mathring{u}^i(c^i(\xi,\eta)(0)) + \int_0^\eta d^i(c^i(\xi,\eta)(t),t,u(c^i(\xi,\eta)(t),t)))\,dt\,.$$

Then we show that on a suitable complete space of functions the operator S is a contraction, so that it has a fixed point.

When we look at the quasi-linear system

$$(1) \qquad u_y(x,y) = A(x,y,u(x,y))\cdot u_x(x,y) + b(x,y,u(x,y)) \ ,$$

we run into a problem at the very first step. For the matrix T which diagonalizes A will depend on u. If we set

$$[T]_x = \frac{\partial T(x,y,u(x,y))}{\partial x} = T_x + \sum_j T_{u^j} u^j_x$$

$$[T]_y = \frac{\partial T(x,y,u(x,y))}{\partial y} = T_y + \sum_j T_{u^j} u^j_y \ ,$$

then the substitution $u(x,y) = T(x,y,u(x,y))\cdot v(x,y)$ leads to equation (b) again, except that now T_x and T_y are replaced by $[T]_x$ and $[T]_y$, which involve u; so we do not even obtain an equation for v. We can reduce our system to one in diagonal form by means of a somewhat more complicated substitution; however, it will be necessary to assume that the matrix $A(x,y,u(x,y))$ in (1) is invertible [in a neighborhood of the points $(x,0,\mathring{u}(x))$].

Suppose we have u satisfying (1), and $T(x,y,u)$ diagonalizes $A(x,y,u)$ for all (x,y,u) in a neighborhood of the points $(x,0,\mathring{u}(x))$. Define v by

$$(2) \qquad u_y = Tv \qquad (\text{i.e.,} \quad u_y(x,y) = T(x,y,u(x,y))\cdot v(x,y)) \ .$$

Then

$$(3) \qquad Tv = Au_x + b \implies (4) \quad u_x = A^{-1}(Tv - b) \ .$$

Differentiating (3) with respect to y, we obtain

$$\begin{aligned} Tv_y + [T]_y v &= Au_{xy} + [A]_y u_x + [b]_y \\ &= A(Tv)_x + [A]_y u_x + [b]_y \qquad \text{by (2)} \\ &= A[T]_x v + ATv_x + [A]_y u_x + [b]_y \ , \end{aligned}$$

so

$$v_y = (T^{-1}AT)v_x + T^{-1}A[T]_x v + T^{-1}[A]_y u_x + T^{-1}[b]_y - T^{-1}[T]_y v \ .$$

Writing out $[T]_x, [A]_y, \ldots,$ substituting for the u_y from (2) and for the u_x from (4), we obtain

$$(5) \qquad \begin{aligned} v_y = (T^{-1}AT)v_x &+ T^{-1}A[T_x + \sum_j T_{u^j}\{A^{-1}(Tv - b)\}^j]v \\ &+ T^{-1}[A_y + \sum_j A_{u^j}(Tv)^j]A^{-1}(Tv - b) \\ &+ T^{-1}[b_y + \sum_j b_{u^j}(Tv)^j] \\ &- T^{-1}[T_y + \sum_j T_{u^j}(Tv)^j]v \ . \end{aligned}$$

If w is the column vector $u^1,\ldots,u^n,v^1,\ldots,v^n,$ then equations (2) and (5) together can be written in the form

$$w_y = Cw_x + d$$

where C is <u>diagonal</u>. We have the initial conditions

$$(6) \quad w(x,0) = \begin{pmatrix} \mathring{u}(x) \\ T^{-1}(x,0,\mathring{u}(x))\cdot[A(x,0,\mathring{u}(x))\cdot\mathring{u}'(x) + b(x,0,\mathring{u}(x))] \end{pmatrix} .$$

Note that if A, b, and u are C^2, and A is C^2 diagonalizable, then C and d are C^1; if the initial condition $\mathring{u}$ is C^2, then the initial condition for w is C^1. Conversely, we have

9. LEMMA. Let $w = u^1,\dots,u^n,v^1,\dots,v^n$ be a C^1 solution of the system (2), (5) with C^1 coefficients C and d and C^1 initial conditions (6). Then u satisfies $u_y = Au_x + b$ (and is C^2).

Proof. Substituting (2) into the last three terms of (5), multiplying by T, and rearranging, we find that

$$(7) \qquad Tv_y + [T]_y v - [b]_y = ATv_x + A[T_x + \sum_j T_{u^j}\{A^{-1}(Tv - b)\}^j]v + [A]_y A^{-1}(Tv - b) .$$

Equation (2) implies that u_y has a continuous partial derivative with respect to x. Hence, by a theorem of calculus, u_x has a continuous first partial derivative with respect to y and $u_{xy} = u_{yx}$. Thus

$$(8) \qquad u_{xy} = u_{yx} = (Tv)_x = Tv_x + [T]_x v .$$

Define

$$s = A^{-1}(Tv - b) - u_x .$$

Then s has a continuous first partial derivative with respect to y, and

$$\begin{aligned} s_y &= [A^{-1}]_y(Tv - b) + A^{-1}(Tv_y + [T]_y v - [b]_y) - u_{xy} \\ &= - A^{-1}[A]_y A^{-1}(Tv - b) + A^{-1}(Tv_y + [T]_y v - [b]_y) - u_{xy} \\ &= - A^{-1}[A]_y(s + u_x) + A^{-1}(Tv_y + [T]_y v - [b]_y) - u_{xy} . \end{aligned}$$

Multiplying by A we have

$$\begin{aligned}
As_y &= -[A]_y s - [A]_y u_x + (Tv_y + [T]_y v - [b]_y) - Au_{xy} \\
&= -[A]_y s - [A]_y u_x + ATv_x + A[T_x + \sum_j T_{u^j}\{A^{-1}(Tv-b)\}^j]v + [A]_y A^{-1}(Tv-b) \\
&\quad - A(Tv_x + [T]_x v) \qquad \text{by (7) and (8)} \\
&= -[A]_y s - [A]_y u_x + A[T_x + \sum_j T_{u^j} s^j + \sum_j T_{u^j} u_x^j]v + [A]_y (s+u_x) - A[T]_x v \\
&= A(\sum_j T_{u^j} s^j)v .
\end{aligned}$$

Thus

$$s_y = \sum_j (T_{u^j} s^j)v .$$

For fixed x, this is a system of ordinary differential equations. But the initial conditions (6) show that $s^j(x,0) = 0$. So by uniqueness of solutions, we have $s = 0$, which means that

$$\begin{aligned}
Au_x + b &= Tv \\
&= u_y \qquad \text{by (2)} .
\end{aligned}$$

Since v is C^1, the partial derivative $u_y = Tv$ is C^1; hence $u_x = A^{-1}(u_y - b)$ is also C^1. Thus u is C^2. ■

Because of Lemma 9, we now restrict our attention to equations

$$u_y(x,y) = C(x,y,u(x,y)) \cdot u_x(x,y) + d(x,y,u(x,y))$$

where $C(x,y,z)$ is a diagonal $n \times n$ matrix in a neighborhood of the points $(x,0,\overset{\circ}{u}(x))$. One further simplification is possible. Introduce two new unknowns u^{n+1}, u^{n+2}, and consider the equations

$$\begin{cases} u_y(x,y) = C(u^{n+1}(x,y),u^{n+2}(x,y),u(x,y))\cdot u_x(x,y) \\ \qquad\qquad + d(u^{n+1}(x,y),u^{n+2}(x,y),u(x,y)) \\ u_y^{n+1}(x,y) = 0 \\ u_y^{n+1}(x,y) = 1 \end{cases}$$

with the initial conditions

$$\begin{cases} u(x,0) = \mathring{u}(x) \\ u^{n+1}(x,0) = x \\ u^{n+2}(x,0) = 0 \;. \end{cases}$$

A solution u, u^{n+1}, u^{n+2} of this system clearly gives a solution u of the original equation, with the initial conditions $u(x,0) = \mathring{u}$. So we might as well consider an equation of the form

$$(1) \qquad u_y(x,y) = C(u(x,y))\cdot u_x(x,y) + d(u(x,y)) \;,$$

with initial conditions

$$(1_0) \qquad u(x,0) = \mathring{u}(x) \;,$$

where the matrix $C = \begin{pmatrix} \lambda^1 & & 0 \\ & \ddots & \\ 0 & & \lambda^n \end{pmatrix}$ is a diagonal matrix in a neighborhood of the points $\mathring{u}(x)$.

Our procedure for solving equation (1) is somewhat more involved than the procedure outlined in the semi-linear case since C now depends on u. For any function u, we define the ith family of <u>characteristic curves of</u> u to be the curves $t \mapsto (c(t),t)$ with

$$\frac{dc(t)}{dt} = -\lambda^i(u(c(t),t)) .$$

Let $c^i(u;\xi,\eta)$ satisfy this equation and the initial condition

$$c^i(u;\xi,\eta)(\eta) = \xi .$$

As before, we find that if u is a solution of (1), with initial conditions (1_0), then

$$(2) \qquad u^i(\xi,\eta) = \mathring{u}^i(c^i(u;\xi,\eta)(0)) + \int_0^\eta d^i(u(c^i(u;\xi,\eta)(t),t))\, dt .$$

And, conversely, if u satisfies (2), then u will satisfy (1), with initial conditions (1_0). We are thus led to define an operator S which takes u to the n-tuple of functions Su given by

$$(Su)^i(\xi,\eta) = \mathring{u}^i(c^i(u;\xi,\eta)(0)) + \int_0^\eta d^i(u(c^i(u;\xi,\eta)(t),t))\, dt .$$

The problem is to show that on a suitable complete metric space of functions, the operator S is a contraction; its fixed point will then be a solution of our equation. The proof is carried out in detail in Courant and Lax [1]. It involves a series of estimates that only an analyst could love, and there doesn't seem to be much point reproducing it here, since the paper is readily accessible, and the sane differential geometer would probably skip it anyway. We would simply like to give a precise statement. Let K be a constant such that all $|\lambda^i| \leq K$ on a region containing $[a,b]\times\{0\}$. We consider the region $\Delta(\delta)$ in $\mathbb{R}^2$ bounded by the lines

$$y = \delta \,, \qquad y = -\delta$$

$$y = \frac{1}{K}(x-a) \qquad y = -\frac{1}{K}(x-a)$$

$$y = -\frac{1}{K}(x-b) \,, \quad y = \frac{1}{K}(x-b)$$

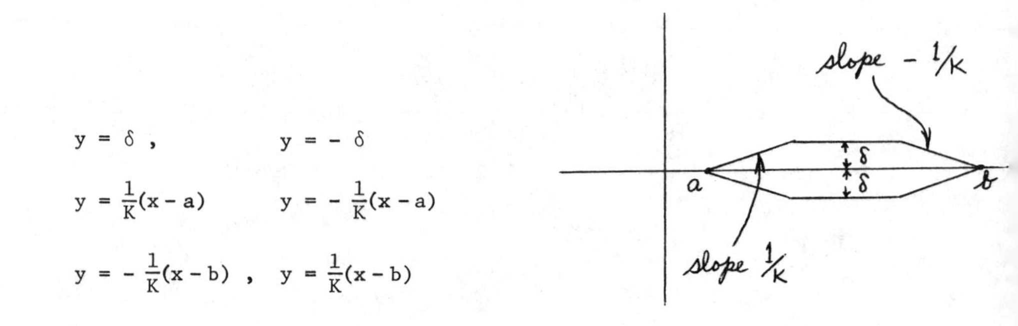

(then the characteristic curves of a function u will have slopes which are larger than the slopes of the sides of $\Delta(\delta)$; so if a characteristic curve begins in $\Delta(\delta)$, it will stay in $\Delta(\delta)$ until it hits the x-axis).

10. THEOREM. Consider a system, with initial conditions,

$$u_y = Au_x + b$$

$$u_x(x,0) = \overset{\circ}{u}(x) \qquad x \in [a,b] \,,$$

such that either

(1) the system is semi-linear, A, b, and $\overset{\circ}{u}$ have continuous partial derivatives satisfying a Lipschitz condition, and A is diagonalizable by a matrix T with the same property

or

(2) the system is quasi-linear, A is invertible, A, b and $\overset{\circ}{u}$ have continuous second partial derivatives satisfying a Lipschitz condition, and A is diagonalizable by a matrix T with the same property.

Then for sufficiently small $\delta > 0$ (which depends on the constants in the Lipschitz conditions), there is a unique solution of the system in the region $\Delta(\delta)$. In case (1), the solution u has continuous partial derivatives satisfying a Lipschitz condition, and in case (2), the solution u has continuous second partial derivatives satisfying a Lipschitz condition.

We are stating this particular theorem simply because it is the most accessible in the literature. Other approaches allow all sorts of improvements. First of all, the differentiability requirements can be weakened. Second of all, the matrix A need not be invertible even in the quasi-linear case -- but then the approach has to be changed considerably. More important, we would like to know that the solutions have a high degree of differentiability if the coefficients and initial conditions do. The proof of this requires considerations like those which are used to prove that the solutions of an ordinary differential equation are differentiable in the initial conditions, considerations which we already omitted in Volume 1. We would also like to consider systems depending on parameters, and show that the solutions are differentiable in the parameters; in section 9 we will use this fact. As in the case of ordinary differential equations, differentiability in the parameters is not very hard, and the reader may work this out for himself, guided by Problem I.5-5.

Despite the somewhat unsatisfactory state in which this section ends, I hope the reader will feel fairly convinced that hyperbolic systems have solutions. In the next 2 sections we will use this fact to show the enormous difference between hyperbolic and elliptic solutions of second order equations.

8. Hyperbolic Second Order Equations in Two Variables

Consider a second order equation

$$(1)\qquad 0 = F(x,y,u,u_x,u_y,u_{xx},u_{xy},u_{yy}) = F(x,y,u,p,q,r,s,t) ,$$

an initial curve, and hyperbolic initial data along this curve such that the curve is free for the initial data. As we saw in section 2, by composing u with a diffeomorphism of the plane, we can assume that the initial curve is a segment $[a,b]$ of the x-axis. The initial data then amount to functions

$$\mathring{u} , \mathring{q} , \mathring{t}$$

on $[a,b]$ satisfying

$$(1_0)\qquad 0 = F(x,0,\mathring{u}(x),\mathring{u}'(x),\mathring{q}(x),\mathring{u}''(x),\mathring{q}'(x),\mathring{t}(x)) ,$$

and the initial data will still be hyperbolic,

$$F_s{}^2 - 4F_rF_t > 0 ,$$

as we remarked in section 5. The requirement that the initial curve $[a,b] \times \{0\}$ be free means that

$$\begin{aligned} 0 &\neq F_t(x,0,\mathring{u}(x),\mathring{u}'(x),\mathring{q}(x),\mathring{u}''(x),\mathring{q}'(x),\mathring{t}(x)) \\ &= F_t(\alpha(x)) , \quad \text{say} \qquad\qquad \text{for } x \in [a,b] . \end{aligned}$$

We claim that we can also assume that $0 \neq F_r(\alpha(x))$. The reason for this is that $F_r = 0$ precisely when the direction of the y-axis is characteristic, and we can always avoid this by an appropriate transformation of the plane.

In detail (compare p. 79), define a new function v by

$$u(x,y) = v(x+\lambda y,y) \ ,$$

where λ is a constant. Note that the map $(x,y) \longmapsto (x+\lambda y,y)$ takes the segment $[a,b]$ of the x-axis into itself. Now

$$u_x = v_x \qquad u_y = \lambda v_x + v_y$$

$$u_{xx} = v_{xx}$$

$$u_{xy} = \lambda v_{xx} + v_{xy}$$

$$u_{yy} = \lambda^2 v_{xx} + 2\lambda v_{xy} + v_{yy}$$

[all partials of u evaluated at (x,y), and all partials of v evaluted at $(x+\lambda y,y)$].

So equation (1) is equivalent to the equation

$$0 = F(x-\lambda y,y,v,v_x,\lambda v_x+v_y,v_{xx},\lambda v_{xx}+v_{xy},\lambda^2 v_{xx}+2\lambda v_{xy}+v_{yy})$$

[all functions $v,\ldots,v_{yy}$ evaluated at (x,y)] .

This can be written

$$0 = G(x,y,v,v_x,v_y,v_{xx},v_{xy},v_{yy}) \ ,$$

where G has the form

$$G(\_\_,r,s,t) = F(\_\_,r,\lambda r+s,\lambda^2 r+2\lambda s+t) \ .$$

Then we have (leaving out the arguments for convenience)

$$G_r = F_r + \lambda F_s + \lambda^2 F_t \ .$$

By choosing λ sufficiently large we can insure that $G_r(\alpha(x)) \neq 0$ for all $x \in [a,b]$.

So we will now consider equation (1) where we have

$$F_r(\alpha(x)),\ F_t(\alpha(x)) \neq 0 \quad \text{for} \quad x \in [a,b]\ .$$

One way of treating equation (1) is to reduce it to an equivalent one by the considerations of section 2. Since $F_t \neq 0$, there is a function f, defined in a neighborhood of all points

$$\beta(x) = (x,0,\mathring{u}(x),\mathring{u}'(x),\mathring{q}(x),\mathring{u}''(x),\mathring{q}'(x))\ ,$$

such that

(a) $$\mathring{t}(x) = f(\beta(x))$$

(b) $$F(x,y,u,p,q,r,s,f(x,y,u,p,q,r,s)) = 0\ .$$

So equation (1) is equivalent to the equation

(1') $$u_{yy} = f(x,y,u,u_x,u_y,u_{xx},u_{xy})\ .$$

Differentiating (b) with respect to r and s we obtain

$$0 = F_r + F_t f_r$$

$$0 = F_s + F_t f_s\ ,$$

so

$$F_r \neq 0 \implies f_r \neq 0$$

$$F_s{}^2 - 4F_rF_t > 0 \implies f_s{}^2 + 4f_r > 0$$

(the latter is just the condition that the solution u of the equation $f(x,y,u,p,q,r,s) - t = 0$ be hyperbolic.) In section 3 we showed that the Cauchy problem for equation (1') is equivalent to a Cauchy problem for a certain system, which in our present notation reads

$$(*)\quad \begin{cases} u_y = q \\ p_y = s \\ q_y = t \\ r_y = s_x \\ s_y = t_x \\ t_y = f_y + f_u \cdot q + f_p \cdot s + f_q \cdot t + f_r s_x + f_s t_x \ . \end{cases}$$

If $\phi = (u,p,q,r,s,t)$, then we can write this system as

$$\phi_y = A\phi_x + \psi \ ,$$

where

$$A = \begin{pmatrix} \bigcirc & \bigcirc \\ & \begin{matrix} 0 & 1 & 0 \end{matrix} \\ \bigcirc & \begin{matrix} 0 & 0 & 1 \\ 0 & f_r & f_s \end{matrix} \end{pmatrix} .$$

The eigenvalues of the matrix

$$A' = \begin{pmatrix} 0 & 1 & 0 \\ 0 & 0 & 1 \\ 0 & f_r & f_s \end{pmatrix}$$

are the roots of

$$-\lambda(\lambda^2 - f_s\lambda - f_r) = 0 .$$

Thus one eigenvalue is 0, and the other two are

$$\frac{f_s \pm \sqrt{f_s^{\,2} + 4f_r}}{2} .$$

These roots are real and distinct, since $f_s^{\,2} + 4f_r > 0$, and they are $\neq 0$, since $f_r \neq 0$. So A' is diagonalizable, and consequently A is. Theorem 10 then tells us that we can solve the system (*) with the appropriate initial conditions; hence we can solve the original equation (1) with the given hyperbolic initial conditions. The only slight problem is that Theorem 10 was stated only for non-singular A (although, as we mentioned, this requirement is not really necessary). We will therefore give another approach which uses Theorem 10 only as stated, and work directly with equation (1), rather than (1'). The new considerations which we will introduce are interesting in their own right, lead to more detailed information about hyperbolic equations, and will also be used in the next section.

For a fixed vector field Z in $\mathbb{R}^2$, and any function g on $\mathbb{R}^2$, we will denote $\nabla_Z g$ simply by g'. In particular, we can write x' and y', by considering x and y to be functions on $\mathbb{R}^2$ (namely, projection on the first and second coordinates). Clearly $Z = (x', y')$. Now for any function u we have

$$r' = r_x \cdot x' + r_y \cdot y'$$

$$s' = s_x \cdot x' + s_y \cdot y' ,$$

or

$$(2) \qquad \begin{cases} x'\cdot r_x + y'\cdot s_x = r' \\ x'\cdot s_x + y'\cdot t_x = s' \ . \end{cases}$$

In addition, if u is a solution of (1), then we have

$$(3) \qquad 0 = \frac{dF(x,\ldots,u_{yy})}{dx} = F_r\cdot r_x + F_s\cdot s_x + F_t\cdot t_x + \{F\}_x \ ,$$

$$\text{where } \{F\}_x = F_x + F_u\cdot u_x + F_p\cdot p_x + F_q\cdot q_x \ .$$

Equations (2) and (3) may be regarded as 3 equations for 3 unknowns r_x, s_x, t_x; in matrix form these equations read

$$(2)\text{-}(3) \qquad \begin{pmatrix} x' & y' & 0 \\ 0 & x' & y' \\ F_r & F_s & F_t \end{pmatrix} \cdot \begin{pmatrix} r_x \\ s_x \\ t_x \end{pmatrix} = \begin{pmatrix} r' \\ s' \\ -\{F\}_x \end{pmatrix} .$$

The condition that the determinant of the matrix of this equation vanish is

$$(4) \qquad (x')^2 F_t - x'y'F_s + (y')^2 F_r = 0 \ .$$

Since our equation is hyperbolic, so that $F_s{}^2 - 4F_rF_t > 0$ in a neighborhood of all points $\alpha(x)$, we can find two continuous everywhere unequal real-valued functions ρ_1, ρ_2 in this neighborhood, each of which is a solution of

$$(4') \qquad F_t - F_s\rho + F_r\rho^2 = 0 \ .$$

The condition $F_r \neq 0$ is used here to guarantee that we have a genuine quadratic equation. The condition $F_t \neq 0$ insures that ρ_1, $\rho_2 \neq 0$, which will be important later on. Notice that the ρ_i are functions on a certain subset of $\mathbb{R}^8$; they depend only on the original equation (1), not on the solution u. If we set $Z = (1,\rho_1)$, and continue the convention that $'$ denotes ∇_Z, then (4) holds in a neighborhood of all points $\alpha(x)$, which means that

$$\operatorname{rank}\begin{pmatrix} x' & y' & 0 \\ 0 & x' & y' \\ F_r & F_s & F_t \end{pmatrix} \leq 2$$

at these points. Now the last column of the matrix

$$\begin{pmatrix} x' & y' & 0 & r' \\ 0 & x' & y' & s' \\ F_r & F_s & F_t & -\{F\}_x \end{pmatrix}$$

is a linear combination of the first three columns [namely the linear combination (r_x,s_x,t_x), by equation (2)-(3)], so this matrix also has rank ≤ 2. Consequently, the determinant of every 3×3 sub-matrix vanishes. In particular,

$$0 = \det\begin{pmatrix} x' & 0 & r' \\ 0 & y' & s' \\ F_r & F_t & -\{F\}_x \end{pmatrix} = -y'F_r r' - x'F_t s' - x'\{F\}_x y' ,$$

or

$$\rho_1 F_r r' + F_t s' + \{F\}_x y' = 0 . \tag{5}$$

If we use $`$ to denote ∇_Z when $Z = (1,\rho_2)$, then we also have

$$\rho_2 F_r r` + F_t s` + \{F\}_x y` = 0 . \tag{6}$$

Exactly the same manipulations may be carried out by differentiating with respect to y, instead of x. We have the equation

$$0 = F_r \cdot r_y + F_s \cdot s_y + F_t \cdot t_y + \{F\}_y ,$$

together with

$$x' \cdot r_y + y' \cdot s_y = s'$$
$$x' \cdot s_y + y' \cdot t_y = t' ,$$

giving 3 equations for the 3 unknowns r_y, s_y, t_y, namely,

$$\begin{pmatrix} x' & y' & 0 \\ 0 & x' & y' \\ F_r & F_s & F_t \end{pmatrix} \cdot \begin{pmatrix} r_y \\ s_y \\ t_y \end{pmatrix} = \begin{pmatrix} s' \\ t' \\ -\{F\}_y \end{pmatrix} .$$

So, as before, we have

$$(7) \qquad \rho_1 F_r s' + F_t t' + \{F\}_y y' = 0 ,$$

as well as an equation involving $\grave{}$.

We will now select 8 equations satisfied by the 8 functions x, y, u, p, q, r, s, t when u is a solution of (1). Our first two equations will be

$$(*1) \qquad y' - \rho_1 x' = 0$$

$$(*2) \qquad y\grave{} - \rho_2 x\grave{} = 0 .$$

The next three will be the "strip conditions"

$$(*3) \qquad u' - px' - qy' = 0$$

$$(*4) \qquad p' - rx' - sy' = 0$$

$$(*5) \qquad q' - sx' - ty' = 0 .$$

Finally, we add in equations (5)-(7):

$$(*6) \qquad \rho_1 F_r r' + F_t s' + (F_x + F_u p + F_p r + F_q s) y' = 0$$

$$(*7) \qquad \rho_1 F_r s' + F_t t' + (F_y + F_u q + F_p s + F_q t) y' = 0$$

$$(*8) \qquad \rho_2 F_r r^{\grave{}} + F_t s^{\grave{}} + (F_x + F_u p + F_p r + F_q s) y^{\grave{}} = 0 .$$

The functions $x, y, \ldots, t$ have initial conditions

$$(*_0) \quad \begin{cases} x(\xi,0) = \xi & y(\xi,0) = 0 \\ u(\xi,0) = \mathring{u}(\xi) & \\ p(\xi,0) = \mathring{u}'(\xi) , & q(\xi,0) = \mathring{q}(\xi) \\ r(\xi,0) = \mathring{u}''(\xi) , & s(\xi,0) = \mathring{q}'(\xi) , \qquad t(\xi,0) = \mathring{t}(\xi) . \end{cases}$$

The beauty of this particular set of 8 equations is the fact that they automatically lead to solutions of equation (1). More precisely, consider 8 functions $x, y, u, \ldots, t$ satisfying the system (*), with the initial conditions $(*_0)$. We are denoting the 8 unknowns of our system by $x, y, \ldots, t$ just for convenience -- we are not assuming that $u_x = p$, etc. Nor do we assume that x and y are projections on the first and second coordinates. We therefore use (ξ, η) as coordinates on $\mathbb{R}^2$, so that we can write x_ξ, x_η, etc.

11. LEMMA. Let x, y, u, p, q, r, s, t be 8 functions satisfying the system (*), with the initial conditions $(*_0)$. Then (x,y) is a coordinate system in a neighborhood of the interval $[a,b]$ of the ξ-axis, and $u\circ(x,y)^{-1}$ is a solution of the PDE (1) with the initial conditions $\overset{\circ}{u}, \overset{\circ}{q}, \overset{\circ}{t}$ along this interval.

Remark. Naturally, we would expect (x,y) to be the identity coordinate system, but we do not assert that right now.

Proof. We claim first that we have

$$(1)\quad 0 = F(x,y,u,p,q,r,s,t) \qquad [= F(x(\xi,\eta),y(\xi,\eta),u(\xi,\eta),\ldots)] .$$

To prove this, we multiply equation (*7) by ρ_1 and add it to equation (*6). Make the substitution

$$\rho_1{}^2F_r + F_t = \rho_1F_s \qquad \text{by (4')} ,$$

divide by $\rho_1 \neq 0$, and use the fact that $y'/\rho_1 = x'$, by (*1). We end up with

$$\begin{aligned} 0 = F_rr' + F_ss' + F_tt' + (F_x + F_up + F_pr + F_qs)x' \\ + (F_y + F_uq + F_ps + F_qt)y' . \end{aligned}$$

Making use of (*3)-(*5) we have

$$\begin{aligned} 0 &= F_rr' + F_ss' + F_tt' + F_xx' + F_yy' + F_uu' + F_pp' + F_qq' \\ &= F' \qquad [\text{where } F \text{ denotes } (\xi,\eta) \mapsto F(x(\xi,\eta),\ldots)] . \end{aligned}$$

On the other hand, since $\mathring{u}$, $\mathring{q}$, $\mathring{t}$ are assumed to satisfy (1_0), the initial conditions $(*_0)$ insure that $F = 0$ on the interval $[a,b]$ of the ξ-axis. Therefore we have $F = 0$ in a whole neighborhood of this interval.

Next note that equations (*1) and (*2),

$$y_\xi + \rho_1 y_\eta = \rho_1(x_\xi + \rho_1 x_\eta)$$
$$y_\xi + \rho_2 y_\eta = \rho_2(x_\xi + \rho_2 x_\eta) ,$$

can be solved for x_η and y_η in terms of x_ξ and y_ξ. Subtracting ρ_2 times the second equation from ρ_1 times the first, we obtain

$$\rho_1\rho_2 x_\eta = - y_\xi .$$

On $[a,b] \times \{0\}$, where we have the initial conditions $(*_0)$, we find that

$$x_\xi = 1 , \qquad x_\eta = 0$$
$$y_\xi = 0 , \qquad y_\eta = 1 .$$

Hence $x', x`, y', y` \neq 0$ on $[a,b] \times \{0\}$. Moreover, (x,y) is a coordinate system in a neighborhood of $[a,b] \times \{0\}$. For any function α, we have

$$\alpha_x = \alpha_\xi \cdot \xi_x + \alpha_\eta \cdot \eta_x$$
$$\alpha_y = \alpha_\xi \cdot \xi_y + \alpha_\eta \cdot \eta_y ,$$

where the partials $\xi_x, \ldots$ are given by

$$\begin{pmatrix} \xi_x & \xi_y \\ \eta_x & \eta_y \end{pmatrix} = \begin{pmatrix} x_\xi & x_\eta \\ y_\xi & y_\eta \end{pmatrix}^{-1} = \frac{1}{x_\xi y_\eta - x_\eta y_\xi} \begin{pmatrix} y_\eta & -x_\eta \\ -y_\xi & x_\xi \end{pmatrix} .$$

Thus

$$\alpha_x = \frac{\alpha_\xi y_\eta - \alpha_\eta y_\xi}{x_\xi y_\eta - x_\eta y_\xi} \qquad \alpha_y = \frac{-\alpha_\xi x_\eta + \alpha_\eta x_\xi}{x_\xi y_\eta - x_\eta y_\xi} .$$

These relations can be combined to give

$$(2) \qquad \begin{cases} x^{\backprime}\alpha^{\prime} - x^{\prime}\alpha^{\backprime} = \alpha_y (x^{\backprime}y^{\prime} - x^{\prime}y^{\backprime}) \\ y^{\backprime}\alpha^{\prime} - y^{\prime}\alpha^{\backprime} = \alpha_x (y^{\backprime}x^{\prime} - y^{\prime}x^{\backprime}) . \end{cases}$$

Now equations (*6) and (*8) can be written in the form

$$\frac{F_r}{x^{\prime}} r^{\prime} + \frac{F_t}{y^{\prime}} s^{\prime} + F_x + F_u p + F_p r + F_q s = 0$$

$$\frac{F_r}{x^{\backprime}} r^{\backprime} + \frac{F_t}{y^{\backprime}} s^{\backprime} + F_x + F_u p + F_p r + F_q s = 0 .$$

Hence

$$\frac{F_r}{x^{\prime}} r^{\prime} + \frac{F_t}{y^{\prime}} s^{\prime} = \frac{F_r}{x^{\backprime}} r^{\backprime} + \frac{F_t}{y^{\backprime}} s^{\backprime} .$$

Using (2), this equation can be written

$$(3) \qquad F_r r_y \left(\frac{y^{\prime}}{x^{\prime}} - \frac{y^{\backprime}}{x^{\backprime}}\right) + F_t s_x \left(\frac{x^{\prime}}{y^{\prime}} - \frac{x^{\backprime}}{y^{\backprime}}\right) = 0 .$$

But

$$\frac{x^{\prime}}{y^{\prime}} \cdot \frac{x^{\backprime}}{y^{\backprime}} = \frac{1}{\rho_1 \rho_2} = \frac{F_r}{F_t} \qquad \text{since } \rho_1, \rho_2 \text{ are the roots of (4') ,}$$

so from (3) we find that

$$r_y = s_x \,. \tag{4}$$

On $[a,b] \times \{0\}$, equation (*4) gives

$$p_\xi + \rho_1 p_\eta = r \cdot 1 + s \cdot \rho_1 \,.$$

Taking into account the initial conditions $(*_0)$, we see that

$$p_\xi = r \,, \quad p_\eta = s \qquad \text{on} \quad [a,b] \times \{0\} \,.$$

It follows that

$$P = p^{\backprime} - rx^{\backprime} - sy^{\backprime} = 0 \qquad \text{on} \quad [a,b] \times \{0\} \,.$$

Moreover

$$P^{\prime} = p^{\prime\backprime} - r^{\prime}x^{\backprime} - s^{\prime}y^{\backprime} - rx^{\prime\backprime} - sy^{\prime\backprime} \,,$$

while equation (*4) gives

$$0 = p^{\prime\backprime} - r^{\backprime}x^{\prime} - s^{\backprime}y^{\prime} - rx^{\prime\backprime} - sy^{\prime\backprime} \,.$$

So we have

$$\begin{aligned} P^{\prime} &= r^{\backprime}x^{\prime} - r^{\prime}x^{\backprime} + s^{\backprime}y^{\prime} - s^{\prime}y^{\backprime} \\ &= (s_x - r_y)(x^{\backprime}y^{\prime} - x^{\prime}y^{\backprime}) && \text{by (2)} \\ &= 0 && \text{by (4)} \,. \end{aligned}$$

Consequently, we have

$$0 = P = p^{\backprime} - rx^{\backprime} - sy^{\backprime} \tag{5}$$

in a neighborhood of $[a,b] \times \{0\}$. From (*4) and (5) we have

$$(6) \qquad \begin{cases} x^{\backprime}p^{\prime} - x^{\prime}p^{\backprime} = s(x^{\backprime}y^{\prime} - x^{\prime}y^{\backprime}) \\ y^{\backprime}p^{\prime} - y^{\prime}p^{\backprime} = r(y^{\backprime}x^{\prime} - y^{\prime}x^{\backprime}) \ . \end{cases}$$

So (2) gives

$$(7) \qquad r = p_x \ , \qquad s = p_y \ .$$

Similarly, set

$$U = u^{\backprime} - px^{\backprime} - qy^{\backprime}$$

$$Q = q^{\backprime} - sx^{\backprime} - ty^{\backprime} \ .$$

Then we have

$$U^{\prime} = u^{\prime\backprime} - p^{\prime}x^{\backprime} - q^{\prime}y^{\backprime} - px^{\prime\backprime} - qy^{\prime\backprime} \ ,$$

and

$$0 = u^{\prime\backprime} - p^{\backprime}x^{\prime} - q^{\backprime}y^{\prime} - px^{\prime\backprime} - qy^{\prime\backprime} \qquad \text{from } (*3) \ ,$$

and thus

$$\begin{aligned} (8) \qquad U^{\prime} &= - p^{\prime}x^{\backprime} - q^{\prime}y^{\backprime} + p^{\backprime}x^{\prime} + q^{\backprime}y^{\prime} \\ &= - s(x^{\backprime}y^{\prime} - x^{\prime}y^{\backprime}) - q^{\prime}y^{\backprime} + q^{\backprime}y^{\prime} && \text{by (6)} \\ &= (q^{\backprime} - sx^{\backprime} - ty^{\backprime})y^{\prime} - (q^{\prime} - sx^{\prime} - ty^{\prime})y^{\backprime} \\ &= Qy^{\prime} && \text{by } (*5) \ . \end{aligned}$$

Likewise,

$$\begin{aligned} Q^{\prime} &= - s^{\prime}x^{\backprime} - t^{\prime}y^{\backprime} + s^{\backprime}x^{\prime} + t^{\backprime}y^{\prime} \\ &= (t_x - s_y)(x^{\backprime}y^{\prime} - x^{\prime}y^{\backprime}) && \text{by (2)} \end{aligned}$$

Now subtract the equation $dF/dx = 0$ (which follows from (1)) from (*6), and subtract the equation $dF/dy = 0$ from (*7). Multiply the resulting equations by x' and y', respectively, and add them. Using (*) we obtain

$$F_t(t_x - s_y)x' + F_q(q_x - s)x' + F_u(u_x - p)x'$$
$$- F_t \frac{x'}{y'}(t_x - s_y)y' + F_q(q_y - t)y' + F_u(u_y - q)y' = 0 .$$

Thus we have

$$\frac{F_t}{y'} Q' + F_q Q + F_u U = 0$$
$$U' = Qy' .$$

Along each integral curve of $Z = (1,\rho_1)$, this is simply a system of ordinary differential equations for U and Q. Since $U = Q = 0$ on $[a,b] \times \{0\}$, we find that $U = Q = 0$ in a neighborhood of $[a,b] \times \{0\}$. As before,

$$U = 0 \text{ and } (*3) \implies p = u_x , \quad q = u_y$$
$$Q = 0 \text{ and } (*5) \implies s = q_x , \quad t = q_y .$$

Thus the first equation (1) in the proof now gives

$$0 = F(x,y,u,u_x,u_y,u_{xx},u_{xy},u_{yy}) ,$$

which is what we wanted to prove. ∎

Now we want to know whether the system (*) does in fact have a solution with the initial conditions $(*_0)$. Our equations are described by the following matrix.

x	y	u	p	q	r	s	t
$-\rho_1$	1	0	0	0	0	0	0
$-\rho_2$	1	0	0	0	0	0	0
*	*	1	0	0	0	0	0
*	*	0	1	0	0	0	0
*	*	0	0	1	0	0	0
0	*	0	0	0	$\rho_1 F_r$	F_t	0
0	*	0	0	0	0	$\rho_1 F_r$	F_t
0	*	0	0	0	$\rho_2 F_r$	F_t	0

We will denote the unknowns by $\phi_1, \ldots, \phi_8$. For convenience, set

$$\rho_1 F_r = A , \qquad \rho_2 F_r = B , \qquad F_t = C .$$

Remembering that

$$\acute{\ } = \frac{\partial}{\partial \xi} + \rho_1 \frac{\partial}{\partial \eta}$$

$$\grave{\ } = \frac{\partial}{\partial \xi} + \rho_2 \frac{\partial}{\partial \eta} ,$$

we see that the system (*) can be written as

$$-\mathcal{P} \cdot \phi_\eta = \mathcal{Q} \cdot \phi_\xi ,$$

where ϕ_η and ϕ_ξ are the column vectors

$$\phi_\eta = \left(\frac{\partial \phi_1}{\partial \eta}, \ldots, \frac{\partial \phi_8}{\partial \eta} \right)^t , \qquad \phi_\xi = \left(\frac{\partial \phi_1}{\partial \xi}, \ldots, \frac{\partial \phi_8}{\partial \xi} \right)^t$$

and $\mathcal{P}$ and $\mathcal{Q}$ are the matrices

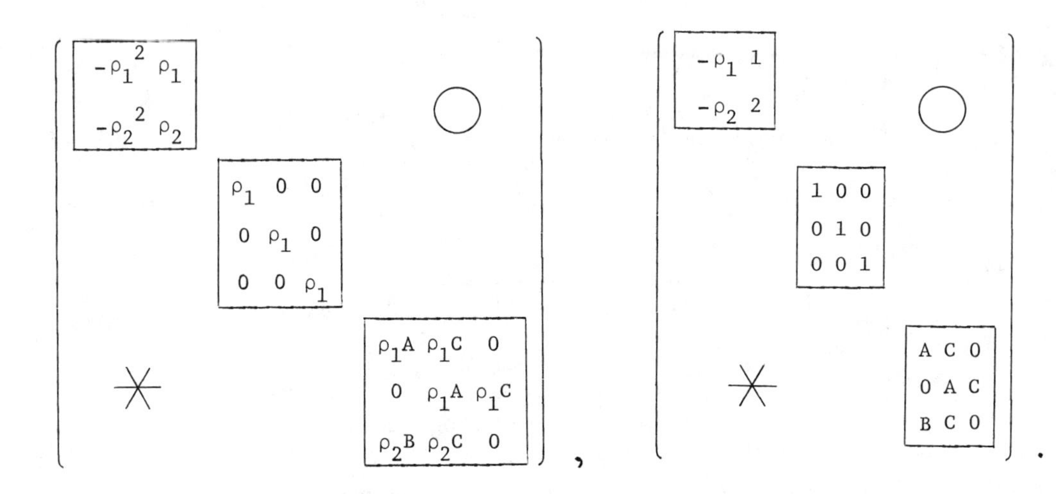

Since

$$Q = \mathcal{D}P \qquad \text{for} \qquad \mathcal{D}^{-1} = \begin{pmatrix} \begin{array}{|cc|} \hline \rho_1 & 0 \\ 0 & \rho_2 \\ \hline \end{array} & & \bigcirc \\ & \begin{array}{|ccc|} \hline \rho_1 & 0 & 0 \\ 0 & \rho_1 & 0 \\ 0 & 0 & \rho_1 \\ \hline \end{array} & \\ \ast & & \begin{array}{|ccc|} \hline \rho_1 & 0 & 0 \\ 0 & \rho_1 & 0 \\ 0 & 0 & \rho_2 \\ \hline \end{array} \end{pmatrix},$$

we can write

$$\phi_\eta = - P^{-1}\mathcal{D}P\phi_\xi \ .$$

The matrix $\mathcal{D}^{-1}$ is diagonalizable, since each box is. So the matrix $-\mathcal{P}^{-1}\mathcal{D}\mathcal{P}$ is diagonalizable, and non-singular. Thus, Theorem 10 shows that we can solve the system (*) with the initial conditions $(*_0)$. So Lemma 11 shows that our original PDE (1) has a solution with the given initial conditions. Moreover, the solution is unique, by the uniqueness part of Theorem 10. (This also shows that the functions x, y in Lemma 11 are actually the usual coordinate system on $\mathbb{R}^2$). We summarize this result in

12. THEOREM. Consider a second order equation

$$F(x,y,u,u_x,u_y,u_{xx},u_{xy},u_{yy}) = 0 \ ,$$

an initial curve, and hyperbolic initial data along this curve such that the curve is free for the initial data. Suppose that the initial curve, initial data, and F have continuous third partial derivatives satisfying a Lipschitz condition. Then in a neighborhood of this initial curve there is a unique solution with this initial data having continuous second partial derivatives satisfying a Lipschitz condition.

Actually, the hypotheses could be weakened considerably here (partly by weakening the hypotheses in Theorem 10, and even more so by using other methods), but we will not worry about this. More important, we would like to know that the solution u has a high order of differentiability if F, the initial curve, and the initial data do, but we must be content with merely asserting this, since we did not prove this for Theorem 10. One thing is clear, however. Even if F and the initial curve are highly differentiable, or even analytic, there may be solutions u which are far less differentiable. In fact, if we have any hyperbolic initial data, then nearby initial data will also be hyperbolic. We can choose this nearby data to be only C^3, and then our solution is at most C^3. As we will see in the next section, this is in marked contrast to the situation for elliptic solutions.

Our approach to Theorem 12 also allows us to consider the case where the curve is everywhere characteristic for the initial data. Again we assume that this curve is a segment $[a,b]$ of the x-axis, but that we now have

$$\text{(a)} \qquad 0 = F_t(x,0,\mathring{u}(x),\mathring{u}'(x),\mathring{q}(x),\mathring{u}''(x),\mathring{q}'(x),\mathring{t}(x)) = F_t(\alpha(x))$$

for all $x \in [a,b]$. We can still assume that $F_r \neq 0$ (for, in the expression for G_r on p. 119 we certainly have $F_s \neq 0$, since $0 < F_s{}^2 - 4F_rF_t = F_s{}^2$). Equation (4') becomes

$$-F_s\rho + F_r\rho^2 = 0 ,$$

with the solutions

$$\rho_1 = 0 , \qquad \rho_2 = F_s/F_r .$$

If there is any solution u of equation (1) with the given initial data $\mathring{u}$, $\mathring{q}$, $\mathring{t}$, then certain conditions must hold. Namely, equations (*1)-(*8) must be true on $[a,b]\times\{0\}$, where

$$\acute{} = \frac{\partial}{\partial\xi} , \qquad \grave{} = \frac{\partial}{\partial\xi} + \rho_2 \frac{\partial}{\partial\eta}$$

$$x = \xi , \qquad y = \eta$$

and

$$r^{`} = \frac{\partial r}{\partial \xi} + \rho_2 \frac{\partial r}{\partial \eta} = \mathring{u}''' + \frac{F_s}{F_r}\mathring{q}'' .$$

All the equations are automatic, except for (*8), which becomes, after dividing by $\rho_2 = y^{`}$,

$$(**) \quad F_r\mathring{u}''' + F_s\mathring{q}'' + (F_x + F_u\mathring{u}' + F_p\mathring{u}'' + F_q\mathring{q}') = 0 \qquad \text{on} \quad [a,b]\times\{0\} .$$

Conversely, suppose we have initial conditions $\mathring{u}$, $\mathring{q}$, $\mathring{t}$, satisfying (a) and (**). We claim that there are infinitely many solutions with these initial conditions. To prove this, we consider a new $\bar{x}$, $\bar{y}$ coordinate system on $\mathbb{R}^2$,

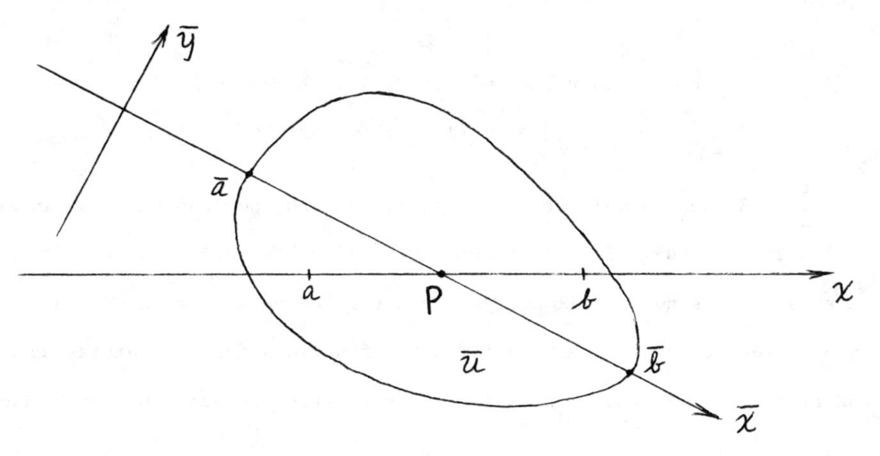

and an interval $[\bar{a},\bar{b}]$ on the $\bar{x}$-axis which intersects the interval $[a,b]$ of the x-axis at a point P. Our equation (1) for u becomes an equation $\bar{F} = 0$ for the function $\bar{u} = u$ expressed in the $(\bar{x},\bar{y})$-coordinate system. At any point, we can determine $\bar{u}_{\bar{x}},\ldots,\bar{u}_{\bar{y}\bar{y}}$ in terms of $u_x,\ldots,u_{yy}$, and conversely. Now choose arbitrary hyperbolic initial data $\mathring{\bar{u}}$, $\mathring{\bar{q}}$, $\mathring{\bar{t}}$ along the interval $[\bar{a},\bar{b}]$ of the $\bar{x}$-axis such that $\mathring{\bar{u}},\mathring{\bar{u}}_{\bar{x}},\mathring{\bar{q}},\ldots,\bar{t}_0$ have the values determined by $u,u_x,q,\ldots,t$ at the point P, and solve the equation $\bar{F} = 0$ with this initial data, by solving the corresponding system $(\bar{*}1)$-$(\bar{*}8)$, with $\bar{\rho}_1, \bar{\rho}_2 \neq 0$. Along the interval $[a,b]$ of the x-axis, the equations $(\bar{*}1)$-$(\bar{*}8)$ are a system of ordinary differential equations for $(\bar{x},\bar{y},\dots,\bar{t})$, and this system agrees with the corresponding system $(*1)$-$(*8)$ for $(x,y,\dots,t)$. By hypothesis, the system $(*1)$-$(*8)$ has the solution $(x,y,\mathring{u},\mathring{u}',\mathring{q},\mathring{u}'',\mathring{q}',\mathring{t})$, and we have arranged for this solution to agree with $(\bar{x},\bar{y},\mathring{\bar{u}},\dots,\mathring{\bar{t}})$ at P. So the solutions agree everywhere. Thus, the solution u of equation (1) corresponding to the solution $\bar{u}$ of $\bar{F} = 0$ has the initial conditions $\mathring{u}, \mathring{q}, \mathring{t}$ along $[a,b] \times \{0\}$.

One particular case will be very important in Chapter 12. We consider an equation

$$(1) \qquad 0 = F(x,y,p,q,r,s,t) = A(rt - s^2) + Br + Cs + Dt + E$$
$$= (Ar + D)t + (Br + Cs + E - As^2) \ ,$$

where $A,\dots,E$ depend only on $x, y, p, q,$ so that our equation is linear in $r, s, t,$ and $rt - s^2$. Such equations are called "Monge-Ampère equations", and they are the kind we will always encounter. Suppose that along the x-axis we have picked $\mathring{u}$ and $\mathring{q}$ as the first two functions for our initial data. We can already see if the x-axis is characteristic, because the condition for this is

$$(2) \qquad 0 = F_t = Ar + D \ ,$$

which only involves $\mathring{u}$, $\mathring{p} = \mathring{u}'$, and $\mathring{q}$ on the x-axis. If equation (2) holds, then there is no hope of selecting $\mathring{t}$ to complete our initial data unless we also have

$$(3) \qquad 0 = Br + Cs + E - As^2$$

along the x-axis. If this equation holds, then any choice of $\overset{\circ}{t}$ will make $\overset{\circ}{u}$, $\overset{\circ}{q}$, $\overset{\circ}{t}$ satisfy (1). Of course, we want to choose $\overset{\circ}{t}$ so that equation (**) holds. This equation now reads

$$
\begin{aligned}
(4) \qquad 0 = {} & (At+B)\overset{\circ}{u}''' + (C-2As)\overset{\circ}{q}'' \\
& + A_x(rt-s^2) + B_x r + C_x s + D_x t + E_x \\
& + [A_p(rt-s^2) + B_p r + C_p s + D_p t + E_p]\overset{\circ}{u}'' \\
& + [A_q(rt-s^2) + B_q r + C_q s + D_q t + E_q]\overset{\circ}{q}' \, .
\end{aligned}
$$

But along the x-axis we have

$$
(2) \quad \Longrightarrow \quad \left\{
\begin{aligned}
A_x r + A\overset{\circ}{u}''' + D_x &= 0 \\
A_p r + D_p &= 0 \\
A_q r + D_q &= 0
\end{aligned}
\right.
$$

and

$$
(3) \quad \Longrightarrow \quad \left\{
\begin{aligned}
B_x r + B\overset{\circ}{u}''' + C_x s + C\overset{\circ}{q}'' + E_x - A_x s^2 - 2As\overset{\circ}{q}'' &= 0 \\
B_p r + C_p s + E_p - A_p s^2 &= 0 \\
B_q r + C_q s + E_q - A_q s^2 &= 0 \, .
\end{aligned}
\right.
$$

From these equations we immediately see that equation (4) is automatic. Thus, in this particular situation, where $\overset{\circ}{u}$ and $\overset{\circ}{q}$ satisfy (2) and (3), we can find solutions of (1) with initial data $\overset{\circ}{u}$, $\overset{\circ}{q}$, $\overset{\circ}{t}$ for arbitrary functions $\overset{\circ}{t}$.

9. Elliptic Solutions of Second Order Equations in Two Variables

The most important topic in the study of elliptic equations is the Dirichlet problem, and if this were a book on PDEs, it would be inexcusable not to devote a great deal of time to this subject. But we won't say another word about it. Instead, we will consider another aspect of elliptic equations, which often isn't even mentioned in a first course in PDEs. We have already noted that every solution of $u_{xx} + u_{yy} = 0$ is automatically analytic. In this section we will prove that every elliptic solution of any second order equation $F(x,y,u,p,q,r,s,t) = 0$ is likewise analytic, provided of course that F is an analytic function of its arguments. This theorem holds for elliptic solutions of second order PDEs in any number of variables, but the proof we will give works only in the two variable case. This deficiency (which doesn't bother us, since we are interested only in the two variable case) is more than compensated for by its conceptual simplicity. Moreover, the proof has one truly beautiful feature -- there isn't a single inequality in it. Since the details of the proof become somewhat complicated, it will probably help to first examine a special case. Consider the equation

$$u_{xx} + u_{yy} = f(x,y,u,p,q) ,$$

where f is a real analytic function of its arguments. We will show that u can be extended to a complex-analytic function from $\mathbb{C}^2$ to $\mathbb{C}$; consequently u must be real analytic.

We recall first that if

$$\alpha(x,y) = \beta(x,y) + i\gamma(x,y)$$

for real-valued β, γ, then the Cauchy-Riemann equations for α are

$$\beta_x = \gamma_y \ , \qquad \beta_y = -\gamma_x \ ;$$

these two equations are equivalent to the single equation

(1) $$\alpha_x = -\,i\alpha_y \ .$$

We will rewrite our equation for u as

(2) $$u_{x_1x_1} + u_{y_1y_1} = f(x_1, y_1, u, p, q) \ .$$

We denote the two coordinates in $\mathbb{R}^2$ by x_1, y_1 and consider $\mathbb{R}^2 \subset \mathbb{C}^2$, where $\mathbb{C}^2$ has coordinates $x = x_1 + ix_2$, $y = y_1 + iy_2$. Thus we think of u as a function such that $u(x_1,0,y_1,0)$ is defined. We first want to find a complex-valued extension $u(x_1,0,y_1,y_2)$ of u which is complex-analytic in $y_1 + iy_2$. Equation (1) shows that our desired extension should satisfy

(3) $$u_{x_1x_1} - u_{y_2y_2} = f(x_1, y_1 + iy_2, u, u_{x_1}, -iu_{y_2})$$

at all points $(x_1,0,y_1,y_2)$. For each fixed y_1, consider equation (3) in the (x_1,y_2)-plane, with the initial conditions

(4) $$u(x_1,0,y_1,0) = \text{the original } u(x_1,0,y_1,0)$$

(5) $$u_{y_2}(x_1,0,y_1,0) = i\cdot u_{y_1}(x_1,0,y_1,0) \ , \qquad \text{for the original } u(x_1,0,y_1,0) \ .$$

Equation (3) is hyperbolic. Actually this statement is misleading, for the right side of (3) is already complex-valued, so we have to allow the solution u to be complex-valued; thus we have to consider (3) as an equation for the real and imaginary parts of u. However, if we replace (3) by a system of quasi-linear equations, and make a new system by looking at the real and imaginary parts of all the functions in the old system, then the new system will in fact be hyperbolic. The reader may check this for himself (we will write things out explicitly later on, for the general case). Then Theorem 10 shows* that we really can solve (3) with initial conditions (4) and (5).

Differentiating (5) gives

$$u_{y_2y_1} = iu_{y_1y_1} \qquad \text{at } (x_1,0,y_1,0) ,$$

while subtracting (3) from (2) gives

$$u_{y_1y_1} + u_{y_2y_2} = 0 \qquad \text{at } (x_1,0,y_1,0) .$$

From these two equations we have

$$u_{y_2y_2} = iu_{y_1y_2} \qquad \text{at } (x_1,0,y_1,0) . \tag{6}$$

Equations (5) and (6) can also be written

$$\left\{\begin{aligned} & u_{y_1} + iu_{y_2} = 0 \\ & \frac{\partial}{\partial y_2}(u_{y_1} + iu_{y_2}) = 0 \end{aligned}\right. \qquad \text{at } (x_1,0,y_1,0) . \tag{7}$$

On the other hand, we can also obtain an equation for $\omega = u_{y_1} + iu_{y_2}$. To equation (3) we apply the operator $\nabla = \partial/\partial y_1 + i\partial/\partial y_2$; in the notation of Addendum 1 to Chapter 9, $\frac{1}{2}\nabla u$ would be written $u_{\bar{y}}$. Then we have

---

*If we use the system of equations derived in the previous section, then Theorem 10 suffices. If we use the system of equations derived in section 3, then we would need the stronger form of Theorem 10 which allows the matrix A to be singular.

$$\omega_{x_1x_1} - \omega_{y_2y_2} = \nabla(u_{x_1x_1} - u_{y_2y_2})$$

$$= f_y\nabla y + f_u\nabla u + f_p\nabla(u_{x_1}) - if_q\nabla(u_{y_2})$$

<u>since</u> f <u>is analytic</u> (compare p. IV.464)

$$= 0 + f_u\omega + f_p\omega_{x_1} - if_q\omega_{y_2} .$$

This is a hyperbolic system for ω. Thus (7) implies that $\omega = 0$, by uniqueness of solutions. Hence

(8) $$u_{y_1}(x_1,0,y_1,y_2) + iu_{y_2}(x_1,0,y_1,y_2) = 0 .$$

Now we will extend u to $\mathbb{R}^4$. We do this by considering the equation

(9) $$u_{y_1y_1} - u_{x_2x_2} = f(x_1 + ix_2, y, u, -iu_{x_2}, u_{y_2}) ;$$

here x_1 and y_2 are the parameters. We use the initial conditions

(10) $u(x_1,0,y_1,y_2)$ = the $u(x_1,0,y_1,y_2)$ already obtained

(11) $u_{x_2}(x_1,0,y_1,y_2) = i\cdot u_{x_1}(x_1,0,y_1,y_2)$,

for the $u(x_1,0,y_1,y_2)$ already obtained .

Again we obtain a hyperbolic system, so we can solve (9), with the initial conditions (10) and (11).

Differentiating (11) gives

$$u_{x_1x_2} = iu_{x_1x_1} \qquad \text{at} \quad (x_1,0,y_1,y_2) ,$$

while differentiating (8) with respect to y_1 and y_2, and then subtracting, gives

$$u_{y_1y_1} + u_{y_2y_2} = 0 \qquad \text{at } (x_1,0,y_1,y_2) .$$

Finally, subtracting (3) from (9) gives

$$u_{y_1y_1} + u_{y_2y_2} - u_{x_1x_1} - u_{x_2x_2} = 0 \qquad \text{at } (x_1,0,y_1,y_2) .$$

From these three equations we obtain

$$(12) \qquad u_{x_2x_2} = iu_{x_1x_2} \qquad \text{at } (x_1,0,y_1,y_2) .$$

Equations (11) and (12) can be written

$$(13) \qquad \left\{ \begin{array}{l} u_{x_1} + iu_{x_2} = 0 \\ \\ \dfrac{\partial}{\partial x_2}(u_{x_1} + iu_{x_2}) = 0 \end{array} \right. \qquad \text{at } (x_1,0,y_1,y_2) .$$

As before, we can also derive an equation for $u_{x_1} + iu_{x_2}$ and conclude that we must have $u_{x_1} + iu_{x_2} = 0$ everywhere. Similarly, we prove that $u_{y_1} + iu_{y_2} = 0$ everywhere. Hence u is complex-analytic, and the real-valued solution of the original equation $u_{xx} + u_{yy} = f(x,y,u,p,q)$ is real analytic.

Now we are ready to tackle the general case. We consider an elliptic solution u of a general second order equation

$$0 = F(x,y,u,u_x,u_y,u_{xx},u_{xy},u_{yy}) = F(x,y,u,p,q,r,s,t) .$$

For convenience, we will often speak as if u were defined on all of $\mathbb{R}^2$, although actually the arguments are entirely local. Ellipticity of u means that

$$0 < 4F_rF_t - F_s^2 \quad [= 4F_r(x,y,u(x,y)\ldots)F_t(x,y,u(x,y)\ldots) - F_s(x,y,u(x,y)\ldots)^2],$$

so that, in particular, F_r, $F_t \neq 0$. In the previous section we used $'$ to denote ∇_Z for a fixed vector field Z on $\mathbb{R}^2$, and noted that $Z = (x', y')$. This holds even if Z is a complex-valued vector field $Z = \alpha \cdot \partial/\partial x + \beta \cdot \partial/\partial y$ for complex-valued functions α, β. Now let ρ_1, ρ_2 be the two continuous everywhere unequal complex-valued functions ρ_1, ρ_2, each of which is a solution of

$$F_t - F_s\rho + F_r\rho^2 = 0 .$$

Denote ∇_Z by $'$ for $Z = (1,\rho_1)$ and ∇_Z by $`$ for $Z = (1,\rho_2)$. Then the considerations of the previous section give us eight equations

$$(*) \qquad \begin{cases} \displaystyle\sum_{j=1}^{8} a_{ij}\phi_j' = 0 & i = 1,\ldots,6 \\ \displaystyle\sum_{j=1}^{8} a_{ij}\phi_j^{`} = 0 & i = 7,8 , \end{cases}$$

where $\phi_1,\ldots,\phi_8$ stand for $x,y,u,\ldots,t$, and the a_{ij} are now complex-valued functions of $\phi_1,\ldots,\phi_8$. Since F is assumed analytic in all its arguments, the functions a_{ij} are complex analytic in some region of $\mathbb{C}^2$ containing the set $R^2 \subset \mathbb{C}^2$ where they are defined. When we arrange our equations in the order used in the previous section, the matrix (a_{ij}) is given on p. 133. Note that

$$\det(a_{ij}) = (\rho_2 - \rho_1)^2 F_r F_t^2 \neq 0 .$$

It will be convenient to use $\xi_1, \xi_2, \eta_1, \eta_2$ as coordinates on $\mathbb{R}^4$. Thus we regard the ϕ_j as functions with $\phi_j(\xi_1,0,\eta_1,0)$ defined; in particular, we have

$$x(\xi_1,0,\eta_1,0) = \phi_1(\xi_1,0,\eta_1,0) = \xi_1 ,$$

$$y(\xi_1,0,\eta_1,0) = \phi_2(\xi_1,0,\eta_1,0) = \eta_1 .$$

The operators $\acute{}$ and $\grave{}$ in the (ξ_1,η_1)-plane = the (x,y)-plane are then given by

$$(*_1) \qquad \begin{cases} \acute{} = \dfrac{\partial}{\partial\xi_1} + \rho_1 \dfrac{\partial}{\partial\eta_1} \\[2ex] \grave{} = \dfrac{\partial}{\partial\xi_1} + \rho_2 \dfrac{\partial}{\partial\eta_1} . \end{cases}$$

We consider the functions a_{ij} as already extended to complex-analytic functions of their eight arguments in a suitable region of $\mathbb{C}^2$. Now for fixed η_1, consider equations (*) as equations in the (ξ_1,η_2)-plane, with the operations $\acute{}$ and $\grave{}$ now being defined by

$$(*_2) \qquad \begin{cases} \acute{} = \dfrac{\partial}{\partial\xi_1} + \dfrac{\partial}{\partial\eta_2} \\[2ex] \grave{} = - \dfrac{\partial}{\partial\xi_1} + \dfrac{\partial}{\partial\eta_2} . \end{cases}$$

This is equivalent to taking $\rho_1 = 1$ and $\rho_2 = -1$. So if we arrange our equations in the order used in the previous section, they become

$$- \mathcal{P}\cdot\phi_{\eta_2} = \mathcal{Q}\cdot\phi_{\xi_1} ,$$

where

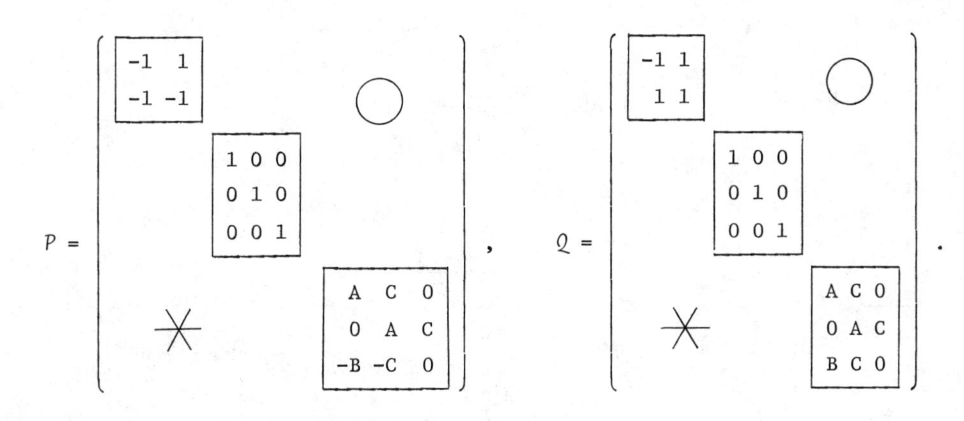

Thus

$$\mathcal{Q} = \mathcal{D}\mathcal{P} \quad \text{for} \quad \mathcal{D} = \begin{pmatrix} \boxed{\begin{matrix} 1 & 0 \\ 0 & -1 \end{matrix}} & & \bigcirc \\ & \boxed{\begin{matrix} 1 & 0 & 0 \\ 0 & 1 & 0 \\ 0 & 0 & 1 \end{matrix}} & \\ \ast & & \boxed{\begin{matrix} 1 & 0 & 0 \\ 0 & 1 & 0 \\ 0 & 0 & -1 \end{matrix}} \end{pmatrix} .$$

Setting

$$\phi_1 = \psi_1 + i\psi_2$$
$$\phi_2 = \psi_3 + i\psi_4 \qquad \psi_i \text{ real-valued,}$$
$$\vdots$$

and writing our equations in terms of the ψ_i, we obtain 16 equations

$$- \mathbf{P}\cdot\psi_{\eta_2} = \mathbf{Q}\cdot\psi_{\xi_1} .$$

Here

$$\mathbf{Q} = \mathbf{D}\,\mathbf{P} \quad ,$$

where the box

$$\boxed{\begin{matrix} 1 & 0 \\ 0 & -1 \end{matrix}} \quad \text{in } \mathcal{D} \text{ is replaced by } \quad \boxed{\begin{matrix} 1 & 0 & 0 & 0 \\ 0 & 1 & 0 & 0 \\ 0 & 0 & -1 & 0 \\ 0 & 0 & 0 & -1 \end{matrix}} \quad \text{in } \mathbf{D} \text{ , etc .}$$

We still have zeros above the boxes, so $\mathbf{D}$ is diagonalizable. By Theorem 10, we can solve (*), with ´ and ` given by $(*_2)$; as our initial conditions we just choose

$$\phi_j(\xi_1,0,\eta_1,0) = \text{the original } \phi_j(\xi_1,0,\eta_1,0) \ .$$

Similarly, we now extend the functions $\phi_j(\xi_1,0,\eta_1,\eta_2)$ to $\mathbb{R}^4$ by fixing ξ_1 and η_2, and considering equations (*) in the (ξ_2,η_1)-plane, with the operations ´ and ` now defined by

$$(*_3) \qquad \left\{ \begin{aligned} ´ &= \frac{\partial}{\partial \xi_2} - \frac{\partial}{\partial \eta_1} \\ ` &= \frac{\partial}{\partial \xi_2} + \frac{\partial}{\partial \eta_1} \ . \end{aligned} \right.$$

Among the extended functions ϕ_j, we have "x" $= \phi_1$ and "y" $= \phi_2$. Since the ϕ_j are now complex-valued, we have four real-valued functions on $\mathbb{R}^4$ defined by

$$x = x_1 + ix_2 \ , \qquad y = y_1 + iy_2 \ .$$

We claim that x_1, x_2, y_1, y_2 is a coordinate system in a neighborhood of any point in the (ξ_1,η_1)-plane. To prove this, we have to compute the Jacobian of (x_1,x_2,y_1,y_2). First of all, since x_1, x_2, y_1, y_2 are simply ξ_1, ξ_2, η_1, η_2 on the (ξ_1,η_1)-plane, at the point in question we have

$$(1)\quad \begin{cases} \dfrac{\partial x_1}{\partial \xi_1} = 1\,, & \dfrac{\partial x_2}{\partial \xi_1} = 0\,, & \dfrac{\partial y_1}{\partial \xi_1} = 0\,, & \dfrac{\partial y_2}{\partial \xi_1} = 0 \\ \dfrac{\partial x_1}{\partial \eta_1} = 0\,, & \dfrac{\partial x_2}{\partial \eta_1} = 0\,, & \dfrac{\partial y_1}{\partial \eta_1} = 1\,, & \dfrac{\partial y_2}{\partial \eta_1} = 0\,. \end{cases}$$

To compute other derivatives, we first write the two complex-conjugate roots ρ_1, ρ_2 of $F_t - F_s\rho + F_r\rho^2 = 0$ as

$$\rho_1 = \sigma_1 + i\sigma_2\,, \qquad \rho_2 = \sigma_1 - i\sigma_2\,, \qquad \sigma_2 \neq 0\,.$$

The equations $y' - \rho_1 x' = 0$ and $y` - \rho_2 x` = 0$, with the two different meanings $(*_2)$ and $(*_3)$ for $'$ and $`$, give the following equations [after making use of (1)]:

$$(2)\quad \begin{cases} \dfrac{\partial y_1}{\partial \eta_2} + i\,\dfrac{\partial y_2}{\partial \eta_2} - (\sigma_1 + i\sigma_2)\left(1 + \dfrac{\partial x_1}{\partial \eta_2} + i\,\dfrac{\partial x_2}{\partial \eta_2}\right) = 0 \\ \dfrac{\partial y_1}{\partial \eta_2} + i\,\dfrac{\partial y_2}{\partial \eta_2} - (\sigma_1 - i\sigma_2)\left(-\,1 + \dfrac{\partial x_1}{\partial \eta_2} + i\,\dfrac{\partial x_2}{\partial \eta_2}\right) = 0 \end{cases}$$

$$(3)\quad \begin{cases} \dfrac{\partial y_1}{\partial \xi_2} + i\,\dfrac{\partial y_2}{\partial \xi_2} - 1 - (\sigma_1 + i\sigma_2)\left(\dfrac{\partial x_1}{\partial \xi_2} + i\,\dfrac{\partial x_2}{\partial \xi_2}\right) = 0 \\ \dfrac{\partial y_1}{\partial \xi_2} + i\,\dfrac{\partial y_2}{\partial \xi_2} + 1 - (\sigma_1 - i\sigma_2)\left(\dfrac{\partial x_1}{\partial \xi_2} + i\,\dfrac{\partial x_2}{\partial \xi_2}\right) = 0\,. \end{cases}$$

Subtracting the first equation of (2) from the second gives

$$2\sigma_1 + 2i\sigma_2\left(\frac{\partial x_1}{\partial \eta_2} + i\,\frac{\partial x_2}{\partial \eta_2}\right) = 0 \quad\Longrightarrow\quad \frac{\partial x_1}{\partial \eta_2} + i\,\frac{\partial x_2}{\partial \eta_2} = i\,\frac{\sigma_1}{\sigma_2}$$

$$\Longrightarrow\quad \frac{\partial x_1}{\partial \eta_2} = 0\ , \quad \frac{\partial x_2}{\partial \eta_2} = \frac{\sigma_1}{\sigma_2}\ .$$

Then we get

$$\frac{\partial y_1}{\partial \eta_2} + i\,\frac{\partial y_2}{\partial \eta_2} = (\sigma_1 + i\sigma_2)\left(1 + i\,\frac{\sigma_1}{\sigma_2}\right) = \frac{i(\sigma_1^{\,2} + \sigma_2^{\,2})}{\sigma_2}$$

$$\Longrightarrow\quad \frac{\partial y_1}{\partial \eta_2} = 0\ , \quad \frac{\partial y_2}{\partial \eta_2} = \frac{\sigma_1^{\,2} + \sigma_2^{\,2}}{\sigma_2}\ .$$

Similarly, from (3) we get

$$2 + 2i\sigma_2\left(\frac{\partial x_1}{\partial \xi_2} + i\,\frac{\partial x_2}{\partial \xi_2}\right) = 0 \quad\Longrightarrow\quad \frac{\partial x_1}{\partial \xi_2} + i\,\frac{\partial x_2}{\partial \xi_2} = \frac{i}{\sigma_2}$$

$$\Longrightarrow\quad \frac{\partial x_1}{\partial \xi_2} = 0\ , \quad \frac{\partial x_2}{\partial \xi_2} = \frac{1}{\sigma_2}\ ,$$

and then

$$\frac{\partial y_1}{\partial \xi_2} + i\,\frac{\partial y_2}{\partial \xi_2} = 1 + (\sigma_1 + i\sigma_2)\frac{i}{\sigma_2} = i\,\frac{\sigma_1}{\sigma_2}$$

$$\Longrightarrow\quad \frac{\partial y_1}{\partial \xi_2} = 0\ , \quad \frac{\partial y_2}{\partial \xi_2} = \frac{\sigma_1}{\sigma_2}\ .$$

So at the point in question, the matrix of derivatives of x_1, x_2, y_1, y_2 with respect to $\xi_1, \xi_2, \eta_1, \eta_2$ is

$$(4) \qquad \begin{array}{c} \\ \xi_1 \\ \xi_2 \\ \eta_1 \\ \eta_2 \end{array} \begin{array}{c} \begin{array}{cccc} x_1 & x_2 & y_1 & y_2 \end{array} \\ \begin{pmatrix} 1 & 0 & 0 & 0 \\ 1 & \frac{1}{\sigma_2} & 0 & \frac{\sigma_1}{\sigma_2} \\ 0 & 0 & 1 & 0 \\ 0 & \frac{\sigma_1}{\sigma_2} & 0 & \frac{\sigma_1^2 + \sigma_2^2}{\sigma_2} \end{pmatrix} \end{array} .$$

The determinant equals 1, so (x_1, x_2, y_1, y_2) is indeed a coordinate system.

Now all partials $\partial/\partial\xi_1, \ldots, \partial/\partial\eta_2$ can be written as certain linear combinations of $\partial/\partial x_1, \ldots, \partial/\partial y_2$. So we can also write

$$(5) \qquad \frac{\partial}{\partial\xi_1} + \frac{\partial}{\partial\eta_2} = A_1 \frac{\partial}{\partial x_1} + B_1 \frac{\partial}{\partial y_1} + C_1\left(\frac{\partial}{\partial x_1} + i \frac{\partial}{\partial x_2}\right) + D_1\left(\frac{\partial}{\partial y_1} + i \frac{\partial}{\partial y_2}\right) .$$

Now the equation $y' - \rho_1 x' = 0$, where $'$ has the significance $(*_2)$, gives

$$(6) \qquad \left(\frac{\partial}{\partial\xi_1} + \frac{\partial}{\partial\eta_2}\right)(y_1 + iy_2) - \rho_1\left(\frac{\partial}{\partial\xi_1} + \frac{\partial}{\partial\eta_2}\right)(x_1 + ix_2) = 0 .$$

But $\partial/\partial x_1 + i\,\partial/\partial x_2$ gives zero when applied to $x_1 + ix_2$ (or any analytic function of x_1, x_2), and similarly for $\partial/\partial y_1 + i\,\partial/\partial y_2$. So when we replace the operator $\partial/\partial\xi_1 + \partial/\partial\eta_2$ in equation (6) by its expression in (5) we end up with

$$B_1 - \rho_1 A_1 = 0 .$$

Note that if we had $A_1 = 0$, then the operator (5) would not be real unless $C_1 = D_1 = 0$, which is impossible, since $\partial/\partial\xi_1 \neq -\,\partial/\partial\eta_2$. So $A_1 \neq 0$. Thus we have

$$(7)\quad \frac{\partial}{\partial\xi_1} + \frac{\partial}{\partial\eta_2} = A_1 \frac{\partial}{\partial x_1} + \rho_1 A_1 \frac{\partial}{\partial y_1} + C_1\left(\frac{\partial}{\partial x_1} + i\,\frac{\partial}{\partial x_2}\right) + D_1\left(\frac{\partial}{\partial y_1} + i\,\frac{\partial}{\partial y_2}\right) ,$$

and similarly

$$(8)\quad -\frac{\partial}{\partial\xi_1} + \frac{\partial}{\partial\eta_2} = A_2 \frac{\partial}{\partial x_1} + \rho_2 A_2 \frac{\partial}{\partial y_1} + C_2\left(\frac{\partial}{\partial x_1} + i\,\frac{\partial}{\partial x_2}\right) + D_2\left(\frac{\partial}{\partial y_1} + i\,\frac{\partial}{\partial y_2}\right) ,$$

$$(9)\quad \frac{\partial}{\partial\xi_2} - \frac{\partial}{\partial\eta_1} = E_1 \frac{\partial}{\partial x_1} + \rho_1 E_1 \frac{\partial}{\partial y_1} + G_1\left(\frac{\partial}{\partial x_1} + i\,\frac{\partial}{\partial x_2}\right) + H_1\left(\frac{\partial}{\partial y_1} + i\,\frac{\partial}{\partial y_2}\right) ,$$

$$(10)\quad \frac{\partial}{\partial\xi_2} + \frac{\partial}{\partial\eta_1} = E_2 \frac{\partial}{\partial x_1} + \rho_2 E_2 \frac{\partial}{\partial y_1} + G_2\left(\frac{\partial}{\partial x_1} + i\,\frac{\partial}{\partial x_2}\right) + H_2\left(\frac{\partial}{\partial y_1} + i\,\frac{\partial}{\partial y_2}\right) ,$$

where $A_1, A_2, E_1, E_2 \neq 0$. All quantities $A_1,\ldots,H_2$ are simply linear combinations of the derivatives of x_1, x_2, y_1, y_2 with respect to $\xi_1, \xi_2, \eta_1, \eta_2$. For example, we obviously have

$$iC_1 = \left(\frac{\partial}{\partial\xi_1} + \frac{\partial}{\partial\eta_2}\right)x_2 , \qquad iD_1 = \left(\frac{\partial}{\partial\xi_1} + \frac{\partial}{\partial\eta_2}\right)y_2 ,$$

$$iC_2 = \left(-\frac{\partial}{\partial\xi_1} + \frac{\partial}{\partial\eta_2}\right)x_2 , \qquad iD_2 = \left(-\frac{\partial}{\partial\xi_1} + \frac{\partial}{\partial\eta_2}\right)y_2 ,$$

$$iG_1 = \left(\frac{\partial}{\partial\xi_2} - \frac{\partial}{\partial\eta_1}\right)x_2 , \qquad iH_1 = \left(\frac{\partial}{\partial\xi_2} - \frac{\partial}{\partial\eta_1}\right)y_2 ,$$

$$iG_2 = \left(\frac{\partial}{\partial\xi_2} + \frac{\partial}{\partial\eta_1}\right)x_2 , \qquad iH_2 = \left(\frac{\partial}{\partial\xi_2} + \frac{\partial}{\partial\eta_1}\right)y_2 .$$

In particular, at a point in the (ξ_1,η_1)-plane we have, from the entries of the matrix (4),

$$
(11) \qquad
\begin{aligned}
iC_1 &= \frac{\sigma_1}{\sigma_2}, & iD_1 &= \frac{\sigma_1^2+\sigma_2^2}{\sigma_2}, \\
iC_2 &= \frac{\sigma_1}{\sigma_2}, & iD_2 &= \frac{\sigma_1^2+\sigma_2^2}{\sigma_2}, \\
iG_1 &= \frac{1}{\sigma_2}, & iH_1 &= \frac{\sigma_1}{\sigma_2}, \\
iG_2 &= \frac{1}{\sigma_2}, & iH_2 &= \frac{\sigma_1}{\sigma_2}.
\end{aligned}
$$

Notice that up till now we have used only the two simplest equations of (*). We will now use the whole set. In the initial plane, the equations (*) hold in three different forms, corresponding to the three meanings of the operators $'$ and $`$, namely

$$
(*_1) \left\{ \begin{aligned} ' &= \frac{\partial}{\partial x_1} + \rho_1 \frac{\partial}{\partial y_1} \\ ` &= \frac{\partial}{\partial x_1} + \rho_2 \frac{\partial}{\partial y_2} \end{aligned} \right. \qquad
(*_2) \left\{ \begin{aligned} ' &= \frac{\partial}{\partial \xi_1} + \frac{\partial}{\partial \eta_2} \\ ` &= -\frac{\partial}{\partial \xi_1} + \frac{\partial}{\partial \eta_2} \end{aligned} \right. \qquad
(*_3) \left\{ \begin{aligned} ' &= \frac{\partial}{\partial \xi_2} - \frac{\partial}{\partial \eta_1} \\ ` &= \frac{\partial}{\partial \xi_2} + \frac{\partial}{\partial \eta_1}. \end{aligned} \right.
$$

From the equations with $(*_2)$ we have, making use of (7) and (11),

$$
(12) \quad \sum_j a_{ij}\left[A_1 \frac{\partial}{\partial x_1} + \rho_1 A_1 \frac{\partial}{\partial y_1} + \frac{\sigma_1}{i\sigma_2}\left(\frac{\partial}{\partial x_1} + i\frac{\partial}{\partial x_2}\right) + \frac{\sigma_1^2+\sigma_2^2}{i\sigma_2}\left(\frac{\partial}{\partial y_1} + i\frac{\partial}{\partial y_2}\right)\right]\phi_j = 0
$$

for $i = 1,\ldots,6$. From the equations with $(*_1)$, we have, after multiplying by A_1,

$$
(13) \qquad \sum_j a_{ij}\left[A_1 \frac{\partial}{\partial x_1} + \rho_1 A_1 \frac{\partial}{\partial y_1}\right]\phi_j = 0
$$

for $i = 1,\ldots,6$. Subtracting (13) from (12) gives us

$$(14) \qquad \sum_j a_{ij}\left[\frac{\sigma_1}{i\sigma_2}\left(\frac{\partial}{\partial x_1} + i\,\frac{\partial}{\partial x_2}\right) + \frac{\sigma_1^{\,2} + \sigma_2^{\,2}}{i\sigma_2}\left(\frac{\partial}{\partial y_1} + i\,\frac{\partial}{\partial y_2}\right)\right]\phi_j = 0 \ .$$

If we do the same thing for $i = 7,8$, except multiply by A_2 instead of A_1, we find that (14) holds also for $i = 7,8$. Since $\det(a_{ij}) \neq 0$, it follows that

$$(15) \qquad \left[\frac{\sigma_1}{\sigma_2}\left(\frac{\partial}{\partial x_1} + i\,\frac{\partial}{\partial x_2}\right) + \frac{\sigma_1^{\,2} + \sigma_2^{\,2}}{\sigma_2}\left(\frac{\partial}{\partial y_1} + i\,\frac{\partial}{\partial y_2}\right)\right]\phi_j = 0 \qquad j = 1,\ldots,8 \ .$$

Similarly, if we start from the equations with $(*_3)$, and then subtract the equations with $(*_1)$, multiplied by E_1 and E_2, we find that

$$(16) \qquad \left[\frac{1}{\sigma_2}\left(\frac{\partial}{\partial x_1} + i\,\frac{\partial}{\partial x_2}\right) + \frac{\sigma_1}{\sigma_2}\left(\frac{\partial}{\partial y_1} + i\,\frac{\partial}{\partial y_2}\right)\right]\phi_j = 0 \qquad j = 1,\ldots,8 \ .$$

For each particular ϕ_j, equations (15) and (16) give two equations for ϕ_j, and since

$$\det\begin{pmatrix} \dfrac{\sigma_1}{\sigma_2} & \dfrac{\sigma_1^{\,2} + \sigma_2^{\,2}}{\sigma_2} \\ \\ \dfrac{1}{\sigma_2} & \dfrac{\sigma_1}{\sigma_2} \end{pmatrix} = -\,1 \neq 0 \ ,$$

we must have

$$(17) \qquad \left(\frac{\partial}{\partial x_1} + i\,\frac{\partial}{\partial x_2}\right)\phi_j = 0 \qquad \text{and} \qquad \left(\frac{\partial}{\partial y_1} + i\,\frac{\partial}{\partial y_2}\right)\phi_j = 0 \ .$$

Thus, the Cauchy-Riemann equations for ϕ_j hold in the plane $x_2 = y_2 = 0$.

Now we want to show that the Cauchy-Riemann equations hold for the y_1, y_2 variables. Let

$$\nabla_x = \frac{\partial}{\partial x_1} + i \frac{\partial}{\partial x_2} , \qquad \nabla_y = \frac{\partial}{\partial y_1} + i \frac{\partial}{\partial y_2} .$$

We denote the partials of the functions a_{ij} with respect to their 8 variables by $\partial a_{ij}/\partial\phi_1$, etc. Because the a_{ij} are <u>analytic</u>, we have

$$\nabla_x a_{ij} = \sum_\ell \frac{\partial a_{ij}}{\partial \phi_\ell} \nabla_x \phi_\ell \qquad \nabla_y a_{ij} = \sum_\ell \frac{\partial a_{ij}}{\partial \phi_\ell} \nabla_y \phi_\ell$$

Consider the first 6 equations (*), with $'$ and $`$ given by $(*_2)$; after division by A_1 they can be written

$$\Sigma\, a_{ij} \frac{\phi_j'}{A_1} = \sum_j a_{ij} \left[\frac{\partial}{\partial x_1} + \rho_1 \frac{\partial}{\partial y_1} + \frac{C_1}{A_1} \nabla_x + \frac{D_1}{A_1} \nabla_y\right] \phi_j = 0 \qquad i = 1,\ldots,6 .$$

Apply ∇_x to this equation. Since we have

$$\nabla_x \frac{\phi_j'}{A_1} = \nabla_x\left(\left(\frac{\partial}{\partial x_1} + \rho_1 \frac{\partial}{\partial y_1} + \frac{C_1}{A_1} \nabla_x + \frac{D_1}{A_1} \nabla_y\right)\phi_j\right)$$

$$= \left(\frac{\partial}{\partial x_1} + \rho_1 \frac{\partial}{\partial y_1} + \frac{C_1}{A_1} \nabla_x + \frac{D_1}{A_1} \nabla_y\right)\nabla_x \phi_j$$

$$+ \left(\nabla_x \frac{C_1}{A_1}\right) \cdot \nabla_x \phi_k + \left(\nabla_x \frac{D_1}{A_1}\right) \cdot \nabla_y \phi_j + \frac{\partial \phi_j}{\partial y_1} \sum_{\ell=1}^{8} \frac{\partial \rho_1}{\partial \phi_\ell} \nabla_x \phi_\ell$$

$$= \frac{1}{A_1}(\nabla_x \phi_j)' + \left(\nabla_x \frac{C_1}{A_1}\right) \cdot \nabla_x \phi_j + \left(\nabla_x \frac{D_1}{A_1}\right) \nabla_y \phi_j + \frac{\partial \phi_j}{\partial y_1} \sum_{\ell=1}^{8} \frac{\partial \rho_1}{\partial \phi_\ell} \nabla_x \phi_\ell ,$$

we obtain an equation of the form

$$(18) \qquad \sum_j a_{ij}(\nabla_x\phi_j)' + \sum_j(b_{ij}\nabla_x\phi_j + c_{ij}\nabla_y\phi_j) = 0 \qquad i = 1,\ldots,6 .$$

Treating the equations for $i = 7,8$ similarly, except dividing by A_2, we obtain

$$(19) \qquad \sum_j a_{ij}(\nabla_x\phi_j)^{\backprime} + \sum_j(b_{ij}\nabla_x\phi_j + c_{ij}\nabla_y\phi_j) = 0 \qquad i = 7,8 .$$

Applying ∇_y similarly to these same equations, we obtain

$$(20) \qquad \sum_j a_{ij}(\nabla_y\phi_j)' + \sum_j(d_{ij}\nabla_x\phi_j + e_{ij}\nabla_y\phi_j) = 0 \qquad i = 1,\ldots,6$$

$$(21) \qquad \sum_j a_{ij}(\nabla_y\phi_j)^{\backprime} + \sum_j(d_{ij}\nabla_x\phi_j + e_{ij}\nabla_y\phi_j) = 0 \qquad i = 7,8 .$$

Equations (18)-(21) are 16 equations for 16 complex-valued functions $\nabla_x\phi_j$, $\nabla_y\phi_j$. The matrix of the system is

$$\begin{pmatrix} (a_{ik}) & 0 \\ 0 & (a_{ik}) \end{pmatrix} .$$

So we easily see that the corresponding system of 32 equations for 32 real-valued functions is hyperbolic. But (17) gives $\nabla_x\phi_j = \nabla_y\phi_j = 0$ for $\eta_2 = 0$. By uniqueness of solutions, it follows that $\nabla_x\phi_j = \nabla_y\phi_j$ for all $(\xi_1,0,\eta_1,\eta_2)$.

In exactly the same way, we show finally that $\nabla_x\phi_j = \nabla_y\phi_j = 0$ for all $(\xi_1,\xi_2,\eta_1,\eta_2)$. Thus all extended ϕ_j, in particular $u = \phi_3$, are complex-analytic. So the original real solution u of our equation is real analytic.

In this proof we need the ϕ_j to have continuous second partial derivatives satisfying a Lipschitz condition (so that the $\nabla_x\phi_j$, $\nabla_y\phi_j$ in the last step will have continuous partials satisfying a Lipschitz condition). Thus we require u to have continuous fourth partial derivatives satisfying a Lipschitz condition. Actually, the result holds even if u is C^3, but that information comes out of other proofs (it might also be derivable from the present proof with enough extra work). We will merely state this stronger result in the summary of all the work of this section:

13. THEOREM. If u is a C^3 elliptic solution of the equation

$$F(x,y,u,u_x,u_y,u_{xx},u_{xy},u_{yy}) = 0 ,$$

where F is a real analytic function of its 8 arguments, then u is real analytic.

Addendum 1. Differential Systems; The Cartan-Kähler Theorem

Suppose we are given everywhere linearly independent 1-forms $\omega_1,\ldots,\omega_\ell$ on an n-manifold M. The Frobenius integrability theorem, in the differential form version (Proposition I.7-14), tells us when every point $p \in M$ lies in some $(n-\ell)$-dimensional manifold $N \subset M$ such that all ω_j restricted to N are zero: this happens if and only if each $d\omega_j$ is in the ideal generated by the $\{\omega_j\}$. Our proof rested on the observation that the $d\omega_j$ have this property if and only if the $(n-\ell)$-dimensional distribution $\Delta = \bigcap \ker \omega_j$ has the property that $[X,Y]$ belongs to Δ whenever X and Y do. On the other hand, simple direct considerations could have shown us that the condition on the $d\omega_j$ is certainly necessary. For suppose that $N \subset M$ is an $(n-\ell)$-dimensional submanifold of M on which all ω_j vanish (i.e. $i^*\omega_j = 0$, where $i\colon N \to M$ is the inclusion map). Then the $d\omega_j$ also vanish on N, since

$$i^*(d\omega_j) = d(i^*\omega_j) = 0 .$$

But the 2-forms $\omega_j \wedge \omega_k$ also vanish on N, and because the ω_j are everywhere linearly independent, at each point $p \in N$ the $\{\omega_j(p) \wedge \omega_k(p)\}$ already span the set of all elements of $\Omega^2(M_p)$ which vanish on N_p. Thus $d\omega_j(p)$ must be a linear combination of the $\{\omega_j(p) \wedge \omega_k(p)\}$.

We could also have given a direct proof that this necessary condition is sufficient, without appealing to the first version of the Frobenius integrability theorem. We will briefly outline this proof, for it not only shows just how the condition on the $d\omega_i$ is related to the classical integrability criterion, but it is also similar in approach to the proof of the main theorem which we will be proving later.

For convenience we set $k = n - \ell$, and number our forms as $\omega_{k+1},\dots,\omega_n$. Since the result is essentially local, we can assume that $M = \mathbb{R}^n$, that the point $p \in M$ in question is $0 \in \mathbb{R}^n$, and, by changing our axes if necessary, that $dx^1,\dots,dx^k,\omega_{k+1},\dots,\omega_n$ span $(\mathbb{R}^n{}_0)^*$. This means that near 0 we can write

$$(1) \qquad dx^\rho = \sum_{h=1}^{k} A_{h\rho} dx^h + \sum_{r=k+1}^{n} B_{r\rho}\omega_r \qquad \rho = k+1,\dots,n .$$

Now take d of equation (1), and consider the coefficient of a term $dx^i \wedge dx^j$ $(i < j \leq k)$, when the right side is expressed in terms of the 2-forms

$$dx^i \wedge dx^j , \qquad dx^i \wedge \omega_r , \qquad \omega_r \wedge \omega_s ,$$

which are linearly independent near 0. When we write $d\omega_r$ in this way, the coefficients of $dx^i \wedge dx^j$ must vanish, since by hypothesis $d\omega_r$ is in the ideal generated by the ω_r. So we obtain

$$\begin{aligned} 0 &= \text{coefficient of } dx^i \wedge dx^j \text{ in } \sum_{h=1}^{k} dA_{h\rho} \wedge dx^h \\ &= \text{coefficient of } dx^i \wedge dx^j \text{ in } \sum_{h=1}^{k} \sum_{\sigma=1}^{n} \frac{\partial A_{h\rho}}{\partial x^\sigma} dx^\sigma \wedge dx^h \\ &= \frac{\partial A_{j\rho}}{\partial x^i} - \frac{\partial A_{i\rho}}{\partial x^j} + \text{coeff. of } dx^i \wedge dx^j \text{ in } \sum_{h=1}^{k} \sum_{\sigma=k+1}^{n} \sum_{\iota=1}^{k} \frac{\partial A_{h\rho}}{\partial x^\sigma} A_{\iota\sigma} dx^\iota \wedge dx^h \\ &\qquad \text{by (1)} \end{aligned}$$

and thus, finally,

$$(2) \qquad 0 = \frac{\partial A_{j\rho}}{\partial x^i} - \frac{\partial A_{i\rho}}{\partial x^j} + \sum_{\sigma=k+1}^{n} \frac{\partial A_{j\rho}}{\partial x^\sigma} A_{i\sigma} - \sum_{\sigma=k+1}^{n} \frac{\partial A_{i\rho}}{\partial x^\sigma} A_{j\sigma} .$$

But now the classical integrability result (Theorem I.6-1) shows that we can find functions $f^{k+1},\ldots,f^n$ in a neighborhood of 0 in $\mathbb{R}^k$ such that

$$(3)\quad \frac{\partial f^\rho}{\partial x^h}(x_1,\ldots,x_k) = A_{h\rho}(x_1,\ldots,x_k,f^{k+1}(x_1,\ldots,x_k),\ldots,f^n(x_1,\ldots,x_k)) .$$

Equation (3) is precisely the condition that the ω_r vanish on the submanifold $\{(x_1,\ldots,x_k,f^{k+1}(x_1,\ldots,x_k),\ldots,f^n(x_1,\ldots,x_k))\}$, so the proof is complete.

Now we want to consider a more general question. Suppose we are given an ideal $\mathcal{I}$ of differential forms on M, not necessarily generated by 1-forms, which satisfies $d\mathcal{I} \subset \mathcal{I}$. When is there a submanifold $N \subset M$ such that all forms of $\mathcal{I}$ vanish on M? We warn right away that everything is going to be much more complicated. The basic information regarding this situation is contained in the Cartan-Kähler theorem (first proved by Cartan when $\mathcal{I}$ is generated by 1-forms and 2-forms, and then generalized by Kähler). We will never use this result, except to give an alternative proof of a theorem, in the Addendum to Chapter 11, but I felt that it should be included here, not only because it is an application of the Cauchy-Kowalewski theorem, but also because it plays such a crucial role in the work of E. Cartan. It enables one to say, in a sense that will be clarified later on, "how many" different submanifolds of $\mathbb{R}^n$ satisfy a given geometric condition, e.g., the condition that H is constant [here we are considering the <u>local</u> theory of submanifolds, without any completeness requirements]; numerous such examples are worked out in E. Cartan { 2 }. The Cartan-Kähler theorem may be thought of as a result about integrability conditions for systems of partial differential equations, of a more complex type than (3). Nevertheless, the systems to be considered are still very special, since they come from differential forms -- one could compare this situation with the Poincaré Lemma, which also involves integrability conditions of a very special sort.

Before we can state the Cartan-Kähler theorem, some preliminary definitions will be required. First we want to be more precise about ideals of differential forms. Let $\Omega^k(M)$ be the vector space of all k-forms on M. Then the direct sum $\Omega(M) = \Omega^0(M) \oplus \cdots \oplus \Omega^n(M)$ is a ring under $\wedge$. For any ideal $\mathcal{I} \subset \Omega(M)$, we set $\mathcal{I}_k = \mathcal{I} \cap \Omega^k(M)$. We will consider only ideals $\mathcal{I}$ which are homogenous, meaning that

$$\mathcal{I} = \mathcal{I}_0 \oplus \mathcal{I}_1 \oplus \cdots \oplus \mathcal{I}_n .$$

Thus, for example, if $\mathcal{I}$ contains $\omega_1 + \omega_2$ where ω_1 is a 1-form and ω_2 is a 2-form, then $\mathcal{I}$ must contain ω_1 and ω_2 (so $\mathcal{I}$ could not be the ideal generated by $\omega_1 + \omega_2$). For a homogeneous ideal $\mathcal{I}$ it is certainly clear what we mean by the condition $d\mathcal{I} \subset \mathcal{I}$: for each k-form $\omega \in \mathcal{I}$, the $(k+1)$-form $d\omega$ must also be in $\mathcal{I}$. A homogenous ideal with this property is called a differential ideal, or sometimes a differential system. For the present we will assume that our differential system $\mathcal{I}$ does not contain functions, i.e., that $\mathcal{I}_0 = 0$.

Let $\mathcal{I}$ be any homogenous ideal with $\mathcal{I}_0 = 0$ (not necessarily satisfying $d\mathcal{I} \subset \mathcal{I}$). An ℓ-dimensional submanifold $N \subset M$, with inclusion map $i\colon N \to M$, is called an integral submanifold of $\mathcal{I}$ if $i^*\omega = 0$ for all forms $\omega \in \mathcal{I}$. It is easy to see that, because $\mathcal{I}$ is an ideal, this condition holds if $i^*\omega = 0$ for all forms $\omega \in \mathcal{I}_\ell$. It is also easy to see that if $\mathcal{I}$ is generated by a set of elements S, then it suffices to have $i^*\omega = 0$ for all $\omega \in S$ of degree $\leq \ell$. In order to analyse integral submanifolds of $\mathcal{I}$, we consider the possible tangent spaces for them. An ℓ-dimensional subspace

$W \subset M_p$ of M_p is called an (ℓ-dimensional) integral element of $\mathcal{I}$ if all $\omega(p)$ are zero when restricted to W, for all $\omega \in \mathcal{I}$; again, it suffices to have this for all $\omega \in \mathcal{I}_\ell$, or for all ω of degree $\leq \ell$ in a generating set S. Notice that a subspace of an integral element is also an integral element. We will also allow the 0-dimensional subspace of M_p, which we will identify with p. It is always an integral element, since we assume that $\mathcal{I}_0 = 0$.

When the ideal $\mathcal{I}$ is generated by 1-forms, we must assume, for the Frobenius integrability theorem, that locally $\mathcal{I}$ is generated by a fixed number of linearly independent 1-forms. The analogous requirements for an arbitrary differential system $\mathcal{I}$ are more involved. Let $W \subset M_p$ be a k-dimension integral element, and let $X_1,\dots,X_k$ be any basis. We define the "polar space"

$$\mathcal{E}(W) = \{X \in M_p\colon \omega(p)(X_1,\dots,X_k,X) = 0 \text{ for all } \omega \in \mathcal{I}_{k+1}\} .$$

[For $k = 0$, this means that $\mathcal{E}(p) = \{X \in M_p\colon \omega(p)(X) = 0 \text{ for all } \omega \in \mathcal{I}_1\}$.] This definition is clearly independent of the basis $X_1,\dots,X_k$, and we have $W \subset \mathcal{E}(W)$. Using the fact that $\mathcal{I}$ is an ideal, we easily see that for all $X \in \mathcal{E}(W)$ and all $h \leq k$ we have

$$\omega(p)(X_{i_1},\dots,X_{i_h},X) = 0 \qquad \text{for all } \omega \in \mathcal{I}_{h+1} .$$

This means that for every $X \in \mathcal{E}(W)$ which is not in W, the space $W \oplus \mathbb{R}\cdot X$ is an extension of W to a $(k+1)$-dimensional integral element; conversely, any $(k+1)$-dimensional integral element extending W is of this form. We will also find it useful to consider explicitly the ordered bases $(X_1,\dots,X_k)$

of integral elements. Let

$$\mathcal{M}_k = \{(p,X_1,\ldots,X_k)\colon X_1,\ldots,X_k \text{ span a } k\text{-dimensional integral element of } M_p\}$$
$$\subset M \times TM \times \cdots \times TM\ .$$

For each $(p,X_1,\ldots,X_k) \in \mathcal{M}_k$, we define

$$\mathcal{E}(p,X_1,\ldots,X_k) = \{X \in M_p\colon \omega(p)(X_1,\ldots,X_k,X) = 0 \text{ for all } \omega \in \mathcal{I}_{k+1}\}$$
$$= \mathcal{E}(k\text{-dimensional integral element spanned by } X_1,\ldots,X_k)\ .$$

We now define <u>regular</u> integral elements inductively as follows. A point p is a <u>regular</u> 0-dimensional integral element if $\dim \mathcal{E}_1(p') = \dim \mathcal{E}_1(p)$ for all p' in a neighborhood of p. A k-dimensional integral element W is <u>regular</u> if

(a) W contains a $(k-1)$-dimensional regular integral element,

(b) $\dim \mathcal{E}_{k+1}(W') = \dim \mathcal{E}_{k+1}(W)$ for all k-dimensional integral elements W' in a neighborhood of W.

In order to talk about a neighborhood of W, we have to specify the topology involved. The k-dimensional integral elements are topologized as a subset of the set of all k-dimensional subspaces of all M_q; locally this looks like $\mathbb{R}^n \times (k\text{-dimensional subspaces of } \mathbb{R}^n)$, and we use the obvious topology on k-dimensional subspaces of $\mathbb{R}^n$ (described in detail in Chapter 13.2). Equivalently, W is regular if it has some basis $X_1,\ldots,X_k$ such that for each $h \le k$ we have $\dim \mathcal{E}_{h+1}(p',X'_1,\ldots,X'_h) = \dim \mathcal{E}_{h+1}(p,X_1,\ldots,X_h)$ for all $(p',X'_1,\ldots,X'_h) \in \mathcal{M}_h$ in a neighborhood of $(p,X_1,\ldots,X_h)$. Notice that the definition does not preclude the possibility that the regular k-dimensional integral element W contains a $(k-1)$-dimensional integral element which is not regular. That is why our second criterion for regularity merely requires the existence of <u>some</u> basis $(X_1,\ldots,X_k)$ with the requisite property -- there may also be bases which do not have this property. A basis $(X_1,\ldots,X_k)$ which does have the required property will be called <u>good</u>.

For a k-dimensional integral element W, consider the codimension

$$c_{k+1}(W) = n - \dim \mathcal{E}_{k+1}(W) ;$$

similarly, for $(p,X_1,\ldots,X_k) \in \mathcal{M}_k$, set

$$c_{k+1}(p,X_1,\ldots,X_k) = n - \dim \mathcal{E}_{k+1}(p,X_1,\ldots,X_k) .$$

Clearly $c_{k+1}(W)$ is the maximum number of $(k+1)$-forms $\omega^{(1)},\omega^{(2)},\ldots \in \mathcal{I}_{k+1}$ such that the $c_{k+1}(W)$ linear functions

$$Y \mapsto \omega^{(\alpha)}(p)(X_1,\ldots,X_k,Y) \qquad Y \in M_p \qquad (X_1,\ldots,X_k \text{ a basis of } W)$$

are linearly independent. It follows that the function $W \mapsto c_{k+1}(W)$ is lower semi-continuous on the set of all k-dimensional integral elements [that is, the value of this function may be greater than $c_{k+1}(W)$ arbitrarily close to W, but it cannot be less than $c_{k+1}(W)$ arbitrarily close to W]. Consequently, the function $W \mapsto \dim \mathcal{E}_{k+1}(W)$ is <u>upper</u> semi-continuous. It follows easily that condition (b) holds on an open dense subset of the set of all k-dimensional integral elements. It certainly holds if $\dim \mathcal{E}_{k+1}(W)$ has the minimum possible value. [In particular, condition (b) holds if $\dim \mathcal{E}_{k+1}(W) = 0$, in which case there is no $(k+1)$-dimensional integral element containing W.] It is easy to see that if M is a connected analytic manifold, and we consider only analytic forms, then condition (b) is equivalent to $\dim \mathcal{E}_{k+1}(W)$ having the minimum possible value.

The appropriateness of the regularity condition is attested to by the following

14. LEMMA. Let $\mathcal{I}$ be a homogeneous ideal with $\mathcal{I}_0 = 0$. If $X_1,\dots,X_k \in M_p$ is a good basis for a regular k-dimensional integral element of $\mathcal{I}$, and $X_{k+1} \in \mathcal{E}_{k+1}(p,X_1,\dots,X_k)$ is linearly independent of $X_1,\dots,X_k$, then near $(p,X_1,\dots,X_{k+1})$, the set $\mathcal{M}_{k+1}$ is a submanifold of $M \times TM \times \cdots \times TM$, of dimension

$$n(k+2) - c_1(p) - c_2(p,X_1) - \cdots - c_{k+1}(p,X_1,\dots,X_k) .$$

Proof. We can assume that $M = \mathbb{R}^n$. Recall that for $Y \in \mathbb{R}^n$, we let $Y_q = (q,Y)$ be the corresponding tangent vector $\in \mathbb{R}^n{}_q$. Choose $A_1,\dots,A_k$ with $X_i = (A_i)_p$. Set

$$\tilde{\mathcal{M}}_{k+1} = \left\{ (q,Y_1,\dots,Y_{k+1}) \in \mathbb{R}^{n(k+2)} : (q,(Y_1)_q,\dots,(Y_{k+1})_q \in \mathcal{M}_{k+1} \right\} .$$

Then $\mathcal{M}_{k+1}$ is the image of $\tilde{\mathcal{M}}_{k+1}$ under an imbedding $\mathbb{R}^{n(k+2)} \to \mathbb{R}^n \times T\mathbb{R}^n \times \cdots \times T\mathbb{R}^n$, so it suffices to prove that $\tilde{\mathcal{M}}_{k+1}$ is a manifold. We will use induction on k, the case $k = 0$ being easy. So suppose that $\tilde{\mathcal{M}}_h \subset \mathbb{R}^{n(h+1)}$ is known to be a submanifold, of dimension

$$(1) \qquad \dim \tilde{\mathcal{M}}_h = n(h+1) - c_1(p) - \cdots - c_h(p,X_1,\dots,X_{h-1}) .$$

For convenience, set

$$c_{h+1} = c_{h+1}(p,X_1,\ldots,X_h) .$$

Choose c_{h+1} $(h+1)$-forms $\omega^{(1)},\omega^{(2)},\ldots \in \mathcal{A}_{h+1}$ such that the c_{h+1} linear functions

$$(*) \qquad Y \longmapsto \omega^{(\alpha)}(p)(X_1,\ldots,X_k,Y_p)$$

are linearly independent. We adopt the convention that if $q,Y_1,\ldots,Y_{h+1} \in \mathbb{R}^n$, then

$$\omega^{(\alpha)}(q,Y_1,\ldots,Y_{h+1}) \quad \text{denotes} \quad \omega^{(\alpha)}(q)((Y_1)_q,\ldots,(Y_{h+1})_q) .$$

Thus we can consider $\omega^{(\alpha)}$ as a function on $\mathbb{R}^{n(h+2)}$. Since $X_1,\ldots,X_k$ is a good basis, we know that for $(q,Y_1,\ldots,Y_h) \in \tilde{\mathcal{M}}_h$ close to $(p,A_1,\ldots,A_h)$, the linear functions

$$Y \longmapsto \omega^{(\alpha)}(q,Y_1,\ldots,Y_h,Y)$$

already span the set of linear functions

$$Y \longmapsto \omega(q,Y_1,\ldots,Y_h,Y) \qquad \text{for } \underline{\text{all}} \ \omega \in \mathcal{A}_{h+1} .$$

This means that near $(p,A_1,\ldots,A_k)$, the set $\tilde{\mathcal{M}}_{h+1}$ is precisely the set of $(q,Y_1,\ldots,Y_h,Y_{h+1})$ such that

$$\begin{cases} (q,Y_1,Y_h) \in \tilde{\mathcal{M}}_h \\ \omega^{(\alpha)}(q,Y_1,\ldots,Y_h,Y_{h+1}) = 0 \qquad \alpha = 1,\ldots,c_{h+1} . \end{cases}$$

Thus, if we define

$$F: \tilde{\mathcal{M}}_h \times \mathbb{R}^n \longrightarrow \mathbb{R}^{c_{h+1}}$$

by

$$F(q, Y_1, \ldots, Y_h, Y_{h+1}) = (\omega^{(1)}(q, Y_1, \ldots, Y_h, Y_{h+1}), \omega^{(2)}(q, Y_1, \ldots, Y_h, Y_{h+1}), \ldots) ,$$

then $\tilde{\mathcal{M}}_{h+1}$ is just $F^{-1}(0)$ near $(p, A_1, \ldots, A_h, A_{h+1})$. Let $Z_1, \ldots, Z_n$

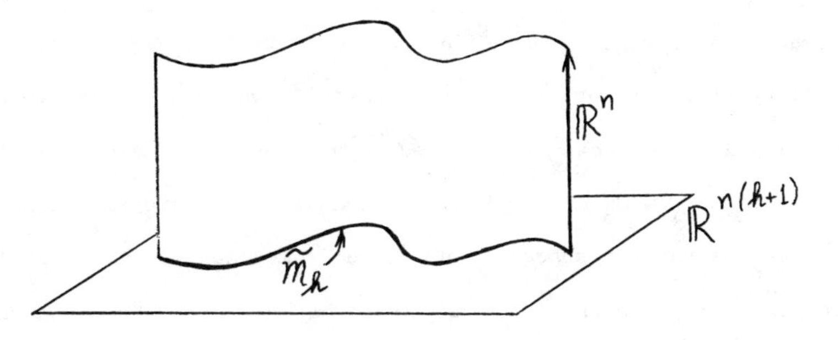

denote the last n basis vectors of $\mathbb{R}^{n(h+2)}$. Then the linear independence of the functions (*) shows that the vectors

$$F_*\left((Z_i)_{(p, A_1, \ldots, A_h, A_{h+1})}\right) \in \mathbb{R}^{c_{h+1}}{}_0$$

are linearly independent. Thus F_* has rank c_{h+1} at $(p, A_1, \ldots, A_h, A_{h+1})$. So in a neighborhood of $(p, A_1, \ldots, A_h, A_{h+1})$ the set

$$\tilde{\mathcal{M}}_{h+1} = F^{-1}(0) \subset \tilde{\mathcal{M}}_h \times \mathbb{R}^n$$

is a manifold, of dimension

$$\dim \tilde{\mathcal{M}}_{h+1} = \dim(\tilde{\mathcal{M}}_h \times \mathbb{R}^n) - c_{h+1}$$

$$= n(h+2) - c_1(p) - \cdots - c_h(p, X_1, \ldots, X_{h-1}) - c_{h+1} \text{ , by (1) .} \blacksquare$$

Our goal is to show that if our ideal $\mathcal{I}$ is a differential system ($d\mathcal{I} \subset \mathcal{I}$), then, at least in the analytic case, a k-dimensional integral element at p which contains a $(k-1)$-dimensional regular integral element (but which need not be regular itself), is the tangent space at p of some k-dimensional integral submanifold of $\mathcal{I}$. We will derive this result as a corollary of a more precise one, which tells when a k-dimensional integral submanifold of $\mathcal{I}$ can be extended to a $(k+1)$-dimensional integral submanifold.

Suppose $W \subset M_p$ is a k-dimensional regular integral element of $\mathcal{I}$ which is the tangent space at p of some k-dimensional integral submanifold N of $\mathcal{I}$. Suppose that $\dim \mathcal{E}_{k+1}(W) > k$, so that there is a vector $X \in \mathcal{E}_{k+1}(W)$ which is not in W; then $W \oplus \mathbb{R}\cdot X$ is a $(k+1)$-dimensional integral element. We will show that there is a $(k+1)$-dimensional integral manifold $N' \supset N$ whose tangent space at p is $W \oplus \mathbb{R}\cdot X$. We can also say precisely how many such integral manifolds N' there are. To do this, we choose a submanifold P of M of dimension

$$\dim P = k + 1 + c_{k+1}(W) ,$$

such that

(a) $$P \supset N$$

(b) $$P_p \cap \mathcal{E}_{k+1}(W) = W \oplus \mathbb{R}\cdot X .$$

We will show that near p there is an (essentially unique) $(k+1)$-dimensional integral submanifold N' of $\mathscr{I}$ with $N \subset N' \subset P$ and $N'_p = W \oplus \mathbb{R}\cdot X$.

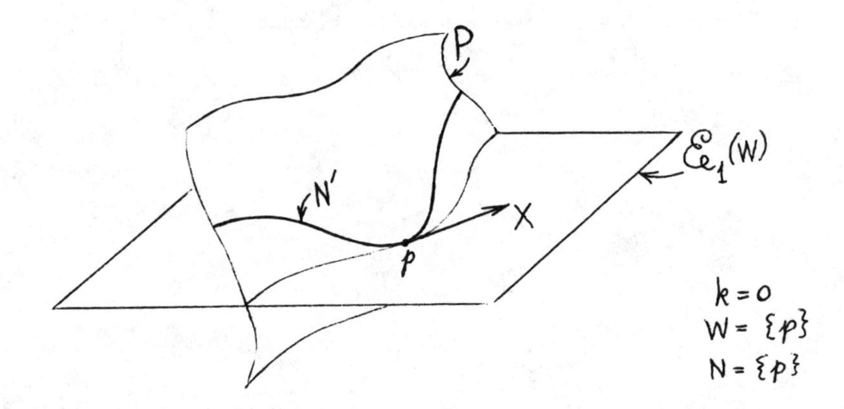

All such submanifolds P can be described locally as follows. Choose $Z \subset M_p$ with $\mathscr{E}_{k+1}(W) \oplus Z = M_p$, so that

$$\dim Z = n - \dim \mathscr{E}_{k+1}(W) = c_{k+1}(W) .$$

Then $W \oplus \mathbb{R}\cdot X \oplus Z$ has dimension

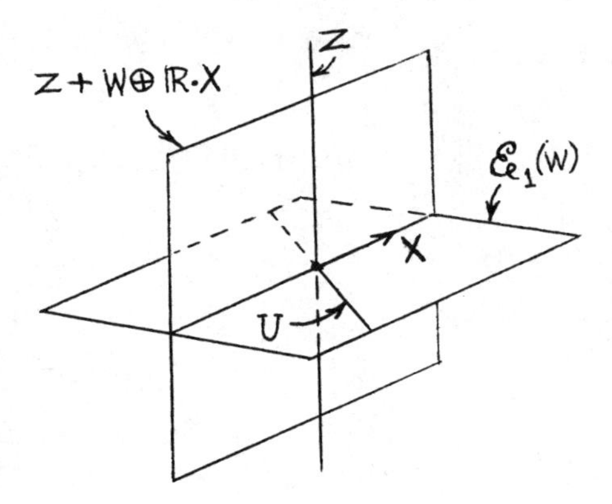

$$\dim(W \oplus \mathbb{R}\cdot X \oplus Z) = k+1+c_{k+1}(W) .$$

Also choose $U \subset M_p$ with $W \oplus \mathbb{R}\cdot X \oplus U = \mathscr{E}_{k+1}(W)$, so that

$$\dim U = \dim \mathscr{E}_{k+1}(W) - k - 1 .$$

Then P can be written as the graph of a function from $W \oplus \mathbb{R}\cdot X \oplus Z$ to U. In classical terminology, the submanifolds P, and hence the desired integral manifolds N', "depend on $\dim \mathcal{E}_{k+1}(W) - k - 1$ arbitrary functions of $k+1+c_{k+1}(W)$ variables."

To prove that N' exists, we can assume without loss of generality that $M = \mathbb{R}^n$, with $p = 0 \in \mathbb{R}^n$, and that

$$
\begin{aligned}
&(e_1)_0,\ldots,(e_k)_0 \text{ is a good basis for } W\\
&X \text{ is } (e_{k+1})_0\\
&\mathcal{E}_{k+1}(W) \text{ is spanned by } (e_1)_0,\ldots,(e_k)_0,(e_{k+1})_0,\ldots,(e_\ell)_0\\
&Z \text{ is spanned by } (e_{\ell+1})_0,\ldots,(e_n)_0\\
&U \text{ is spanned by } (e_{k+2})_0,\ldots,(e_\ell)_0 .
\end{aligned}
$$

By Theorem I.2-10(2), we can assume, by composing $\mathbb{R}^n$ with a diffeomorphism, that

$$(1) \qquad P = \{(x^1,\ldots,x^{k+1},0,\ldots,0,x^{\ell+1},\ldots,x^n)\} .$$

Let N be

$$N = \{(x^1,\ldots,x^k,f_{k+1}(x^1,\ldots,x^k),\ldots,f_n(x^1,\ldots,x^k))\} ,$$

for certain functions $f_{k+1},\ldots,f_n$ with

$$(2) \qquad D_i f_t = 0 \qquad i = 1,\ldots,k\ ; \quad t = k+1,\ldots,n .$$

In order to have $P \supset N$ we must have

$$f_{k+1}(x^1,\ldots,x^k) = x^{k+1}$$

$$f_{k+2}(x^1,\ldots,x^k) = \cdots = f_\ell(x^1,\ldots,x^k) = 0$$

$$f_\nu(x^1,\ldots,x^k) = x^\nu \qquad \nu = \ell+1,\ldots,n\ .$$

Now the map

$$(x^1,\ldots,x^n) \mapsto (x^1,\ldots,x^\ell,\ x^{\ell+1} - f_{\ell+1}(x^1,\ldots,x^k),\ldots,x^n - f_n(x^1,\ldots,x^k))$$

has Jacobian matrix equal to the identity at 0, by (2); so by another application of Theorem I.2-10(2) we can assume that

$$(3) \qquad N = \{(x^1,\ldots,x^k,0,\ldots,0)\}\ .$$

The required N' must be of the form

$$(4) \quad N' = \{(x^1,\ldots,x^{k+1},0,\ldots,0,g_{\ell+1}(x^1,\ldots,x^{k+1}),\ldots,g_n(x^1,\ldots,x^{k+1}))\}\ ,$$

where the functions g_ν satisfy

$$(5) \qquad g_\nu(x^1,\ldots,x^k,0) = 0 \qquad \nu = \ell+1,\ldots,n\ .$$

If

$$\omega = \sum_{i_1<\cdots<i_{k+1}} \omega_{i_1\cdots i_{k+1}}\, dx^{i_1}\wedge\cdots\wedge dx^{i_{k+1}}$$

is any $(k+1)$-form, then ω restricted to N' is zero if and only if the coefficient of $dx^1\wedge\cdots\wedge dx^{k+1}$ is zero when we replace

$\omega_{i_1 \cdots i_{k+1}}$ by

$$(x^1,\ldots,x^{k+1}) \mapsto \omega_{i_1 \cdots i_{k+1}}(x^1,\ldots,x^{k+1},0,\ldots,0,g_{\ell+1}(x^1,\ldots,x^{k+1}), \ldots,g_n(x^1,\ldots,x^{k+1}))$$

dx^j by 0, $\quad j = k+2,\ldots,\ell$

dx^ν by $\sum_{i=1}^{k+1} \frac{\partial g_\nu}{\partial x^i} dx^i$ $\quad \nu = \ell+1,\ldots,n$.

Thus ω restricted to N' is zero if and only if

$$\sum_{\mu=\ell+1}^{n} \omega_{1,2,\ldots,p,\mu}(x^1,\ldots,x^{k+1},0,\ldots,0,g_{\ell+1}(x^1,\ldots,x^{k+1}), \ldots,g_n(x^1,\ldots,x^{k+1}))\frac{\partial g_\mu}{\partial x^{k+1}}$$

$$= \text{certain terms involving the } \partial g_\rho/\partial x^h , \qquad h \le k .$$

We can write this as

$$(6) \quad \sum_{\mu=\ell+1}^{n} C_\mu(x^1,\ldots,x^{k+1},g_{\ell+1},\ldots,g_n)\cdot \frac{\partial g_\mu}{\partial x^{k+1}} = D\left(x^1,\ldots,x^{k+1},\ldots.g_\rho\ldots\frac{\partial g_\rho}{\partial x^h}\ldots\right)$$

[all g_ρ and $\partial g_\rho/\partial x^h$ evaluated at $(x^1,\ldots,x^{k+1})$] ,

where

$$(7) \quad C_\mu(x^1,\ldots,x^{k+1},g_{\ell+1},\ldots,g_n) = \omega_{1,2,\ldots,p,\mu}(x^1,\ldots,x^{k+1},0,\ldots,0,g_{\ell+1},\ldots,g_n) .$$

Choose $n-\ell$ $(k+1)$-forms $\omega^{\ell+1},\ldots,\omega^n \in \mathscr{I}_{k+1}$ so that, with the conventions of the proof of Lemma 14, the $n-\ell$ linear functions

$$Y \longmapsto \omega^{(\nu)}(0,e_1,\ldots,e_k,Y)$$

are linearly independent. This means that

$$0 \neq \det(\omega^{(\nu)}(0,e_1,\ldots,e_k,e_\mu)) \qquad \ell+1 \leq \mu,\ \nu \leq n\ .$$

So if we write $\omega^{(\nu)}$ as

$$\omega^{(\nu)} = \sum_{i_1<\cdots<i_{k+1}} \omega^{(\nu)}_{i_1\cdots i_{k+1}}\, dx^{i_1} \wedge \cdots \wedge dx^{i_{k+1}}\ ,$$

then

$$(8) \qquad 0 \neq \det(\omega^{(\nu)}_{1,2,\ldots,k,\mu}(0)) \qquad \ell+1 \leq \mu,\ \nu \leq n\ .$$

Consider the equations (6) for each $\omega^{(\nu)}$:

$$(9) \qquad \sum_{\mu=\ell+1}^{n} C^{(\nu)}_{\mu}(x^1,\ldots,x^{k+1},g_{\ell+1},\ldots,g_n)\frac{\partial g_\mu}{\partial x^{k+1}} = D\left(x^1,\ldots,x^{k+1},\ldots.g_\rho\ldots\frac{\partial g_\rho}{\partial x^h}\ldots\right).$$

Equation (7), together with (8), shows that

$$0 \neq \det(C^{(\nu)}_{\mu}(0)) \qquad \ell+1 \leq \mu,\ \nu \leq n\ .$$

So equations (9) can be written, near 0, as

$$(10) \qquad \frac{\partial g_\nu}{\partial x^{k+1}} = E_\nu\left(x^1,\ldots,x^{k+1},\ldots.g_\rho\ldots\frac{\partial g_\rho}{\partial x^h}\ldots\right).$$

Now we have arrived at a familiar looking problem.

15. THEOREM (THE CARTAN-KÄHLER THEOREM). Let M be an analytic manifold, and let $\mathcal{I}$ be a differential system (of analytic forms) with $\mathcal{I}_0 = 0$. Let $W \subset M_p$ be a regular k-dimensional integral element, and let N be a k-dimensional integral submanifold of $\mathcal{I}$ with $N_p = W$. Let $X \in \mathcal{E}_{k+1}(W)$ be a vector not in W, and let P be an analytic submanifold of M of dimension $k+1+c_{k+1}(W)$ such that $P \supset N$ and $P_p \cap \mathcal{E}_{k+1}(W) = W \oplus \mathbb{R}\cdot X$. Then there is a unique analytic $(k+1)$-dimensional integral submanifold N' of $\mathcal{I}$ with $N \subset N' \subset P$ and $N'_p = W \oplus \mathbb{R}\cdot X$.

Proof. The previous considerations show that the existence of N' is equivalent to the existence of functions g_ν satisfying

$$g_\nu(x^1,\ldots,x^k,0) = 0 \qquad \nu = \ell+1,\ldots,n$$

and also equations (6) for all $\omega \in \mathcal{I}_{k+1}$. In particular, the functions g_ν must satisfy (10), with the above initial conditions. The Cauchy-Kowalewski theorem (together with the considerations at the end of section 3) shows that there are unique analytic functions g_ν with this property. This already proves uniqueness, and proves the existence of N', with inclusion map $i\colon N' \to \mathbb{R}^n$, satisfying

$$i^*\omega^{\ell+1} = \cdots = i^*\omega^n = 0 .$$

To complete the proof of existence we must show that $i^*\omega = 0$ for <u>all</u> $\omega \in \mathcal{I}_{k+1}$. Here is where the regularity of W is required.

<u>We will continue to use the convention in the proof of Lemma 14</u>. For each $h \le k$, choose $(h+1)$-forms $\omega^{(\alpha)}_{h+1}$ such that the linear functions

$$Y \longmapsto \omega_{h+1}^{(\alpha)}(0,e_1,\ldots,e_h,Y) \qquad \alpha = 1,\ldots,c_{h+1} = c_{h+1}(0,e_1,\ldots,e_h)$$

are linearly independent. Thus the forms $\omega_{k+1}^{(\alpha)}$ are the forms $\omega^{\ell+1},\ldots,\omega^n$ introduced previously. Consider the $(k+1)$-forms

$$\text{(I)} \quad \left\{ \begin{array}{ll} \omega_1^{(\alpha)} \wedge dx^2 \wedge dx^3 \wedge \ldots \wedge dx^{k+1} & \alpha = 1,\ldots,c_1 \\ \quad \omega_2^{(\alpha)} \wedge dx^3 \wedge \ldots \wedge dx^{k+1} & \alpha = 1,\ldots,c_2 \\ \quad \vdots & \\ \qquad \omega_{k+1}^{(\alpha)} & \alpha = 1,\ldots,c_{k+1} . \end{array} \right.$$

We use all of these forms to construct a map $G\colon \mathbb{R}^{n(k+2)} \longrightarrow \mathbb{R}^{c_1+\cdots+c_{k+1}}$, defined by

$$G(q,Y_1,\ldots,Y_{k+1}) = ((\omega_1^{(1)} \wedge dx^2 \wedge \ldots \wedge dx^{k+1})(q,Y_1,\ldots,Y_{k+1}), \ldots, \omega_{k+1}^{(c_{k+1})}(q,Y_1,\ldots,Y_{k+1})) .$$

Since

$$\omega_i^{(\alpha)} \in \mathcal{I}_i \implies (\omega_i^{(\alpha)} \wedge dx^{i+1} \wedge \ldots \wedge dx^{k+1})(0,e_1,\ldots,e_i,\ldots,Y_h,\ldots,e_{k+1}) = 0 ,$$

the Jacobian matrix of G has the form

$$
\begin{array}{c|cccccc|}
 & q\quad Y_1 & Y_2 & Y_3 & \cdots & Y_{k+1} \\
c_1\ \{ & \boxed{A_1} & 0 & 0 & \cdots & 0 \\
c_2\ \{ & & \boxed{A_2} & 0 & \cdots & 0 \\
 & & & \boxed{A_3} & & \vdots \\
 & & & & \ddots & 0 \\
c_{k+1}\ \{ & & & & & \boxed{A_{k+1}}
\end{array}
\quad \text{at} \quad (0,e_1,\ldots,e_{k+1}) .
$$

By our choice of the $\omega_{h+1}^{(\alpha)}$, the block A_{h+1} has rank c_{h+1}. So the whole matrix has maximal rank $c_1+\cdots+c_{k+1}$. Thus $G^{-1}(0)$ is an (analytic) submanifold of $\mathbb{R}^{c_1+\cdots+c_{k+1}}$ near $(0,e_1,\ldots,e_{k+1})$, of dimension $n(k+2)-c_1-\cdots-c_{k+1}$. But the forms (I) are all in the ideal $\mathcal{I}$, so $G^{-1}(0)$ contains the manifold $\tilde{\mathcal{M}}_{k+1}$ in the proof of Lemma 14. It also has the same dimension as this manifold, so it equals this manifold near $(0,e_1,\ldots,e_{k+1})$. We will write the forms in (I) as

$$
\text{(II)} \qquad \begin{cases} \eta^{(\beta)} \wedge dx^{k+1} & \beta = 1,\ldots,d = c_1+\cdots+c_k \\ \omega_{k+1}^{(\alpha)} & \alpha = 1,\ldots,c_{k+1} , \end{cases}
$$

where the forms $\eta^{(\beta)}$ are all in $\mathcal{I}_k$.

Now consider an arbitrary $(k+1)$-form $\omega \in \mathcal{I}_{k+1}$. Since W is a regular integral element, we know that we can write

$$
\omega(q,Y_1,\ldots,Y_k,Y) = \sum_{\alpha=1}^{c_{k+1}} B_\alpha(q,Y_1,\ldots,Y_k)\cdot\omega_{k+1}^{(\alpha)}(q,Y_1,\ldots,Y_k,Y)
$$

for all $(q,Y_1,\ldots,Y_k) \in \tilde{\mathcal{M}}_k$ close to $(0,e_1,\ldots,e_k)$. The functions B_α can be solved for explicitly by Cramer's rule, so they are actually analytic functions in a whole neighborhood of $(0,e_1,\ldots,e_k)$ in $\mathbb{R}^{n(k+1)}$, even though the equation need hold only for $(q,Y_1,\ldots,Y_k) \in \tilde{\mathcal{M}}_k$. We may express this situation as follows:

the function

$$\text{(a)} \qquad \omega - \sum_{\alpha=1}^{c_{k+1}} B_\alpha \omega^{(\alpha)}$$

on $\mathbb{R}^{n(k+2)}$ vanishes on the submanifold $\tilde{\mathcal{M}}_k \times \mathbb{R}^n$, and hence on the (analytic) submanifold $\tilde{\mathcal{M}}_{k+1}$, defined by the equations

$$\text{(b)} \qquad \eta^{(\beta)} \wedge dx^{k+1} = 0 , \qquad \omega_{k+1}^{(\alpha)} = 0 .$$

It follows easily (Problem 1) that locally the function (a) is a sum of analytic functions times the functions in (b). Consequently, we can write

$$\begin{aligned}\omega(q,Y_1,\ldots,Y_{k+1}) = &\sum_{\alpha=1}^{c_{k+1}} C_\alpha(q,Y_1,\ldots,Y_{k+1})\cdot\omega^{(\alpha)}(q,Y_1,\ldots,Y_{k+1}) \\ &+ \sum_{\beta=1}^{d} D_\beta(q,Y_1,\ldots,Y_{k+1})\cdot(\eta^{(\beta)} \wedge dx^{k+1})(q,Y_1,\ldots,Y_{k+1}) ,\end{aligned}$$

for analytic C_α and D_β. This implies that if $q \in N'$, and $Y_1,\ldots,Y_{k+1}$ are tangent to N', then

$$(i^*\omega)(q,Y_1,\ldots,Y_{k+1}) = 0 + \sum_{\beta=1}^{d} D_\beta(q,Y_1,\ldots,Y_{k+1})\, i^*(\eta^{(\beta)} \wedge dx^{k+1})(q,Y_1,\ldots,Y_{k+1}) .$$

So it suffices to show that

$$i^*(\eta^{(\beta)} \wedge dx^{k+1}) = 0 \qquad \beta = 1,\dots,d .$$

From the form of N' (equation (4) on p. 171) it is clear that $x^1,\dots,x^{k+1}$ is a coordinate system on N'. So we write each $i^*\eta^{(\beta)}$ as

$$i^*\eta^{(\beta)} = \sum_{j=1}^{k+1} (-1)^{j+1} h_j^{(\beta)}\, dx^1 \wedge \dots \wedge \widehat{dx^j} \wedge \dots \wedge dx^{k+1} .$$

Now the above analysis for the $(k+1)$-form $\omega \in \mathscr{I}_{k+1}$ can be applied, in particular, for $\omega = \eta^{(\beta)} \wedge dx^j$. Thus each $i^*(\eta^{(\beta)} \wedge dx^j)$ is a linear combination, with analytic coefficients, of the forms $i^*(\eta^{(\beta)} \wedge dx^{k+1})$. Since

$$i^*(dx^j \wedge \eta^{(\beta)}) = dx^j \wedge i^*\eta^{(\beta)} = h_j^{(\beta)}\, dx^1 \wedge \dots \wedge dx^{k+1} \qquad j \leq k$$

$$\begin{aligned} i^*(dx^{k+1} \wedge \eta^{(\beta)}) &= h_{k+1}^{(\beta)}\, dx^1 \wedge \dots \wedge dx^{k+1} \\ &= H^{(\beta)}\, dx^1 \wedge \dots \wedge dx^{k+1}, \qquad \text{say,} \end{aligned}$$

this shows that we can write $h_j^{(\beta)}$ for $j \leq k$ as an analytic linear combination of the $H^{(\beta)}$,

$$h_j^{(\beta)} = \sum_{\gamma=1}^{d} E_{j\beta\gamma} H^{(\gamma)} .$$

Since $d\mathscr{I} \subset \mathscr{I}$, we can also write each $i^*d\eta^{(\beta)}$ as a linear combination of the $i^*(\eta^{(\beta)} \wedge dx^{k+1})$,

$$i^*d\eta^{(\beta)} = \sum_{\gamma=1}^{d} F_{\beta\gamma} H^{(\gamma)}\, dx^1 \wedge \dots \wedge dx^{k+1} .$$

But

$$i^*d\eta^{(\beta)} = di^*\eta^{(\beta)} = d\left(\sum_{j=1}^{k+1}(-1)^{j+1}h_j^{(\beta)}\,dx^1\wedge\ldots\wedge\widehat{dx^j}\wedge\ldots\wedge dx^{k+1}\right)$$

$$= \left[\frac{\partial h_1^\beta}{\partial x^1}+\cdots+\frac{\partial h_{k+1}^{(\beta)}}{\partial x^{k+1}}\right]dx^1\wedge\ldots\wedge dx^{k+1}$$

$$= \left[\frac{\partial(\sum_\gamma E_{1\beta\gamma}H^{(\gamma)})}{\partial x^1}+\cdots+\frac{\partial(\sum_\gamma E_{k\beta\gamma}H^{(\gamma)})}{\partial x^k}+\frac{\partial H^{(\beta)}}{\partial x^{k+1}}\right]dx^1\wedge\ldots\wedge dx^{k+1}\ .$$

Comparing with the original expression for $i^*d\eta^{(\beta)}$, we see that we have a system of equations

$$(*)\qquad \frac{\partial H^{(\beta)}}{\partial x^{k+1}} = \sum_{\gamma=1}^{d} F_{\beta\gamma}H^{(\gamma)} + \sum_{j=1}^{k} G_{j\beta\gamma}\frac{\partial H^{(\gamma)}}{\partial x^j}\,,\qquad \beta = 1,\ldots,d\ ,$$

with everything in sight being analytic. Finally, we have to use the fact that the original manifold N (equation (3) on p. 171) is an integral submanifold of $\mathcal{I}$. This implies that all forms $\eta^{(\beta)}$ vanish on N, which means that

$$(*_0)\qquad H^{(\beta)}(x^1,\ldots,x^k,0) = 0\ ,\qquad \beta = 1,\ldots,d\ .$$

The uniqueness part of the Cauchy-Kowalewski theorem shows that the only solutions $H^{(\beta)}$ of $(*)$ with the initial conditions $(*_0)$ is $H^{(\beta)} = 0$. Thus all $h_j^{(\beta)} = 0$, so all $i^*\eta^{(\beta)} = 0$. ■

As an immediate consequence we obtain

16. COROLLARY. Let M be an analytic manifold, and let $\mathcal{I}$ be a differential system (of analytic forms) with $\mathcal{I}_0 = 0$. Let $W \subset M_p$ be a k-dimensional integral element which contains a regular $(k-1)$-dimensional integral element. Then there is a k-dimensional analytic integral submanifold N of $\mathcal{I}$ with $N_p = W$.

Proof. Choose a good basis $X_1, \ldots, X_k$ of W, and consider the subspaces $W_1 \subset W_2 \subset \cdots \subset W_k$ with W_i the subspace spanned by $X_1, \ldots, X_i$. The desired result then follows by induction from Theorem 15, starting with p as a 0-dimensional integral submanifold. ∎

The reader may easily check for himself that if $\mathcal{I}$ is an ideal generated by linearly independent 1-forms $\omega_1, \ldots, \omega_\ell$, then for every k-dimensional integral element W we have $c_{k+1}(W) = n - \ell$. Consequently, every integral element is regular. Thus the Frobenius theorem follows, in the analytic case, from the Cartan-Kähler theorem.

As a final remark, we point out that it is not hard to take care of the case $\mathcal{I}_0 \neq 0$. One merely has to assume that $\{q \in M: f(q) = 0 \text{ for all } f \in \mathcal{I}_0\}$ is a submanifold $M' \subset M$ near p, and then apply the previous considerations to $\mathcal{I}|M'$.

Addendum 2. An Elementary Maximum Principal

It is well-known that if u is harmonic $(\partial^2 u/\partial x^2 + \partial^2 u/\partial y^2 = 0)$, then u cannot have a relative maximum at an interior point of an open set. A more general principal holds, and its proof, although tricky, is elementary.

On an open set $U \subset \mathbb{R}^n$, consider the second order differential operator L defined by

$$(*) \qquad Lu = \sum_{i,j=1}^{n} a_{ij} \frac{\partial^2 u}{\partial x_i \partial x_j} + \sum_{i=1}^{n} b_i \frac{\partial u}{\partial x_i} + cu \,,$$

for certain functions a_{ij}, b_i, c on U. We assume that $a_{ij} = a_{ji}$, and that the matrix $A = (a_{ij})$ is everywhere definite. [Thus the equation $Lu = 0$ is the most general second order linear elliptic equation.] To be more specific, we will assume that $A = (a_{ij})$ is <u>positive</u> definite. Thus $\sum_{i,j} a_{ij}\xi_i\xi_j > 0$ for $0 \neq \xi \in \mathbb{R}^n$; equivalently, the 1×1 matrix

$$\xi \cdot A \cdot \xi^t > 0 \qquad \text{for } 0 \neq \xi \in \mathbb{R}^n .$$

An elementary observation about definite matrices will be needed. Suppose that B is also positive definite, so that

$$\xi \cdot B \cdot \xi^t > 0 \qquad \text{for } 0 \neq \xi \in \mathbb{R}^n .$$

For any non-singular matrix P we then have

$$\xi \cdot PBP^t \cdot \xi^t = (\xi P)B(\xi P)^t > 0 \qquad \text{for } 0 \neq \xi \in \mathbb{R}^n ,$$

so $PBP^t = C = (c_{ij})$ is also positive definite. Now the symmetric matrix A can be diagonalized -- there is an orthogonal matrix P such that

$$PAP^{-1} = PAP^t = \begin{pmatrix} \lambda_1 & & \bigcirc \\ & \ddots & \\ \bigcirc & & \lambda_n \end{pmatrix} \qquad \lambda_i > 0 .$$

Then

$$\begin{aligned} \text{trace } AB &= \text{trace } PABP^t = \text{trace}(PAP^t)(PBP^t) \\ &= \text{trace} \begin{pmatrix} \lambda_1 & & \bigcirc \\ & \ddots & \\ \bigcirc & & \lambda_n \end{pmatrix} C \\ &= \text{trace}(\lambda_i c_{ij}) \\ &= \sum_i \lambda_i c_{ii} > 0 . \end{aligned}$$

Similarly, we have trace $AB \geq 0$ if B is positive semi-definite, and trace $AB \leq 0$ if B is negative semi-definite.

Now consider the operator (*), where we assume that

$$\text{(i)} \qquad c \leq 0 \quad \text{in} \quad U .$$

Suppose that $u: U \to \mathbb{R}$ is a twice differentiable function with a relative maximum at some point $p \in U$. Assume, moreover, that

$$\text{(ii)} \qquad u(p) \geq 0 .$$

From (i) and (ii) we have

$$\text{(iii)} \qquad \sum_{i,j=1}^{n} a_{ij} \frac{\partial^2 u}{\partial x_i \partial x_j}(p) = (Lu)(p) - c(p)u(p) \geq Lu(p) .$$

On the other hand, since u has a maximum at p, the matrix

$$B = \left(\frac{\partial^2 u}{\partial x_i \partial x_j}(p) \right)$$

is negative semi-definite. Hence we have

$$\text{(iv)} \qquad 0 \geq \text{trace } A(p) \cdot B = \sum_{i,j=1}^{n} a_{ij} \frac{\partial^2 u}{\partial x_i \partial x_j}(p) .$$

Since (iii) and (iv) imply that $Lu(p) \leq 0$, we find

(A) If the operator (*) has (a_{ij}) positive definite on U and $c \leq 0$ on U, and the twice differentiable function u satisfies $Lu > 0$ on U, then u cannot have a non-negative relative maximum on U.

The significant fact is that we can replace the condition $Lu > 0$ by $Lu \geq 0$.

17. THEOREM (E. HOPF). Consider a second order differential operator

$$Lu = \sum_{i,j=1}^{n} a_{ij} \frac{\partial^2 u}{\partial x_i \partial x_j} + \sum_{i=1}^{n} b_i \frac{\partial u}{\partial x_i} + cu \qquad c \leq 0$$

on a connected open set $U \subset \mathbb{R}^n$. Assume that the functions b_i and c are locally bounded, and that in a neighborhood of any point of U there are constants $\varepsilon, M > 0$ such that the matrix (a_{ij}) satisfies

$$\varepsilon \cdot \sum_{i=1}^{n} \xi_i^{\,2} \leq \sum_{i,j=1}^{n} a_{ij} \xi_i \xi_j \leq M \cdot \sum_{i=1}^{n} \xi_i^{\,2} \qquad \xi \in \mathbb{R}^n .$$

Suppose that u is a twice differentiable function on U satisfying

$$Lu \geq 0 .$$

Then u cannot have a non-negative maximum on U, unless u is a constant.

Proof. Suppose u has a maximum at $p \in U$, with $u(p) \geq 0$. If u is not constant, then there is clearly a point $q \in U$ and an open ball B centered at q with $\bar{B} \subset U$ such that $u(q) < u(p)$, but $u(p^*) = u(p)$ for some $p^* \in$ boundary B. Moreover, by choosing the smallest ball B with this

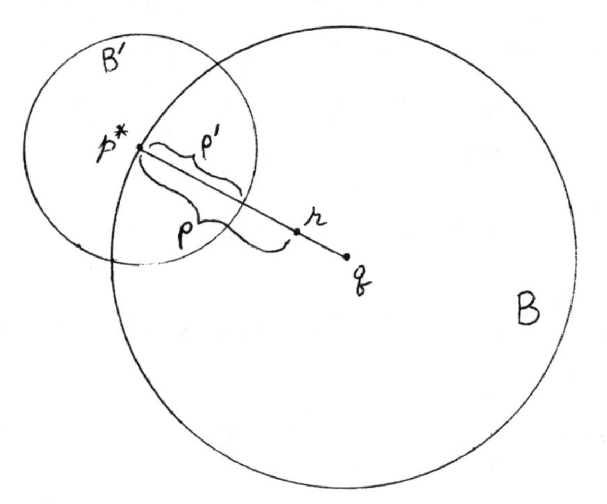

property, we can assume that $u < u(p)$ in B. Let r be a point on the open segment $\overline{qp^*}$, set

$$\rho = d(r,p^*) \qquad \text{and choose} \qquad 0 < \rho' < \rho .$$

Let B' be the open ball of radius ρ' around p^*; assume ρ' chosen sufficiently small so that $\bar{B}' \subset U$.

Now consider the function

$$v(x) = e^{-kd(r,x)^2} - e^{-k\rho^2}$$

where k is a constant. We find that

$$Lv(x) = e^{-kd(r,x)^2}\left[4k^2 \sum_{i,j} a_{ij}(x_i - r_i)(x_j - r_j) - 2k \sum_i b_i(x_i - r_i)\right] + c\left[e^{-k\rho^2} - e^{-kd(r,x)^2}\right].$$

By choosing k sufficiently large, we will obviously have $Lv > 0$ at all points of $\bar{B}'$. So for all $\lambda > 0$ we will have

(1) $$L(u+\lambda v) = Lu + \lambda Lv > 0 \quad \text{in } \bar{B}' .$$

We also have

(2) $$v(p^*) = 0 \implies (u+\lambda v)(p^*) = u(p^*) = u(p) .$$

Now consider S' = boundary B'. It is the union of a set C and a closed set D such that

$$x \in C \implies d(x,r) > \rho$$
$$\implies v(x) < 0$$
$$x \in D \implies x \in B \implies u(x) < u(p) .$$

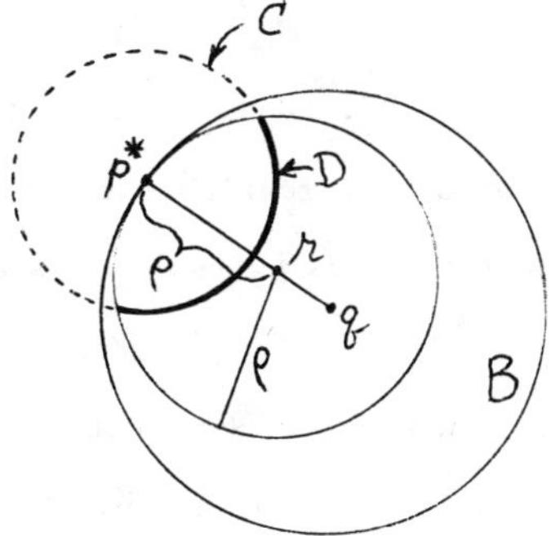

This clearly implies that for sufficiently small $\lambda > 0$ we have

(3) $$u + \lambda v < u(p) \quad \text{on } S' .$$

But (2) and (3) imply that $u+\lambda v$ has a maximum $\geq u(p) \geq 0$ on B'. Together with (1), this contradicts (A). ■

The example

$$u = -(x^2+y^2) - 4$$

$$Lu = \frac{\partial^2 u}{\partial x^2} + \frac{\partial^2 u}{\partial y^2} - u = x^2 + y^2 \geq 0$$

shows that the function u in Theorem 17 may well have a negative maximum on U. The example

$$u = -(x^2+y^2) + 5$$

$$Lu = \frac{\partial^2 u}{\partial x^2} + \frac{\partial^2 u}{\partial y^2} + u = 1 - (x^2+y^2) \geq 0 \qquad \text{for } x^2+y^2 \leq 1$$

shows that the hypothesis $c \leq 0$ is essential. If we assume $c = 0$, then we get a stronger conclusion:

18. COROLLARY. Consider the operator L of Theorem 17, with $c = 0$. If u is a twice differentiable function on U with $Lu \geq 0$, then u cannot have a maximum on U unless u is a constant.

Proof. Suppose u has a maximum at p. Let $v = u - u(p)$. Then v has a maximum of 0 at p. Moreover,

$$Lv = Lu \geq 0 .$$

So Theorem 17 implies that v is a constant. ■

As an application, consider a function $f: M \to \mathbb{R}$ on a Riemannian manifold M. Then we have the Laplacian Δf, defined in Addendum 1 to Chapter 7. In a coordinate system $(x^1,\ldots,x^n)$ on M, the formula for Δf (p. IV.193) is precisely of the form considered in Corollary 18. So if $\Delta f \geq 0$, then Δf cannot have a maximum on M, unless f is a constant function. This gives another proof of Bochner's Lemma (7-60).

In contrast to Corollary 18, where we assume $c = 0$, there is another result where c is arbitrary.

19. COROLLARY. Consider the operator L of Theorem 17, with arbitrary c. If u is a twice differentiable function on U with $Lu \geq 0$ and $u \leq 0$, then u cannot have the value 0 anywhere on U unless u is identically 0.

Proof. Let

$$Pu = \sum_{i,j=1}^{n} a_{ij} \frac{\partial^2 u}{\partial x_i \partial x_j} + \sum_{i=1}^{n} b_i \frac{\partial u}{\partial x_i} .$$

Then we have

$$Pu + cu = Lu \geq 0 .$$

Hence

$$Pu + \min(c,0)u = [\min(c,0) - c]u + Lu .$$

Now

$$\min(c,0) \leq 0$$

$$\min(c,0) - c \leq 0 \implies [\min(c,0) - c]u \geq 0 \qquad \text{since } u \leq 0$$

$$\implies [\min(c,0) - c]u + Lu \geq 0 .$$

Applying Theorem 17 to the operator $Pu + \min(c,0)u$, we conclude that u cannot have a non-negative maximum unless it is a constant. ■

There is also a version of Theorem 17 when u has its maximum at a boundary point of U.

20. THEOREM. Consider a second order differential operator L as in Theorem 17. Let u be a twice differentiable function on U satisfying $Lu \geq 0$ and such that at some point $p \in$ boundary U the function u has a maximum $u(p) \geq 0$ on $U \cup \{p\}$. Suppose moreover that there is some closed ball $\overline{B_\rho(r)} \subset U \cup \{p\}$ containing p on its boundary, and that the directional

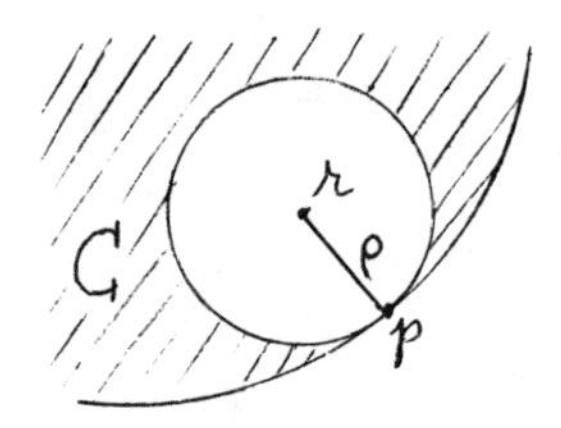

derivative of u at p in the direction from p to r is ≥ 0 (all directional derivatives in directions tangent to the boundary of $B_\rho(r)$ are clearly equal to 0). Assume, finally, that the functions a_{ij}, b_i, c have the same properties as in Theorem 17, but in $U \cup \{p\}$. Then u is a constant function.

Proof. Suppose u is not a constant function. Choose $0 < \rho_1 < \rho = d(p,r)$, and let

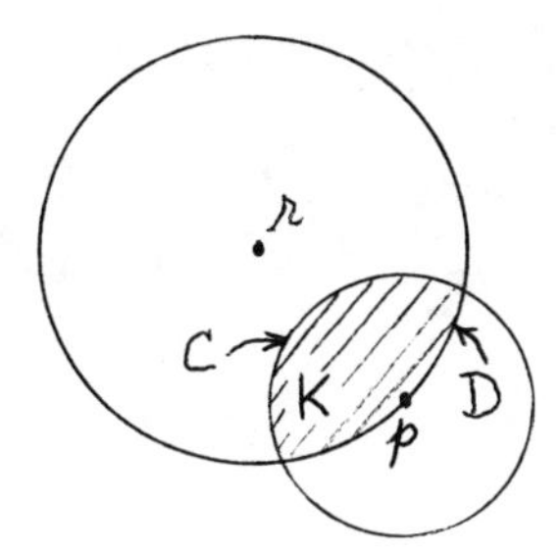

$$K = \{x\colon d(x,r) \leq \rho \text{ and } d(x,p) \leq \rho_1\} .$$

Again define

$$v(x) = e^{-kd(r,x)^2} - e^{-k\rho^2} .$$

The boundary of K is the union of a closed set $C \subset U$ and a closed set D on which $v = 0$. Theorem 17 implies (choosing ρ smaller if necessary) that $u(x) < u(p)$ for $x \in C$. So for sufficiently small $\lambda > 0$ we have

$$-\lambda v(x) \geq u(x) - u(p) \qquad \text{for } x \in \text{boundary } K .$$

But for sufficiently large k, we have $Lv > 0$ in K. So

$$L(u - u(p) + \lambda v) = Lu - cu(p) + \lambda Lv > 0 .$$

It follows from (A) that

$$-\lambda v(x) \geq u(x) - u(p) \qquad \text{for all } x \in K .$$

So the directional derivative of u at p, in the direction from p to r, is less than or equal to this directional derivative at p of $-\lambda v$. But we easily compute that the latter directional derivative is

$$-2\lambda k\rho e^{-k\rho^2} < 0 ,$$

contradicting the hypotheses. ■

Naturally, there are analogous versions of Corollaries 18 and 19.

PROBLEM

1. (a) Let $f\colon \mathbb{R}^n \to \mathbb{R}$ be a C^∞ [respectively, C^ω] function such that $f = 0$ on the points $(0,\ldots,0, x^{n-k+1},\ldots,x^n)$ near 0. Show that there are $C^\infty[C^\omega]$ functions h_i near 0 such that

$$f = \sum_{i=1}^{k} h_i x^i .$$

[The C^ω case is actually trivial; the C^∞ case can be proved by generalizing the argument in the proof of Lemma I.3-2.]

(b) Let $g\colon \mathbb{R}^n \to \mathbb{R}^k$ be a C^∞ $[C^\omega]$ function whose Jacobian has rank k on $g^{-1}(0)$, and let $f\colon \mathbb{R}^n \to \mathbb{R}$ be a C^∞ $[C^\omega]$ function which vanishes on $g^{-1}(0)$. Then near any point of $g^{-1}(0)$ there are C^∞ $[C^\omega]$ functions h_i such that

$$f = \sum_{i=1}^{k} h_i g^i .$$

Chapter 11. Existence and Non-existence of Isometric Imbeddings

In the past we have had some very special results about the non-existence of isometric imbeddings of certain Riemannian manifolds in other Riemannian manifolds. For example, a compact surface of everywhere negative curvature cannot be isometrically imbedded, or even immersed in $\mathbb{R}^3$, nor can a complete surface of constant negative curvature be isometrically immersed in $\mathbb{R}^3$. Ideally, differential geometry should be replete with such results, so that we could have a reasonable chance of finding the smallest dimensional Euclidean space into which a given Riemannian manifold can be isometrically imbedded. But at present only quite isolated facts are known, and a general theory can hardly be said to exist

There are, first of all, purely topological, or at any rate differential-topological, questions which have to be considered in any imbedding problem -- for there is no point trying to isometrically immerse or imbedd a Riemannian manifold in $\mathbb{R}^m$ unless its underlying differentiable manifold has some differentiable immersion or imbedding in $\mathbb{R}^m$. Generally speaking, the methods used to settle such questions are of little interest to differential geometry *per se*. We note, however, that one special result of this sort has already been proved in Volume I: a compact hypersurface imbedded in $\mathbb{R}^m$ is always orientable (Theorem I.11-14). Thus, for example, there is no imbedding of the projective plane $\mathbb{P}^2$ in $\mathbb{R}^3$. We can supplement this result with a simple differential geometric one: If $\langle\ ,\ \rangle$ is a metric on $\mathbb{P}^2$ with $K > 0$, then $(\mathbb{P}^2, \langle\ ,\ \rangle)$ cannot even be isometrically *immersed* in $\mathbb{R}^3$; this follows directly from Hadamard's Theorem (2-11).

At the other extreme from these global topological restrictions, there are certain purely local results. For example, if $n \geq 3$, and M^n has all sectional curvatures < 0, then M^n cannot be locally isometrically imbedded in $\mathbb{R}^{n+1}$; for the principal curvatures $k_1,\ldots,k_n$ would have to satisfy $k_i k_j < 0$ for all i, j, while some pair k_i, k_j must have the same sign. Similarly, Theorem 7-50 shows that if $n \geq 3$, and $M^n \subset \mathbb{R}^{n+1}$ has Ricci tensor $\text{Ric} = 0$, then M is flat; so for $n \geq 3$, a non-flat M^n with $\text{Ric} = 0$ is not isometrically imbeddable in $\mathbb{R}^{n+1}$. Historically, this was first used to show that the 4-dimensional Schwartzschild metric of general relativity is not imbeddable in $\mathbb{R}^5$.

To obtain purely local results for higher codimension, we need to replace the trivial algebraic considerations used before by something more substantial.

1. LEMMA. Let $s\colon \mathbb{R}^n \times \mathbb{R}^n \to \mathbb{R}^k$ be a symmetric bilinear map, and let $\langle\!\langle\, ,\, \rangle\!\rangle$ be a positive definite inner product on $\mathbb{R}^k$. Let $S^{n-1} \subset \mathbb{R}^n$ be the unit sphere (with respect to the usual inner product $\langle\, ,\, \rangle$ on $\mathbb{R}^n$), and consider the function $f(x) = \|s(x,x)\|^2$ for $x \in S^{n-1}$.

(1) If $x \in S^{n-1}$ is a critical point of f, then

$$\langle\!\langle s(x,x), s(x,y) \rangle\!\rangle = 0 \qquad \text{for all } y \text{ with } \langle x,y\rangle = 0 .$$

Consequently, if $f(x) \neq 0$, then

$$s(x,y) = 0 \implies \langle x,y\rangle = 0 .$$

(2) If $x \in S^{n-1}$ is a minimum point of f, then for all $y \in S^{n-1}$ with $\langle x,y\rangle = 0$ we have

$$\langle\!\langle s(x,x), s(y,y) \rangle\!\rangle + 2\langle\!\langle s(x,y), s(x,y) \rangle\!\rangle \geq \langle\!\langle s(x,x), s(x,x) \rangle\!\rangle .$$

Proof. (1) Using the fact that the derivative $DA(x)$ of a linear transformation A is always A itself, we easily see that the map $S\colon \mathbb{R}^n \to \mathbb{R}^k$ defined by $S(x) = s(x,x)$ has derivative $DS(x)\colon \mathbb{R}^n \to \mathbb{R}^k$ given by

$$(DS)(x)(y) = 2s(x,y) .$$

It follows that

$$(Df)(x)(y) = 2\langle S(x)(y), S(x)\rangle = 4\langle s(x,y), s(x,x)\rangle .$$

Since $x \in S^{n-1}$ is a critical point for f, we must have $(Df)(x)(y) = 0$ for all $y \in S^{n-1}_x$, i.e., for $\langle x,y\rangle = 0$.

Now suppose $s(x,y) = 0$ and $s(x,x) \neq 0$. Writing $y = \lambda x + y'$, with

$$\langle x,y'\rangle = 0 \implies \langle s(x,x), s(x,y')\rangle = 0 \qquad \text{by the above paragraph,}$$

we have

$$0 = \langle s(x,x), s(x,y)\rangle = \langle s(x,x), s(x,\lambda x)\rangle + \langle s(x,x), s(x,y')\rangle = \langle s(x,x), s(x,\lambda x)\rangle .$$

Since $s(x,x) \neq 0$, this implies that $\lambda = 0$.

(2) Let c be the curve in S^{n-1} defined by

$$c(t) = (\cos t)x + (\sin t)y .$$

Since x is a minimum of f we have

$$0 \leq \left.\frac{d^2}{dt^2}\right|_{t=0} f(c(t)) .$$

A short computation shows that the right side is

$$4[-\langle s(x,x),s(x,x)\rangle+\langle s(x,x),s(y,y)\rangle+2\langle s(x,y),s(x,y)\rangle] . \blacksquare$$

From this we derive, first of all, a purely local result.

<u>2. PROPOSITION</u>. Let N be a manifold of dimension $2n-2$ with all sectional curvatures $\geq K_0$, and let M be a manifold of dimension n with all sectional curvatures $< K_0$. Then M cannot be isometrically immersed in N. The result also holds if all sectional curvatures of N are $> K_0$ and all sectional curvatures of M are $\leq K_0$.

<u>Proof</u>. Suppose we could isometrically immerse M in N, and let s be the second fundamental form. For any $p \in M$, and orthonormal $X, Y \in M_p$, Gauss' equation gives, under the first hypothesis,

$$\begin{aligned} K_0 \leq \langle R'(X,Y)Y,X\rangle &= \langle R(X,Y)Y,X\rangle + \langle s(X,Y),s(X,Y)\rangle - \langle s(X,X),s(Y,Y)\rangle \\ &< K_0 + \langle s(X,Y),s(X,Y)\rangle - \langle s(X,X),s(Y,Y)\rangle . \end{aligned}$$

Under the second hypothesis the $\leq$ and $<$ are interchanged. In either case, we obtain

$$(1) \quad \langle s(X,X),s(Y,Y)\rangle - \langle s(X,Y),s(X,Y)\rangle < 0 , \qquad X, Y \in M_p \text{ independent.}$$

(for this final inequality we do not need X, Y to be orthonormal). Choose $X \in M_p$ to be a minimum point of $X \mapsto |s(X,X)|$ on the unit sphere of M_p. Since $\{Y: s(X,Y) = 0\}$ has dimension ≥ 2, there is a unit vector Y linearly independent of X with

$$(2) \qquad s(X,Y) = 0 .$$

From (1) and (2) we see that we must have $s(X,X) \neq 0$. But then Lemma 1(1) implies that $\langle X,Y\rangle = 0$, so Lemma 1(2) gives

$$\langle s(X,X),s(Y,Y)\rangle + 0 \geq \langle s(X,X),s(X,X)\rangle \geq 0 ,$$

contradicting (1) and (2). ■

In particular, an n-manifold M of constant curvature $K < K_0$ cannot be locally isometrically immersed in a $(2n-2)$-manifold N of constant curvature K_0. An example of a (non-complete) n-manifold of constant negative curvature in $\mathbb{R}^{2n-1}$ is given in Problem 1. It seems reasonable to conjecture that there is no immersion of a complete n-manifold of constant negative curvature in $\mathbb{R}^{2n-1}$, but this has not been proved (whether one can be found in $\mathbb{R}^{2n}$ is any body's guess). It is known, however, that no such immersion exists if M is compact. This follows from the next theorem, whose proof combines the local information from Lemma 1 with just a smidgin of globalness.

3. PROPOSITION. Let N be the complete simply-connected $(2n-1)$-dimensional manifold of constant curvature $K_0 \leq 0$, and let M be a compact n-manifold with all sectional curvatures ≤ 0. Then M cannot be isometrically immersed in N.

Proof. Suppose there were an isometric immersion $f\colon M \to N$. Let $q_0 \in N$ be a fixed point, and choose $p \in M$ so that $f(p)$ is furthest from q_0. Then (p.IV.118) there is $\xi \in M_p{}^{\perp}$ with

(1) $\langle s(X,X),\xi\rangle > \sqrt{-K_0} \implies \langle s(X,X),s(X,X)\rangle > -K_0$ for $X \in M_p$.

Choose $X \in M_p$ to be the minimum point of $X \mapsto |s(X,X)|^2$ on the unit sphere in M_p. Since $\{Y: s(X,Y) = 0\}$ has dimension ≥ 1, there is a unit vector $Y \in M_p$ with

(2) $s(X,Y) = 0 \implies \langle X,Y\rangle = 0$ by Lemma 1(1) .

Then Lemma 1(2) gives

(3) $\langle s(X,X),s(Y,Y)\rangle \geq \langle s(X,X),s(X,X)\rangle > -K_0$, by (1).

Moreover, applying Gauss' equation to the plane $P \subset M_p$ spanned by the orthonormal vectors X, Y, we have

$$K(P) = \langle s(X,X),s(Y,Y)\rangle - \langle s(X,Y),s(X,Y)\rangle + K_0$$
$$> -K_0 + 0 + K_0 \qquad \text{by (2) and (3).}$$

This contradicts the assumption that $K(P) \leq 0$. ■

Remark. More generally, if each tangent space M_p contains an ℓ-dimensional subspace on which all sectional curvatures are ≤ 0, and the compact manifold M can be isometrically immersed in the complete simply-connected $(n+k)$-dimensional manifold of sectional curvature $K_0 \leq 0$, then we must have $k \geq \ell$. Proposition 2 can be generalized similarly.

4. COROLLARY (THOMPKINS). An n-dimensional compact flat Riemannian manifold cannot be isometrically immersed in $\mathbb{R}^{2n-1}$.

Obviously $2n-1$ is the best possible dimension here, since the flat n-torus $S^1 \times \cdots \times S^1$ is isometrically imbedded in $\mathbb{R}^{2n}$. It is also isometrically imbedded in $H^{2n}(K_0)$ for any $K_0 < 0$, since it is isometrically imbedded in S^{2n-1}, and these are spheres of all curvatures in $H^{2n}(K_0)$. I do not know if there is a non-flat compact n-manifold in $\mathbb{R}^{2n}$ or $H^{2n}(K_0)$ with all sectional curvatures ≤ 0.

The proof of Proposition 3 breaks down if N is a sphere, with constant curvature $K_0 > 0$, since we cannot guarantee the existence of the requisite point $p \in M$ unless we know that M lies in a hemisphere. Indeed, the n-dimensional flat torus $S^1 \times \cdots \times S^1$ can be isometrically imbedded in S^{2n-1}. It seems reasonable to assume that M^n cannot be isometrically immersed in S^{2n-1} if all sectional curvatures of M are > 0 and < 1. For the special case where M has constant curvature, see Problem 2.

Now we are going to consider some more elaborate algebraic results. Let V be a real vector space, and let $\Phi\colon V \times V \to \mathbb{R}$ be bilinear. With Φ we can associate a linear transformation $\tilde{\Phi}\colon V \to V^*$ by

$$\tilde{\Phi}(v)(w) = \Phi(v,w) .$$

A collection $\Phi^1,\ldots,\Phi^n$ of bilinear forms on V is called <u>exteriorly orthogonal</u> if for all $v_1, v_2 \in V$ we have

$$\sum_{i=1}^{n} \tilde{\Phi}^i(v_1) \wedge \tilde{\Phi}^i(v_2) = 0 \in \Omega^2(V) ;$$

equivalently,

$$\sum_{i=1}^{n} [\Phi^i(v_1,w_1)\Phi^i(v_2,w_2) - \Phi^i(v_1,w_2)\Phi^i(v_2,w_1)] = 0$$

for all v_1, v_2, w_1, $w_2 \in V$. Before stating the main result about exteriorly orthogonal bilinear forms, we make a simple observation. If $\Phi \neq 0$ is $\Phi = \phi \otimes \phi$ for some $\phi \in V^*$, so that $\Phi(v,w) = \phi(v)\cdot\phi(w)$, then $\tilde{\Phi}(v) = \phi(v)\cdot\phi$, and consequently range $\tilde{\Phi} \subset V^*$ is 1-dimensional. Conversely, if range $\tilde{\Phi} \subset V^*$ is 1-dimensional, then $\tilde{\Phi}$ must be of the form

$$\tilde{\Phi}(v)(w) = \psi(v)\cdot\phi(w)$$

for ϕ, $\psi \in V^*$. If, moreover, $\tilde{\Phi}$ is symmetric, so that $\psi(v)\cdot\phi(w) = \psi(w)\cdot\phi(v)$ for all v, w, then we have $\phi = \psi$ [since $(\phi-\psi)(w) = 0$ for some $w \neq 0$], so Φ is of the form $\Phi = \phi \otimes \phi$.

5. THEOREM (E. CARTAN). Let V be a real vector space of dimension n, and let $\Phi^1,\ldots,\Phi^n$ be n exteriorly orthogonal symmetric bilinear forms on V. Suppose that

$$(*) \quad 0 = \bigcap_{i=1}^{n} \ker \tilde{\Phi}^i = \{v \in V\colon \Phi^i(v,w) = 0 \text{ for all } v \in V \text{ and all } i\} .$$

Then there is an orthogonal $n \times n$ matrix A and linearly independent $\phi^1,\ldots,\phi^n \in V^*$ such that

$$\Phi^i = \sum_{j=1}^{n} A^i_j \, \phi^j \otimes \phi^j .$$

Proof. We claim that there is a vector $v \in V$ such that the $\tilde{\Phi}^i(v) \in V^*$ are linearly independent. To prove this, let $v_0 \in V$ be a vector such that the subspace

$$[\tilde{\Phi}^1(v_0),\ldots,\tilde{\Phi}^n(v_0)] \subset V^*$$

spanned by the $\tilde{\Phi}^i(v_0)$ has maximal dimension $d \leq n$, and suppose that $d < n$. Replacing the $\{\tilde{\Phi}^i\}$ by an orthogonal linear combination of them changes neither the hypotheses nor the conclusion of the Theorem, so without loss of generality we can assume that

$$\tilde{\Phi}^1(v_0),\ldots,\tilde{\Phi}^d(v_0) \quad \text{are linearly independent ,}$$
$$\tilde{\Phi}^{d+1}(v_0) = \cdots = \tilde{\Phi}^n(v_0) = 0 \ .$$

Then for any vector $v \in V$ we have

$$\sum_{i=1}^{d} \tilde{\Phi}^i(v_0) \wedge \tilde{\Phi}^i(v) = 0 \ .$$

Cartan's Lemma thus implies that for $i = 1,\ldots,d$, the $\tilde{\Phi}^i(v)$ are a linear combination of the $\tilde{\Phi}^i(v_0)$, $i = 1,\ldots,d$. Consequently

$$\mathcal{V} = \{\tilde{\Phi}^i(v) \colon v \in V,\ 1 \leq i \leq d\} \subset V^*$$

also has dimension exactly d. Since $d < n$, there is a vector $0 \neq w \in V$ such that $\phi(w) = 0$ for all $\phi \in \mathcal{V}$. But by (*) there is some i and some $v \in V$ such that $\Phi^i(v,w) \neq 0$; clearly $i > d$. Now consider the vector $v_0 + \varepsilon v$. If $\varepsilon > 0$ is sufficiently small, then

$$\dim[\tilde{\Phi}^1(v_0+\varepsilon v),\ldots,\tilde{\Phi}^d(v_0+\varepsilon v)] = \dim[\tilde{\Phi}^1(v_0),\ldots,\tilde{\Phi}^d(v_0)] = d$$

$$\Downarrow$$

$$[\tilde{\Phi}^1(v_0+\varepsilon v),\ldots,\tilde{\Phi}^d(v_0+\varepsilon v)] = \mathcal{V}\,, \qquad \text{since } \dim \mathcal{V} = d\,.$$

But $\tilde{\Phi}^i(v_0+\varepsilon v) \notin \mathcal{V}$, by the choice of v and i. So

$$\dim[\tilde{\Phi}^1(v_0+\varepsilon v),\ldots,\tilde{\Phi}^n(v_0+\varepsilon v)] > d\,,$$

contradicting the definition of d. This establishes the claim.

Now choose a basis $v_1,\ldots,v_n$ of V such that $\tilde{\Phi}^1(v_1),\ldots,\tilde{\Phi}^n(v_1)$ are linearly independent. Since

$$\sum_{i=1}^{n} \tilde{\Phi}^i(v_1) \wedge \tilde{\Phi}^i(v_j) = 0\,,$$

Cartan's Lemma implies that there is a symmetric matrix $C(j)$, with $C(1) =$ identity, such that

$$\tilde{\Phi}^i(v_j) = \sum_{h=1}^{n} C(j)^i_h\, \tilde{\Phi}^h(v_1)\,.$$

The equation

$$\sum_h \tilde{\Phi}^h(v_j) \wedge \tilde{\Phi}^h(v_k) = 0$$

implies that $C(j)$ and $C(k)$ commute. Then a well-known theorem of linear algebra (Problem 3) states that there is an orthogonal matrix B such that $B\cdot C(i)\cdot B^t$ is a diagonal matrix for all i. If we set

$$\Psi^i = \sum_h B_h^i \Phi^h ,$$

then $\tilde{\Psi}^i(v_j)$ is a constant times $\tilde{\Psi}^i(v_1)$. Thus range $\tilde{\Psi}^i$ is 1-dimensional, so $\Psi^i = \phi^i \otimes \phi^i$ for some $\phi^i \in V^*$. We choose $A = B^{-1}$. The ϕ^i must be linearly independent, for otherwise there is $0 \neq v \in V$ such that $\phi^i(v) = 0$ for all i, contradicting (*). ■

The hypothesis (*) in Theorem 5 may be interpreted as saying that the set $\{\Phi^i\}$ "depends on n variables" -- we cannot find $\phi^1,\ldots,\phi^{n-1} \in V^*$ such that each Φ^i is a linear combination of the $\phi^j \otimes \phi^k$, $1 \leq j, k \leq n-1$. More generally, if $\Phi^1,\ldots,\Phi^k$ are bilinear forms on V, then the set $\{\Phi^i\}$ <u>depends on</u> q <u>variables</u> if the subspace $\bigcap_{i=1}^{k} \ker \tilde{\Phi}^i$ has codimension q.

<u>6. COROLLARY</u>. Let V be a real vector space of dimension n, and let $\Phi^1,\ldots,\Phi^k$ be k exteriorly orthogonal symmetric bilinear forms on V which depend on $\ell \geq k$ variables (so necessarily $k \leq n$). Then $\ell = k$ and there is an orthogonal $k \times k$ matrix A and linearly independent $\phi^1,\ldots,\phi^k \in V^*$ such that

$$\Phi^i = \sum_{j=1}^{k} A_j^i \, \phi^j \otimes \phi^j .$$

In particular, k exteriorly orthogonal symmetric bilinear forms always depend on $\leq k$ variables.

<u>Proof</u>. Without loss of generality, we can assume that $\ell = n$ [by applying the

result to a subspace of V complementary to $\bigcap_{i=1}^{k} \ker \tilde{\Phi}^i$]. If $k < \ell = n$, we set

$$\Phi^{k+1} = \cdots = \Phi^n = 0 .$$

The n bilinear forms $\Phi^1,\ldots,\Phi^n$ are then exteriorly orthogonal and $\bigcap_{i=1}^{n} \ker \tilde{\Phi}^i = 0$. By the Theorem, there is an orthogonal $n \times n$ matrix A, and linearly independent $\phi^1,\ldots,\phi^n \in V^*$ with

$$\Phi^i = \sum_{j=1}^{n} A^i_j \, \phi^j \otimes \phi^j .$$

So we cannot have $\Phi^i = 0$ for any i, so actually $k = n = \ell$. ■

These algebraic results were used by Cartan for a systematic local study of n-dimensional manifolds M of constant curvature K isometrically imbedded in an $(n+k)$-dimensional manifold N of constant curvature $K_0 > K$. For an adapted orthonormal moving frame $X_1,\ldots,X_m$ on $M \subset N$ we have, as in Chapter 1,

$$\psi^r_j = \sum_i s^r_{ij}\theta^i , \qquad s^r_{ij} = s^r_{ji} ;$$

the second fundamental forms II^r are given by

$$II^r = \sum_i \psi^r_i \otimes \theta^i .$$

We also have

$$K_0\, \theta^i \wedge \theta^j = \Omega^i_j - \sum_r \psi^r_i \wedge \psi^r_j$$

$$= K\, \theta^i \wedge \theta^j - \sum_r \psi^r_i \wedge \psi^r_j \ ,$$

or equivalently

$$(1) \qquad \sum_r (s^r_{ij} s^r_{k\ell} - s^r_{i\ell} s^r_{kj}) = (K - K_0)(\delta_{ij}\delta_{k\ell} - \delta_{i\ell}\delta_{kj}) \ .$$

If we define

$$\Psi = \sqrt{K_0 - K}\,(\sum_i \theta^i \otimes \theta^i) \ ,$$

then equation (1) says that the $k+1$ bilinear forms $\{II^r, \Psi\}$ are exteriorly orthogonal. The collection $\{II^r, \Psi\}$ certainly depends on all n variables, since Ψ alone does. So Corollary 6 implies that $n \leq k+1$, showing once again that M^n cannot be isometrically imbedded in N^{2n-2}. Cartan showed, using his theory of exterior differential systems (Chapter 10, Addendum 1) that the analytic local imbeddings of M^n in N^{2n-1} depend upon $n(n-1)$ functions of one variable.

Another consequence of Corollary 6 depends on two definitions, one intrinsic and one extrinsic. For a point p of a Riemannian manifold M^n, we define the *index of nullity* at p to be

$$\mu(p) = \dim\{X \in M_p : R(X,Y) = 0 \text{ for all } Y \in M_p\} \ .$$

Equivalently, $n - \mu(p)$ is the minimum number of 1-forms in terms of which we can express the collection of 2-forms $\{\Omega^i_j(p)\}$. For $M^n \subset \mathbb{R}^{n+k}$, with second fundamental form s, we define the index of relative nullity at p to be

$$\nu(p) = \dim\{X \in M_p : A_\xi(X) = 0 \text{ for all } \xi \in M_p{}^\perp\}$$
$$= \dim\{X \in M_p : s(X,Y) = 0 \text{ for all } Y \in M_p\} .$$

Equivalently, $n - \nu(p)$ is the minimum number of 1-forms in terms of which we can express the collection of forms $\{II^r(p)\}$, for an orthonormal set $\nu^{n+1},\dots,\nu^{n+k} \in M_p{}^\perp$.

7. PROPOSITION. For $M^n \subset \mathbb{R}^{n+k}$ we have

$$\nu(p) \le \mu(p) \le \nu(p) + \text{rank } s(p) \le \nu(p) + k .$$

Proof. The first inequality follows from Gauss' equation, which shows that

$$\{X \in M_p : s(X,Y) = 0 \text{ for all } Y \in M_p\} \subset \{X \in M_p : R(X,Y) = 0 \text{ for all } Y \in M_p\} .$$

For the second we can assume, by choosing the ν^r appropriately, that $II^{n+1}(p),\dots,II^{n+d}(p)$ are linearly independent, for $d = \text{rank } s(p)$, while the other $II^r(p)$ are all 0. Gauss' equation shows that $II^{n+1}(p),\dots,II^{n+d}(p)$ are exteriorly orthogonal restricted to $\{X \in M_p : R(X,Y) = 0 \text{ for all } Y \in M_p\}$. Let W be a subspace such that

$$\{X \in M_p : R(X,Y) = 0 \text{ for all } Y \in M_p\} = W \oplus \{X \in M_p : s(X,Y) = 0 \text{ for all } Y \in M_p\}.$$

Then $II^{n+1}(p),\dots,II^{n+d}(p)$ are exteriorly orthogonal on W, and depend on all variables of W. Corollary 6 implies that $d \ge \dim W = \nu(p) - \mu(p)$. ∎

Remark. We have a similar result for $M^n \subset N^{n+k}$, where N^{n+k} has constant curvature K_0, provided we re-define

$$\mu(p) = \dim\{X \in M_p : R(X,Y)Z = K_0[\langle Y,Z\rangle X - \langle X,Z\rangle Y]\} .$$

8. COROLLARY. If M^n is a compact manifold immersed in $\mathbb{R}^{n+k}$, then

$$k \geq \min_{p \in M} \mu(p) .$$

Proof. Proposition 7-30 shows that for some $p \in M$ we have $\nu(p) = 0$. ∎

Note that Corollary 4 is a special case (admittedly the only reasonably general consequence we can give). Recently Corollary 6 has been used to prove results of quite another sort, which we will mention in the next Chapter.

This ends our treatment of non-imbeddability theorems, and pretty much exhausts the subject in its present state (a few other special results are mentioned in the Bibliography). Now we will take a more positive approach to life and try to prove that under certain circumstances isometric imbeddings do exist. We first consider the purely local problem of isometrically imbedding a surface in $\mathbb{R}^3$. So we assume that we are given functions g_{ij} (= E, F, G) on a neighborhood of $0 \in \mathbb{R}^2$, with $\det(g_{ij}) > 0$, and we want to find a function $f\colon U \to \mathbb{R}^3$, on some smaller neighborhood U of $0 \in \mathbb{R}^2$, such that $I_f = f^*\langle\ ,\ \rangle$ has components g_{ij}. This means that the component functions f^α of f must satisfy

$$g_{ij} = \sum_{\alpha=1}^{3} \langle f^\alpha_i, f^\alpha_j\rangle ,$$

so that we have three (non-linear) partial differential equations in three unknowns. We also know that f can be found once we have functions ℓ_{ij} satisfying Gauss' equation and the Codazzi-Mainardi equation, which again gives us three equations in three unknowns. There is also a way of introducing a single second order equation, which was used classically. Suppose that the required f exists; let n be its normal map, and let ℓ_{ij} be the components of II_f. Then for each component function f^α of f we have the Gauss formulas

$$
\begin{aligned}
&f^\alpha_{11} - \Gamma^1_{11}f^\alpha_1 - \Gamma^2_{11}f^\alpha_2 = \ell_{11}n^\alpha \\
(1) \qquad &f^\alpha_{12} - \Gamma^1_{12}f^\alpha_1 - \Gamma^2_{12}f^\alpha_2 = \ell_{12}n^\alpha \\
&f^\alpha_{22} - \Gamma^1_{22}f^\alpha_1 - \Gamma^2_{22}f^\alpha_2 = \ell_{22}n^\alpha .
\end{aligned}
$$

If we denote these component functions of f by $u, v, w,$ then

$$
\begin{aligned}
n = \frac{f_1 \times f_2}{\sqrt{\det(g_{ij})}} &= \frac{(u_1,v_1,w_1) \times (u_2,v_2,w_2)}{\sqrt{\det(g_{ij})}} \\
&= \frac{(v_1w_2 - v_2w_1,\ w_1u_2 - u_1w_2,\ u_1v_2 - u_2v_1)}{\sqrt{\det(g_{ij})}} .
\end{aligned}
$$

So we have, for example,

$$
\begin{aligned}
\det(g_{ij})\cdot(n^3)^2 &= (u_1v_2 - u_2v_1)^2 \\
&= ({u_1}^2 + {v_1}^2)({u_2}^2 + {v_2}^2) - (u_1u_2 + v_1v_2)^2 \\
&= (g_{11} - {w_1}^2)(g_{22} - {w_2}^2) - (g_{12} - w_1w_2)^2 \\
&= \det(g_{ij}) - (g_{22}{w_1}^2 - 2g_{12}w_1w_2 + g_{11}{w_2}^2) .
\end{aligned}
$$

Using equations (1) for $\alpha = 3$, we obtain

$$(*)\quad (w_{11} - \Gamma^1_{11}w_1 - \Gamma^2_{11}w_2)(w_{22} - \Gamma^1_{22}w_1 - \Gamma^2_{22}w_2) - (w_{12} - \Gamma^1_{12}w_1 - \Gamma^2_{22}w_2)^2$$

$$= (\ell_{11}\ell_{22} - {\ell_{12}}^2) \cdot \frac{\{\det(g_{ij}) - (g_{22}{w_1}^2 - 2g_{12}w_1w_2 - g_{11}{w_2}^2)\}}{\det(g_{ij})}$$

$$= K\{\det(g_{ij}) - (g_{22}{w_1}^2 - 2g_{12}w_1w_2 + g_{11}{w_2}^2)\} ,$$

where the Γ's and K are all computable in terms of the g_{ij}. We thus have a certain non-linear second order partial differential equation (*) for w. Notice that this equation does not contain w explicitly. If w is any solution, then w + constant is also a solution, so we can always specify $w(0)$ arbitrarily.

It is easily checked (and is *a priori* clear on symmetry grounds) that u and v also satisfy equation (*). On the other hand, it is by no means true that (u,v,w) is a solution of our problem whenever u, v, w each satisfy (*), even if (u,v,w) is an immersion. In order to obtain more precise information, we must use a different procedure, due to Darboux. Suppose first that we are given an immersion $f = (u,v,w)$ such that I_f has components E, F, G; thus

$$(1)\quad du \otimes du + dv \otimes dv + dw \otimes dw = E\, dx \otimes dx + F[dx \otimes dy + dy \otimes dx] + G\, dy \otimes dy ,$$

where (x,y) is the standard coordinate system on $\mathbb{R}^2$. By composing f with a Euclidean motion, if necessary, we can assume that

$$(2)\quad w_1(0,0) = w_2(0,0) = 0$$

$$\Downarrow$$

$$(3)\quad \begin{pmatrix} u_1 & u_2 \\ v_1 & v_2 \end{pmatrix} \text{ is nonsingular at } (0,0) .$$

Consider the tensor

$$\begin{aligned} <\ ,\ >' &= E\, dx\otimes dx + F[dx\otimes dy + dy\otimes dx] + G\, dy\otimes dy - dw\otimes dw \\ &= (E-{w_1}^2)dx\otimes dx + (F-w_1w_2)[dx\otimes dy + dy\otimes dx] + (G-{w_2}^2)dy\otimes dy\ . \end{aligned}$$

Using (1) we can write

$$(4) \qquad <\ ,\ >' = du\otimes du + dv\otimes dv\ .$$

This is positive definite at $(0,0)$ by (2), and hence positive definite in a neighborhood of $(0,0)$. Moreover, (u,v) is a coordinate system for $\mathbb{R}^2$ in a neighborhood of $(0,0)$, by (3). So equation (4) says that $<\ ,\ >'$ is flat, and thus has curvature $K' = 0$.

Recall (p.III.343) that the metric with coefficients E, F, G has curvature K given by

$$(5) \qquad K(EG-F^2)^2 = \det\begin{pmatrix} -\frac{1}{2}G_{11}+F_{12}-\frac{1}{2}E_{22} & \frac{1}{2}E_1 & F_1-\frac{1}{2}E_2 \\ F_2-\frac{1}{2}G_1 & E & F \\ \frac{1}{2}G_2 & F & G \end{pmatrix}$$

$$- \det\begin{pmatrix} 0 & \frac{1}{2}E_2 & \frac{1}{2}G_1 \\ \frac{1}{2}E_2 & E & F \\ \frac{1}{2}G_1 & F & G \end{pmatrix}.$$

To obtain the condition $K' = 0$, we set the right side equal to 0 after replacing E by $E-{w_1}^2$, etc. With the standard notation

$$p = w_1 , \qquad q = w_2$$

$$r = w_{11} , \qquad s = w_{12} , \qquad t = w_{22} ,$$

the $(1,1)$ term in the first matrix becomes

$$-\frac{1}{2}(G - w_2{}^2)_{11} + (F - w_1 w_2)_{12} - \frac{1}{2}(E - w_1{}^2)_{22} = -\frac{1}{2}G_{11} + F_{12} - \frac{1}{2}E_{22} + (s^2 - rt) ,$$

<u>all third derivatives cancelling</u>. We thus obtain a second order equation for w, the "Darboux equation," which written out explicitly becomes

$$\begin{aligned}(**) \quad 0 = &- 4(EG - F^2)(rt - s^2)\\
&+ 2pr[2GF_2 - GG_1 - FG_2] + 2qr[EG_2 + FG_1 - 2FF_2]\\
&+ 4ps[FG_1 - GE_2] + 4qs[FE_2 - EG_1]\\
&+ 2pt[GE_1 + FE_2 - 2FF_1] + 2qt[2EF_1 - EE_2 - FE_1]\\
&+ (E - p^2)[E_2G_2 - 2F_1G_2 + (G_1)^2]\\
&+ (F - pq)[E_1G_2 - E_2G_1 - 2E_2F_2 - 2G_1F_1 + 4F_1F_2]\\
&+ (G - q^2)[G_1E_1 - 2F_2E_1 + (E_2)^2]\\
&+ 2[EG - F^2 - Gp^2 - Eq^2 + 2Fpq]\cdot[2F_{12} - E_{22} - G_{11}] .\end{aligned}$$

Brute force computations will show that equations (*) and (**) are, in fact, the same (a somewhat more refined approach is given in Problem 4). But our derivation of (**) now enables us to relate solutions of (**) with functions $f = (u,v,w)$ satisfying (1). For suppose that w is a solution of (**) satisfying (2). Then $\langle\ ,\ \rangle'$ is positive definite, and has curvature $K' = 0$. So there is a coordinate system (u,v) satisfying (4), which implies that (u,v,w) satisfies (1). The possible coordinate systems (u,v) for the flat metric $\langle\ ,\ \rangle'$ all differ by a Euclidean motion of $\mathbb{R}^2$, so (u,v) is determined by specifying

$$u(0)\ ,\quad v(0)$$
$$u_1(0)\ ,\quad u_2(0)\ ,\quad v_1(0)\ ,\quad v_2(0)\ ,$$

where the $u_i(0)$ and $v_i(0)$ have to be chosen so that (1) holds at $(0,0)$. Notice that u and v will automatically satisfy (**), since this equation is equivalent to (*), which is satisfied by all component functions $f = (u,v,w)$ satisfying (1). We thus have the paradoxical situation that u, v, w all satisfy (**) $\equiv$ (*), but that once we pick the initial conditions w_1, w_2 along the x-axis which determine w, then we have almost no choice left for the initial conditions for u and v; of course, we could just as well pick the initial conditions for u, say, and then be stuck with those for v and w.

The Darboux equation is not linear, but it is linear in $(rt-s^2)$, r, s, t; as we mentioned in Chapter 10, section 8, equations of this sort are called "Monge-Ampère equations." We can write our equation as

$$t(r+Ap+Bq) + C = 0\ ,$$

where A, B, C do not involve t, and thus we can solve for t in terms of the other quantities,

$$t = \frac{-C}{r+Ap+Bq} = g(x,y,p,q,r,s)\ . \tag{6}$$

More precisely, if we are given initial conditions along the x-axis such that $Ap+Bq+r \neq 0$ at $(0,0)$, then we can write our equation in this form near $(0,0)$. In Chapter 10, Part 4, we showed that the Cauchy problem for such an equation is equivalent to the Cauchy problem for a quasi-linear first order system, which can always be solved, by the Cauchy-Kowalewski theorem, if all functions in the equation and the initial data are analytic. Thus we see that the required isometric imbedding f exists locally if E, F, G are analytic.

Naturally we would like to know to what extent this restriction to analytic E, F, G and analytic initial data is necessary. Recall that a solution w of a second order PDE

$$F(x,y,w,p,q,r,s,t) = 0$$

is elliptic [respectively, hyperbolic] if and only if

$$4F_r F_t - F_s{}^2 > 0 \qquad [\text{respectively}, \ < 0] .$$

We consider the Darboux equation in the form (*) on p. 208. Our condition becomes

$$4(w_{22} - \Gamma_{22}^1 w_1 - \Gamma_{22}^2 w_2)(w_{11} - \Gamma_{11}^1 w_1 - \Gamma_{11}^2 w_2) - 4(w_{12} - \Gamma_{12}^1 w_1 - \Gamma_{22}^2 w_2)^2 > 0$$
$$[\text{respectively}, \ < 0] .$$

Using equations (1) on p. 207, this becomes

$$(n^3)^2(\ell_{22}\ell_{11} - \ell_{12}{}^2) > 0 \qquad [\text{respectively}, \ < 0] .$$

So at all points where $n^3 \neq 0$, the solution w is elliptic [hyperbolic] if and only if the corresponding surface (obtained by choosing u, v as before) has $K > 0$ $[K < 0]$. Thus the cases $K > 0$ and $K < 0$ require separate treatment.

We note first that Theorem 10-13 shows that if $K > 0$ everywhere and E, F, G are analytic, then every imbedding f such that I_f has components E, F, G is automatically analytic, so there is no point considering initial data which are not analytic. Expressed somewhat differently, if a surface $M \subset \mathbb{R}^3$ has $K > 0$ everywhere and a metric which is analytic in some coordinate system, then M is actually an analytic submanifold of $\mathbb{R}^3$. In particular, if M is any (C^3) surface of constant positive curvature, then M is automatically analytic. When E, F, G are not analytic, there is no known criterion on the initial data which will guarantee the existence of a solution of the Darboux equation (**). However, we might simply ask if there is some solution of the Darboux equation (without specifying initial conditions), and hence some imbedding f such that I_f has components E, F, G. The fact that a second order equation like (**) actually has solutions has been "known" for a long time -- an actual proof may be found in Jacobowitz [1].

When $K < 0$ there is no problem. Theorem 10-12 shows that for any initial conditions with $F_t = r + Ap + Bq \neq 0$, there is always a solution w of (**) in a neighborhood of $(0,0)$, and we can obtain solutions less differentiable than the functions E, F, G. (In the next Chapter we will have occasion to examine the case where the initial conditions are such that $F_t = 0$.) We see, in particular, that there are surfaces of constant negative curvature in $\mathbb{R}^3$ which are C^∞, but not analytic.

By the way, it is interesting to note that a surface of constant mean curvature H is always analytic (the sign of H couldn't be relevant, since it is not even well-determined). For suppose that $M \subset \mathbb{R}^3$ is a surface with $H = C$, given locally as the graph of a function h. Then formula (B') on p. III.201 gives

$$0 = F(x,y,h,p,q,r,s,t)$$
$$= (1+q^2)r - 2pqs + (1+p^2)t - 2C(1+p^2+q^2)^{3/2} ,$$

so

$$4F_rF_t - F_s{}^2 = 4[(1+q^2)(1+p^2) - p^2q^2] = 4(1+p^2+q^2) > 0 ,$$

and h is analytic by Theorem 10-13.

We now consider the general problem of locally imbedding an n-manifold in $\mathbb{R}^m$. We are given g_{ij} on a neighborhood of $0 \in \mathbb{R}^n$, and we seek $f: U \longrightarrow \mathbb{R}^m$, on some smaller neighborhood U, such that

$$(1) \qquad g_{ij} = \langle f_i, f_j \rangle = \sum_{\alpha=1}^{m} \langle f_i^\alpha, f_j^\alpha \rangle = \sum_{\alpha=1}^{m} \frac{\partial f^\alpha}{\partial x_i} \cdot \frac{\partial f^\alpha}{\partial x_j} .$$

Since this is a set of $s_n = n(n+1)/2$ equations, it seems unlikely that we can always find f if $m < s_n$. In fact, if (1) is to hold, then all equations obtained from (1) by partial differentiation must also hold. If we evaluate these equations at 0, we obtain polynomial formulas expressing the derivatives

$$(2) \qquad \frac{\partial^{r_1+\cdots+r_n} g_{ij}}{\partial^{r_1}x_1 \cdots \partial^{r_n}x_n}(0) \qquad 0 \le r_1 + \cdots + r_n \le r-1$$

in terms of the derivatives

$$(3) \qquad \frac{\partial^{r_1+\cdots+r_n} f^\alpha}{\partial^{r_1}x_1 \cdots \partial^{r_n}x_n}(0) \qquad 1 \le r_1 + \cdots + r_n \le r .$$

For each g_{ij}, the number of derivatives in (2) is the binomial coefficient $\binom{n+r-1}{n}$ = the number of ways of picking n things from $n+r-1$ things, for we can associate to each set $\alpha_1 < \alpha_2 < \cdots < \alpha_n$ of integers from 1 to $n+r-1$ the numbers

$$r_1 = \alpha_1 - 1 , \quad r_2 = \alpha_2 - \alpha_1 - 1 , \ldots, \quad r_n = \alpha_n - \alpha_{n-1} - 1 .$$

Thus,

$$\# \text{ of derivatives in (2) is } a = s_n \cdot \binom{n+r-1}{n} .$$

Similarly,

$$\# \text{ of derivatives in (3) is } b = m \cdot \left[\binom{n+r}{n} - 1\right] .$$

Now if $m < s_n$, then the first of these numbers will be greater than the second for large enough r. In fact,

$$(s_n - 1) \cdot \frac{(n+r)!}{n!r!} = s_n \cdot \frac{(n+r-1)!}{n!(r-1)!} \qquad \text{for} \qquad r = n(s_n - 1) .$$

But this means that the set of all possible derivatives (2), considered as a point in $\mathbb{R}^a$, is the image of a polynomial map defined on a lower dimensional space $\mathbb{R}^b$, so the derivatives (2) cannot be assigned arbitrarily for a map $f: \mathbb{R}^n \to \mathbb{R}^m$, $m < s_n$. In other words, not every g_{ij} can be obtained from some f.

It also seems reasonable to conjecture that we can always find an appropriate f when $m = s_n$. With the proper handling of subsidiary considerations, the proof of this conjecture can be reduced to the Cauchy-Kowalewski theorem (which means that we will have to assume that the g_{ij} are analytic). First a preliminary definition. Given $f\colon U \to \mathbb{R}^m$, consider the space spanned by the vectors

$$\frac{\partial f}{\partial x^i} \quad 1 \le i \le n\,, \qquad\qquad \frac{\partial^2 f}{\partial x^i \partial x^j} \quad 1 \le i,\, j \le n$$

at a point $p \in U$ [this is the direct sum of the tangent space of $f(U)$ at $f(p)$, and the first normal space at $f(p)$, in the terminology of Addendum 4 to Chapter 7]. The map f is called non-degenerate if these $n + s_n$ vectors are linearly independent, for each $p \in U$. For example, a curve in $\mathbb{R}^2 \subset \mathbb{R}^m$ is non-degenerate if its curvature is nowhere zero.

9. THEOREM (BURSTIN-JANET-CARTAN). Let g_{ij} be the components of an analytic Riemannian metric in a neighborhood of $0 \in \mathbb{R}^n$. Then there is an analytic isometric imbedding $f\colon U \to \mathbb{R}^{s_n}$ (defined on some smaller neighborhood U).

Proof. Let V_i be i-dimensional subspaces of $\mathbb{R}^n{}_0$ with

$$V_1 \subset \cdots \subset V_n = \mathbb{R}^n{}_0\,,$$

and let

$$H_i = \exp_0(V_i) \subset \mathbb{R}^n$$

(the exponential being defined with respect to the metric given by the g_{ij}). Since H_1 is a curve, we can clearly find an analytic isometric imbedding

$f_1\colon H_1 \longrightarrow \mathbb{R}^{s_n}$; moreover, we can arrange for f_1 to be non-degenerate. We will now show that if $f_k\colon H_k \longrightarrow \mathbb{R}^{s_n}$ is a free analytic isometric imbedding, then f_k can be extended to an analytic isometric imbedding $f_{k+1}\colon H_{k+1} \longrightarrow \mathbb{R}^{s_n}$ (defined perhaps in a smaller neighborhood of 0). Moreover, for $k+1 < n$ we will show that f_{k+1} can be chosen to be non-degenerate [note that $\ell < n \Longrightarrow \ell + s_\ell < s_n$]. This will clearly prove the theorem.

Step 1. By changing our coordinate system $x_1,\ldots,x_k$, $y = x_{k+1}$ on H_{k+1} we can assume that

$$H_k = \{(x,y)\colon y = 0\}$$

$$g_{i,k+1} = 0 \qquad (1 \le i \le k)\ , \qquad\qquad g_{k+1,k+1} = 1\ .$$

Then the equations $g_{ij} = \langle f_i, f_j \rangle$ become

$$(1) \qquad \begin{aligned} \langle f_{x_i}, f_{x_j} \rangle &= g_{ij} \\ \langle f_{x_i}, f_y \rangle &= 0 \\ \langle f_y, f_y \rangle &= 1\ . \end{aligned}$$

Differentiating the first equation with respect to y, and the second with respect to x_j, we find that if f satisfies (1), then it also satisfies the following set of equations, which are of first order with respect to the y variable:

$$(2)\qquad \begin{aligned} &(a) \quad \langle f_y, f_{x_i x_j}\rangle = -\tfrac{1}{2}(g_{ij})_y \\ &(b) \quad \langle f_y, f_{x_i}\rangle = 0 \\ &(c) \quad \langle f_y, f_y\rangle = 1 \,. \end{aligned}$$

Similar manipulations show that f also satisfies the equations

$$(3)\qquad \begin{aligned} &(a) \quad \langle f_{yy}, f_{x_i}\rangle = 0 \\ &(b) \quad \langle f_{yy}, f_y\rangle = 0 \\ &(c) \quad \langle f_{yy}, f_{x_i x_j}\rangle = -\tfrac{1}{2}(g_{ij})_{yy} + \langle f_{yx_i}, f_{yx_j}\rangle \,, \end{aligned}$$

which are of second order with respect to the y variables.

Conversely, suppose that f satisfies (3), and also satisfies (2) on H_k. We claim that f satisfies (1). First of all, since (3b) says that $\langle f_y, f_y\rangle_y = 0$, equation (2c) on H_k implies that $\langle f_y, f_y\rangle = 1$ everywhere. Consequently, we also have $\langle f_y, f_{x_i y}\rangle = 0$. So (3a) says that $\langle f_{x_i}, f_y\rangle_y = 0$, and then (2b) on H_k implies that $\langle f_{x_i}, f_y\rangle = 0$. Thus we have the last two equations of (1). Now from

$$\langle f_{x_i}, f_y\rangle = 0 \,, \qquad \langle f_{x_j}, f_y\rangle = 0 \,, \qquad \langle f_y, f_y\rangle = 1$$

we obtain

$$\text{(i)} \quad \langle f_{x_i x_j}, f_y\rangle + \langle f_{x_i}, f_{x_j y}\rangle = 0$$

$$\text{(ii)} \quad \langle f_{x_j x_i}, f_y\rangle + \langle f_{x_j}, f_{x_i y}\rangle = 0$$

(iii) $\langle f_{x_i y}, f_y\rangle = 0$,

and then

(iv) $\langle f_{x_j x_i y}, f_y\rangle + \langle f_{x_j x_i}, f_{yy}\rangle + \langle f_{x_j y}, f_{x_i y}\rangle + \langle f_{x_j}, f_{x_i yy}\rangle = 0$ from (ii)

(v) $\langle f_{x_i x_j y}, f_y\rangle + \langle f_{x_i y}, f_{x_j y}\rangle = 0$ from (iii).

Equations (iv) and (v) give

$$\langle f_{x_i yy}, f_{x_j}\rangle = -\langle f_{x_i x_j}, f_{yy}\rangle ,$$

so we have

$$\begin{aligned}\langle f_{x_i}, f_{x_j}\rangle_{yy} &= \langle f_{x_i yy}, f_{x_j}\rangle + 2\langle f_{x_i y}, f_{x_j y}\rangle + \langle f_{x_i}, f_{x_j yy}\rangle \\ &= -2\langle f_{x_i x_j}, f_{yy}\rangle + 2\langle f_{x_i y}, f_{x_j y}\rangle \\ &= (g_{ij})_{yy} \qquad \text{by (3c).}\end{aligned}$$

On the other hand, (i) and (ii) give

$$\begin{aligned}\langle f_{x_i}, f_{x_j}\rangle_y &= -2\langle f_{x_i x_j}, f_y\rangle \\ &= (g_{ij})_y \quad \text{on } H_k, \qquad \text{by 2(a).}\end{aligned}$$

It follows that $\langle f_{x_i}, f_{x_j}\rangle_y = (g_{ij})_y$ everywhere. Since we have $\langle f_{x_i}, f_{x_j}\rangle = g_{ij}$ on H_k, we conclude that $\langle f_{x_i}, f_{x_j}\rangle = g_{ij}$ everywhere, as desired.

Step 2. Having established this, we now claim that there is an analytic function χ on H_k such that

$$
(2') \qquad \begin{aligned}
&\langle\chi, f_{x_i x_j}\rangle = -\tfrac{1}{2}(g_{ij})_y \\
&\langle\chi, f_{x_i}\rangle = 0 \qquad\qquad \text{on } H_k . \\
&\langle\chi,\chi\rangle = 1 \\
&\chi \text{ is linearly independent of } f_{x_i}, f_{x_i x_j}
\end{aligned}
$$

The reason for this is the following. At 0, we have $(g_{ij})_y = 0$ (Proposition II. 4-1). So $\chi(0)$ is just a unit vector in $\mathbb{R}^{s_n}$ which is perpendicular to all $f_{x_i}(0)$ and $f_{x_i x_j}(0)$ [such a vector exists, since $k + s_k < s_n$]. In general, we first pick a linear combination χ_1 of the (linearly independent) vectors f_{x_i}, $f_{x_i x_j}$ so that the first two conditions in (2') hold for χ_1. Near 0, this makes χ_1 a vector of small norm. Then we add on an appropriate vector orthogonal to the f_{x_i} and $f_{x_i x_j}$ so that the norm becomes 1. There is no problem arranging for χ to be analytic.

Consider the following system of equations for functions $f, q\colon H_{k+1} \to \mathbb{R}^{s_n}$:

$$
(*) \quad \left\{
\begin{aligned}
&f_y = q \\
&\langle q_y, f_{x_i}\rangle = 0 \qquad i = 1,\dots,k \\
&\langle q_y, q\rangle = 0 \\
&\langle q_y, f_{x_i x_j}\rangle = -\tfrac{1}{2}(g_{ij})_{yy} + \langle q_{x_i}, q_{x_j}\rangle \qquad i, j = 1,\dots,k ,
\end{aligned}
\right.
$$

with the initial conditions

$$(*_0) \qquad \left\{\begin{array}{l} f(x,0) = f_k(x) \\ q(x,0) = \chi(x) \end{array}\right. \qquad \text{on } H_k .$$

If we have a solution (f,q), then f will be a solution of (3) such that $\chi(x) = f_y(x,0)$ satisfies (2) on H_k, so f will be an analytic isometric imbedding extending f_k in a neighborhood of 0. Now (*) is rather like the equations considered in the Cauchy-Kowalewski theorem, expressing the partials of the $2s_n$ functions $f^1,\ldots,f^{s_n},q^{s_1},\ldots,q^{s_n}$ with respect to y in terms of their partials with respect to $x_1,\ldots,x_k$. However, we have more unknowns than equations (except for $k = n-1$), and the q_y are not explicitly solved for. Such a problem is handled as follows.

Step 3. Write the last three sets of equations of (*) as a matrix equation

$$\underbrace{\begin{pmatrix} f_{x_i} \\ \\ q \\ \\ f_{x_ix_j} \end{pmatrix}}_{\substack{(k+1+s_k)\times s_n \\ \text{matrix}}} \cdot \underbrace{\begin{pmatrix} q_y^1 \\ \cdot \\ \cdot \\ \cdot \\ q_y^{s_n} \end{pmatrix}}_{\substack{s_n\times 1 \\ \text{matrix}}} = \underbrace{\begin{pmatrix} 0 \\ \vdots \\ 0 \\ \vdots \\ -\frac{1}{2}(g_{ij})_{yy} + \langle q_{x_i}, q_{x_j}\rangle \\ \vdots \end{pmatrix}}_{\substack{(k+1+s_k)\times 1 \\ \text{matrix}}} ,$$

or for short as

$$B(f_{x_i},q,f_{x_ix_j})\cdot q_y = g .$$

On H_k, the rows of $B(f_{x_i},q,f_{x_ix_j})$ are linearly independent, because f_k is non-degenerate, and by the choice (2') of χ. So this matrix has a right inverse. Moreover, we can pick this inverse analytically -- that is, for any $r \leq s$ there is an analytic map $B \mapsto \tilde{B}$ from the $r \times s$ matrices of rank r to the $s \times r$ matrices such that

$$B \cdot \tilde{B} = r \times r \text{ identity matrix} .$$

[One specific way to define $\tilde{B}$ is as follows. Write B as a collection of row vectors,

$$B = \begin{pmatrix} v_1 \\ \vdots \\ v_r \end{pmatrix}$$

for $v_i \in \mathbb{R}^s$. There is a unique decomposition

$$v_i = w_i + z_i ,$$

where

$$w_i \in \text{subspace } W_i \subset \mathbb{R}^s \text{ spanned by } v_1, \dots, \hat{v}_i, \dots, v_r$$

$$z_i \perp W_i ;$$

clearly $z_i \neq 0$. Then

$$\langle v_i, z_j \rangle = \begin{cases} 0 & i \neq j \\ \langle z_i, z_i \rangle & i = j , \end{cases}$$

so we can choose $\tilde{B}$ to be the matrix whose columns are $z_i/\langle z_i,z_i\rangle$.]

Now consider the system of equations

$$(**) \qquad \begin{cases} f_y = q \\ q_y = \tilde{B}(f_{x_i},q,f_{x_ix_j})\cdot g \,, \end{cases}$$

with the initial conditions $(*_0)$. This equation makes sense in a neighborhood of H_k, since the rows of $B(f_{x_i},q,f_{x_ix_j})$ are linearly independent there. Any solution of (**) will be a solution of (*), since

$$q_y = \tilde{B}(f_{x_i},q,f_{x_ix_j})\cdot g \implies B(f_{x_i},q,f_{x_ix_j})\cdot q_y = g \,.$$

But (**) is a set of equations to which the Cauchy-Kowalewski theorem applies [more precisely, it can be reduced to such a set by the method of section 4 of Chapter 10]. Thus we have established the existence of the extension f.

Step 4. We still have to arrange for f to be non-degenerate when $k+1 < n$. We claim first that we can choose χ on H_k so that χ satisfies (2'), and also so that the vectors χ, χ_{x_i}, f_{x_i}, $f_{x_ix_j}$ are linearly independent at 0. To do this we again first choose χ_1 to be a linear combination of the f_{x_i}, $f_{x_ix_j}$ which satisfies the first two conditions of (2'); near 0 we have $|\chi_1| < 1$. We next choose $\alpha\colon H_k \longrightarrow \mathbb{R}^{s_n}$ to be an analytic map with α perpendicular to f_{x_i}, $f_{x_ix_j}$ and $|\alpha|$ small. Then there is a constant λ_0 with $|\chi_1 + \lambda_0\alpha| = 1$ at 0. We can assume, by renumbering, that the vectors of the set

$$A = \left\{\alpha, (\chi+\lambda_0\alpha)_{x_1}, \dots, (\chi+\lambda_0\alpha)_{x_h}, f_{x_i}, f_{x_ix_j}\right\}$$

are linearly independent at 0, and that $h \leq k$ is the largest integer with this property ($h = 0$ is a possibility, i.e., there may be no vectors $(x+\lambda_0\alpha)_{x_i}$ in our set). If $h = k$ we are done. Otherwise, pick non-zero analytic functions $\beta_{h+1},\ldots,\beta_k$ which are orthogonal to the vectors of A, and also mutually orthogonal. Then determine the analytic function λ so that

$$|\chi(x)| = \left|\chi_1(x) + \lambda(x)\cdot(\alpha(x) + \sum_{i=h+1}^{k} \beta_i(x)\cdot x_i)\right| = 1 .$$

Thus $\lambda_0 = \lambda(0)$. Suppose some linear combination of χ, χ_{x_i}, f_{x_i}, $f_{x_i x_j}$ vanishes at 0. I.e., suppose that at 0 we have

$$\begin{aligned} 0 = a(\chi_1+\lambda\alpha) &+ \sum_{i=1}^{h} a_i((\chi_1)_{x_i} + \lambda\alpha_{x_i} + \lambda_{x_i}\alpha) \\ &+ \sum_{i=h+1}^{k} a_i((\chi_1)_{x_i} + \lambda\alpha_{x_i} + \lambda_{x_i}\alpha + \lambda\beta_i) \\ &+ \sum_{i=1}^{k} b_i f_{x_i} + \sum_{i,j=1}^{k} b_{ij} f_{x_i x_j} . \end{aligned}$$

Take the inner product with some $\beta_\iota(0)$ $(\iota = h+1,\ldots,k)$. The β_i are mutually orthogonal, and β_ι is perpendicular to all vectors of A; moreover, each $(\chi_1)_{x_i} + \lambda\alpha_{x_i}$ $(i = h+1,\ldots,k)$ is a linear combination of elements of A, by the maximality property of A. So we end up with

$$\lambda(0)a_\iota|\beta_\iota(0)|^2 = 0 \quad\Longrightarrow\quad a_\iota = 0 .$$

Then we also have $a = a_i = b_i = b_{ij} = 0$, since the vectors of A are linearly independent. Thus χ has all the required properties.

Now in a neighborhood of the $s_k + 2k + 1$ linearly independent vectors

$\chi(0)$, $\chi_{x_i}(0)$, $f_{x_i}(0)$, $f_{x_ix_j}(0)$ we can choose an analytic map $(v,v_i,w_i,w_{ij}) \mapsto h(v,v_i,w_i,w_{ij}) \neq 0$ such that

$$h(v,v_i,w_i,w_{ij}) \quad \text{is perpendicular to} \quad v,\ v_i,\ w_i,\ w_{ij}\ .$$

Consider the equations

$$(**) \qquad \begin{cases} f_y = q \\ q_y = \tilde{B}(f_{x_i},q,f_{x_ix_j})\cdot g + h(q,q_{x_i},f_{x_i},f_{x_ix_j}) \ , \end{cases}$$

with the initial condition $(*_0)$. A solution will again be a solution of $(*)$, since our choice of h gives

$$B(f_{x_i},q,f_{x_ix_j})\cdot h(q,q_{x_i},f_{x_i},f_{x_ix_j}) = 0\ .$$

The vectors f_y, f_{yx_i}, f_{x_i}, $f_{x_ix_j}$ are linearly independent near 0, by our choice of χ. Then $f_{yy} = q_y$ will be independent of f_y, f_{yx_i}, f_{x_i}, $f_{x_ix_j}$, since $h(f_y,f_{yx_i},f_{x_i},f_{x_ix_j})$ is linearly independent of all these vectors, while our explicit construction of $\tilde{B}$ makes its columns span the same space as the rows of B, which implies that $\tilde{B}(f_{x_i},q,f_{x_ix_j})\cdot g$ is a linear combination of the q, f_{x_i}, $f_{x_ix_j}$. So f is non-degenerate. ■

This proof can naturally be applied to the case $n = 2$, and then there is only the step from $k = 1$ to $k+1 = 2$. In this case, the matrix $B(f_x,q,f_{xx})$ is a 3×3 invertible matrix, so we just consider the equations

$$f_y = q$$
$$q_y = B(f_x, q, f_{xx})^{-1} \cdot g \; .$$

One can check that this is a hyperbolic system when $K < 0$, so that Theorem 10-12 can be applied, with the initial choice f_1: (x-axis) $\longrightarrow \mathbb{R}^3$ being C^∞ rather than analytic; thus we can obtain the same results as we got by looking at the Darboux equation previously. Perhaps one could even try to analyse higher dimensional cases similarly, when the given metric has all sectional curvatures < 0. There is not much interest in doing this, however, for although analyticity was required to obtain the "best possible" local result of Theorem 9, there are global results where it is not needed. These results are essentially theorems in analysis, rather than geometry [with certain significant exceptions], and generally require rather involved techniques, some of which were created precisely for this problem. So we will merely indicate what these results are, and our discussion will be particularly brief since there are now several research reports which cover the field quite well.

One class of global results give very strong information about the special case of surfaces in $\mathbb{R}^3$. The first such question was raised by Hermann Weyl, who asked whether every metric $\langle\ ,\ \rangle$ on S^2 with $K > 0$ comes from an isometric immersion in $\mathbb{R}^3$ (necessarily an imbedding as a convex surface, by Hadamard's Theorem). Although Weyl indicated an approach to this problem, the first affirmative solution, for analytic metrics, was given by H. Lewy [2]. A proof for C^k metrics, $k \geq 4$, was given by Nirenberg [1], and the cases $k = 2, 3$ were later handled by Heinz [1] . Already in 1942, A.D. Alexandrov had considered Weyl's problem from a completely different, totally geometric approach, involving polyhedral approximations to the surface. He was able to solve Weyl's problem for C^2 metrics, although his result did not indicate how differentiable the resulting surface would be when the metric was more differentiable. But this was established by later research, especially that of Pogorelov. At the same time, this pioneering work of Alexandrov led him to investigate arbitrary convex surfaces (which need not be smooth at all); although such surfaces may not have Riemannian metrics, we can still define an isometry between two such surfaces to be a homeomorphism preserving lengths of curves, and there are suitable generalizations of other differential geometric concepts like curvature (which may exist only almost everywhere). In consequence, there has developed an entirely disjoint school of differential geometry, whose practitioners are almost exclusively Russian mathematicians, which proves certain results in far greater generality than classical differential geometry, and has sometimes proved results from this field which are still inaccessible to the classical methods. Some examples of this will be mentioned in the next Chapter, and the Bibliography gives further references to the Russian school.

In contrast to Weyl's problem, which arises by considering the metric $\langle\ ,\ \rangle$ induced on S^2 by some imbedding of S^2 as a convex set in $\mathbb{R}^3$, we now consider a strictly convex surface $M \subset S^3$ and define a function $k > 0$ on S^2 by

$$K(p) = k(\nu(p)) ,$$

where $K > 0$ is the curvature of M, and the diffeomorphism $\nu\colon M \to S^2$ is the normal map. This function k always satisfies certain integral equalities. To derive them, we note that we have

$$0 = \int_M \nu^i \, dA = \int_M \left\langle \nu, \frac{\partial}{\partial x^i} \right\rangle dA \; ;$$

this follows from the Divergence Theorem (Problem I.9-13 or Theorem 7-57), applied to the region D bounded by M. Consequently, if da is the volume element of S^2, and $x^i/k(x)$ denotes the function $x \mapsto x^i/k(x)$ on S^2, then

$$(*) \qquad \int_{S^2} \frac{1}{k(x)} \cdot x^i \, da = \int_M \nu^*\left(\frac{1}{k(x)} \cdot x^i \, da\right)$$
$$= \int_M \frac{1}{k \circ \nu} \cdot \nu^i \cdot \nu^*(da)$$
$$= \int_M \frac{1}{K} \cdot \nu^i \cdot K \, dA = \int_M \nu^i \, dA$$
$$= 0 \; .$$

"Minkowski's problem" is to show that any function $k > 0$ on S^2 which satisfies the conditions (*) is $K \circ \nu^{-1}$ for some convex $M \subset \mathbb{R}^3$. This problem was solved by Lewy [3] in the analytic case, and by Nirenberg [1] in the C^2 case. It should also be mentioned that generalizations of Minkowski's problem have been given in higher dimensions, in the style of the Russian school, by A.D. Alexandrov [3] and Fenchel and Jessen [1], but for the higher dimensional cases little is known about the differentiability of the hypersurfaces obtained.

Less delicate, but much more general, results are now available for the problem of isometrically imbedding arbitrary Riemannian manifolds in some Euclidean space. The first results along this line were by Nash [1], supplemented by Kuiper [1]. For a compact n-dimensional Riemannian manifold M, their results show that if M has any imbedding in $\mathbb{R}^q$, with $q \geq n+1$, then it also has a C^1 isometric imbedding. Thus compact orientable surfaces always have a C^1 isometric imbedding in $\mathbb{R}^3$; in particular, even the flat torus can be C^1 isometrically imbedded in $\mathbb{R}^3$! The most important isometric imbedding theorems stem from a second paper of Nash [2], where he proved that every C^∞ Riemannian manifold can be C^∞ isometrically imbedded in some Euclidean space. We will not give the dimensions of the Euclidean spaces involved; for this the reader may consult Gromov and Rokhlin [1], which gives a very complete discussion of the results known up to 1970. We merely mention that almost nothing is known about the lowest dimensional Euclidean space in which the imbedding is possible.

Addendum. The Embedding Problem via Differential Systems

Although the general line of argument for the proof of Theorem 9 was proposed by Janet, it was Burstin who gave the first rigorous proof. The result is often known as the Cartan-Janet theorem because E. Cartan gave another (completely different) rigorous proof, using his theory of differential systems (Chapter 10, Addendum 1). We will give this proof here; so we assume that we have an analytic Riemannian metric in a neighborhood of $0 \in \mathbb{R}^n$, and we seek a local isometric imbedding into $\mathbb{R}^{s_n}$, $s_n = n(n+1)/2$.

Let $O(\mathbb{R}^{s_n})$ be the bundle of orthonormal frames of $\mathbb{R}^{s_n}$, on which we have the dual forms ϕ^α and connection forms ψ^α_β $(1 \le \alpha, \beta \le s_n)$; for simplicity we do not use bold-face letters for these forms on $O(\mathbb{R}^{s_n})$. The forms ϕ^α, ψ^α_β give a basis for the dual space of the tangent space $O(\mathbb{R}^{s_n})_u$ for any $u \in O(\mathbb{R}^{s_n})$. Also let $Z_1,\dots,Z_n$ be some fixed orthonormal moving frame on $\mathbb{R}^n$, with dual forms θ^i, connection forms ω^i_j, and curvature forms Ω^i_j. Suppose that $f\colon U \to \mathbb{R}^{s_n}$ is an isometry, for some neighborhood U of 0 in $\mathbb{R}^n$. Let $s = (Y_1,\dots,Y_n,Y_{n+1},\dots,Y_{s_n})$ be any orthonormal moving frame on $f(U)$ with $Y_i = f_* Z_i$ for $i = 1,\dots,n$; then $s^*\phi^\alpha$ and $s^*\psi^\alpha_\beta$ are its dual forms and connection forms. Since f is an isometry, we clearly have

$$\theta^i = f^*(s^*\phi^i) = (s\circ f)^*\phi^i \qquad i = 1,\dots,n$$

$$0 = f^*(s^*\phi^r) = (s\circ f)^*\phi^r \qquad r = n+1,\dots,s_n\,.$$

Conversely, if $F\colon U \to O(\mathbb{R}^{s_n})$ is a map which can be written as $F = s\circ f$, and

$$\theta^i = F^*\phi^i \qquad i = 1,\dots,n$$

$$0 = F^*\phi^r \qquad r = n+1,\dots,s_n \;,$$

then f is an isometry. We will look for F, and hence f, by looking for its graph in $\mathbb{R}^n \times O(\mathbb{TR}^{s_n})$. We have two projections

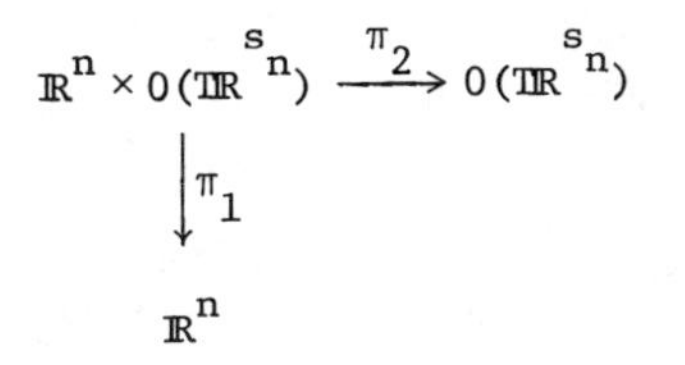

and for simplicity we will denote

$$\pi_1{}^*\theta^i \text{ by } \theta^i, \qquad \pi_1{}^*\omega^i_j \text{ by } \omega^i_j, \qquad \pi_1{}^*\Omega^i_j \text{ by } \Omega^i_j,$$

$$\pi_2{}^*\phi^\alpha \text{ by } \phi^\alpha, \qquad \pi_2{}^*\psi^\alpha_\beta \text{ by } \psi^\alpha_\beta .$$

We easily see that our problem is solved if there is an n-dimensional manifold $\Gamma \subset \mathbb{R}^n \times O(\mathbb{TR}^{s_n})$ through some point $(0,u)$, such that $\pi_{1*}\colon \Gamma_{(0,u)} \to \mathbb{R}^n{}_0$ is one-one, and such that

$$\phi^i - \theta^i = 0 \text{ on } \Gamma \qquad i = 1,\dots,n$$

$$\phi^r = 0 \text{ on } \Gamma \qquad r = n+1,\dots,s_n .$$

We want to find Γ as an integral manifold for an appropriate differential system $\mathscr{I}$. So, first of all, we want $\mathscr{I}$ to contain the $\phi^i - \theta^i$ and the ϕ^r. Now

$$
\begin{aligned}
d(\phi^i - \theta^i) &= -\sum_{\alpha=1}^{s_n} \psi^i_\alpha \wedge \phi^\alpha + \sum_{j=1}^{n} \omega^i_j \wedge \theta^j \\
&= -\sum_{j=1}^{n} (\psi^i_j - \omega^i_j) \wedge \theta^j - \sum_{j=1}^{n} \psi^i_j \wedge (\phi^j - \theta^j) - \sum_{r=n+1}^{s_n} \psi^i_r \wedge \phi^r ,
\end{aligned}
$$

and

$$
\begin{aligned}
d\phi^r &= -\sum_{\alpha=1}^{s_n} \psi^r_\alpha \wedge \phi^\alpha \\
&= -\sum_{j=1}^{n} \psi^r_j \wedge \theta^j - \sum_{j=1}^{n} \psi^r_j \wedge (\phi^j - \theta^j) - \sum_{t=n+1}^{s_n} \psi^r_t \wedge \theta^t ,
\end{aligned}
$$

so in order to have $d\mathcal{I} \subset \mathcal{I}$ we also want the $\psi^i_j - \omega^i_j$ and the $\sum_j \psi^r_j \wedge \theta^j$ to be in $\mathcal{I}$. Similarly, since

$$
\begin{aligned}
d(\psi^i_j - \omega^i_j) &= -\sum_{\alpha=1}^{s_n} \psi^i_\alpha \wedge \psi^\alpha_j - \Omega^i_j + \sum_{h=1}^{n} \omega^i_h \wedge \omega^h_j \\
&= -\sum_{h=1}^{n} \psi^i_h \wedge (\psi^h_j - \omega^h_j) - \sum_{h=1}^{n} (\psi^i_h - \omega^i_h) \wedge \omega^h_j - \sum_{r=n+1}^{s_n} \psi^i_r \wedge \psi^r_j - \Omega^i_j ,
\end{aligned}
$$

we also want the $\sum_r \psi^i_r \wedge \psi^j_r - \Omega^i_j$ to be in $\mathcal{I}$. Moreover, we easily see that if $\mathcal{I}$ is generated by

(a) $\phi^i - \theta^i$ $\qquad i = 1,\ldots,n$

(b) ϕ^r $\qquad r = n+1,\ldots,s_n$

(c) $\psi^i_j - \omega^i_j$ $\qquad i, j = 1,\ldots,n$

(d) $\sum_{j=1}^{n} \psi^r_j \wedge \theta^j$ $\qquad r = n+1,\ldots,s_n$

(e) $\sum_{r=n+1}^{s_n} \psi^i_r \wedge \psi^j_r - \Omega^i_j$ $\qquad i, j = 1,\ldots,n$,

then we have $d\mathscr{I} \subset \mathscr{I}$. The Cartan-Kähler Theorem (10-15) tells us that the desired n-dimensional manifold $\Gamma \subset \mathbb{R}^n \times O(\mathbb{T}\mathbb{R}^{s_n})$ exists if for some $u \in O(\mathbb{T}\mathbb{R}^{s_n})$ there is an n-dimensional integral element $W \subset O(\mathbb{T}\mathbb{R}^{s_n})_{(0,u)}$ of $\mathscr{I}$ which contains a regular $(n-1)$-dimensional integral element of $\mathscr{I}$, and for which $\pi_{1*}\colon W \longrightarrow \mathbb{R}^n{}_0$ is one-one. We assume $n \geq 2$, since the case $n = 1$ is trivial.

We claim, first of all, that every point of $\mathbb{R}^n \times O(\mathbb{T}\mathbb{R}^{s_n})$ is a regular 0-dimensional element of $\mathscr{I}$. To prove this we have to consider each $\mathscr{E}_1((x,u))$ for $(x,u) \in \mathbb{R}^n \times O(\mathbb{T}\mathbb{R}^{s_n})$. By definition, $\mathscr{E}_1((x,u))$ is the set of all vectors (X,Y) [with $X \in \mathbb{R}^n{}_x$ and $Y \in O(\mathbb{T}\mathbb{R}^{s_n})_u$] such that the forms (a)-(c) vanish on (X,Y), that is:

$$(1)\qquad \begin{cases} \phi^i(u)(Y) = \theta^i(x)(X) & i = 1,\ldots,n \\ \phi^r(u)(Y) = 0 & r = n+1,\ldots,s_n \\ \psi^i_j(u)(Y) = \omega^i_j(x)(X) & i, j = 1,\ldots,n\,. \end{cases}$$

Because the $\phi^\alpha(u)$, $\psi^\alpha_\beta(u)$ are a basis for $O(\mathbb{T}\mathbb{R}^{s_n})_u$, the dimension of $\mathscr{E}_1((x,u))$ is always exactly the dimension of $\mathbb{R}^n \times O(\mathbb{T}\mathbb{R}^{s_n})$ minus the number of forms (a)-(c), and thus a (non-zero) constant. So each (x,u) is a regular 0-dimensional integral element.

For any tangent vector $Y \in O(\mathbb{T}\mathbb{R}^{s_n})_u$, it will be convenient to consider n vectors $Y^{(i)}$ in $\mathbb{R}^{s_n - n} = \mathbb{R}^{n(n-1)/2}$, defined by

$$Y^{(i)} = (\psi_i^{n+1}(u)(Y),\ldots,\psi_i^{s_n}(u)(Y)) .$$

Note that we can always choose $Y \in O(\mathbb{T}\mathbb{R}^{s_n})_u$ satisfying (1) [i.e., with $(X,Y) \in \mathcal{E}_1((x,u))$] such that the $Y^{(i)}$ are any given n vectors in $\mathbb{R}^{n(n-1)/2}$.

Now at any point (x,u), pick $(X_1,Y_1) \in \mathcal{E}_1((x,u))$ with X_1 a unit vector so that the $n-1$ vectors

$$Y_1^{(1)},\ldots,Y_1^{(n-1)} \in \mathbb{R}^{n(n-1)/2}$$

are linearly independent [this is possible since $n(n-1)/2 \geq n-1$ for $n \geq 2$], and consider $\mathcal{E}_2((x,u),(X_1,Y_1))$. It is the set of all (X_2,Y_2) such that (1) holds, and such that the forms (d) and (e) vanish on the pair (X_1,Y_1), (X_2,Y_2). If X_2 is a multiple of X_1, then (1) implies that Y_2 is the same multiple of Y_1, so we will assume that X_2 is linearly independent of X_1. Since $(X_2,Y_2) \in \mathcal{E}_2((x,u),(X_1,Y_1))$ implies that any linear combination of (X_1,Y_1) and (X_2,Y_2) is also in $\mathcal{E}_2((x,u),(X_1,Y_1))$, in computing the dimension of this space we can restrict our attention to (X_2,Y_2) with X_1, X_2 orthonormal. Extend X_1, X_2 to an orthonormal basis $X_1,\ldots,X_n$ at u. Then (d) and (e) vanish on the pair (X_1,Y_1), (X_2,Y_2) if and only if

$$Y_2^{(1)} = Y_1^{(2)} \tag{2}$$

$$Y_2^{(j)}\cdot Y_1^{(i)} - Y_2^{(i)}\cdot Y_1^{(j)} - \langle R(X_1,X_2)X_j,X_i\rangle = 0 \qquad 1 \leq i < j \leq n , \tag{3}$$

where $\cdot$ denotes the usual inner product in $\mathbb{R}^{n(n-1)/2}$. Equation (2) determines $Y_2^{(1)}$; then equation (3) for $i = 1$, $j = 2$ determines a hyperplane

$H_2 \subset \mathbb{R}^{n(n-1)/2}$ in which $Y_2^{(2)}$ must lie; then equations (3) for $i = 1, j = 3$ and $i = 2, j = 3$ determine a plane $H_3 \subset \mathbb{R}^{n(n-1)/2}$ of codimension 2 in which $Y_2^{(3)}$ must lie; etc. [we use here the fact that $Y_1^{(1)},\ldots,Y_1^{(n-1)}$ are linearly independent]. In particular, the dimension of $\mathcal{E}_2((x,u),(X_1,Y_1))$ is the minimum possible. Thus (X_1,Y_1) generates a regular 1-dimensional integral element. Notice that $\mathcal{E}_2((x,u),(X_1,Y_1))$ does contain some (X_2,Y_2) [with X_1, X_2 orthonormal], since each H_α $(\alpha = 2,\ldots,n)$ has dimension

$$\frac{n(n-1)}{2} - (\alpha - 1) \geq \frac{n(n-1)}{2} - (n-1) \geq 0 \qquad \text{for } n \geq 2 .$$

In the case $n = 2$, we have just shown that there is some 2-dimensional integral element containing the regular 1-dimensional integral element generated by (X_1,Y_1), which completes the proof. In the case $n \geq 3$ we claim that we can choose Y_2 so that

$$Y_1^{(1)},\ldots,Y_1^{(n-1)},Y_2^{(2)},\ldots,Y_2^{(n-1)}$$

are linearly independent. We choose $Y_2^{(\alpha)}$ successively for $\alpha = 2,\ldots,n-1$. We want $Y_2^{(\alpha)}$ to be linearly independent of the vectors in the set

$$A = \{Y_1^{(1)},\ldots,Y_1^{(n-1)},Y_2^{(2)},\ldots,Y_2^{(\alpha-1)}\} .$$

Now $Y_2^{(\alpha)}$ must lie in the plane H_α, which is perpendicular to the vectors in the set

$$B = \{Y_1^{(1)},\ldots,Y_1^{(\alpha-1)}\} .$$

So we just need to have $\dim H_\alpha$ greater than the number of vectors in the set

$$A - B = \{Y_1^{(\alpha)},\ldots,Y_1^{(n-1)},Y_2^{(2)},\ldots,Y_2^{(\alpha-1)}\} .$$

Thus we need

$$\dim H_\alpha = \frac{n(n-1)}{2} - (\alpha - 1) > (n-\alpha) + (\alpha - 2) ,$$

or

$$\frac{n(n-1)}{2} > (n-1) + (\alpha - 2) .$$

But for $\alpha \le n-1$ we have

$$\begin{aligned} (n-1) + (\alpha-2) &\le (n-1) + (n-3) \\ &< (n-1) + (n-2) \\ &\le 1 + \cdots + n-1 = \frac{n(n-1)}{2} . \end{aligned}$$

This proves the claim.

Now suppose that for some $k \le n-1$ we have found $(X_1,Y_1),\ldots,(X_k,Y_k)$ with $X_1,\ldots,X_k$ orthonormal such that

(i) $(X_1,Y_1),\ldots,(X_k,Y_k)$ generate a k-dimensional integral element,

(ii) $(X_1,Y_1),\ldots,(X_{k-1},Y_{k-1})$ generate a regular $(k-1)$-dimensional integral element,

(iii) the $(n-1) + (n-2) + \cdots + (n-k)$ vectors

$$Y_1^{(1)},\ldots,Y_1^{(n-1)},Y_2^{(2)},\ldots,Y_2^{(n-1)},\ldots,Y_k^{(k)},\ldots,Y_k^{(n-1)}$$

are linearly independent.

[We have just done this for $k = 2$.] We claim that $(X_1,Y_1),\dots,(X_k,Y_k)$ generate a <u>regular</u> k-dimensional integral element, which can be extended to a $(k+1)$-dimensional integral element generated by $(X_1,Y_1),\dots,(X_{k+1},Y_{k+1})$ [with $X_1,\dots,X_{k+1}$ orthonormal]; moreover, if $k+1 \le n-1$, then Y_{k+1} can be picked so that $Y_{k+1}^{(k+1)},\dots,Y_{k+1}^{(n-1)}$ are linearly independent of the vectors in (iii). Once we have proved this claim, it will clearly follow that there is some n-dimensional integral element spanned by $(X_1,Y_1),\dots,(X_n,Y_n)$, with $X_1,\dots,X_n$ orthonormal, such that $(X_1,Y_1),\dots,(X_{n-1},Y_{n-1})$ span a regular $(n-1)$-dimensional integral element. Thus the proof will be complete.

To calculate the dimension of $\mathcal{E}_{k+1}((x,u),(X_1,Y_1),\dots,(X_k,Y_k))$ we consider (X_{k+1},Y_{k+1}) with $X_1,\dots,X_{k+1}$ orthonormal, and again extend $X_1,\dots,X_{k+1}$ to an orthonormal basis $X_1,\dots,X_n$ at x. Then (d) and (e) vanish on the pairs (X_h,Y_h), (X_{k+1},Y_{k+1}) $(1 \le h \le k)$ if and only if

$$(2') \qquad Y_{k+1}^{(h)} = Y_h^{(k+1)} \qquad h = 1,\dots,k$$

$$(3') \qquad Y_{k+1}^{(j)}\cdot Y_h^{(i)} - Y_{k+1}^{(i)}\cdot Y_h^{(j)} - \langle R(X_h,X_{k+1})X_j,X_i\rangle = 0 \qquad \begin{matrix} h = 1,\dots,k \\ 1\le i\le j\le n. \end{matrix}$$

Equations (2') determine $Y_{k+1}^{(h)}$ for $h \le k$. We claim that with these values of $Y_{k+1}^{(h)}$, equation (3') holds for $i, j \le k$. In fact, by hypothesis (i) we have

$$(a) \qquad Y_h^{(i)} = Y_i^{(h)}, \qquad Y_h^{(j)} = Y_j^{(h)} \qquad i, j, h \le k,$$

as well as

$$\text{(b)} \qquad Y_i^{(\ell)} \cdot Y_j^{(\lambda)} - Y_i^{(\lambda)} \cdot Y_j^{(\ell)} - \langle R(X_j, X_i)X_\ell, X_\lambda \rangle = 0 \qquad \begin{array}{l} i,\ j \le k \\ \text{all } \ \ell,\ \lambda \le n \ . \end{array}$$

Choose $\ell = h$ and $\lambda = k+1$ in (b), and substitute (2') and (a) into the equation. Using the identity $\langle R(X_j, X_i)X_h, X_{k+1} \rangle = \langle R(X_h, X_{k+1})X_j, X_i \rangle$, we obtain (3') for $i,\ j \le k$.

Thus, we need to consider (3') only for i or $j \ge k+1$. Since we are choosing $i < j$, we have $j \ge k+1$ in either case. Moreover, we claim that for each $j \ge k+1$, and $h \le k$, we need only consider the cases $h \le i$. For if we have all these cases, and $\iota < h$, then by choosing ι as our h, and h as our i, we have

$$\text{(c)} \qquad Y_{k+1}^{(j)} \cdot Y_\iota^{(h)} - Y_{k+1}^{(h)} \cdot Y_\iota^{(j)} - \langle R(X_\iota, X_{k+1})X_j, X_h \rangle = 0 \ .$$

Moreover, by (b) we also have

$$\text{(d)} \qquad Y_h^{(k+1)} \cdot Y_\iota^{(j)} - Y_h^{(j)} \cdot Y_\iota^{(k+1)} - \langle R(X_\iota, X_h)X_{k+1}, X_j \rangle = 0 \ .$$

In addition, (2') and (a) give

$$Y_h^{(k+1)} = Y_{k+1}^{(h)} \ , \qquad Y_\iota^{(h)} = Y_h^{(\iota)} \ , \qquad Y_\iota^{(k+1)} = Y_{k+1}^{(\iota)} \ .$$

So adding (c) and (d) gives

$$Y_{k+1}^{(j)} \cdot Y_h^{(\iota)} - Y_{k+1}^{(\iota)} \cdot Y_h^{(j)} - [\langle R(X_\iota, X_h)X_{k+1}, X_j \rangle + \langle R(X_\iota, X_{k+1})X_j, X_h \rangle] = 0 \ .$$

Using the identities for the curvature tensor, we obtain finally

$$Y_{k+1}^{(j)} \cdot Y_h^{(\iota)} - Y_{k+1}^{(\iota)} \cdot Y_h^{(j)} - \langle R(X_h, X_{k+1})X_j, X_\iota \rangle = 0 \ ,$$

which is indeed just the required identity for i, j.

So consider now the equations

$$\text{(3')} \quad Y_{k+1}^{(j)} \cdot Y_h^{(i)} - Y_{k+1}^{(i)} \cdot Y_h^{(j)} - \langle R(X_h, X_{k+1})X_j, X_i \rangle = 0 \qquad \begin{cases} h \le k & i \ge h \\ i < j & j \ge k+1 . \end{cases}$$

For $j = k+1$, there is one equation for each of the vectors

$$Y_1^{(1)}, \ldots, Y_1^{(k)}, Y_2^{(2)}, \ldots, Y_2^{(k)}, \ldots, Y_k^{(k)} ,$$

which are linearly independent, by (iii). Thus $Y_{k+1}^{(k+1)}$ is restricted to lie in some plane $H_{k+1} \subset \mathbb{R}^{n(n-1)/2}$ of codimension $1 + \cdots + k$. For $j = k+2$, there is then one equation for each of the linearly independent vectors

$$Y_1^{(1)}, \ldots, Y_1^{(k+1)}, Y_2^{(2)}, \ldots, Y_2^{(k+1)}, \ldots, Y_k^{(k)}, Y_k^{(k+1)} .$$

So $Y_{k+1}^{(k+2)}$ is restricted to lie in some plane H_{k+2} of codimension $2 + \cdots + (k+1)$. Etc. We see right away that

$$\dim \mathcal{E}_{k+1}((x,u), (X_1, Y_1), \ldots, (X_k, Y_k))$$

is the minimum possible, so $(X_1, Y_1), \ldots, (X_k, Y_k)$ do generate a regular k-dimensional integral element. Moreover, it can be extended to a $(k+1)$-dimensional integral element, by choosing an appropriate (X_{k+1}, Y_{k+1}) [with $X_1, \ldots, X_{k+1}$ orthonormal], since each H_α $(\alpha = k+1, \ldots, n)$ has dimension

$$\frac{n(n-1)}{2} - [(\alpha - k) + \cdots + (\alpha - 1)] = 1 + \cdots + (n-1) - [(\alpha - k) + \cdots + (\alpha - 1)] \ge 0.$$

We claim, finally, that if $k+1 \leq n-1$, then Y_{k+1} can be picked so that $Y_{k+1}^{(k+1)},\dots,Y_{k+1}^{(n-1)}$ are linearly independent of the vectors in (iii). We pick $Y_{k+1}^{(\alpha)}$ successively, for $\alpha = k+1,\dots,n-1$. The vector $Y_{k+1}^{(\alpha)}$ has to be picked linearly independent of the vectors in the set

$$A = \{Y_1^{(1)},\dots,Y_1^{(n-1)},\dots,Y_k^{(k)},\dots,Y_k^{(n-1)},Y_{k+1}^{(k+1)},\dots,Y_{k+1}^{(\alpha-1)}\} ,$$

with cardinality $(n-k) + \cdots + (n-1) + (\alpha-k-1)$

Equations (3') say that $Y_{k+1}^{(\alpha)}$ must lie in a plane H_α perpendicular to the vectors of the set

$$B = \{Y_1^{(1)},\dots,Y_1^{(\alpha-1)},Y_2^{(2)},\dots,Y_2^{(\alpha-1)},\dots,Y_k^{(k)},\dots,Y_k^{(\alpha-1)}\} ,$$

with cardinality r, say.

This is possible if $\dim H_\alpha$ is greater than the number of vectors in the difference set $A-B$. Since r is just the codimension of H_α, we thus need to have

$$\frac{n(n-1)}{2} - r > (n-k) + \cdots + (n-1) + (\alpha-k-1) - r .$$

But for $\alpha \leq n-1$ we have

$$\begin{aligned}(n-k)+\cdots+(n-1)+(\alpha-k-1) &\leq (n-k)+\cdots+(n-1)+(n-k-2)\\ &< (n-k)+\cdots+(n-1)+(n-k-1)\\ &\leq 1+\cdots+(n-1) = \frac{n(n-1)}{2} ,\end{aligned}$$

as required.

PROBLEMS

1. Let $a_i \neq 0$ for $1 < i < n-1$ with $\Sigma\, a_i{}^2 = 1$. Define an immersion

$$f\colon \{x \in \mathbb{R}^n : x_n < 0\} \longrightarrow \mathbb{R}^{2n-1}$$

by

$$\left.\begin{aligned} f^{2i-1}(x) &= a_i e^{x_n} \cos(x_i/a_i) \\ f^{2i}(x) &= a_i e^{x_n} \sin(x_i/a_i) \end{aligned}\right\} \qquad 1 \le i \le n-1$$

$$f^{2n-1}(x) = \int_0^{x_n} \sqrt{1 - e^{2t}}\, dt\ .$$

Calculate that the induced metric has constant negative curvature.

2. Let M^n be a manifold of constant curvature K isometrically immersed in a manifold N^{2n-1} of constant curvature $K_0 > K$.

(a) Using equation (1) on p. 204, generalize the argument in the second proof of Lemma 5-10 to prove that the bracket of two unit asymptotic vector fields on M is zero.

(b) Also use an argument from this proof to show that if M is complete, then its universal covering space must be $\mathbb{R}^n$.

(c) Conclude that we cannot have $K > 0$.

3. (a) Let $T: V \longrightarrow V$ be a self-adjoint linear transformation, and let $V = V_1 \oplus \ldots \oplus V_k$ where the V_i are the mutually orthogonal eigenspaces for the distinct eigenvalues $\lambda_1, \ldots, \lambda_k$. Let $P_i: V \longrightarrow V_i$ be the corresponding orthogonal projections, $P_i(\Sigma\, a_j v_j) = a_i v_i$. Show that P_i is a polynomial in T, namely,

$$\frac{(T - \lambda_1)\ldots(T - \lambda_{i-1})(T - \lambda_{i+1})\ldots(T - \lambda_k)}{(\lambda_i - \lambda_1)\ldots(\lambda_i - \lambda_{i-1})(\lambda_i - \lambda_{i+1})\ldots(\lambda - \lambda_k)} \;.$$

Thus we have $T = \Sigma\, \lambda_i P_i$ where the P_i are polynomials in T.

(b) Let $S: V \longrightarrow V$ be another self-adjoint linear transformation with $S = \Sigma\, \mu_j Q_j$ for polynomials Q_j in S. If S and T commute, then all P_i and Q_j commute. Let $A = \Sigma\, a_{ij} P_i Q_j$ for distinct $a_{ij} \in \mathbb{R}$. Show that A is a self-adjoint transformation, that S and T are both polynomials in A, and that any linear transformation that commutes with S and T also commutes with A.

(c) If $T_1, \ldots, T_r$ are commuting self-adjoint operators, then there is a self-adjoint transformation A such that each T_i is a polynomial in A.

(d) Consequently, $T_1, \ldots, T_r$ can be simultaneously diagonalized.

4. (a) Let $f\colon M \to \mathbb{R}$, where $(M, \langle\ ,\ \rangle)$ is a Riemannian manifold, and consider the symmetric covariant tensor $\nabla(df)$ of order 2, with components $f_{i;j}$ in a coordinate system. Each $\nabla(f)(p)\colon M_p \times M_p \to \mathbb{R}$ can be regarded as a linear transformation $M_p \to M_p$, by using the inner product on M_p. So we can form $\mathscr{D}f(p) = \det(\nabla(df)(p))$. Equivalently, $\mathscr{D}f(p) = \det(\nabla(df)(p)(X_i, X_j))$, where $X_1, \ldots, X_n$ is any orthonormal basis of M_p. In a coordinate system we have

$$\mathscr{D}f = \frac{\det(f_{i;j})}{\det g_{ij}} .$$

Show that equation (*) on p. 208 can be written

$$\mathscr{D}w = K(1 - \Delta_1 w).$$

(b) Check that equations (*) and (**) are the same when $F = 0$, and conclude that they are always the same.

Chapter 12. Rigidity

In Chapter 7 we proved a result (Theorem 7-47) which is but a special case of the following more general

1. THEOREM. Let M and $\bar{M}$ be immersed hypersurfaces in $\mathbb{R}^{n+1}$, and let $\phi\colon M \to \bar{M}$ be an isometry. Suppose that $d\nu\colon M_p \to M_p$ has rank ≥ 3. Then $(\phi^*\overline{II})(p) = \pm\, II(p)$. [This equation makes sense even though ν and $\bar{\nu}$ may be defined only locally, and then only up to sign.]

Consequently, if M and $\bar{M}$ are connected (not necessarily complete) hypersurfaces and $d\nu\colon M_p \to M_p$ has rank ≥ 3 for all $p \in M$, then ϕ is the restriction of a Euclidean motion.

Proof. To deduce the second part of the theorem from the first part, we recall (p.IV.92) that there is an inner product preserving bundle isomorphism $\tilde{\phi}\colon \text{Nor } M \to \text{Nor } \bar{M}$ covering ϕ. The first part of the theorem shows that if $p \in M$, then

$$\text{either} \qquad \bar{s}(\phi_*X,\phi_*Y) = \tilde{\phi}(s(X,Y)) \qquad \text{for all } X, Y \in M_p$$

$$\text{or} \qquad \bar{s}(\phi_*X,\phi_*Y) = -\,\tilde{\phi}(s(X,Y)) \qquad \text{for all } X, Y \in M_p .$$

Moreover, only one alternative can hold at each p, since $d\nu\colon M_p \to M_p$ is non-singular. It follows that one of the alternatives holds for all $p \in M$. Then Theorem 7-21 shows that f is the restriction of a Euclidean motion.

To prove the first part, let $X_1,\dots,X_n \in M_p$ be an orthonormal basis, and define the $n \times n$ symmetric matrix S by $S_{ij} = II(X_i,X_j)$. Similarly, let $\bar{X}_i = \phi_*X_i \in \bar{M}_{f(p)}$ and define the symmetric $n \times n$ matrix $\bar{S}$ by $\bar{S}_{ij} = \overline{II}(\bar{X}_i,\bar{X}_j)$. Gauss' equation shows that

$$\begin{aligned} S_{i_1j_1}S_{i_2j_2} - S_{i_1j_2}S_{i_2j_1} &= \langle R(X_{i_1},X_{i_2})X_{j_2},X_{j_1}\rangle \\ &= \langle \bar{R}(\bar{X}_{i_1},\bar{X}_{i_2})\bar{X}_{j_2},\bar{X}_{j_1}\rangle , \qquad \text{since } \phi \text{ is an isometry} \\ &= \bar{S}_{i_1j_1}\bar{S}_{i_2j_2} - \bar{S}_{i_1j_2}\bar{S}_{i_2j_1} . \end{aligned}$$

We now use an algebraic

2. LEMMA. Let S and $\bar{S}$ be symmetric $n \times n$ matrices, with rank $S \geq 3$. Suppose that the determinant of every 2×2 submatrix of S equals the determinant of the corresponding 2×2 submatrix of $\bar{S}$. Then $\bar{S} = \pm S$.

Proof. To isolate the main idea of the proof, we first consider

Case 1. The matrices S and $\bar{S}$ are non-singular. (Then the hypothesis on the rank just means that $n \geq 3$.) Let $T, \bar{T}\colon \mathbb{R}^n \to \mathbb{R}^n$ be the non-singular linear transformations with matrices $S, \bar{S}$. Then $T^*, \bar{T}^*\colon \mathbb{R}^{n*} \to \mathbb{R}^{n*}$ are also non-singular. We can also consider the linear transformations

$$T^*, \bar{T}^*\colon \Omega^2(\mathbb{R}^n) \to \Omega^2(\mathbb{R}^n) .$$

If $\phi_1,\ldots,\phi_n \in \mathbb{R}^{n*}$ is the dual basis to the standard basis of $\mathbb{R}^n$, then

$$\begin{aligned} T^*(\phi_{i_1} \wedge \phi_{i_2}) &= T^*\phi_{i_1} \wedge T^*\phi_{i_2} \\ &= \sum_{j_1} S_{i_1j_1}\phi_{j_1} \wedge \sum_{j_2} S_{i_2j_2}\phi_{j_2} \\ &= \sum_{j_1<j_2} (S_{i_1j_1}S_{i_2j_2} - S_{i_1j_2}S_{i_2j_1}) \cdot \phi_{j_1} \wedge \phi_{j_2} , \end{aligned}$$

and similarly for $\bar{T}^*$. Thus we see that the hypotheses on the determinants of S and $\bar{S}$ is equivalent to the assertion that

$$(1) \qquad T^* = \bar{T}^* \colon \Omega^2(\mathbb{R}^n) \longrightarrow \Omega^2(\mathbb{R}^n) .$$

Now given any $\phi \in \mathbb{R}^{n*}$, we claim that $T^*\phi$ and $\bar{T}^*\phi$ must be linearly dependent. For otherwise we could choose $\psi \in \mathbb{R}^{n*}$ with $T^*\phi$, $\bar{T}^*\phi$, $T^*\psi$ linearly independent. Then we would have

$$\begin{aligned} 0 \neq T^*\phi \wedge T^*\psi \wedge \bar{T}^*\phi &= T^*(\phi \wedge \psi) \wedge \bar{T}^*\phi \\ &= \bar{T}^*(\phi \wedge \psi) \wedge \bar{T}^*\phi \qquad \text{by (1)} \\ &= \bar{T}^*\phi \wedge \bar{T}^*\psi \wedge \bar{T}^*\phi \\ &= 0 , \end{aligned}$$

a contradiction. Thus $\bar{T}^*\phi = c \cdot T^*\phi$ for some $c \in \mathbb{R}$. If we choose linearly independent $\phi_1, \phi_2 \in \mathbb{R}^{n*}$, and apply this result to $\phi_1, \phi_2, \phi_1 + \phi_2$, we find that there are constants c_1, c_2, c with

$$\bar{T}^*\phi_1 = c_1 \cdot T^*\phi_1 , \qquad \bar{T}^*\phi_2 = c_2 \cdot T^*\phi_2$$
$$\bar{T}^*\phi_1 + \bar{T}^*\phi_2 = c \cdot (T^*\phi_1 + T^*\phi_2) .$$

It follows that $c_1 = c_2$. So $\bar{T}^* = c \cdot T^*$ for some $c \in \mathbb{R}$. From (1) we see that $c = \pm 1$.

<u>Case 2. General Case</u>. Since S is symmetric, the map $T \colon \mathbb{R}^n \to \mathbb{R}^n$ is self-adjoint with respect to the usual metric on $\mathbb{R}^n$; that is, $\langle Tv, w \rangle = \langle v, Tw \rangle$ for $v, w \in \mathbb{R}^n$. Similarly, if we give $\mathbb{R}^{n*}$ the inner product with $\langle \phi_i, \phi_j \rangle = \delta_{ij}$, then

(2) $$\langle T^*\phi,\psi\rangle = \langle\phi,T^*\psi\rangle \qquad \text{for} \quad \phi,\ \psi \in \mathbb{R}^{n*} .$$

Let $(\ker T^*)^\perp \subset \mathbb{R}^{n*}$ be the orthogonal complement of $\ker T^* \subset \mathbb{R}^{n*}$ with respect to this inner product. We easily see from (2) that T^* takes $(\ker T^*)^\perp$ into itself and that

(3) $$T^*\colon (\ker T^*)^\perp \longrightarrow (\ker T^*)^\perp \quad \text{is one-one} .$$

We now claim:

(4) $$\ker T^* = \{\phi\colon T^*\phi \wedge T^*\psi = 0 \text{ for all } \psi \in \mathbb{R}^{n*}\} = W , \qquad \text{say} .$$

It is clear that $\ker T^* \subset W$. Conversely, given $\phi \in W$, write

$$\phi = \phi_1 + \phi_2 \qquad \text{with} \qquad \phi_1 \in \ker T^* \subset W , \quad \phi_2 \in (\ker T^*)^\perp .$$

Then

(5) $$\phi_2 \in (\ker T^*)^\perp \cap W .$$

We want to show that $\phi_2 = 0$. Note that

$$\dim (\ker T^*)^\perp = \operatorname{rank} T^* = \operatorname{rank} S \geq 3 ,$$

so if $\phi_2 \neq 0$, then there is $\psi \in (\ker T^*)^\perp$ with ϕ_2 and ψ linearly independent. By (3), this means that $T^*\phi_2$ and $T^*\psi$ are linearly independent, so

$$0 \neq T^*\phi_2 \wedge T^*\psi ,$$

contradicting the fact that $\phi_2 \in W$ [by (5)]. Thus ϕ_2 must be 0, and we have demonstrated (4).

Notice that in proving (4) we really used only the fact that rank $S \geq 2$. Now we also have rank $\bar{S} \geq 2$ (otherwise every 2×2 submatrix of $\bar{S}$ would have determinant 0). So we also have

$$(\bar{4}) \qquad \ker \bar{T}^* = \{\phi\colon \bar{T}^*\phi \wedge \bar{T}^*\psi = 0 \text{ for all } \psi \in \mathbb{R}^{n*}\} .$$

Then equation (1) shows that $\ker T^* = \ker \bar{T}^*$. Now we can apply Case 1 to

$$T^*, \bar{T}^*\colon (\ker T^*)^{\perp} \to (\ker T^*)^{\perp} ,$$

for these maps are one-one by (3) and the corresponding $(\bar{3})$, and moreover $\dim(\ker T^*)^{\perp} = \text{rank } T^* = \text{rank } S \geq 3$. ∎

The rank of $d\nu\colon M_p \to M_p$ is called the type number $t(p)$ of M at p; it is the number of non-zero principal curvatures at p. The hypothesis that $t(p) \geq 3$ says, roughly speaking, that M curves in at least 3 different directions at p. The hypersurfaces with $t(p) = 0$ or 1 for all p are precisely the flat hypersurfaces, with curvature tensor $R = 0$, while the hypersurfaces with $t(p) = 2$ for all p may be regarded as a sort of generalization of this class. They have been classified into three different types by E. Cartan [1], but the classification suffers the same defect as the classical classification of flat surfaces, for there is no discussion of the manner in which hypersurfaces of different types can be joined together. We will have a little more to say about this later on.

Theorem 1 is often expressed by saying that a hypersurface in $\mathbb{R}^{n+1}$ which bends enough is "rigid." The first precise proof was by Killing {1}, although the result had been stated by Beez [1], who found it so astounding that he could barely cease discussing it, and practically regarded it as a proof that space can't be 4-dimensional. While we might not be willing to go quite so far as that, it is nevertheless true that because of this result most of the interest in rigidity phenomena has centered on the case of surfaces in $\mathbb{R}^3$, where intuition tells us that a small piece of surface is not rigid, and at the same time suggests that compact surfaces should be rigid. Actually, there are several different senses in which a surface can be rigid. Books written in English often consider only one possible sense, or tend to be rather sloppy about distinguishing the various possibilities. I therefore propose to introduce some terminology which, although it may not be especially appealing aesthetically, and suffers the disadvantage of not being standard, at least has the virtue of being unambiguous.

Consider a C^∞ imbedding $f: M \to \mathbb{R}^3$. The strongest sense in which $f(M)$ can be "bent" corresponds to the ordinary conception of the word, whereby $f(M)$ passes continuously from one shape to another, without being stretched. To express this idea precisely, we define a <u>bending</u> of the imbedding $f: M \to \mathbb{R}^3$ to be a C^∞ map $\alpha: [0,1] \times M \to \mathbb{R}^3$ such that

(a) each map $\bar{\alpha}(t): M \to \mathbb{R}^3$, given by $p \mapsto \alpha(t,p)$, is an imbedding,

(b) $\bar{\alpha}(0) = f$,

(c) $\bar{\alpha}(t)^*\langle\ ,\ \rangle = \bar{\alpha}(0)^*\langle\ ,\ \rangle$ for all $t \in [0,1]$.

Thus α is a "variation" of f, in the terminology of Chapter 9. To be a little more precise, α should be called a <u>bending through imbeddings</u>, and we can also define a bending through immersions. The bending $\alpha: [0,1] \times M \to \mathbb{R}^3$ is called <u>trivial</u> if each $\bar{\alpha}(t)$ is $A_t \circ f$ for some Euclidean motion A_t; it is called non-trivial if at least one $\bar{\alpha}(t)$ is not of this form. We say that the imbedding $f\colon M \to \mathbb{R}^3$ is bendable if there is a non-trivial bending of f; otherwise it is called unbendable. To be precise, we must speak of bendability and unbendability through imbeddings or through immersions. We can also define when an immersion $f\colon M \to \mathbb{R}^3$ is bendable or unbendable; in this case, of course, only bendings through immersions can be relevant. (It is also possible to consider C^ℓ bendings of C^k imbeddings and immersions, for $1 \le \ell \le k \le \omega$; but we shall hardly ever stray from the case $k = \ell = \infty$.) Finally, a submanifold $M \subset \mathbb{R}^3$ is called bendable or unbendable (through imbeddings or immersions) according as whether the inclusion map $i\colon M \to \mathbb{R}^3$ is bendable or unbendable.

One way of modifying the concept of a bending is by taking a discrete analogue: We will call an imbedding $f\colon M \to \mathbb{R}^3$ warpable if there is an imbedding $g\colon M \to \mathbb{R}^3$ such that $f^*\langle\ ,\ \rangle = g^*\langle\ ,\ \rangle$, but such that g is not $A \circ f$ for any Euclidean motion A. If f is not warpable, it will be called un-warpable. We can also define a warpable immersion. It is conceivable that there is an imbedding $f\colon M \to \mathbb{R}^3$ such that

(i) there exists an immersion $g\colon M \to \mathbb{R}^3$ with $f^*\langle\ ,\ \rangle = g^*\langle\ ,\ \rangle$

(ii) there does not exist an imbedding $g\colon M \to \mathbb{R}^3$ with $f^*\langle\ ,\ \rangle = g^*\langle\ ,\ \rangle$ except for g of the form $A \circ f$ for some Euclidean motion A;

we can express this by saying that f is warpable as an immersion, but not as an imbedding (an actual example of this phenomenon will be mentioned later). A surface $M \subset \mathbb{R}^3$ is called warpable or unwarpable (as an imbedding or immersion) according as whether the inclusion map $i\colon M \to \mathbb{R}^3$ is warpable or unwarpable. An unwarpable surface $M \subset \mathbb{R}^3$ is sometimes called "uniquely determined" (for M is then uniquely determined, up to a Euclidean motion, by its induced metric); similarly, we can speak of an imbedding or immersion $f\colon M \to \mathbb{R}^3$ being "uniquely determined." A bendable surface is obviously warpable, but it is not *a priori* clear whether there are any warpable surfaces which are not bendable.*

We can also consider an infinitesimal analogue of a bending α, by looking at its "variation vector field" Z. This is the vector field along f defined by

$$Z(p) = \text{tangent vector at } 0 \text{ of } t \mapsto \alpha(t,p) \in \mathbb{R}^3{}_{f(p)} .$$

Consider for the moment the case where $M \subset \mathbb{R}^3$ and f = inclusion map. Since α satisfies

$$\langle X,Y\rangle = \langle \bar{\alpha}(t)_*(X),\ \bar{\alpha}(t)_*(Y)\rangle$$

for all t, and $X, Y \in M_p$, we have

$$(1) \qquad \begin{aligned} 0 &= \frac{d}{dt} \langle \bar{\alpha}(t)_*(X),\ \bar{\alpha}(t)_*(Y)\rangle \\ &= \left\langle \frac{D'}{\partial t}\, \bar{\alpha}(t)_*(X),\ \bar{\alpha}(t)_*(Y)\right\rangle + \left\langle \bar{\alpha}(t)_*(X),\ \frac{D'}{\partial t}\, \bar{\alpha}(t)_*(Y)\right\rangle , \end{aligned}$$

where $D'/\partial t$ denotes covariant differentiation in the ambient space $\mathbb{R}^3$, as usual. Now if c is a curve in M with $c'(0) = X$, then

---

*As if matters were not already sufficiently complicated, one more possibility must be mentioned, which for the sake of simplicity we shall describe in terms of submanifolds, rather than imbeddings. Let $M \subset \mathbb{R}^3$ be a surface. It is conceivable that M is warpable, so that there is an isometry $\phi\colon M \to \bar{M} \subset \mathbb{R}^3$ which is not the restriction of a Euclidean motion, but that whenever we have an isometry $\phi\colon M \to \bar{M}$ then there is also another isometry $\psi\colon M \to \bar{M}$ which *is* the restriction of a Euclidean motion. Thus M might be warpable, but only into surfaces which happen to be congruent to M. No example of such a phenomenon is known, however.

$$\left.\frac{D'}{\partial t}\right|_{t=0} \bar{\alpha}(t)_*(X) = \left.\frac{D'}{\partial t}\right|_{t=0} \left.\frac{\partial}{\partial s}\right|_{s=0} \alpha(t,c(s))$$

$$= \left.\frac{D'}{\partial s}\right|_{s=0} \left.\frac{\partial}{\partial t}\right|_{t=0} \alpha(t,c(s)) \qquad \text{by Proposition II.6-9 (or simply equality of mixed partials)}$$

$$= \left.\frac{D'}{\partial s}\right|_{s=0} Z(c(s))$$

$$= \nabla'_X Z \ .$$

So equation (1) becomes

$$(2) \qquad 0 = \langle \nabla'_X Z, Y \rangle + \langle X, \nabla'_Y Z \rangle \qquad \text{for all } X,\ Y \text{ tangent to } M\ .$$

This is equivalent, by polarization, to

$$(2') \qquad 0 = \langle \nabla'_X Z, X \rangle \qquad \text{for all } X \text{ tangent to } M\ ,$$

and these equations can also be written

$$(2'') \qquad \begin{cases} 0 = \langle dZ(X), X \rangle & \text{for all } X \text{ tangent to } M \\ 0 = \langle dZ(X), Y \rangle + \langle X, dZ(Y) \rangle & \text{for all } X,\ Y \text{ tangent to } M\ , \end{cases}$$

where Z is considered as an $\mathbb{R}^3$-valued function on M, and the tangent vector X of $\mathbb{R}^3$ is identified with an element of $\mathbb{R}^3$.

For the general case of an immersion $f\colon M \to \mathbb{R}^3$, the vector field Z along f still satisfies (2), except that now the term

$$\text{"}\nabla'_X Z\text{"} = \left.\frac{D'}{\partial s}\right|_{s=0} Z(c(s)) = \left(\left.\frac{dZ^1(c(s))}{ds}\right|_{s=0}, \ldots, \left.\frac{dZ^3(c(s))}{ds}\right|_{s=0} \right)$$

denotes a "covariant derivative of a vector field along f." Equation (2") becomes

$$(2''') \qquad 0 = \langle dZ(X), f_*(X) \rangle$$
$$= \langle dZ(X), df(X) \rangle \qquad \text{for all } X \text{ tangent to } M ,$$

where Z is considered as an $\mathbb{R}^3$-valued function on M, and $f_*(X)$ is identified with an element of $\mathbb{R}^3$, or f is considered as an $\mathbb{R}^3$-valued function on M. This equation is sometimes written simply

$$\langle dZ, df \rangle = 0 \qquad \text{or} \qquad dZ \cdot df = 0 .$$

[Note: In some books, X (or German $\mathfrak{X}$) is used to denote the immersion $X: M \longrightarrow \mathbb{R}^3$, and this equation appears as $dZ \cdot dX = 0$.]

A vector field Z along an immersion $f: M \longrightarrow \mathbb{R}^3$ will be called an infinitesimal bending of f if it satisfies equation (2'''). Clearly this equation will be satisfied by the variation vector field of a variation α which merely "preserves lengths up to first order," equation (1) being, in fact, the analytic expression of this condition. Of course, we can always find infinitesimal bendings by taking the variation vector field Z of a bending α by means of Euclidean motions,

$$\alpha(t,p) = f(p) \cdot B(t) + v(t) ,$$

where $B(t) \in O(3)$ with $B(0) = I$, and $v(t) \in \mathbb{R}^3$ with $v(0) = 0$ [and $f(p) \cdot B(t)$ denotes the product of the 1×3 matrix $f(p)$ with the 3×3 matrix $B(t)$]. In this case we have

$$Z(p) = f(p) \cdot B'(0) + v'(0) ,$$

where $B'(t) \in \mathfrak{o}(3) = \{3 \times 3 \text{ skew-symmetric matrices}\}$. Conversely, if Z is an infinitesimal bending of f of the form

$$(3) \qquad Z(p) = f(p)\cdot C + w\,, \qquad C \in \mathfrak{o}(3)\,,$$

then Z is the variation vector field of the bending α through Euclidean motions defined by

$$\alpha(t,p) = f(p)\cdot e^{tC} + tw\,.$$

So we will call an infinitesimal bending Z trivial if it is of the form (3). An immersion $f\colon M \to \mathbb{R}^3$ will be called infinitesimally bendable if there is a non-trivial infinitesimal bending of f; otherwise it will be called infinitesimally rigid. (The word "rigid" is sometimes used to mean infinitesimally rigid, but unfortunately it is also sometimes sloppily used to mean un-warpable, or unbendable.)

Notice that the product of a vector by a skew-symmetric matrix,

$$(x,y,z)\cdot\begin{pmatrix}0 & -a & -b\\ a & 0 & -c\\ b & c & 0\end{pmatrix} = (ay+bz,\ -ax+cz,\ -bx-cy)\,,$$

can also be written as a cross-product

$$(x,y,z)\times(c,-b,a) = (ay+bz,\ -ax+cz,\ -bx-cy)\,.$$

So equation (3) can also be written

$$(3') \qquad Z(p) = f(p)\times Y + w\,, \qquad Y, w \in \mathbb{R}^3\,.$$

As an easy consequence of the triviality condition (3') we have

$$dZ(X) = df(X) \times Y \qquad \text{all } X \text{ tangent to } M .$$

Now the same formula holds for an arbitrary infinitesimal bending Z, provided that we allow Y to vary:

3. LEMMA. If Z is an infinitesimal bending of $f\colon M \to \mathbb{R}^3$, then for each $p \in M$, there is a unique $Y(p) \in \mathbb{R}^3$ such that

$$dZ(X) = df(X) \times Y(p) \qquad \text{for all } X \in M_p .$$

Proof. Let $X_1, X_2 \in M_p$ be linearly independent. Since $\langle dZ(X_i), df(X_i)\rangle = 0$, there are certainly some vectors $Y_i \in \mathbb{R}^3$ with

$$dZ(X_i) = df(X_i) \times Y_i .$$

Moreover,

$$\begin{aligned} 0 &= \langle dZ(X_2), df(X_1)\rangle + \langle dZ(X_1), df(X_2)\rangle \\ &= \langle df(X_2) \times Y_2, df(X_1)\rangle + \langle df(X_1) \times Y_1, df(X_2)\rangle \\ &= \langle Y_2, df(X_1) \times df(X_2)\rangle - \langle Y_1, df(X_1) \times df(X_2)\rangle \\ &= \langle Y_2 - Y_1, df(X_1) \times df(X_2)\rangle . \end{aligned}$$

Thus $Y_2 - Y_1 \in df(M_p)$, so we can write

$$Y_2 - Y_1 = a\, df(X_1) + b\, df(X_2) .$$

If we set

$$Y(p) = Y_2 - a\, df(X_1) = Y_1 + b\, df(X_2)$$

then we have

$$dZ(X_i) = df(X_i) \times Y_i = df(X_i) \times Y(p) \ .$$

Uniqueness is obvious. ■

The vector field $p \longmapsto Y(p)$ of Lemma 3 is called the (infinitesimal) rotation field of the infinitesimal bending Z. We know that Y is constant when Z is trivial, and conversely,

4. LEMMA. If the rotation field Y of the infinitesimal bending Z is constant, then Z is trivial.

Proof. By assumption, there is a vector $Y_0 \in \mathbb{R}^3$ with

$$dZ(X) = df(X) \times Y_0$$

for all X tangent to M. Let c be a curve in M, with $c(0) = p_0 \in M$. Then

$$\frac{dZ(c(t))}{dt} = dZ(c'(t)) = df(c'(t)) \times Y_0 = \frac{df(c(t))}{dt} \times Y_0 \ .$$

Therefore

$$Z(c(t)) - Z(p_0) = [f(c(t)) - f(p_0)] \times Y_0 \ ,$$

or

$$Z(c(t)) = f(c(t)) \times Y_0 + w_0 ,$$

where $w_0 \in \mathbb{R}^3$ does not depend on c. So for all $p \in M$ we have

$$Z(p) = f(p) \times Y_0 + w_0 ,$$

and Z is trivial. ■

At first sight it might seem that every bendable surface must also be infinitesimally bendable. As a matter of fact, we certainly do have

5. LEMMA. Let $\alpha\colon [0,1] \times M$ be a bending, and let Z_t be the variation vector field of α at time t. If each Z_t is trivial, then the bending α is trivial.

Proof. By definition,

$$(1) \qquad Z_t(p) = \frac{d}{dt}\,\alpha(t,p) ,$$

and since each Z_t is trivial we have

$$(2) \qquad Z_t(p) = \alpha(t,p) \times Y_t + w_t \qquad Y_t,\ w_t \in \mathbb{R}^3 .$$

Then for all $p_1, p_2 \in M$ we have

$$\begin{aligned}
\frac{d}{dt}|\alpha(t,p_1) - \alpha(t,p_2)|^2 &= 2\langle \alpha(t,p_1) - \alpha(t,p_2),\ Z_t(p_1) - Z_t(p_2)\rangle \qquad \text{by (1)} \\
&= 2\langle \alpha(t,p_1) - \alpha(t,p_2),\ [\alpha(t,p_1) - \alpha(t,p_2)] \times Y_t\rangle \\
&= 0 .
\end{aligned}$$

So $|\alpha(t,p_1) - \alpha(t,p_2)|$ is constant in t. In particular, $|\alpha(t,p_1) - \alpha(t,p_2)|$ $= |\alpha(0,p_1) - \alpha(t,p_2)|$. This implies that each α_t differs by a Euclidean motion from α_0. ■

Nevertheless, it is conceivable that $f\colon M \to \mathbb{R}^3$ is bendable, yet that every bending α of f has trivial variation vector field Z_0 at time $t = 0$, so that f is not infinitesimally bendable. No example of such a weird phenomenon is known, but there are also no positive results along this line, except for the obvious fact that if f is an analytic immersion which is analytically infinitesimally rigid, then f is analytically unbendable.

There are a couple of other surprising facts about infinitesimal bendings. First of all, there are non-trivial infinitesimal bendings Z of a plane which vanish outside a compact set. If our plane is the (x,y)-plane, we can choose Z to be $h \cdot \frac{\partial}{\partial z}$, where h is any C^∞ function vanishing outside a compact set. For any tangent vector X of the (x,y)-plane we have

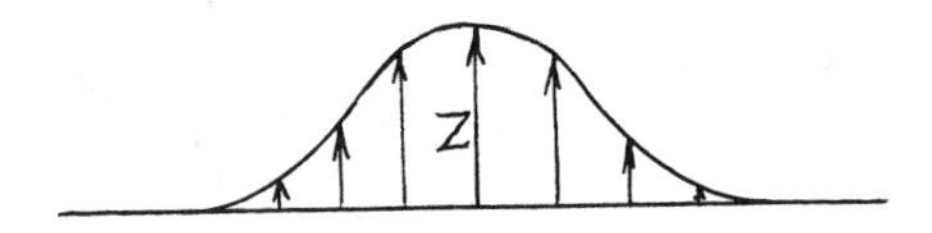

$$dZ(X) = (X(0),X(0),X(h)) = X(h) \cdot \frac{\partial}{\partial z},$$

so the infinitesimal rotation vector field of Z is

$$Y = X(h) \cdot \frac{\partial}{\partial z}.$$

Since Y is not constant, Z is non-trivial. Even more surprising, perhaps, is an immediate consequence of this fact: any surface containing a portion of a plane is infinitesimally bendable.

Notice that the infinitesimal bending Z constructed above is everywhere perpendicular to the surface $M = (x,y)$-plane. This is essentially the only possibility:

6. LEMMA. (1) If Z is an infinitesimal bending of an open subset M of the (x,y)-plane, and Z is always tangential to the (x,y)-plane, then Z is trivial,

$$Z(p) = p \cdot C + v$$

for a 2×2 skew-symmetric matrix C, and $v \in (x,y)$-plane.

(2) More generally, Z is an infinitesimal bending of $M \subset (x,y)$-plane if and only if the tangential component $\mathbf{T}Z$ of Z is an infinitesimal bending, and hence trivial. In particular, any vector field Z normal to M is an infinitesimal bending.

(3) Let $M \subset \mathbb{R}^3$ be a surface and let Z be any infinitesimal bending of M which is everywhere normal to M. Then at every point $p \in M$ where $Z(p) \neq 0$, the second fundamental form $II(p) = 0$. (So if $Z(p) \neq 0$ for all p in an open set $U \subset M$, then U lies in a plane.)

Proof. (1) Let

$$Z(x,y) = (a(x,y), b(x,y)) = a(x,y)\frac{\partial}{\partial x} + b(x,y)\frac{\partial}{\partial y} \quad .$$

Then

$$dZ(\frac{\partial}{\partial x}) = \frac{\partial a}{\partial x}\cdot\frac{\partial}{\partial x} + \frac{\partial b}{\partial x}\cdot\frac{\partial}{\partial y} \qquad dZ(\frac{\partial}{\partial y}) = \frac{\partial a}{\partial y}\cdot\frac{\partial}{\partial x} + \frac{\partial b}{\partial y}\cdot\frac{\partial}{\partial y} .$$

Thus

$$0 = \left\langle\frac{\partial}{\partial x}, dZ(\frac{\partial}{\partial x})\right\rangle \implies \frac{\partial a}{\partial x} = 0$$

$$0 = \left\langle\frac{\partial}{\partial y}, dZ(\frac{\partial}{\partial y})\right\rangle \implies \frac{\partial b}{\partial y} = 0 ,$$

so we can write

$$a(x,y) = \bar{a}(y) , \qquad b(x,y) = \bar{b}(x) .$$

Moreover,

$$0 = \left\langle\frac{\partial}{\partial x}, dZ(\frac{\partial}{\partial y})\right\rangle + \left\langle\frac{\partial}{\partial y}, dZ(\frac{\partial}{\partial x})\right\rangle \implies \bar{a}'(y) + \bar{b}'(x) = 0 .$$

Since this is true for all x, y, the derivatives $\bar{a}'(y)$ and $\bar{b}'(x)$ must be constants. So we must have

$$a(x,y) = \bar{a}(y) = \alpha y + \beta$$

$$b(x,y) = \bar{b}(x) = -\alpha x + \delta .$$

Then

$$Z(x,y) = (x,y)\cdot\begin{pmatrix}0 & -\alpha\\ \alpha & 0\end{pmatrix} + (\beta,\delta) .$$

This completes the proof of (1).

Now for any $M \subset \mathbb{R}^3$, with unit normal ν, consider a vector field

$$Z = \mathbf{T}Z + \phi \cdot \nu \ .$$

Then for $X \in M_p$ we have

$$\nabla'_X Z = \nabla_X \mathbf{T}Z + II(X, \mathbf{T}Z) \cdot \nu + X(\phi) \cdot \nu - \phi(p) d\nu(X) \ .$$

So $0 = \langle \nabla'_X Z, X \rangle$ if and only if

$$(*) \qquad \langle \nabla_X \mathbf{T}Z, X \rangle = \phi(p) \cdot II(X,X) \ .$$

If $M \subset (x,y)$-plane, then $II(X,X) = 0$ and $II(X, \mathbf{T}Z) = 0$, so (*) says that Z is an infinitesimal bending if and only if $0 = \langle \nabla_X \mathbf{T}Z, X \rangle = \langle \nabla'_X \mathbf{T}Z, X \rangle$ for all X, so that $\mathbf{T}Z$ is an infinitesimal bending. This proves (2).

On the other hand, if Z is an infinitesimal bending with $\mathbf{T}Z = 0$, and $\phi(p) \neq 0$, then (*) shows that $II(X,X) = 0$ for all $X \in M_p$. ∎

There also turns out to be a relationship between warpability and infinitesimal bendability, which at first sight seem to have nothing to do with each other.

7. LEMMA. Let Z be an infinitesimal bending of an immersion $f\colon M \to \mathbb{R}^3$. Define $\alpha_t\colon M \to \mathbb{R}^3$ by

$$\alpha_t(p) = f(p) + t \cdot Z(p) \ ,$$

where $Z(p)$ is considered as an element of $\mathbb{R}^3$ as usual. Then in a neighborhood of any point $p \in M$, the map α_t is an immersion for sufficiently small t, and the induced metric $\alpha_t{}^*\langle\ ,\ \rangle$ on M is related to the metric $f^*\langle\ ,\ \rangle$ by

$$[\alpha_t{}^*\langle\ ,\ \rangle](X,Y) = [f^*\langle\ ,\ \rangle](X,Y) + t^2\langle dZ(X), dZ(Y)\rangle .$$

In particular, the metrics $\alpha_t{}^*\langle\ ,\ \rangle$ and $\alpha_{-t}{}^*\langle\ ,\ \rangle$ on M are the same.

Proof. If X is a tangent vector on M, with $X = c'(0)$ for some curve c in M, then

$$\begin{aligned}\alpha_{t*}X &= \left.\frac{d}{ds}\right|_{s=0} \alpha_t(c(s)) \\ &= \left.\frac{d}{ds}\right|_{s=0} c(s) + tZ(c(s)) \\ &= c'(0) + tdZ(c'(0)) \\ &= X + tdZ(X) .\end{aligned}$$

This immediately leads to the desired formula, and this formula shows that α_t is an immersion for small t, in any neighborhood of p on which Z is bounded. ∎

An illustration of this phenomenon is provided by the infinitesimal bending Z of the (x,y)-plane given previously. The map taking

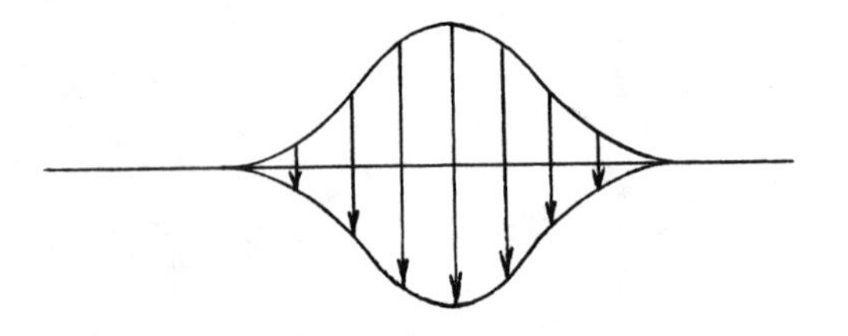

$p + tZ(p) \mapsto p - tZ(p)$ is the restriction of a Euclidean motion, namely reflection through the (x,y)-plane. But this is the _only_ case in which this happens:

8. LEMMA. Let Z be a non-trivial infinitesimal bending of a surface $M \subset \mathbb{R}^3$ which is not part of a plane. Then for $t \neq 0$, the map

$$p + tZ(p) \mapsto p - tZ(p)$$

[which is an isometry by Lemma 7] is not the restriction of a Euclidean motion.

Proof. If this map is the restriction of a length preserving map of $\mathbb{R}^3$, then for all $p, q \in M$ we have

$$|p + tZ(p) - \{q + tZ(q)\}| = |p - tZ(p) - \{q - tZ(q)\}| ,$$

that is,

$$|p - q + t\{Z(p) - Z(q)\}| = |p - q - t\{Z(p) - Z(q)\}| .$$

This implies that

$$(1) \qquad \langle p - q, \ Z(p) - Z(q)\rangle = 0 .$$

Without loss of generality, we may assume that M contains the point 0, and that $Z(0) = 0$. Then equation (1) gives

$$\langle p, Z(p)\rangle = 0 \qquad p \in M ,$$

from which we further deduce that

$$(2) \qquad \langle p, Z(q)\rangle + \langle q, Z(p)\rangle = 0 , \qquad p, q \in M .$$

Since M is not contained in a plane, there are three linearly independent points $r_1, r_2, r_3 \in M$. Now if λ_i are numbers with $\Sigma\, \lambda_i r_i \in M$, then

$$\begin{aligned}
\langle Z(\Sigma\, \lambda_i r_i),\, r_j\rangle &= -\langle \Sigma\, \lambda_i r_i,\, Z(r_j)\rangle && \text{by (2)}\\
&= -\Sigma\, \lambda_i \langle r_i,\, Z(r_j)\rangle \\
&= \Sigma\, \lambda_i \langle Z(r_i),\, r_j\rangle && \text{by (2)}\\
&= \langle \Sigma\, \lambda_i Z(r_i),\, r_j\rangle .
\end{aligned}$$

This implies that

$$Z(\Sigma\, \lambda_i r_i) = \Sigma\, \lambda_i Z(r_i) .$$

So Z is the restriction to M of a linear transformation T. Since a linear transformation is its own derivative, equation (2") shows that

$$0 = \langle TX,Y\rangle + \langle X,TY\rangle$$

for all pairs of vectors X, Y which are in some M_p (when M_p is identified with a subspace of $\mathbb{R}^3$ in the usual way). Since M is not contained in a plane, there are three distinct subspaces $M_{p_1}, M_{p_2}, M_{p_3}$ (we regard these as vector subspaces of $\mathbb{R}^3$). Choose

$$\begin{aligned}
X_3 &\in M_{p_1} \cap M_{p_2}\\
X_2 &\in M_{p_1} \cap M_{p_3}\\
X_1 &\in M_{p_2} \cap M_{p_3} .
\end{aligned}$$

Then the X_i are linearly independent, and $0 = \langle TX_i,X_j\rangle + \langle X_i,TX_j\rangle$ for each

i, j. This implies that $0 = \langle TX,Y\rangle + \langle X,TY\rangle$ for all $X, Y \in \mathbb{R}^3$. Thus T is skew-adjoint, and its matrix C is skew-symmetric. In other words,

$$Z(p) = p \cdot C \qquad C \in \mathfrak{o}(3) ,$$

and hence Z is trivial. ■

In order to obtain some deeper results about rigidity, we will find it useful to consider various $\mathbb{R}^3$-valued differential forms on a surface M. Many of these forms will be defined in terms of other $\mathbb{R}^3$-valued forms and functions on M by means of the inner product and cross product on $\mathbb{R}^3$. If $g, h\colon M \longrightarrow \mathbb{R}^3$ are two functions, then there is only one reasonable meaning for $f \times g$, namely the function

$$p \longmapsto f(p) \times g(p) \in \mathbb{R}^3 .$$

But if ω and η are $\mathbb{R}^3$-valued 1-forms on M, then $\omega \times \eta$ might mean any of the following:

$X \longmapsto \omega(X) \times \eta(X)$ (a quadratic function on tangent vectors)

$(X,Y) \longmapsto \omega(X) \times \eta(Y)$ (a bilinear function on tangent vectors)

$(X,Y) \longmapsto \omega(X) \times \eta(Y) - \omega(Y) \times \eta(X)$ (a 2-form) .

To distinguish these possibilities, we might write the last two as

$$\omega \overset{\otimes}{\times} \eta \qquad \text{and} \qquad \omega \overset{\wedge}{\times} \eta .$$

Since we shall, in fact, only be interested in the last case, we will introduce the simpler symbol $\mathbf{x}$ and define

$$\omega \times \eta(X,Y) = \omega(X) \times \eta(Y) - \omega(Y) \times \eta(Y) .$$

The present situation is actually just a special case of the one already considered at the end of Chapter I.10; see p. I.547 and especially Problems I.10-20, 21. In general, if ω and η are $\mathbb{R}^3$-valued forms of degree k and ℓ, respectively, then we define a $(k+\ell)$-form $\omega \times \eta$ by

$$\omega \times \eta(X_1,\ldots,X_k,X_{k+1},\ldots,X_{k+\ell}) = \frac{1}{k!\ell!} \sum_{\sigma \in S_{k+\ell}} \operatorname{sgn}\sigma \cdot \omega(X_{\sigma(1)},\ldots,X_{\sigma(k)}) \times \eta(X_{\sigma(k+1)},\ldots,X_{\sigma(k+\ell)}) .$$

[Actually, since we will be dealing with a surface M, only the cases $k, \ell \leq 1$ are relevant]. In an exactly analogous way, we define $\omega \bullet \eta$, using the product $v \cdot w = \langle v,w \rangle$ in $\mathbb{R}^3$. Then we have (Problem I.10-21(a))

$$d(\omega \times \eta) = d\omega \times \eta + (-1)^k \omega \times d\eta$$
$$d(\omega \bullet \eta) = d\omega \bullet \eta + (-1)^k \omega \bullet d\eta .$$

Notice that since $\times$ is not commutative, $\omega \times \omega$ need not be zero. In fact, for a 1-form ω we have

$$\omega \times \omega(X,Y) = 2\omega(X) \times \omega(Y) .$$

More generally,

$$\omega \times \eta = (-1)^{k\ell+1} \eta \times \omega$$
$$\omega \bullet \eta = (-1)^{k\ell} \eta \bullet \omega .$$

It is also easy to see that the formula

$$v \cdot (w \times z) = - w \cdot (v \times z) \qquad v, w, z \in \mathbb{R}^3$$

leads to the relation

$$\omega \bullet (\eta \times \lambda) = (-1)^{k\ell+1}\eta \bullet (\omega \times \lambda) .$$

Pure notational fiddling would lead us to write $\omega \bullet (\eta \times \lambda)$ in the form $\mathbf{det}(\omega,\eta,\lambda)$, which can also be defined directly: if $\omega = (\omega^1,\omega^2,\omega^3)$ for ordinary 1-forms ω^i, and similarly for η and λ, then $\mathbf{det}(\omega,\eta,\lambda)$ denotes

$$\det\begin{pmatrix} \omega^1 & \omega^2 & \omega^3 \\ \lambda^1 & \lambda^2 & \lambda^3 \\ \eta^1 & \eta^2 & \eta^3 \end{pmatrix} ,$$

where the determinant is expanded out as usual, with all multiplications being replaced by $\wedge$, and care being taken to write products in the correct order (namely, the same order as the columns they appear in). One easily checks, either from this definition, or from the alternative form $\omega \bullet (\eta \times \lambda)$, that

$$\begin{aligned} d\ \mathbf{det}(\omega,\eta,\lambda) = {} & \mathbf{det}(d\omega,\eta,\lambda) + (-1)^k\ \mathbf{det}(\omega,d\eta,\lambda) \\ & + (-1)^{k+\ell}\ \mathbf{det}(\omega,\eta,d\lambda) . \end{aligned}$$

We will frequently use the various formulas given here without specific comment.

Now consider an immersion $f\colon M \to \mathbb{R}^3$ of an oriented surface M, and the corresponding normal map $n\colon M \to \mathbb{R}^3$. If dA is the volume element of M for the metric $f^*\langle\ ,\ \rangle$, then we have the following identities among $\mathbb{R}^3$-valued 2-forms on M:

(I) $df \times df = 2n\ dA$

(II) $df \times dn = -\ 2Hn\ dA$

(III) $dn \times dn = 2Kn\ dA$.

To prove these simple relations, pick vectors $X_1, X_2 \in M_p$ with (X_1,X_2) positively oriented. Now $df(X_i)$ is just $f_*(X_i)$, considered as an element of $\mathbb{R}^3$. So

$$\begin{aligned} df \mathbin{\boldsymbol{\times}} df(X_1,X_2) &= 2\ df(X_1) \times df(X_2) \\ &= 2\ n(p) \cdot \text{area of parallelogram spanned by } df(X_1),\ df(X_2) \\ &= 2\ n(p) \cdot dA(X_1,X_2)\ . \end{aligned}$$

Moreover, if

$$\begin{aligned} dn(X_1) &= \alpha\ df(X_1) + \beta\ df(X_2) \\ dn(X_2) &= \gamma\ df(X_1) + \delta\ df(X_2)\ , \end{aligned}$$

then

$$\begin{aligned} df \mathbin{\boldsymbol{\times}} dn(X_1,X_2) &= \delta\ df(X_1) \times df(X_2) - \alpha\ df(X_2) \times df(X_1) \\ &= (\alpha + \delta)\ df(X_1) \times df(X_2) \\ &= (\alpha + \delta) n(p)\ dA(X_1,X_2) \\ &= -\ 2H(p) n(p)\ dA(X_1,X_2)\ , \end{aligned}$$

and

$$\begin{aligned} dn \mathbin{\boldsymbol{\times}} dn(X_1,X_2) &= 2\ dn(X_1) \times dn(X_2) \\ &= 2[\alpha\ df(X_1) + \beta\ df(X_2)] \times [\gamma\ df(X_1) + \delta\ df(X_2)] \\ &= 2(\alpha\delta - \beta\gamma)\ df(X_1) \times df(X_2) \\ &= 2K(p) n(p)\ dA(X_1,X_2)\ . \end{aligned}$$

Naturally, all these formulas can be applied when $M \subset \mathbb{R}^3$, and f is just the inclusion map $i\colon M \to \mathbb{R}^3$. And, in fact, we shall usually apply them to imbedded, rather than immersed, surfaces. But it nevertheless seems conceptually easier always to regard M as an abstract surface sitting off in the void, so that f and n can be thought of simply as certain $\mathbb{R}^3$-valued functions on M without worrying about the geometry they induce; most of the geometric information in question is already presented in formulas (I)-(III).

We will also need to recall the support function $h = -f \cdot n \colon M \to \mathbb{R}^3$, which is defined in Problem 3-7. As we saw, $h(p)$ is the signed distance from the origin to the tangent plane of $f(M)$ at $f(p)$; when f is an imbedding with $f(M)$ star-shaped with respect to 0, and n is inward pointing, it is precisely this distance. This happens, in particular, when $f(M)$

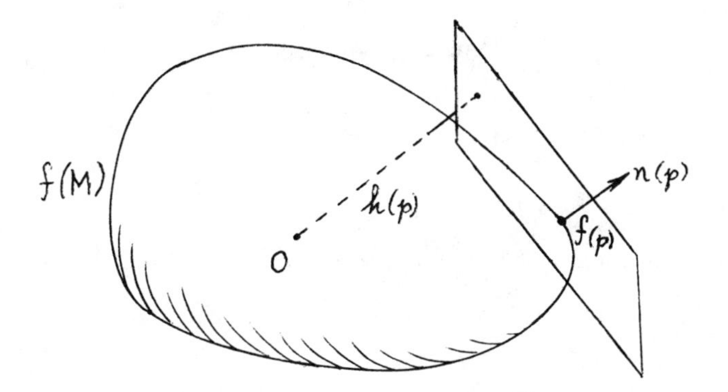

is convex, and 0 lies inside it.

Now consider the $\mathbb{R}^3$-valued 1-form α on M defined by

$$\alpha = (f \times n) \cdot df .$$

We have

$$d\alpha = (df \times n) \cdot df + (f \times dn) \cdot df$$

$$= - n\cdot(df \times df) + f\cdot(dn \times df)$$ [the dots do not have to be bold face since f and n are functions]

$$= - 2\, dA - 2H(f\cdot n)\, dA$$ by (I) and (II)

$$= - 2\, dA + 2hH\, dA .$$

If M is compact, then $\int_M d\alpha = \int_{\partial M} \alpha = 0$, so we obtain

(IV) $$\text{area}(M) = \int_M hH\, dA .$$

Similarly, for the 1-form

$$\beta = (f \times n) \cdot dn ,$$

we have

$$d\beta = (df \times n) \cdot dn + (f \times dn) \cdot dn$$

$$= - n\cdot(df \times dn) + f\cdot(dn \times dn)$$

$$= 2H\, dA - 2hK\, dA$$ by (II) and (III) .

Hence for compact M we have

(V) $$\int_M H\, dA = \int_M hK\, dA .$$

Equations (IV) and (V) are sometimes called "Minkowski's formulas."

As a first application of these formulas, we reprove a rigidity result which appeared a long time ago. The theorem that a compact surface with constant $K > 0$ must be a (standard) sphere in $\mathbb{R}^3$ can also be stated as follows:

a sphere is unwarpable.* To prove this from our present formulas, we consider any compact surface $M \subset \mathbb{R}^3$ with constant $K > 0$. It is convex, by Hadamard's theorem, and we can assume that $0 \in \mathbb{R}^3$ lies in its interior, so that the support function h is always positive for the inward pointing n. There is no loss of generality in assuming that $K = k_1 k_2 = 1$. Since for $x > 0$ we always have

$$x + \frac{1}{x} \geq 2 ,$$

with equality only if $x = 1$, this implies that

$$H = \frac{1}{2}\left(k_1 + \frac{1}{k_1}\right) \geq 1 ,$$

with equality only if $k_1 = k_2$. So

$$\begin{aligned} \int_M h \, dA &= \int_M H \, dA && \text{by (V)} \\ &\geq \int_M dA \\ &= \int_M hH \, dA && \text{by (IV)} , \end{aligned}$$

and consequently

$$\int_M h(1 - H) \, dA \geq 0 .$$

---

*Actually, the statement that a sphere is unwarpable is formally stronger than the statement that any compact surface isometric to a sphere is a sphere -- see the footnote on p. 251 . But the complete equivalence of the two statements follows easily from the fact that any isometry of a sphere onto itself is the restriction of a Euclidean motion.

Since $h > 0$ everywhere, and $1 - H \leq 0$ everywhere, it follows that $H = 1$ everywhere. This implies that $k_1 = k_2$ everywhere.

Similarly, we can reprove Theorem 5-3, and in a little greater generality.

9. PROPOSITION. The only star-shaped surfaces of constant mean curvature H are spheres.

Proof. We still have $h > 0$, and there is no loss of generality in assuming that

$$1 = H = \frac{k_1 + k_2}{2} .$$

Then

$$K = k_1 k_2 = k_1(2 - k_1) = 2k_1 - (k_1)^2 \leq 1 ,$$

with equality only if $k_1 = 1$. Now

$$\begin{aligned} \int_M h \, dA &= \int_M dA && \text{by (IV)} \\ &= \int hK \, dA && \text{by (V)} , \end{aligned}$$

so

$$\int_M h(1 - K) \, dA = 0 .$$

Thus we must have $K = 1$ everywhere, hence $k_1 = 1$ everywhere, hence $k_2 = 1 = k_1$ everywhere. ■

This proof is mainly a curiosity, since, as we showed in Chapter 9 (Addendum 2 or 3), a much stronger result actually holds. By considering another $\mathbb{R}^3$-valued 1-form, however, we obtain a real theorem, one of the first in the subject.

10. THEOREM. Let $M \subset \mathbb{R}^3$ be any compact convex surface which does not contain a portion of a plane. Then M is infinitesimally rigid.

Remarks. (1) We have already pointed out that the conclusion fails if M does contain a portion of a plane.
(2) We could replace convexity with the hypothesis $K \geq 0$ (see the remark after Theorem 2-11 or Proposition 7-32).

Proof. For simplicity, we first consider the case where $K > 0$ everywhere. Let Z be an infinitesimal bending of the inclusion map $f\colon M \to \mathbb{R}^3$, and let Y be its rotation field, so that

$$dZ(X_p) = df(X_p) \times Y(p)$$

for all $X_p \in M_p$. This relation can be written in terms of the $\mathbb{R}^3$-valued 1-forms dZ and df, and $\mathbb{R}^3$-valued function Y, as

$$dZ = df \times Y .$$

Now

$$(1) \qquad 0 = d(dZ) = -\, df \times dY .$$

This means that for $X_1, X_2 \in M_p$ we have

$$(2) \qquad df(X_1) \times dY(X_2) - df(X_2) \times dY(X_1) = 0 .$$

Taking the dot product of this equation with $df(X_1)$ and $df(X_2)$, we find that $dY(X_i)$ lies in the plane spanned by $df(X_1)$ and $df(X_2)$, which is nothing but the tangent plane M_p moved over to the origin. In other words, we can consider dY as a map $dY\colon M_p \to M_p$.

Now choose a moving frame X_1, X_2 in a neighborhood of p. We can write

$$(3) \qquad \begin{aligned} dY(X_1) &= \alpha\, df(X_1) + \beta\, df(X_2) \\ dY(X_2) &= \gamma\, df(X_1) + \delta\, df(X_2) \end{aligned}$$

for some functions $\alpha, \beta, \gamma, \delta$. Equation (1) implies that

$$(4) \qquad \alpha + \delta = 0 .$$

Remembering that f is just the inclusion map, so that X_1, X_2 are vector fields on M, we can write

$$\begin{aligned} (5) \qquad 0 = d(dY)(X_1,X_2) &= X_1(dY(X_2)) - X_2(dY(X_1)) - dY([X_1,X_2]) \\ &= \gamma\, \nabla'_{X_1}X_1 + \delta\, \nabla'_{X_1}X_2 - \alpha\, \nabla'_{X_2}X_1 - \beta\, \nabla'_{X_2}X_2 \\ &\qquad + \text{something tangent to } M . \end{aligned}$$

Taking the inner product with n, and using (4), we get

$$\begin{aligned} 0 &= \gamma\, II(X_1,X_1) - 2\alpha\, II(X_1,X_2) - \beta\, II(X_2,X_2) \\ &= \gamma \ell - 2\alpha m - \beta n , \qquad \text{say.} \end{aligned}$$

In particular, suppose that we choose X_1, X_2 to be principal vectors at some point p, so that at p we have $m = 0$. Then our equation is simply

(6) $$0 = \gamma\ell - \beta n \ .$$

Since $K = \ell n > 0$, this shows that γ and β have the same sign. So

$$0 \leq \beta\gamma \quad \text{and} \quad 0 = \beta\gamma \quad \text{only if} \quad \beta\gamma = 0 \ .$$

Hence at each point p we have

(7) $$0 \leq \alpha^2 + \beta\gamma = - \det dY$$

with equality only if $\alpha = \beta = \gamma = 0 \implies dY = 0$.

Consider the 1-form

$$\omega = (f \times Y) \bullet dY$$

(closed related to the 1-form β considered previously). We have

$$\begin{aligned} d\omega &= (df \times Y) \bullet dY + (f \times dY) \bullet dY \\ &= - Y\cdot(df \times dY) + f\cdot(dY \times dY) \\ &= 0 + f\cdot(dY \times dY) \qquad \text{by (1)} \ . \end{aligned}$$

But we also have

$$dY \times dY = 2(\det dY) n \, dA \ ,$$

by the very same argument which proved formula (III). Hence

$$d\omega = 2h(\det dY) \, dA \ .$$

So for our compact manifold M we have the integral formula of Blaschke:

(*) $$\int_M h(\det dY) \, dA = 0 \ .$$

Since $h > 0$, and $\det dY \leq 0$ by (7), we must have $\det dY = 0$ everywhere. Then (7) also shows that $dY = 0$ everywhere. Therefore Y is constant. So Z is trivial by Lemma 4.

Now we consider the case where $K \geq 0$, but M contains no portion of a plane, so that the planar points of M are nowhere dense. At a parabolic point we have $n = 0$ and $\ell \neq 0$, say. Equation (6) then shows that $\gamma = 0$. Hence we still have

$$0 \leq \alpha^2 + \beta\gamma = -\det dY .$$

Thus we have $\det dY \leq 0$ at all non-planar points, which implies that $\det dY \leq 0$ everywhere. So we can still conclude from (*) that $\det dY = 0$ everywhere. We want to show that consequently $dY = 0$ everywhere; it obviously suffices to show that $dY(p) = 0$ when p is a parabolic point.

If the parabolic point p lies in the closure of $\{q \in M\colon K(q) > 0\}$, then clearly $dY(p) = 0$. So consider a parabolic point p which has a neighborhood on which $K = 0$. By Proposition 5-4 and Corollary 5-6, M contains a ruled surface

$$(s,t) \longmapsto c(s) + td(s) \qquad |d| = 1 \implies \langle d,d'\rangle = 0$$

around p such that the ruling through p has its endpoints in the closure of $\{q \in M\colon K(q) > 0\}$, and consequently $dY = 0$ at the endpoints of this ruling. We can choose the curve c to be the intersection of the ruled surface with

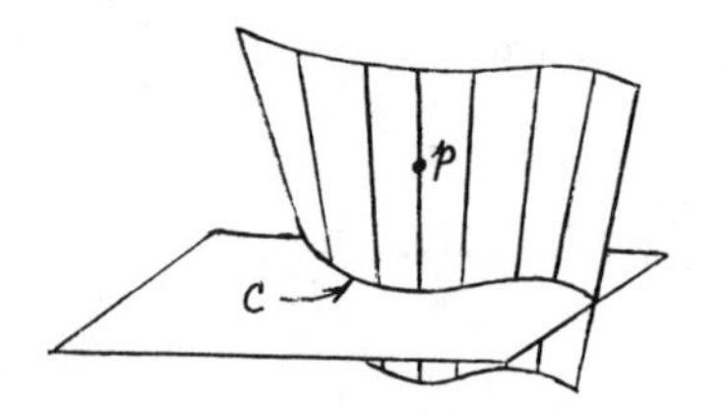

a plane perpendicular to the ruling through p. So if X_1, X_2 are the coordinate vector fields

$$X_1(s,t) = c'(s) + td'(s)$$
$$X_2(s,t) = d(s) ,$$

then along the ruling through p we have

$$\langle X_1, X_2 \rangle = \langle c' + td', d \rangle = \langle c', d \rangle = 0 .$$

Now X_2 is the principal vector with principal curvature $n = 0$. So along the ruling through p, the vector field X_1 is the other principal vector, with principal curvature $\ell \neq 0$. So we have [by (6)]

$$(8) \qquad \gamma = 0 \qquad \text{along the ruling through } p .$$

Since we have $0 = -\det Y = \alpha^2 + \beta\gamma$ everywhere, we also have

$$(9) \qquad \alpha = 0 \qquad \text{along the ruling through } p .$$

So equations (3) and (4) give

$$\left.\begin{aligned} dY(X_1) &= \beta \, df(X_2) \\ dY(X_2) &= 0 \end{aligned}\right\} \quad \text{along the ruling through } p .$$

Then equation (5) becomes simply

$$\begin{aligned} 0 &= 0 - X_2(\beta \, df(X_2)) - 0 \\ &= - X_2(\beta) \, df(X_2) - \beta \, \nabla'_{X_2} X_2 \\ &= - X_2(\beta) \, df(X_2) . \end{aligned}$$

Thus $0 = X_2(\beta)$ along the ruling through p, so that β is constant on this ruling. But $\beta = 0$ at the endpoints, since $dY = 0$ at the endpoints. It follows that also $\beta = 0$ at p; together with (8) and (9) we now have $dY(p) = 0$. ∎

At this point it might be nice to have some non-trivial examples of surfaces which <u>are</u> infinitesimally bendable. Consider a surface given as a graph,

$$f(x,y) = (x,y,u(x,y)) ;$$

we introduce the standard notation

$$p = u_1 = \frac{\partial u}{\partial x}, \qquad q = \frac{\partial u}{\partial y}$$

$$r = \frac{\partial^2 u}{\partial x^2}, \qquad s = \frac{\partial^2 u}{\partial x \partial y}, \qquad t = \frac{\partial^2 u}{\partial y^2}.$$

Suppose that $Z = (\ , \ ,\zeta)$ is an infinitesimal bending, with rotation field $Y = (\alpha,\beta,\psi)$. Then

$$\begin{aligned} (\ , \ ,\zeta_1) = dZ\left(\frac{\partial}{\partial x}\right) &= df\left(\frac{\partial}{\partial x}\right) \times (\alpha,\beta,\psi) \\ &= (1,0,p) \times (\alpha,\beta,\psi) \\ &= (\ , \ ,\beta) \end{aligned}$$

and similarly

$$(\ , \ ,\zeta_2) = (\ , \ , -\alpha) .$$

So Y must be of the form

$$Y = (-\zeta_2,\zeta_1,\psi) .$$

Using equation (2) in the proof of Theorem 10, we see that we must have

$$(1,0,p)\times(-\zeta_{22},\zeta_{12},\psi_2) - (0,1,q)\times(-\zeta_{12},\zeta_{11},\psi_1) = 0 ,$$

which is equivalent to

$$(*)\qquad \begin{cases} \psi_1 = q\zeta_{11} - p\zeta_{12} \\ \psi_2 = -q\zeta_{12} + p\zeta_{22} . \end{cases}$$

Conversely, if ψ_i, ζ_{ij} satisfy these equations, and we define the vector valued 1-form W by

$$W(X) = df(X)\times(-\zeta_2,\zeta_1,\psi) ,$$

then W will satisfy $dW = 0$, so on any simply-connected portion of the (x,y)-plane there will be Z with $dZ = W$. Now equations (*) can be solved for ψ if and only if

$$(q\zeta_{11} - p\zeta_{12})_2 = (-q\zeta_{12} + p\zeta_{22})_1 ,$$

which leads to an equation for ζ:

$$(**)\qquad r\zeta_{22} - 2s\zeta_{12} + t\zeta_{11} = 0 .$$

As a particular case, we consider the paraboloid $u(x,y) = \frac{1}{2}(x^2+y^2)$. We obtain the equation

$$\zeta_{11} + \zeta_{22} = 0 ,$$

whose solutions are the real part of any entire function on $\mathbb{C} = \mathbb{R}^2$. Thus there are non-trivial infinitesimal bendings of the complete convex surface

$\{(x,y,u(x,y))\}$.

As we have already pointed out, it is now known whether infinitesimal rigidity generally implies unbendability. But we can deduce this further property in the special situation considered in Theorem 10.

11. COROLLARY. Let $M \subset \mathbb{R}^3$ be a compact convex surface with $K > 0$ everywhere. Then M is unbendable.

Proof. Let $\alpha\colon [0,1] \times M$ be any bending of the inclusion map $i\colon M \to \mathbb{R}^3$. Then all $\bar{\alpha}(t)(M)$ have $K > 0$ everywhere, so all $\bar{\alpha}(t)(M)$ are infinitesimally rigid, by Theorem 10. So the variation vector field Z_t of α at time t is trivial. Then by Lemma 5, the bending α is trivial. ∎

In this Corollary, the case $K \geq 0$ eluded us, but we aren't going to worry very much about it, because we are now going to prove a much better result anyway, the famous theorem of Cohn-Vossen that any convex surface is unwarpable. This result is the uniqueness part of Weyl's Problem, mentioned in the previous Chapter; the present proof stems from the work of Herglotz.

12. THEOREM (COHN-VOSSEN). If $M \subset \mathbb{R}^3$ is a compact convex surface, then M is unwarpable.

Proof. As in the proof of Theorem 11, for simplicity we first consider the case where $K > 0$ everywhere. So we consider two imbeddings $f, \bar{f}\colon M \to \mathbb{R}^3$ with $f^*\langle\ ,\ \rangle = \bar{f}^*\langle\ ,\ \rangle$, such that the curvature $K = \bar{K}$ for this metric is

> 0 everywhere. Let $n, \bar{n}$ be the inward pointing normals, for the convex surfaces $f(M)$ and $\bar{f}(M)$, and orient these surfaces so that n and $\bar{n}$ are the normals determined by the orientations. We can assume that M has an orientation which makes both maps $f\colon M \longrightarrow f(M)$ and $\bar{f}\colon M \longrightarrow \bar{f}(M)$ orientation preserving (by composing $\bar{f}$ with a reflection if necessary). For each $p \in M$ we have two subspaces $df(M_p)$, $d\bar{f}(M_p) \subset \mathbb{R}^3$, which are just $f_*(M_p)$ and $\bar{f}_*(M_p)$ moved over to the origin. So we can consider

$$\iota = d(f \circ \bar{f}^{-1})\colon d\bar{f}(M_p) \longrightarrow df(M_p) .$$

The magic 1-form which we want to consider is

$$\omega = (f \times n) \bullet (\iota \circ d\bar{n}) .$$

Figuring out $d\omega$ will be quite a bit harder than in the previous theorem. First we will get an expression for ω in terms of moving frames. We choose a moving frame X_1, X_2 on M which is orthonormal for $f^*\langle\ ,\ \rangle = \bar{f}^*\langle\ ,\ \rangle$, and let θ^1, θ^2 and $\omega_1^2 = -\omega_2^1$ be its dual forms and connection forms. We can consider (f_*X_1, f_*X_2, n) and $(\bar{f}_*X_1, \bar{f}_*X_2, \bar{n})$ as adapted orthonormal moving frames on $f(M)$ and $\bar{f}(M)$; let ψ_1^3, ψ_2^3 be f^* of the corresponding forms on $f(M)$, and define $\bar{\psi}_1^3, \bar{\psi}_2^3$ similarly. Then for X tangent to M we have

$$(1) \qquad \begin{gathered} d\bar{n}(X) = \bar{\psi}_3^1(X)\cdot d\bar{f}(X_1) + \bar{\psi}_3^2(X)\cdot d\bar{f}(X_2) \\ \Downarrow \\ \iota \circ d\bar{n}(X) = \bar{\psi}_3^1(X)\cdot df(X_1) + \bar{\psi}_3^2(X)\cdot df(X_2) . \end{gathered}$$

We will also express $f(p)$ as a linear combination

$$(2) \qquad f(p) = y_1(p)\cdot df(X_1(p)) + y_2(p)\cdot df(X_2(p)) + y_3(p)\cdot n(p) ,$$

where, in particular,

$$(3) \qquad y_3 = f\cdot n = -\, h \ .$$

Then for $X \in M_p$ we have

$$\begin{aligned}
\omega(X) &= [f(p)\times n(p)]\cdot[\bar\psi_3^1(X)\cdot df(X_1) + \bar\psi_3^2(X)\cdot df(X_2)] && \text{by (1)}\\
&= [y_1(p)\cdot df(X_1)\times n(p) + y_2(p)\cdot df(X_2)\times n(p)]\ \cdot\\
&\quad \cdot\ [\bar\psi_3^1(X)\cdot df(X_1) + \bar\psi_3^2(X)\cdot df(X_2)] && \text{by (2)}\\
&= y_1(p)\cdot\bar\psi_3^2(X) - y_2(p)\cdot\bar\psi_3^1(X) \ ,
\end{aligned}$$

and consequently

$$(4) \qquad \omega = -\, y_1\bar\psi_2^3 + y_2\bar\psi_1^3 \ .$$

Now equation (2) implies that for $X \in M_p$ we have

$$\begin{aligned}
df(X) &= dy_1(X)\cdot df(X_1(p)) + y_1(p)\nabla'_{f_*X}f_*X_1 + \cdots\\
&= dy_1(X)\cdot df(X_1(p)) + y_1(p)\omega_1^2(X)\cdot df(X_2(p)) + y_1(p)\psi_1^3(X)n(p) + \cdots\\
&= [dy_1(X) + y_2(p)\omega_2^1(X) + y_3(p)\psi_3^1(X)]df(X_1) + \cdots \ .
\end{aligned}$$

But also

$$df(X) = \theta^1(X)\cdot df(X_1) + \theta^2(X)\cdot df(X_2) \ .$$

Hence we have

$$(5) \qquad \begin{aligned} dy_1 &= \theta^1 + y_2\omega_1^2 + y_3\psi_1^3 \\ dy_2 &= \theta^2 + y_1\omega_2^1 + y_3\psi_2^3 \,. \end{aligned}$$

Now we can compute

$$\begin{aligned} d\omega &= d(-\, y_1\bar{\psi}_2^3 + y_2\bar{\psi}_1^3) \qquad \text{by (4)} \\ &= -\, (\theta^1 + y_2\omega_1^2 + y_3\psi_1^3) \wedge \bar{\psi}_2^3 \;+\; y_1(\bar{\psi}_1^3 \wedge \omega_2^1) \\ &\quad + (\theta^2 + y_1\omega_2^1 + y_3\psi_2^3) \wedge \bar{\psi}_1^3 \;-\; y_2(\bar{\psi}_2^3 \wedge \omega_1^2) \\ &\qquad \text{by (5) and the structural equations} \\ &= -\, \theta^1 \wedge \bar{\psi}_2^3 \;+\; \theta^2 \wedge \bar{\psi}_1^3 \;-\; y_3(\psi_1^3 \wedge \bar{\psi}_2^3 - \psi_2^3 \wedge \bar{\psi}_1^3) \\ &= -\, \{(\bar{\ell}_{11} + \bar{\ell}_{22}) + y_3(\ell_{11}\bar{\ell}_{22} + \bar{\ell}_{11}\ell_{22} - 2\ell_{12}\bar{\ell}_{12})\} \, dA \,, \end{aligned}$$

where

$$\ell_{ij} = II_f(X_i, X_j) \,, \qquad \bar{\ell}_{ij} = II_{\bar{f}}(X_i, X_j) \,, \qquad dA = \theta^1 \wedge \theta^2 \,.$$

Now one has to observe that

$$\begin{aligned} \ell_{11}\bar{\ell}_{22} - 2\ell_{12}\bar{\ell}_{12} + \ell_{11}\bar{\ell}_{12} &= (\ell_{11}\ell_{22} - \ell_{12}{}^2) + (\bar{\ell}_{11}\bar{\ell}_{22} - \bar{\ell}_{12}{}^2) \\ &\qquad - \det\begin{pmatrix} \bar{\ell}_{11} - \ell_{11} & \bar{\ell}_{12} - \ell_{12} \\ \bar{\ell}_{12} - \ell_{12} & \bar{\ell}_{22} - \ell_{22} \end{pmatrix} \\ &= 2K - \det(d\bar{n} - dn) \,, \end{aligned}$$

where we now regard dn and $d\bar{n}$ as maps $dn, d\bar{n}\colon M_p \longrightarrow M_p$. We obtain finally

$$d\omega = -\, \{2\bar{H} - h(2K - \det(d\bar{n} - dn))\} \, dA \,.$$

So for our compact M we have

$$(6) \qquad 2\int_M \bar{H}\, dA - 2\int_M hK\, dA = -\int_M h \det(d\bar{n} - dn)\, dA .$$

Using formula (V), we obtain the <u>Herglotz integral formula</u>:

$$(7) \qquad 2\int_M \bar{H} - H\, dA = -\int_M h \det(d\bar{n} - dn)\, dA .$$

[Note that formula (V) also follows from (6) by taking $f = \bar{f}$.] Now we need an algebraic

<u>13. LEMMA</u>. Let A and B be two self-adjoint linear transformations on $\mathbb{R}^2$ which are positive semi-definite (i.e., have eigenvalues ≥ 0). Suppose that $\det A = \det B$. Then

$$\det(A - B) \leq 0 .$$

Moreover, if A and B are positive definite, then equality holds only if $A = B$; and if A and B are positive semi-definite, then equality holds only if A and B are proportional.

<u>Proof</u>. Consider A and B as symmetric matrices, and suppose first that A is positive definite. Since A is self-adjoint, there is an orthogonal matrix P with

$$PAP^t = PAP^{-1} = \begin{pmatrix} \lambda_1 & 0 \\ 0 & \lambda_2 \end{pmatrix} \qquad \lambda_1, \lambda_2 > 0 .$$

If we set

$$Q = CP \ , \qquad C = \begin{pmatrix} \frac{1}{\sqrt{\lambda_1}} & 0 \\ 0 & \frac{1}{\sqrt{\lambda_2}} \end{pmatrix} ,$$

then

$$\text{(a)} \qquad QAQ^t = CPAP^tC = I \ .$$

Now consider QBQ^t. It is also symmetric, so there is an orthogonal R with

$$\text{(b)} \qquad (RQ)B(RQ)^t = R(QBQ^t)R^t = \begin{pmatrix} \mu_1 & 0 \\ 0 & \mu_2 \end{pmatrix} , \qquad \mu_1, \mu_2 > 0 \ .$$

Moreover, equation (a) gives

$$\text{(c)} \qquad (RQ)A(RQ)^t = R(QAQ^t)R^{-1} = I \ .$$

[We have simply reproved the well-known result that two positive definite quadratic forms can be simultaneously diagonalized.] So for $S = RQ$ we have

$$(\det S)^2 \det B = \mu_1\mu_2 \qquad \text{by (b)} ,$$

$$(\det S)^2 \det A = 1 \qquad \text{by (c)} .$$

If $\det A = \det B$, then

$$\mu_1\mu_2 = 1 \ .$$

Moreover,

$$(\det S)^2 \det(A-B) = \det\begin{pmatrix} 1-\mu_1 & 0 \\ 0 & 1-\mu_2 \end{pmatrix} = 2 - (\mu_1+\mu_2) .$$

Since for $x > 0$ we always have

$$x + \frac{1}{x} \geq 2 , \qquad \text{with equality only if } x = 1 ,$$

it follows that

$$2 - (\mu_1+\mu_2) \leq 0 ,$$

with equality only if $\mu_1 = \mu_2 = 1 \implies A = B$.

Now suppose that A and B are positive semi-definite, with $A \neq 0$, say. We can now obtain Q with

(a') $$QAQ^t = \begin{pmatrix} 1 & 0 \\ 0 & 0 \end{pmatrix} ,$$

and R with

(b') $$(RQ)B(RQ)^t = \begin{pmatrix} \mu_1 & 0 \\ 0 & \mu_2 \end{pmatrix} \qquad \mu_1, \mu_2 \geq 0 , \quad \mu_1\mu_2 = 0 .$$

As before, we also have

(c') $$(RQ)A(RQ)^t = \begin{pmatrix} 1 & 0 \\ 0 & 0 \end{pmatrix} .$$

So for $S = RQ$ we have

$$(\det S)^2 \det(A-B) = \det\begin{pmatrix} 1-\mu_1 & 0 \\ 0 & -\mu_2 \end{pmatrix} = -\mu_2 + \mu_1\mu_2 = -\mu_2 .$$

Thus $\det(A-B) \leq 0$, and equality holds only if $\mu_2 = 0 \implies B = \mu_1 \cdot A$. Q.E.D.

Applying the Lemma to the positive definite maps dn, $d\bar{n}\colon M_p \to M_p$, with the same determinant $K(p)$, we conclude from equation (7) that

$$\int_M \bar{H}\, dA - \int_M H\, dA \geq 0 .$$

But we can interchange f and $\bar{f}$ in this inequality to obtain

$$\int_M H\, dA - \int_M \bar{H}\, dA \geq 0 .$$

Hence

$$\int_M H\, dA = \int_M \bar{H}\, dA .$$

Then equation (7) gives

$$(*) \qquad \int_M h \det(d\bar{n} - dn)\, dA = 0 .$$

Now (*) implies that $\det(d\bar{n} - dn) = 0$ everywhere. Then Lemma 13 implies that $d\bar{n} = dn$ everywhere. So the fundamental theorem of surface theory implies that f and $\bar{f}$ differ by a Euclidean motion.

Now we consider the case $K \geq 0$. We can still obtain (*) and thus conclude that $\det(d\bar{n} - dn) = 0$ everywhere. We have to show that $d\bar{n}(p) = dn(p)$ for points p with $K(p) = 0$, and it is only necessary to consider points p with $K = 0$ in a whole neighborhood of p. If $f(p)$ and $\bar{f}(p)$ are both planar points, there is nothing to prove. So suppose that $f(p)$, say, is a parabolic point. Then, as in the previous proof, the point $f(p)$ is on some line segment $\Gamma \subset f(M)$, whose endpoints Q_1, Q_2 are in the closure of the set where $K > 0$. Let $\bar{\Gamma} \subset \bar{f}(M)$ be the image of Γ under the isometry $\bar{f} \circ f^{-1}\colon f(M) \to \bar{f}(M)$. Then $\bar{\Gamma}$ is a geodesic in $\bar{f}(M)$. Now Γ is also an asymptotic curve, $II(X,X) = 0$ for tangent vectors X pointing along Γ. The last part of Lemma 13 then shows that we must have $\overline{II}(Y,Y) = 0$ for tangent vectors Y pointing along $\bar{\Gamma}$. So $\bar{\Gamma}$ has normal curvature $\bar{\kappa}_n = 0$; since $\bar{\Gamma}$ has geodesic curvature $\bar{\kappa}_g = 0$, it follows that $\bar{\Gamma}$ has curvature $\bar{\kappa} = \sqrt{\bar{\kappa}_n{}^2 + \bar{\kappa}_g{}^2} = 0$. Hence $\bar{\Gamma}$ is also a straight line segment, with endpoints $\bar{Q}_1, \bar{Q}_2$, say. Lemma 5-5 says that the non-zero principal curvature k along Γ is of the form

$$k(s) = \frac{1}{As + B},$$

where $k(s)$ is the value of k at the point on Γ at distance s from $f(p)$. In particular, k cannot approach zero at Q_1 or Q_2, so Q_1 and Q_2 are not planar points. Since Q_1, Q_2 are in the closure of the set where $K > 0$ we have $dn(Q_i) = d\bar{n}(\bar{Q}_i)$, so $\bar{Q}_1, \bar{Q}_2$ are also not planar points. So by Corollary 5-6, $\bar{f}(p)$ is not a planar point. Thus the non-zero principal curvature $\bar{k}$ along $\bar{\Gamma}$ is of the form

$$\bar{k}(s) = \frac{1}{\bar{A}s + \bar{B}},$$

where s now measures the distance from $\bar{f}(p)$. But since $dn(Q_i) = d\bar{n}(\bar{Q}_i)$, we have $k(Q_i) = \bar{k}(\bar{Q}_i)$. It follows that $A = \bar{A}$ and $B = \bar{B}$. Hence $d\bar{n}(p) = dn(p)$. ∎

For later use, we insert here a form of the Herglotz integral formula for compact surfaces-with-boundary.

14. LEMMA. Let M be a compact oriented surface with boundary ∂M, and let $f, \bar{f}: M \longrightarrow \mathbb{R}^3$ be immersions with $f^*\langle\ ,\ \rangle = \bar{f}^*\langle\ ,\ \rangle$. Let $n, \bar{n}: M \longrightarrow \mathbb{R}^3$ be the normals determined by the orientation, let dA be the volume form on $(M, f^*\langle\ ,\ \rangle)$, and let ds be the volume form on ∂M. For $p \in \partial M$ let $\mathbf{t}(p)$, $\mathbf{u}(p)$ be the first two vectors of the Darboux frame at $f(p)$ for the curve $f(\partial M)$ on $f(M)$; we regard $\mathbf{t}$ and $\mathbf{u}$ as elements of $\mathbb{R}^3$, as usual. Let κ_n and τ_g be the normal curvature and geodesic torsion for this curve, and let $\bar{\kappa}_n$ and $\bar{\tau}_g$ be the corresponding quantities for the curve $\bar{f}(\partial M)$ on $\bar{f}(M)$. Then

$$\int_{\partial M} (\bar{\tau}_g - \tau_g)\langle f, \mathbf{t}\rangle + (\kappa_n - \bar{\kappa}_n)\langle f, \mathbf{u}\rangle \; ds$$

$$= \int_M h \det(d\bar{n} - dn)\; dA \;+\; 2\int_M \bar{H} - H \; dA \;.$$

Proof. We consider the 1-form

$$\omega = (f \times n) \bullet (\iota \circ d\bar{n})$$

of the previous proof, for which we have

$$(1) \qquad d\omega = -\;\{2\bar{H} - h(2K - \det(d\bar{n} - dn))\}\; dA \;.$$

If $\mathbf{s}$ is the unit tangent vector of the curve ∂M on $(M, f^*\langle\ ,\ \rangle)$, so that $df(\mathbf{s}) = \mathbf{t}$, then

$$dn(\mathbf{s}) = -\kappa_n \mathbf{t} - \tau_g \mathbf{u} \quad ,$$

by definition of κ_n and τ_g. Similarly,

$$(\iota \circ d\bar{n})(\mathbf{s}) = -\bar{\kappa}_n \mathbf{t} - \bar{\tau}_g \mathbf{u} \quad .$$

Therefore

$$\begin{aligned} (2) \qquad \omega(\mathbf{s}) &= f \cdot n \times (\iota \circ d\bar{n})(\mathbf{s}) = f \cdot n \times [-\bar{\kappa}_n \mathbf{t} - \bar{\tau}_g \mathbf{u}] \\ &= f \cdot (\bar{\tau}_g \mathbf{t} - \bar{\kappa}_n \mathbf{u}) \\ &= \bar{\tau}_g \langle f, \mathbf{t} \rangle - \bar{\kappa}_n \langle f, \mathbf{u} \rangle \quad . \end{aligned}$$

Substituting (1) and (2) into Stokes' Theorem,

$$\int_{\partial M} \omega = \int_M d\omega \quad ,$$

we obtain

$$\begin{aligned} (3) \qquad \int_{\partial M} \bar{\tau}_g \langle f, \mathbf{t} \rangle - \bar{\kappa}_n \langle f, \mathbf{u} \rangle \; ds = -2 \int_M \bar{H} \; dA + 2 \int_M hK \; dA \\ + \int_M h \det(d\bar{n} - dn) \; dA \quad . \end{aligned}$$

We cannot use formula (V) for $\int_M hK \; dA$, since M is not compact. But choosing $\bar{f} = f$ in (3), we obtain

$$\int_{\partial M} \tau_g \langle f, \mathbf{t} \rangle - \kappa_n \langle f, \mathbf{u} \rangle \; ds = -2 \int_M H \; dA + 2 \int_M hK \; dA \quad .$$

Substituting this into (3), we obtain the desired result. ■

The proofs of Theorems 10 and 12 have a formal correspondence which is even more complete than their superficial resemblance. Indeed, suppose that the two imbeddings f, $\bar{f}$ of Theorem 12 are part of a variation α of f, with variation vector field Z. Then each $\bar{\alpha}(t)$ has a normal field n_t, and the integrand $h\det(d\bar{n}_t - dn)$ of the Herglotz integral formula can be expanded in powers of t; the terms up to second order in t turn out to be exactly the integrand in Blaschke's integral formula.

More to the point, perhaps, is the fact that the proofs of Theorems 10 and 12 are equally mysterious. They depend on discovering 1-forms ω for which $d\omega = (\text{something interesting})\cdot dA$; these 1-forms ω are suggested by the geometry in only the vaguest way, and one simply has to carry out the computations explicitly to see what $d\omega$ really is. In this connection, however, the following may be mentioned. The requirement that $\bar{f}$ be an imbedding with the same metric as f is a system of partial differential equations (in 2 variables) for the components of $\bar{f}$. (The requirement that α be a bending of f is an even more complicated equation in 3 variables; the basic aim of introducing infinitesimal bendings is to reduce the problem to one in only 2-variables. This "linearization" of the problem leads to a system of linear partial differential equations.) Theorems 10 and 12 may be regarded as uniqueness theorems for partial differential equations on M. As Stoker {1} points out, "the proofs of uniqueness theorems for boundary-value problems involving other partial differential equations also usually require the invention of special tricks and devices, above all if the problems are nonlinear, and such devices commonly involve integrals over the domains in question (e.g., energy integrals in problems having their origin in mathematical physics)."

This is perhaps an opportune moment to describe briefly the original, quite geometric, proofs of these rigidity results. Theorem 10 was first proved by Liebmann, and the crux of his proof was the following observation. Let Z be an infinitesimal bending of $M \subset \mathbb{R}^3$. Regarding each $Z(p)$ as an element of $\mathbb{R}^3$, we obtain a surface $N = \{Z(p) \colon p \in M\} \subset \mathbb{R}^3$. Of course, this may not really be an immersed surface; indeed we hope to prove that it contains only one point when M is compact with $K > 0$. Liebmann showed that at points $p \in M$ where $p \mapsto Z(p)$ is an immersion, the curvature of N is < 0 at $Z(p)$ when the curvature of M is > 0 at p. This immediately shows that $p \mapsto Z(p)$ cannot be an immersion everywhere when M is compact with $K > 0$, since N would then be a compact surface with $K < 0$. Liebmann showed that even when N has singularities, it is nevertheless true that if N is not a point, then N has the character of a surface of negative curvature, in the sense that no point q of N has a support plane (a plane containing q and all points of N on one side of it); this property again contradicts the compactness of N. Since Liebmann's proof involves the investigation of singularities, it is hardly surprising that it works only in the analytic case.

Cohn-Vossen's proof was also originally restricted to the analytic case, and is quite similar to the proof of Hopf's Theorem (9-33) on surfaces of constant mean curvature, which was obviously inspired by it. Given $M, \bar{M} \subset \mathbb{R}^3$, with normals ν and $\bar{\nu}$, and an isometry $\phi \colon M \to \bar{M}$, we call $p \in M$ a "congruence point" if $\phi^*(\overline{II}(\phi(p)) = II(p)$. If ϕ is not the restriction of a Euclidean motion, then by analyticity the congruence points are isolated. At all other points p, Lemma 13 shows that

$$\det[\phi^*(\overline{II}(\phi(p))) - II(p)] < 0 \ ,$$

where the bilinear functions $II(p)$ and $\phi^*(\overline{II}(\phi(p)))$ are regarded as linear transformations on M_p by means of the metric on M_p. It follows that the linear transformation corresponding to the difference $\phi^*(\overline{II}(\phi(p)) - II(p))$ has two eigenspaces, one with a positive eigenvalue, and one with a negative eigenvalue. By picking the one with the positive eigenvalue, say, we obtain a 1-dimensional distribution defined everywhere on M except at the congruence points. At each congruence point we can define the index of the distribution, and the sum of these indices is 2 if M is homeomorphic to S^2 (Theorem 4-21). On the other hand, Cohn-Vossen showed that if M has positive curvature, then the index would have to be negative. The Bibliography will guide the interested reader to descriptions of Cohn-Vossen's argument, as well as alternative arguments and refinements introduced later. We merely mention here that the assumption of analyticity can be dropped by using appropriate results about partial differential equations, and that the whole argument can be formalized to yield an "index method," which has been successfully used in studying certain questions in surface theory; it is one of the few methods which has never yet been generalized to higher dimensions.

We will now return to the use of integral formulas, and prove a result similar to Cohn-Vossen's, which although not strictly speaking a rigidity theorem, nevertheless seems to belong in this chapter since it is the uniqueness part of Minkowski's Problem (p. 228). The original proof was based on the general "Brunn-Minkowski inequality" for the "mixed volumes" of convex sets (see Bonnesen-Fenchel { 1}). The present proof, obviously inspired by Herglotz' proof of Cohn-Vossen's theorem, is due to Chern.

15. THEOREM (MINKOWSKI). Let M be a compact surface with $K > 0$ everywhere, and let $f, \bar{f}: M \to \mathbb{R}^3$ be two isometric imbeddings with $n = \bar{n}$. Then f and

$\bar{f}$ differ by a translation. (Alternatively stated, if two compact convex surfaces in $\mathbb{R}^3$ with everywhere positive curvatures have the same curvatures at points where the normals are parallel, then one surface is a translate of the other.)

<u>Proof</u>. Since $K > 0$, the maps $n, \bar{n}: M \to S^2 \subset \mathbb{R}^3$ are diffeomorphisms, so we can consider the imbeddings

$$g = f \circ n^{-1}: S^2 \to \mathbb{R}^3$$
$$\bar{g} = \bar{f} \circ \bar{n}^{-1}: S^2 \to \mathbb{R}^3 .$$

These imbeddings have the property that $p \in S^2$ is normal to the tangent plane of $g(S^2)$ at $g(p)$, and similarly for $\bar{g}$. Thus the normal maps $\xi, \bar{\xi}: S^2 \to \mathbb{R}^3$ for g and $\bar{g}$ are both the identity map $\mathrm{id}: S^2 \to S^2$. Since $n = \bar{n}$, and $f, \bar{f}$ are isometries by hypothesis, the maps $g, \bar{g}$ induce the same metric on S^2. It clearly suffices to show that g and $\bar{g}$ differ by a translation.

Consider the 1-form

$$\omega = (g \times \bar{g}) \bullet d\bar{g}$$

on S^2. We have

$$(1) \qquad d\omega = (dg \times \bar{g}) \bullet d\bar{g} + (g \times d\bar{g}) \bullet d\bar{g}$$
$$= -\bar{g} \bullet (dg \times d\bar{g}) + g \bullet (d\bar{g} \times d\bar{g}) .$$

To calculate $d\omega$ explicitly we consider a positively oriented moving frame on S^2 which is orthonormal for the usual metric on S^2. If we move the vectors $X_i(p)$ over to $g(p)$, then the translated vectors, $Y_i(g(p))$, will be tangent to

$g(S^2)$ at $g(p)$, since $\xi(p) = p$. Thus we obtain an orthonormal moving frame Y_1, Y_2 on $g(S^2)$. Let θ^1, θ^2 be the dual forms for Y_1, Y_2 and let ψ^α_β be the connection forms for the moving frame (Y_1, Y_2, ν) on $g(S^2)$, where ν is the unit normal field on $g(S^2)$. Define $\bar{\theta}^i$ and $\bar{\psi}^\alpha_\beta$ similarly, using $\bar{g}$. If we set

$$\eta^i = g^*\theta^i , \qquad \bar{\eta}^i = \bar{g}^*\bar{\theta}^i ,$$

then for X tangent to S^2 we have

$$\begin{aligned} (2) \qquad & dg(X) = \eta^1(X)\cdot Y_1 + \eta^2(X)\cdot Y_2 = \eta^1(X)\cdot X_1 + \eta^2(X)\cdot X_2 , \\ & d\bar{g}(X) = \bar{\eta}^1(X)\cdot X_1 + \bar{\eta}^2(X)\cdot X_2 , \end{aligned}$$

where the X_i and Y_i are now regarded as $\mathbb{R}^3$-valued functions. It follows that we have

$$\begin{aligned} dg \times d\bar{g} &= (\eta^1 \wedge \bar{\eta}^2 - \eta^2 \wedge \bar{\eta}^1)\cdot \mathrm{id} , \\ d\bar{g} \times d\bar{g} &= 2(\bar{\eta}^1 \wedge \bar{\eta}^2)\cdot \mathrm{id} , \end{aligned}$$

so that (1) becomes

$$(3) \qquad d\omega = \bar{h}(\eta^1 \wedge \bar{\eta}^2 - \eta^2 \wedge \bar{\eta}^1) - 2h(\bar{\eta}^1 \wedge \bar{\eta}^2) ,$$

where h and $\bar{h}$ are the support functions of g and $\bar{g}$, respectively.

Now we have to relate the forms $\eta^i, \bar{\eta}^i$ to the dual forms for the moving frame X_1, X_2 on S^2. On $g(S^2)$ we have

$$\psi^3_i = \Sigma\, \ell_{ij}\theta^j$$

where $\ell_{ij} = II(Y_i, Y_j)$; hence

$$g^*\psi_i^3 = \Sigma(\ell_{ij} \circ g) \cdot \eta^j \ .$$

But

$$\begin{aligned} g^*\psi_i^3(X) &= \psi_i^3(g_* X) = -\psi_3^i(g_* X) \\ &= \langle -\nu_* g_* X, Y_i \rangle \\ &= \langle -d(\nu \circ g)(X), Y_i \rangle \\ &= \langle -d\xi(X), Y_i \rangle \\ &= \langle -X, Y_i \rangle = \langle -X, X_i \rangle \ . \end{aligned}$$

So if ζ^1, ζ^2 are the dual forms for X_1, X_2, so that $\zeta^1 \wedge \zeta^2 = dA$, the volume element of S^2, then

$$\zeta^i = -\Sigma(\ell_{ij} \circ g) \cdot \eta^j \ .$$

If λ is the 2×2 matrix of functions on S^2 defined by

$$\lambda(p) = (\ell_{ij}(g(p)))^{-1} \ ,$$

then

$$\eta^i = -\Sigma \lambda_{ij} \cdot \zeta^j \ .$$

Similarly,

$$\bar{\eta}^i = -\Sigma \bar{\lambda}_{ij} \zeta^j \ ,$$

where $\bar{\lambda}$ is the inverse of the matrix $(\bar{\ell}_{ij}\circ\bar{g})$. Since

$$\det(\ell_{ij}\circ g) = \det(\bar{\ell}_{ij}\circ\bar{g}) = K ,$$

where K is the curvature for the metrics $g^*\langle\ ,\ \rangle$ and $\bar{g}^*\langle\ ,\ \rangle$ on S^2, we likewise have

$$\det(\lambda_{ij}) = \det(\bar{\lambda}_{ij}) = K^{-1} .$$

Now we have

$$\bar{\eta}^1 \wedge \bar{\eta}^2 = (\det \bar{\lambda})\ \zeta^1 \wedge \zeta^2 = K^{-1}\ dA$$

and

$$\eta^1 \wedge \bar{\eta}^2 - \eta^2 \wedge \bar{\eta}^1 = (\lambda_{11}\bar{\lambda}_{22} + \bar{\lambda}_{11}\lambda_{22} - 2\lambda_{12}\bar{\lambda}_{12})\ dA .$$

Calculating as in the preceeding proof, we see that

$$\eta^1 \wedge \bar{\eta}^2 - \eta^2 \wedge \bar{\eta}^1 = [2K^{-1} - \det((d\bar{\nu})^{-1} - (d\nu)^{-1})]\ dA .$$

So equation (3) becomes

$$d\omega = -\ \{\bar{h}\ \det((d\bar{\nu})^{-1} - (d\nu)^{-1}) + 2K^{-1}(\bar{h} - h)\}\ dA .$$

Hence

$$\int_{S^2} \bar{h}\ \det((d\bar{\nu})^{-1} - (d\nu)^{-1})\ dA = -\int_{S^2} 2K^{-1}(\bar{h} - h)\ dA .$$

Then Lemma 13 shows that the left side is ≤ 0, so that the right side is also

≤ 0. By symmetry we conclude that the right side is 0, and then by Lemma 13 again that in fact $d\nu = d\bar{\nu}$. Thus there is a Euclidean motion A such that $\bar{g} = A \circ g$ and $\bar{\nu} = A_* \nu$. The latter implies that A is a translation. ■

Sometimes this result is expressed in a way that looks quite different: Let $M, \bar{M} \subset \mathbb{R}^3$ be compact surfaces with $K, \bar{K} > 0$ everywhere, and let $\phi\colon M \to \bar{M}$ be a map such that $\bar{K}(\phi(p)) = K(p)$ and such that ϕ preserves the third fundamental forms. Then ϕ is the restriction of a Euclidean motion. To see the equivalence of this statement and Theorem 15, just note that by Proposition 2-7, the normal maps of M and $\bar{M}$ are congruent, so after rotating $\bar{M}$ suitably we have two surfaces satisfying the hypothesis of Theorem 15.

Unlike the last few results, Theorem 15 is not true if we allow $K \geq 0$; a counterexample is shown below. Notice, however, that the components of the

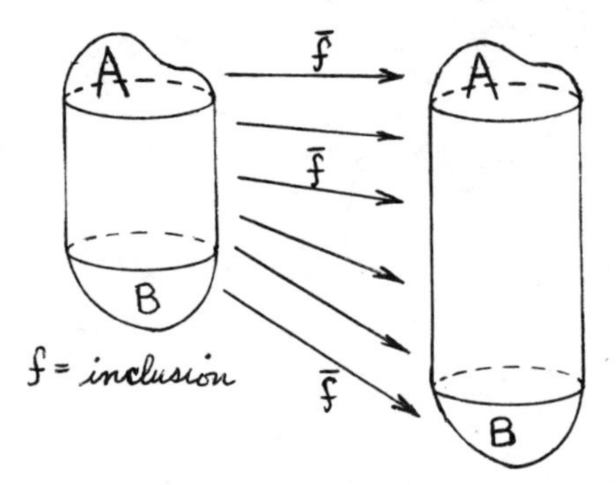

set where $K \neq 0$ on one surface do differ from the corresponding components in the other surface by a translation. It is not hard to see that this is always the case, and that Theorem 15 remains true if the set where $K = 0$ is nowhere dense. (For the alternative statement of the theorem surely the only reasonable situation is that in which n is one-one.)

In our next result, the condition $K > 0$ is more crucial, for we are going to consider the sum

$$\frac{1}{k_1} + \frac{1}{k_2} = \frac{H}{K}$$

of the reciprocals of the principal curvatures; these reciprocals are classically called the "radii of principal curvature." After constructing a proof of Minkowski's Theorem by means of integral formulas, Chern then succeeded in constructing a similar proof for the following result, which is actually older than any yet considered.

16. THEOREM (CHRISTOFFEL). Let M be a compact surface with $K > 0$ everywhere, and let $f, \bar{f}\colon M \to \mathbb{R}^3$ be two imbeddings with $n = \bar{n}$ such that

$$\frac{1}{k_1} + \frac{1}{k_2} = \frac{1}{\bar{k}_1} + \frac{1}{\bar{k}_2} ,$$

where the functions k_i and $\bar{k}_i$ on M are the principal curvatures at the corresponding points of $f(M)$ and $\bar{f}(M)$, respectively. Then f and $\bar{f}$ differ by a translation.

Proof. We introduce the imbeddings $g, \bar{g}\colon S^2 \to \mathbb{R}^3$ of the previous proof, and we will use all the notation and formulas from that proof. In addition we note that for the normal map $\xi = \text{id}$ of g we have, of course,

$$d\xi(X) = X = \zeta^1(X)\cdot X_1 + \zeta^2(X)\cdot X_2 ,$$

from which it follows that

$$(1) \qquad d\xi \times dg = (\zeta^1 \wedge \eta^2 - \zeta^2 \wedge \eta^1)\cdot\xi$$
$$= -(\lambda_{22} + \lambda_{11})\xi\ dA$$
$$= -(k_1^{-1} + k_2^{-1})\xi\ dA .$$

Similarly, since $\xi = \bar{\xi}$, we have

$$(2) \qquad d\xi \times d\bar{g} = -(\bar{k}_1^{-1} + \bar{k}_2^{-1})\xi\ dA .$$

Consider first the 1-form

$$\omega_1 = (g \times \xi) \bullet d\bar{g} .$$

We have

$$d\omega_1 = (dg \times \xi) \bullet d\bar{g} + (g \times d\xi) \bullet d\bar{g}$$
$$= -\xi \bullet (dg \times d\bar{g}) + g \bullet (d\xi \times d\bar{g})$$
$$= -\xi \bullet (dg \times d\bar{g}) - (\bar{k}_1^{-1} + \bar{k}_2^{-1}) g\cdot\xi\ dA \qquad \text{by (2)}$$
$$= -\xi \bullet (dg \times d\bar{g}) + h(\bar{k}_1^{-1} + \bar{k}_2^{-1})\ dA .$$

Similarly, for the 1-form

$$\omega_2 = (g \times \xi) \bullet dg$$

we have

$$d\omega_2 = -\xi \bullet (dg \times dg) + h(k_1^{-1} + k_2^{-1})\ dA .$$

Since $k_1^{-1} + k_2^{-1} = \bar{k}_1^{-1} + \bar{k}_2^{-1}$, we obtain the integral formula

$$(3) \qquad \int_{S^2} \xi \bullet (dg \times d\bar{g}) - \xi \bullet (dg \times dg)\ dA = 0 .$$

By interchanging the roles of g and $\bar{g}$ we also obtain

$$(4) \qquad \int_{S^2} \xi \bullet (d\bar{g} \times dg) - \xi \bullet (d\bar{g} \times d\bar{g})\ dA = 0\ .$$

Adding (3) and (4) we obtain an integral formula

$$(*) \qquad \int_{S^2} I\ dA = 0\ .$$

Now we have

$$\xi \bullet (dg \times d\bar{g}) - \xi \bullet (dg \times dg) = (\eta^1 \wedge \bar{\eta}^2 - \eta^2 \wedge \bar{\eta}^1)\xi\cdot\xi - 2(\eta^1 \wedge \eta^2)\xi\cdot\xi$$
$$= (\lambda_{11}\bar{\lambda}_{22} + \bar{\lambda}_{11}\lambda_{22} - 2\lambda_{12}\bar{\lambda}_{12} - 2(\lambda_{11}\lambda_{22} - \lambda_{12}^2)\ dA\ ,$$

$$\xi \bullet (d\bar{g} \times dg) - \xi \bullet (d\bar{g} \times d\bar{g}) =$$
$$(\bar{\lambda}_{11}\lambda_{22} + \lambda_{11}\bar{\lambda}_{22} - 2\lambda_{12}\bar{\lambda}_{12} - 2(\bar{\lambda}_{11}\bar{\lambda}_{22} - \bar{\lambda}_{12}^2)\ dA\ .$$

So the integrand I in (*) is

$$I = 2(\lambda_{11}\bar{\lambda}_{22} + \bar{\lambda}_{11}\lambda_{22} - 2\lambda_{12}\bar{\lambda}_{12}) - 2(\lambda_{11}\lambda_{22} - \lambda_{12}^2) - 2(\bar{\lambda}_{11}\bar{\lambda}_{22} - \bar{\lambda}_{12}^2)\ dA$$
$$= -2[(\bar{\lambda}_{11} - \lambda_{11})(\bar{\lambda}_{22} - \lambda_{22}) - (\bar{\lambda}_{12} - \lambda_{12})^2]\ dA\ .$$

Since $\lambda_{11} + \lambda_{22} = \bar{\lambda}_{11} + \bar{\lambda}_{22}$ by hypothesis, we can write

$$I = -2[(\lambda_{22} - \bar{\lambda}_{22})(\bar{\lambda}_{22} - \lambda_{22}) - (\bar{\lambda}_{12} - \lambda_{12})^2]\ dA$$
$$= 2[(\bar{\lambda}_{22} - \lambda_{22})^2 + (\bar{\lambda}_{12} - \lambda_{12})^2]\ dA\ .$$

So the integrand I in (*) is everywhere ≥ 0. Hence it must be everywhere $= 0$. Hence $\lambda_{ij} = \bar{\lambda}_{ij}$, so $d\nu = d\bar{\nu}$, and the proof is complete, as before. ∎

In Christoffel's time the radii of principal curvature were regarded as the fundamental entities (which is pretty awkward when $K = 0$), so Theorem 16 was the natural result to try to prove. Nowadays, of course, the result looks rather weird and we would like to formulate it for $H = k_1 + k_2$ instead. Oddly enough, this more reasonable looking problem hasn't been solved. The pair of surfaces pictured on p. 298 give a counterexample of sorts, but I do not know of any counterexample which is strictly convex. This same pair of surfaces illustrates the need for the final hypothesis appearing in the following result along these lines, which replaces the conditions on the normal maps by one on the imbeddings themselves (as some sort of compensation for the stringency of this hypothesis, notice that no hypothesis on K is required).

17. THEOREM (HOPF AND VOSS). Let M be a compact surface, and $f, \bar{f}\colon M \to \mathbb{R}^3$ two imbeddings with $H = \bar{H}$ such that $\bar{f}(p) - f(p)$ is always parallel to a fixed vector $v \in \mathbb{R}^3$. Suppose, moreover, that $f(M)$ and $\bar{f}(M)$ do not contain a portion of a cylinder with generators parallel to v. Then f and $\bar{f}$ differ by a translation in the direction of v.

Proof. Write

$$\bar{f} = f + \alpha \cdot v$$

for some function α on M. For any $X_1, X_2 \in M_p$ we have

$$\begin{aligned} d\bar{f}(X_1) \times d\bar{f}(X_2) &= [df(X_1) + d\alpha(X_1)\cdot v] \times [df(X_2) + d\alpha(X_2)\cdot v] \\ &= df(X_1) \times df(X_2) + [d\alpha(X_1)\cdot v \times df(X_2) - d\alpha(X_2)\cdot v \times df(X_1)] \,. \end{aligned}$$

So

$$d\bar{f} \times d\bar{f} = df \times df + 2(d\alpha \cdot v \times df) .$$

By (I), this is equivalent to

$$(1) \qquad \bar{n}\, d\bar{A} = n\, dA + (d\alpha \cdot v \times df) .$$

Consider the 1-form

$$\omega_1 = (\alpha \cdot v \times n) \bullet df .$$

We have

$$\begin{aligned} (2) \qquad d\omega_1 &= (d\alpha \cdot v \times n) \bullet df + (\alpha \cdot v \times dn) \bullet df \\ &= - n \bullet (d\alpha \cdot v \times df) + \alpha \cdot v \bullet (dn \times df) \\ &= dA - (n \cdot \bar{n})\, d\bar{A} + 2\alpha H (v \cdot n)\, dA \qquad \text{by (1) and (I)} . \end{aligned}$$

Similarly, for the 1-form

$$\omega_2 = (\alpha \cdot v \times \bar{n}) \bullet df ,$$

we have

$$\begin{aligned} (3) \qquad d\omega_2 &= (d\alpha \cdot v \times \bar{n}) \bullet df + (\alpha \cdot v \times d\bar{n}) \bullet df \\ &= (d\alpha \cdot v \times \bar{n}) \bullet df + (\alpha \cdot v \times d\bar{n}) \bullet d\bar{f} , \\ &\qquad \text{since } d\bar{f} = df + d\alpha \cdot v \\ &= - \bar{n} \bullet (d\alpha \cdot v \times df) + \alpha \cdot v \bullet (d\bar{n} \times d\bar{f}) \\ &= n \cdot \bar{n}\, dA - d\bar{A} + 2\alpha \bar{H} (v \cdot \bar{n})\, d\bar{A} \qquad \text{by (1) and (I)} \\ &= n \cdot \bar{n}\, dA - d\bar{A} + 2\alpha \bar{H} (v \cdot n)\, dA \qquad \text{by (1)} . \end{aligned}$$

From (2) and (3) we derive the integral formula

$$2\int_M \alpha(\bar{H}-H)(v\cdot n)\ dA = \int_M (1-n\cdot\bar{n})(dA+d\bar{A}) \ .$$

Since $H = \bar{H}$, we thus have

$$\int_M (1-n\cdot\bar{n})(dA+d\bar{A}) = 0 \ .$$

But $n\cdot\bar{n} \le 1$, so we must have $n\cdot\bar{n} = 1$ everywhere, and hence $n = \bar{n}$ everywhere. We will use this to show that α is constant.

For simplicity assume that v points along the z-axis. The final hypothesis implies that $n(p)$ has non-zero z-component for all points p in a dense open set. If p is such a point, then $f(M)$ and $\bar{f}(M)$ can be represented near p as the graphs of two functions, g and $\bar{g}$, say. Then the normals are given by

$$\frac{(g_1,g_2,-1)}{\sqrt{1+{g_1}^2+{g_2}^2}} \qquad \text{and} \qquad \frac{(\bar{g}_1,\bar{g}_2,-1)}{\sqrt{1+{\bar{g}_1}^2+{\bar{g}_2}^2}} \ .$$

Since these normals are everywhere equal, we must have $g_i = \bar{g}_i$, so α is constant in a neighborhood of p. ∎

Before returning to rigidity theory proper, we ought to mention that Minkowski's Theorem and Christoffel's Theorem have generalizations to compact hypersurfaces $M \subset \mathbb{R}^{n+1}$ with all principal curvatures $k_1,\ldots,k_n > 0$. For each $i = 1,\ldots,n$ we can consider the elementary symmetric polynomial

$$P_i = \sigma_i\left(\frac{1}{k_1},\ldots,\frac{1}{k_n}\right) \ .$$

If M and $\bar{M}$ are two such hypersurfaces, and for some i the functions P_i and $\bar{P}_i$ agree at points of M and $\bar{M}$ where the normals are parallel, then one hypersurface is a translate of the other. A proof may be found in Chern [1]. (Our proofs of Theorems 15 and 16 are taken from that paper, and are pretty representative of the sort of argument which is used. In fact, the proof of Theorem 15 is a special case of Chern's proof for all $i > 1$, while the proof of Theorem 16 is a special case of Chern's proof for $i = 1$, which needs a separate argument. Chern remarks that that distinction is significant, since the case $i = 1$ involves linear partial differential equations, while the case $i > 1$ involves non-linear ones.)

Another remark is necessary to put Theorem 16 on an equal footing with Theorems 12 and 15. The latter two results give the uniqueness of imbeddings whose existence was discussed in Chapter 11 (Weyl's problem and Minkowski's problem). The corresponding existence result for Christoffel's problem is much more involved, for there are complicated relations which must be satisfied by a given function on S^2 in order for it to be $1/k_1 + 1/k_2$ for some imbedded surface. Many incomplete treatments of this problem have been given, and the correct necessary and sufficient conditions (in all dimensions) were discovered only in 1967 by Firey [1]. One could also seek the conditions on a function in order that it be P_i $(1 < i < n)$ for some convex hypersurface, but this problem is perhaps hopelessly complicated.

Finally, we want to point out that the higher dimensional generalization of the Minkowski and Christoffel theorems can also be expressed as follows: Let $M, \bar{M} \subset \mathbb{R}^{n+1}$ be compact hypersurfaces with all principal curvatures > 0, and let $\phi: M \to \bar{M}$ be a map such that $\bar{P}_i(\phi(p)) = P_i(p)$ and such that ϕ preserves the third fundamental forms. Then ϕ is the restriction of a Euclidean motion. The argument is just the same as on p. 298 , using Problem 7-18 in place of Proposition 2-7. In this connection it is interesting that E. Cartan [2] raised the possibility of studying surfaces by means of their second fundamental form, rather than their first. For example, he showed that if II is positive definite, then the curvature of (M,II) can be written in terms of the ordinary principal curvatures and their derivatives. Grove [1] used integral formulas to prove that a diffeomorphism of compact convex surfaces which preserves II and the Gaussian curvature $K = k_1k_2$ is the restriction of a Euclidean motion, and Walden [1] used the index method to prove the same result if either $k_1 + k_2$ or ${k_1}^{-1} + {k_2}^{-1}$ is preserved (as well as if ${k_1}^2 + {k_2}^2$ or ${k_1}^{-2} + {k_2}^{-2}$ is preserved). On the other hand, it is perfectly conceivable that a diffeomorphism of compact convex surfaces is the restriction of a Euclidean motion if it merely preserves II, and perhaps the surfaces need not even be convex. In higher dimensions, the only result is that of Gardner [1], proved using integral formulas, which generalizes Grove's result to the case where II and $k_1 \cdots k_n$ are preserved.

All our rigidity results have required the hypothesis of convexity, so it is only natural to wonder what happens in the case of non-convex surfaces. There is a standard example, illustrated below, of two compact rotation surfaces

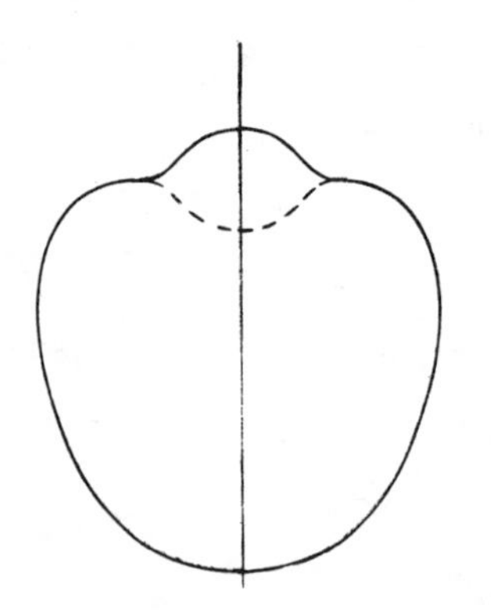

which are isometric but not congruent. To be sure, this example is rather unsatisfying, since it is merely C^∞, and cannot be modified to be analytic. Moreover, the surface consists of two parts which are individually kept rigid, but which are glued together in two different ways along a plane curve where $K = 0$. Finally, this example is merely a reflection of the fact that the plane has infinitesimal bendings which vanish outside a compact set. It could have been obtained by starting with a surface containing a portion of a plane, finding

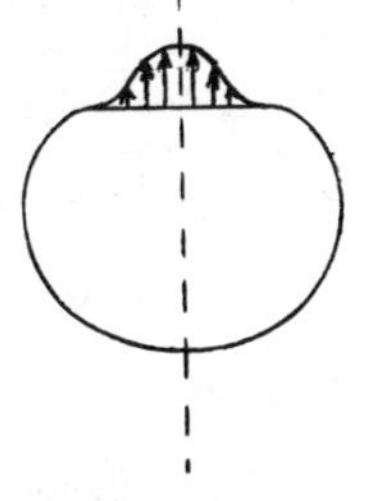

an infinitesimal bending Z which vanishes outside the planar region, and then mapping $p + Z(p)$ to $p - Z(p)$, which is an isometry, by Lemma 7, but not the restriction of a Euclidean motion, by Lemma 8. We can at least show that there is no bending connecting our two non-congruent isometric C^∞ surfaces. In fact, there is not even a bending taking the small region A of positive curvature pictured below to its corresponding region A' in the other surface, the

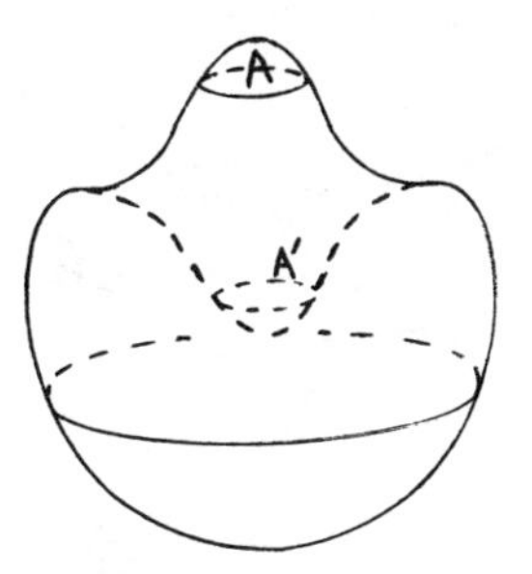

region A' being simply the mirror image of A. To prove this, we simply recall that any surface of positive curvature has a natural orientation, which has to be preserved during the bending.

Long before such C^∞ trickiness was in vogue, Cohn-Vossen [1] had investigated infinitesimally bendable rotation surfaces by quite different methods, and he found C^∞ rotation surfaces with non-trivial C^2 infinitesimal bendings. Afterwards, Rembs [1] managed to obtain an example where the rotation surface is analytic; but the infinitesimal bending is still only C^2 (a point which is by no means made clear in the paper). Applying Lemmas 7 and 8 in this case we merely obtain two C^2 isometric non-congruent surfaces, which seems a lot worse than C^∞; but at least this example is not a bald-faced trick like the previous one. Since the example is interesting, but nevertheless rather disappointing, it has been relegated to an Addendum.

It seems that at present it is simply unknown whether every analytic compact surface is unwarpable. *A fortiori* it is unknown whether every analytic compact surface is unbendable; it certainly seems likely that even C^∞ compact surfaces are unbendable.

There is only one crumb of comfort which we can offer in this dismal situation. It *is* known that a torus of revolution is unwarpable. Such a torus M

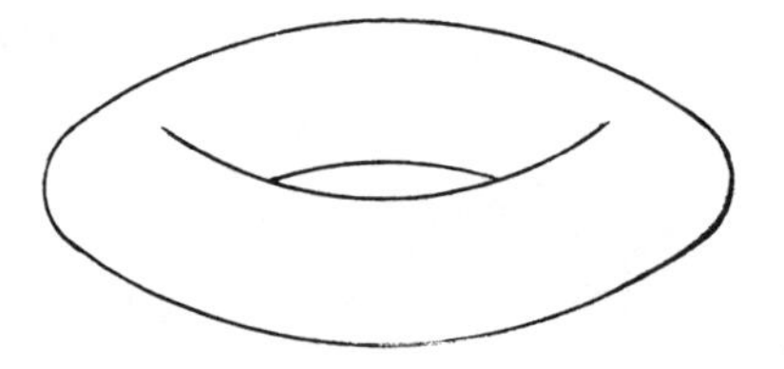

has the property that $\int_{M^+} K\, dA = 4\pi$, where $M^+ = \{p \in M\colon K(p) > 0\}$; the closure $C(M^+)$ of M^+ is a compact surface with boundary such that $\partial C(M^+)$ is the union of plane curves along which the tangent space of M is constant. Suppose that $\phi\colon M \to \bar{M}$ is an isometry. Then also $\int_{\bar{M}^+} K\, dA = 4\pi$, so it follows from Theorem 6-16 that $C(\bar{M}^+)$ is a compact surface of the same sort as $C(M^+)$. Now apply Lemma 14 to f = inclusion map of $C(M^+)$ and $\bar{f} = \phi\colon C(M^+) \to C(\bar{M}^+)$. The terms $\tau_g, \kappa_n, \bar{\tau}_g, \bar{\kappa}_n$ are all 0, since the normal is constant along each component of $\partial C(M^+)$ and $\partial C(\bar{M}^+)$. So we have simply

$$0 = \int_{C(M^+)} h \det(d\bar{n} - dn)\, dA + 2\int_{C(M^+)} \bar{H} - H\, dA ,$$

the same equality which we used in the proof of Theorem 12. Then the same argument which was used in this proof shows that $\phi\colon C(M^+) \to C(\bar{M}^+)$ is the restriction of a Euclidean motion. This already shows that any <u>analytic</u> surface of minimal total absolute curvature is unwarpable in the class of <u>analytic</u> surfaces, a result originally due to Alexandrov [4]. A proof that ϕ is also the restriction of a Euclidean motion on the part of the surface with $K < 0$ has been given by Nirenberg [2]. The proof, which involves a discussion of hyperbolic equations, requires some additional, rather unsatisfactory, hypothesis, but these hypotheses are satisfied at least in the special case of a torus of revolution.

We can also ask about the rigidity of complete convex non-compact surfaces. It is not hard to see (compare the pictures on p.IV.125) that the normal map ν of such a surface M always lie in a hemisphere, so that

$$\int_M K\, dA \leq 2\pi .$$

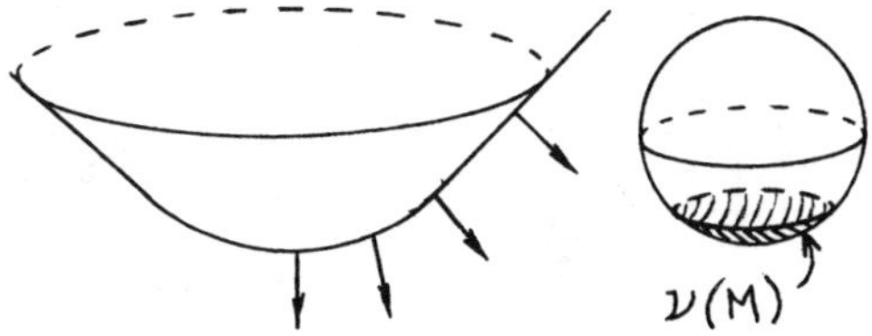

$\nu(M)$

The first result on complete convex non-compact surfaces was the surprising theorem of Olowjanischnikow [1] that M is <u>warpable</u> if $\int_M K\,dA < \pi$. Olowjanischnikow's proof uses the methods of the Russian school of differential geometry, which was briefly discussed in Chapter 11. As we have already mentioned, these methods, though intricate and difficult, allow one to prove certain results for surfaces which are merely continuous. For example, Pogorelov {1} has proved Cohn-Vossen's theorem for arbitrary convex surfaces: if M and $\bar{M}$ are the boundaries of compact convex sets (with non-empty interiors) in $\mathbb{R}^3$, and $\phi: M \to \bar{M}$ is a map which preserves lengths of curves, then ϕ is a congruence. Similarly, Olowjanischnikow's result holds whenever M is the boundary of a closed non-compact convex set (with non-empty interior) in $\mathbb{R}^3$. Pogorelov {2; p.114} also showed that any surface isometric to such a surface M may be joined to M by a continuous bending. To my knowledge, no one has ever provided simpler proofs of these results when the surfaces considered are C^∞.

When our complete convex surface M has $\int_M K\,dA = 2\pi$, it is unwarpable. The proof of this is due to Pogorelov [2]. We have already seen that such surfaces, although unwarpable, and hence unbendable, may nevertheless be infinitesimally bendable; and I think that this is the only known instance of such a phenomenon.

As opposed to the complete convex surfaces, consider what happens when we delete a set with non-empty interior from a convex surface. Is it bendable, or

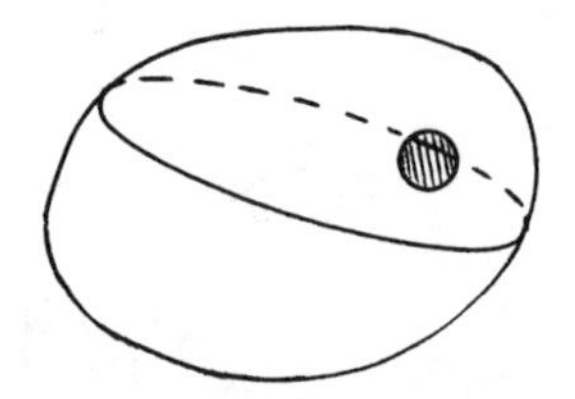

warpable, or infinitesimally bendable? The promptings of intuition seem to vary from person to person, and historically there was considerable confusion on the question. We claim first of all that any open set $U \subset S^2$ whose closure $\bar{U}$ is contained in an open hemisphere of S^2 is warpable. For this purpose we consider the rotation surface M_a of constant curvature 1 given on p.III.237, with $a > 1$. Locally there is an isometry $S^2 \to M_a$ taking the

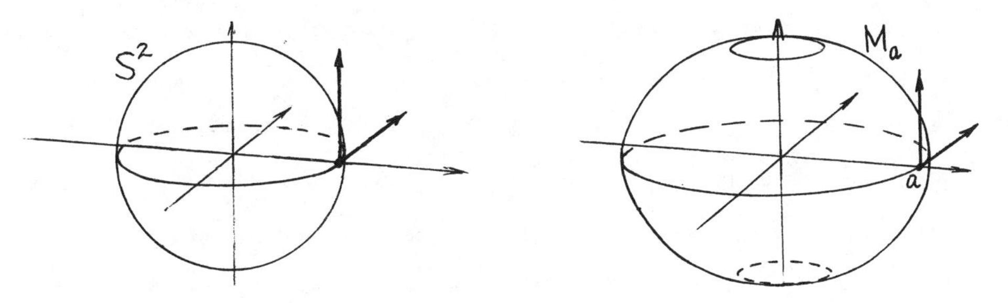

tangent vectors $(0,1,0)$ and $(0,0,1)$ at $(1,0,0) \in S^2$ to the tangent vectors $(0,1,0)$ and $(0,0,1)$ at $(a,0,0) \in M_a$. If a is sufficiently close to 1, then this isometry can be extended to cover $\bar{U} \subset$ the hemisphere $\{p \in S^2 : p^1 > 0\}$. The image of U is contained in the open set $V \subset M_a$ which is obtained by deleting the profile curve of M_a in the left half (x,z)-plane, as well as a neighborhood of the top and bottom boundary curves. We also claim that this

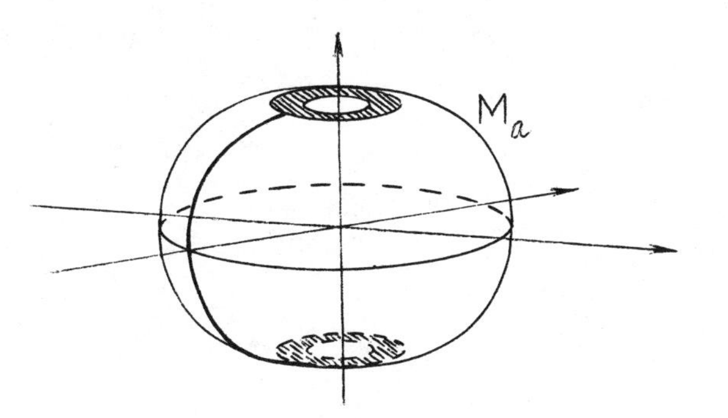

open set V is bendable. To prove this we consider all the rotation surfaces $M_{a'}$ for a' close to a. Then there will be an isometry $f_{a'}: V \to M_{a'}$ which takes the tangent vectors $(0,1,0)$ and $(0,0,1)$ at $(a,0,0) \in M_a$ to the tangent vectors $(0,1,0)$ and $(0,0,1)$ at $(a',0,0) \in M_{a'}$. The 1-parameter family of isometries $\{f_{a'}\}$ gives us the bending.

The warpability of U was noted in a paper by Liebmann [1] in 1900, which also offered up a proof that U is not warpable if it contains a closed hemisphere. Fifteen years later, Blaschke [1] observed that the proof, "as simple as this assertion may appear," was incorrect. In 1919 Liebmann [2] showed that in fact the sphere minus any disc, no matter how small, is bendable. He did this by specifically constructing the bending, using other classical examples of open surfaces of constant curvature 1. A physically intuitive argument, involving soap bubbles, is given in Hilbert and Cohn-Vossen {1}. Liebmann was then willing to conjecture that any convex surface with $K > 0$ everywhere is bendable after a small disc is removed. The infinitesimal bendability of such surfaces was proved by Cohn-Vossen [2] in 1927, and bendability was proven by Hellwig [1] in 1955; both proofs require facts about partial differential equations, with more difficult theorems required for bendability. (This question has also been treated by the Russian school; see Pogorelov {2; p.104}.) It should be noted that the bendability of a convex surface M with a set $A \subset M$ removed does not necessarily imply the bendability of $M - B$ for $A \subset B$. For conceivably the bending of $M - A$ might always be constant

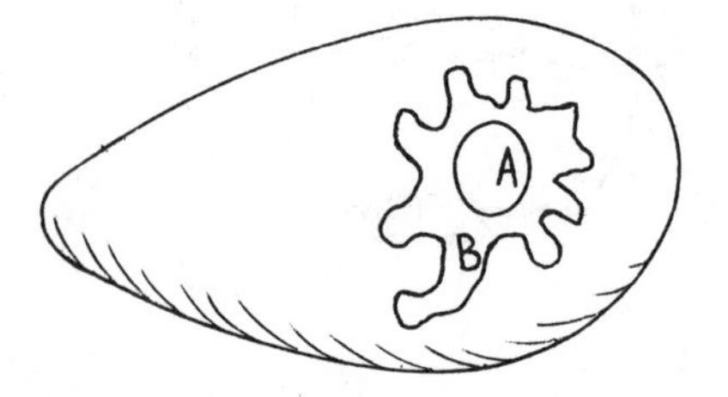

on $M-B$! But at least we do not have to worry about this anomaly if M is analytic, with $K > 0$ everywhere. For it then follows from the results of Chapter 11 that all warpings of $M-A$ are also analytic.

We can also ask what happens when we delete even smaller sets from a convex surface M. Hilbert and Cohn-Vossen {1} claim that the sphere minus any segment of a great circle is bendable, but I have never seen a reference to such a result. I am almost certain that nothing similar is known when the sphere is replaced by an arbitrary convex surface M. As recently as 1971, Greene and Wu [1] showed that if only finitely many points are removed from a compact surface M with $K \geq 0$ everywhere, then the resulting surface $M' = M - \{p_1, \ldots, p_k\}$ is unwarpable. On the other hand, Pogorelov [1] had already shown that if $k \geq 2$, then M' is warpable as an immersion: there is an isometric immersion of M' which is not an imbedding (in fact, there are infinitely many inequivalent isometric immersions). In one special case this is easy to see: Take M' to be $S^2 - \{p_1, p_2\}$, where p_1 and p_2 are the north and south poles, and consider the surface of revolution M_a on p. III.237

with $a = 1/2$; it is not hard to compute that the area of M_a is just $1/2$ the area of S^2. Then there will be an isometry of a closed hemisphere of $S^2 - \{p_1, p_2\}$ onto M_a which takes the two semi-circles on the boundary onto the same curve in M_a, namely its profile curve in the left-hand (x,z)-plane. By using a similar map on the other hemisphere of $S^2 - \{p_1, p_2\}$ we obtain a local isometry $S^2 - \{p_1, p_2\} \longrightarrow M_a$ which is a double covering.

To end this discussion of holey surfaces we mention one more indication of the abysmal state of our ignorance: It is not known whether the standard torus minus a disc is bendable, or even warpable.

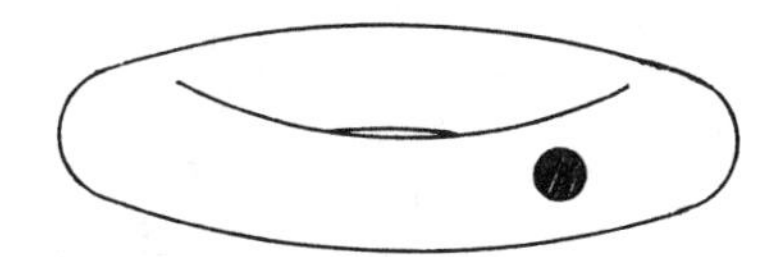

We now turn our attention to purely local results about rigidity, where further surprises are in store for us. We will begin by examining some of the classical results along this line, partly to give an idea of the sort of questions which used to be investigated, and partly because some of these questions throw great light on the geometric aspects of the Darboux equation.

Consider a surface $M \subset \mathbb{R}^3$ and an arc-length parameterized curve $\tilde{c}\colon [a,b] \to M$. Given another arc-length parameterized curve $c\colon [a,b] \to \mathbb{R}^3$, we ask whether there is a neighborhood $\tilde{V}$ of $\tilde{c}([a,b])$ in M and an isometry $\phi\colon \tilde{V} \to V$ of $\tilde{V}$ onto a surface $V \subset \mathbb{R}^3$ such that $\phi \circ \tilde{c} = c$. In other words, we want to know to what extent a curve $\tilde{c}$ on M can be changed in a local warping. We might as well assume that M is the image of an isometric immersion $\tilde{f}\colon (U,(g_{ij})) \to \mathbb{R}^3$, where $U \subset \mathbb{R}^2$ is an open set containing $[a,b] \times \{0\}$, and $\tilde{c}(x) = \tilde{f}(x,0)$ for $x \in [a,b]$. On $[a,b] \times \{0\}$ we can compute the geodesic curvature $\tilde{\kappa}_g(x)$ of $\tilde{c}$. If the isometry ϕ exists, then the geodesic curvature κ_g of c on V must be the same as $\tilde{\kappa}_g$. On the other hand, if c is a curve on any surface V whatsoever, then its geodesic curvature κ_g on V always has absolute value less than or equal to its curvature κ (which is a known function). Thus we see that the isometry ϕ cannot exist unless

$$|\tilde{\kappa}_g| \le \kappa .$$

Consider the case where we have the strict inequality $|\tilde{\kappa}_g(x)| < \kappa(x)$ for all x. If the surface V exists, and $\nu(c(x))$ is its normal at $c(x)$, then equation (9) on p. III.277 shows that we must have

$$\tilde{\kappa}_g = \kappa_g = \kappa \cdot \sin\phi ,$$

where ϕ is the angle from the principal normal $\mathbf{n}(x)$ of c to $\nu(c(x))$. This equation determines two possible choices for ϕ, and hence two possible choices for the tangent space $V_{c(x)}$. Consider either of the two possible continuous choices of $V_{c(x)}$ along c. We claim that we can find V with this

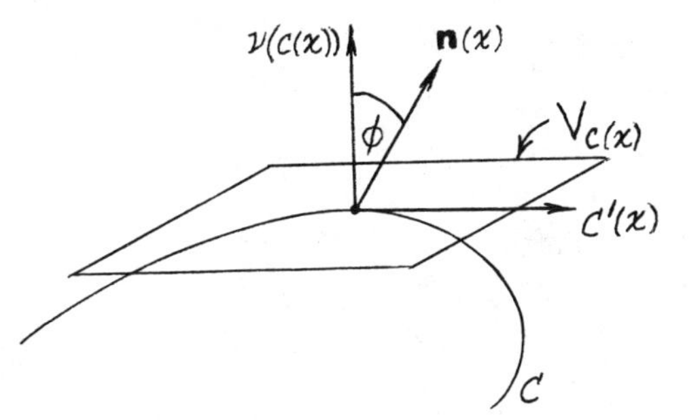

choice of $V_{c(x)}$. We will look for V as the image of an isometric immersion $f\colon U \to \mathbb{R}^3$ with $c(x) = f(x,0)$. By changing our coordinate system on U, we can assume that

$$g_{12} = 0 , \quad g_{22} = 1 ; \quad \text{and} \quad g_{11} = 1 \quad \text{along} \quad [a,b] \times \{0\} .$$

We easily compute that in this case the curve $x \mapsto (x,0)$ has geodesic curvature

$$(1) \qquad \tilde{\kappa}_g(x) = -\tfrac{1}{2}(g_{11})_y .$$

Now the vector $\mathbf{v} = (0,1)_{(x,0)} \varepsilon U_{(x,0)}$ is a unit vector perpendicular to $(1,0)_{(x,0)} \varepsilon U_{(x,0)}$. Obviously we want the vector

$$f_*(\mathbf{v}) = \chi(x) , \qquad \text{say}$$

to be a unit vector in the known vector space $V_{c(s)}$, perpendicular to the known tangent vector $c'(x)$. Thus $\chi(x)$ is determined along $[a,b] \times \{0\}$. For the values of $\chi(x)$ so determined we have

$$(2) \qquad \begin{cases} \langle\chi(x), c'(x)\rangle = 0 \\ \langle\chi(x), \chi(x)\rangle = 1 \\ \langle\chi(x), c''(x)\rangle = \kappa(x) \cdot \langle\chi(x), \mathbf{n}(x)\rangle \\ \qquad = \kappa(x) \cdot \sin\phi \\ \qquad = \tilde{\kappa}_g(x) = -\frac{1}{2}(g_{11})_y , \qquad \text{by (1)} . \end{cases}$$

Now we can use the arguments in the proof of Theorem 11-9. We have the map f defined on $[a,b] \times \{0\}$ by $f(x,0) = c(x)$, and we want to extend f to a neighborhood of $[a,b] \times \{0\}$ in U. The submanifold $[a,b] \times \{0\}$ isn't necessarily of the form $\exp_0(V_1)$, but that is irrelevant, because we have already determined χ satisfying (2), which are just the equations (2') on p. 220. Moreover, χ is linearly independent of f_x, f_{xx} on $[a,b] \times \{0\}$, as required. Thus we can solve equation (*) in the proof of Theorem 11-9 with the initial conditions $f(x,0) = c(x)$ and $q(x,0) = \chi(x)$.

This argument requires that the original surface M and the curves $\tilde{c}$, c be analytic, in order to apply the Cauchy-Kowalewski theorem. But analyticity is not needed if M has curvature $K < 0$, for in this case the system involved is hyperbolic, as we mentioned on p. 226.

For other purposes, it is also important that we examine the classical treatment of this problem, by means of the Darboux equation. We continue to use the special coordinate system on U, with $g_{12} = 0$ and $g_{22} = 1$; and $g_{11} = 1$ along $[a,b] \times \{0\}$. We would like to find $f = (u,v,w)$ by using the three components of the equations

$$f(x,0) = c(x)$$
$$f_y(x,0) = \chi(x)$$

as the initial conditions for the solutions u, v, w of the Darboux equation. Unfortunately that simple procedure won't work, since, as we have already seen, there is not that much arbitrariness permitted in the choice of u, v, w. What we have to do is first find a solution w of the Darboux equation with

$$(3) \qquad w(x,0) = c^3(x) , \qquad w_y(x,0) = \chi^3(x) ,$$

and then choose u and v so that we at least have

$$(4) \qquad \begin{aligned} u(0,0) &= c^1(0) , & u_x(0,0) &= c^{1\prime}(0) , & u_y(0,0) &= \chi^1(0) \\ v(0,0) &= c^2(0) , & v_x(0,0) &= c^{2\prime}(0) , & v_y(0,0) &= \chi^2(0) , \end{aligned}$$

and so that $f = (u,v,w)$ is an isometry, and consequently satisfies

$$(5) \qquad \left\{ \begin{aligned} g_{11} &= \langle f_x, f_x \rangle = u_x \cdot u_x + v_x \cdot v_x + w_x \cdot w_x \\ 0 &= \langle f_x, f_y \rangle = u_x \cdot u_y + v_x \cdot v_y + w_x \cdot w_y \\ 1 &= \langle f_y, f_y \rangle = u_y \cdot u_y + v_y \cdot v_y + w_y \cdot w_y \\ &\Downarrow \text{ (as in the proof of Theorem 11-9)} \\ -\tfrac{1}{2}(g_{11})_y &= \langle f_{xx}, f_y \rangle = u_{xx} \cdot u_y + v_{xx} \cdot v_y + w_{xx} \cdot w_y . \end{aligned} \right.$$

We claim that u and v will automatically have the desired initial conditions on $[a,b]\times\{0\}$. To see this, consider the $\mathbb{R}^2$-valued functions

$$\alpha(x) = (c^{1\prime}(x), c^{2\prime}(x))$$
$$\beta(x) = (\chi^1(x), \chi^2(x)) .$$

By substituting equations (3) into equations (2), we find that α and β satisfy

$$\text{(i)} \qquad \langle\alpha,\beta\rangle = -\, w_y(x,0)\cdot w_x(x,0)$$

$$\text{(ii)} \qquad \langle\beta,\beta\rangle = 1 - w_y(x,0)\cdot w_y(x,0)$$

$$\text{(iii)} \qquad \langle\alpha',\beta\rangle = -\frac{1}{2}(g_{11})_y - w_y(x,0)\cdot w_{xx}(x,0) ,$$

while we also have

$$\text{(iv)} \qquad \langle\alpha,\alpha\rangle = 1 - w_x(x,0)\cdot w_x(x,0) = g_{11}(x,0) - w_x(x,0)\cdot w_x(x,0) .$$

Simple arguments (Problem 1) show that the solution of the system of equations (i)-(iv) is completely determined once $\alpha(0)$, $\beta(0)$ are known. But equations (5) show that

$$(u_x(x,0), v_x(x,0))$$
$$(u_y(x,0), v_y(x,0))$$

satisfy this system, while equations (4) insure that

$$\alpha(0) = (u_x(0,0), v_x(0,0))$$
$$\beta(0) = (u_y(0,0), v_y(0,0)) .$$

It follows that for all $x \in [a,b]$ we have

$$(c^{1\prime}(x),c^{2\prime}(x)) = \alpha(x) = (u_x(x,0),v_x(x,0))$$

$$\Rightarrow (c^1(x),c^2(x)) = (u(x,0),v(x,0)) \text{ by (4)}$$

$$(\chi^1(x),\chi^2(x)) = \beta(x) = (u_y(x,0),v_y(x,0)) \ .$$

Thus u and v will indeed have the initial conditions which we would like, and $f = (u,v,w)$ will be the required isometric imbedding.

We have already noted in the previous Chapter that the Darboux equation is hyperbolic when $K < 0$, so we do not need the Cauchy-Kowalewski theorem in that case. In this case there is still one problem remaining, however, for in order to solve the Darboux equations for w with the initial conditions (3), we need to know that the interval $[a,b]$ of the x-axis is free for these initial conditions. Here is the place where the geometry links up beautifully with the analysis. To begin with, suppose we have an isometric immersion $f = (u,v,w)\colon (U,(g_{ij})) \to \mathbb{R}^3$, and a curve γ in U. Then w is a solution of the Darboux equation, and we would like to know when γ is free for the initial conditions which we obtain by restricting w to γ. Recall that for the second order PDE

$$F(x,y,w,p,q,r,s,t) = 0 \ ,$$

this means that

$$\frac{\partial F}{\partial r}\cdot(\nu^1)^2 + \frac{\partial F}{\partial s}\cdot\nu^1\nu^2 + \frac{\partial F}{\partial t}\cdot(\nu^2)^2 \neq 0 \qquad \text{at } \gamma(t) \ ,$$

where (ν^1,ν^2) is the normal to γ at t. Since (ν^1,ν^2) is proportional to $(-\gamma_2{}',\gamma_1)$, this means that

$$\frac{\partial F}{\partial r}\cdot(\gamma_2{}')^2 - \frac{\partial F}{\partial s}\cdot\gamma_1{}'\gamma_2{}' + \frac{\partial F}{\partial t}\cdot(\gamma_2{}')^2 \neq 0 \ .$$

Consider the Darboux equation in the form (*) on p. 208 . Our condition becomes

$$(6) \qquad (w_{22} - \Gamma^1_{22}w_1 - \Gamma^2_{22}w_2)(\gamma_2{}')^2 + 2(w_{12} - \Gamma^1_{12}w_1 - \Gamma^2_{22}w_2)\gamma_1{}'\gamma_2{}' + (w_{11} - \Gamma^1_{11}w_1 - \Gamma^2_{11}w_2)(\gamma_1{}')^2 \neq 0 \ ,$$

which by equation (1) on p. 207 becomes

$$0 \neq n^3[\ell_{22}(\gamma_2{}')^2 + 2\ell_{12}\gamma_1{}'\gamma_2{}' + \ell_{11}(\gamma_1{}')^2] = n^3 II_f(\gamma',\gamma') \ .$$

Thus the curve γ is free for the initial conditions determined by w if and only if the curve $f\circ\gamma$ on $f(U)$ is nowhere asymptotic, and the tangent plane for $f(U)$ is nowhere vertical along γ. On the other hand, γ is characteristic for these initial conditions if and only if $f\circ\gamma$ is an asymptotic curve, except perhaps at points where the tangent plane of $f(U)$ is vertical. These conditions have the paradoxical feature customarily associated with the Darboux equation: the condition on a single component w of f is stated in terms of the whole map f. When we are merely given initial conditions along a curve, rather than a solution, we simply write out equation (6) as stated. In the situation we are considering, our initial curve is just the interval $[a,b]$ of the x-axis, so that $\gamma_1{}' = 1$ and $\gamma_2{}' = 0$, and we compute that in our special coordinate system on U we have

$$\begin{cases} [11,2] = -\frac{1}{2}(g_{11})_y \\ [12,1] = [21,1] = \frac{1}{2}(g_{11})_y \\ \text{other } [ij,k] = 0 \end{cases} \qquad \text{along the } x\text{-axis}$$

$$\begin{cases} \Gamma^1_{12} = \Gamma^1_{21} = \frac{1}{2}(g_{11})_y \\ \Gamma^2_{11} = -\frac{1}{2}(g_{11})_y \\ \text{other } \Gamma^k_{ij} = 0 \end{cases} \qquad \text{along the } x\text{-axis.}$$

Then equation (6) becomes

$$(6') \qquad w_{11} + \frac{1}{2}(g_{11})_y w_2 \neq 0 , \qquad \text{or} \qquad (c^3)'' - \tilde{\kappa}_g \chi^3 \neq 0 .$$

Now in our situation, χ is not a multiple of c''. Simply by rotating everything, we can then insure that $(c^3)'' - \tilde{\kappa}_g \chi^3 \neq 0$ on $[a,b] \times \{0\}$. Thus we really can solve the Darboux equation and obtain an isometry $\phi\colon \tilde{V} \longrightarrow V$ with $\phi \circ \tilde{c} = c$ (suitably rotated). Naturally we can then obtain a new isometry ϕ' with $\phi' \circ \tilde{c} = c$.

Things work out quite differently when we try to find an isometry $\phi\colon \tilde{V} \longrightarrow V$ with $\phi \circ \tilde{c} = c$ in the case where $\tilde{\kappa}_g(x) = \kappa(x)$ for all x. If ϕ exists, then we must have $\kappa_g = \tilde{\kappa}_g = \kappa$, so c must be an asymptotic curve on V, which means that V must have $K \le 0$ along c so M must have $K \le 0$ along $\tilde{c}$. We will actually assume $K < 0$, so that the Darboux equations are hyperbolic. We first suppose that $\kappa(x) > 0$ for all x. Then the Beltrami-Enneper Theorem (4-7) shows that the torsion τ of c must satisfy

$$\tau(x) = \pm \sqrt{-K(c(x))} = \pm \sqrt{-K(\tilde{c}(x))} .$$

Thus there is, up to Euclidean motions, only one possibility for c. If we are given this curve c, and $\phi\colon \tilde{V} \longrightarrow V$ exists, then $V_{c(x)}$ must be the osculating plane of c at x, so our choice for $\chi(x) = f_y(x,0)$ must be the principal normal $\mathbf{n}(x)$ of c at χ. In this situation we have $c'' = \kappa\,\mathbf{n} = \kappa\chi = \tilde{\kappa}_g\chi$, so equation (6') is <u>not</u> true; our initial curve is characteristic for the initial conditions. Fortunately, we have complete information about this situation, since we are dealing with a Monge-Ampère equation. Along the x-axis our Darboux equation (from p. 208) is

$$[w_{11}+\tfrac{1}{2}(g_{11})_y w_2]\cdot w_{22} - [w_{12}-\tfrac{1}{2}(g_{11})_y w_1]^2 = K(1-w_1{}^2-w_2{}^2)\ .$$

As we pointed out at the end of section 8 of Chapter 10, there is no hope of solving for w_{22} along the x-axis unless we also have

$$(7) \qquad - [w_{12}-\tfrac{1}{2}(g_{11})_y w_1]^2 = K(1-w_1{}^2-w_2{}^2)$$

along the x-axis. Moreover, if this equation does hold, then we can choose w_{22} arbitrarily, and there will be a solution with these intial conditions. We claim that equation (7) holds as a consequence of our choice of c. For,

$$\begin{aligned}[]
[w_{12}-\tfrac{1}{2}(g_{11})_y w_1]^2 &= \{(\mathbf{n}'+\kappa c')^3\}^2 && \text{[the 3 denotes third component]}\\
&= \{(-\kappa c'+\tau\,\mathbf{b}+\kappa c')^3\}^2 && \text{by Serret-Frenet}\\
&= \{(\tau\,\mathbf{b})^3\}^2 = \tau^2(\mathbf{b}^3)^2\\
&= -K(\mathbf{b}^3)^2\ ,
\end{aligned}$$

so we just have to show that

$$(\mathbf{b}^3)^2 = 1 - w_1{}^2 - w_2{}^2 = 1 - \{(c')^3\}^2 - \{\mathbf{n}^3\}^2\ .$$

This is elementary: we have

$$e_3 = \langle e_3, c'\rangle c' + \langle e_3, \mathbf{n}\rangle \mathbf{n} + \langle e_3, \mathbf{b}\rangle \mathbf{b} ,$$

and when we take the inner product with e_3 we get

$$1 = \{(c')^3\}^2 + \{\mathbf{n}^3\}^2 + \{\mathbf{b}^3\}^2 ,$$

as desired.

Thus, when c is a curve with $\kappa = \tilde{\kappa}_g$ and $\tau^2 = -K$, we can find infinitely many isometries $\phi\colon \tilde{V} \to V$ with $\phi \circ \tilde{c} = c$; all the surfaces V are tangent to each other along c. In particular, if $\tilde{c}$ is an asymptotic curve on M, with $K < 0$ along $\tilde{c}$, then we can take c to be $\tilde{c}$, and we see that a neighborhood of $\tilde{c}$ can be continuously bent keeping $\tilde{c}$ fixed; all surfaces in the bending are tangent to M along $\tilde{c}$. As opposed to this, if $\tilde{c}$ satisfies $\tilde{\kappa}_g < \tilde{\kappa}$ everywhere, then there is only one other surface containing $\tilde{c}$ which is isometric to a neighborhood of $\tilde{c}$ in M, and it is nowhere tangent to M along $\tilde{c}$.

The case where $\kappa = \tilde{\kappa} = 0$ (both $\tilde{c}$ and c are straight lines) is similar, except that now there is even complete leeway in the choice of the tangent space of V along c.

The discovery that asymptotic lines of a surface are precisely the curves along which the surface may be bent leads one to formulate all sorts of other questions. For example, when is there an isometry $\phi\colon M \to \bar{M}$ which takes both families of asymptotic lines of M to asymptotic lines of $\bar{M}$? It is easy to see that this happens essentially only when ϕ is the restriction of a Euclidean motion. In fact, if $f\colon U \to M$ is an imbedding for which the parameter lines are asymptotic curves, so that $\ell = n = 0$, and we define $\bar{f} = \phi\circ f$, then also $\bar{\ell} = \bar{n} = 0$. But we have, in addition,

$$\ell n - m^2 = \bar{\ell}\bar{n} - \bar{m}^2 \implies m = \pm\,\bar{m}\,.$$

If we restrict our attention to surfaces with $K < 0$, then we must have $m = \bar{m}$ or $m = -\bar{m}$ everywhere, and we can assume $m = \bar{m}$ by suitable choice of the normal. Hence ϕ is the restriction of a Euclidean motion.

Since this question turned out to be rather uninteresting, we modify it by investigating isometries $\phi\colon M \to \bar{M}$ which take the asymptotic curves of only one family of asymptotic lines on M to asymptotic lines on $\bar{M}$. Choose an orthonormal moving frame X_1, X_2 on M such that $II(X_1,X_1) = 0$, so that the integral curves of X_1 are the given family of asymptotic curves. Then we have

$$(1)\qquad \begin{aligned} \psi_1^3 &= \phantom{m\theta^1 + {}} m\theta^2 \\ \psi_2^3 &= m\theta^1 + n\theta^2\,, \end{aligned}$$

where $0 = \ell = II(X_1,X_1)$ and $m = II(X_1,X_2)$ and $n = II(X_2,X_2)$. Let $\bar{X}_1, \bar{X}_2$ be the orthonormal moving frame $\bar{X}_i = \phi_* X_i$ on $\bar{M}$, and let barred forms (e.g. $\bar{\psi}_1^3$) actually denote ϕ^* of the corresponding forms on $\bar{M}$. Then $\bar{\theta}^i = \theta^i$ and $\bar{\omega}_2^1 = \omega_2^1$. Now $\bar{\ell} = \overline{II}(\bar{X}_1,\bar{X}_1) = 0$ by the hypothesis that our family of asymptotic curves is taken into asymptotic curves. Moreover,

$$\ell n - m^2 = \bar{\ell}\bar{n} - \bar{m}^2 \implies m = \pm\,\bar{m}\,;$$

again we consider only the case $K < 0$ everywhere, so we might as well assume that $m = \bar{m}$, by suitable choice of the normals. Then we have

$$(2) \qquad \begin{aligned} \bar{\psi}_1^3 &= \quad m\theta^2 \\ \bar{\psi}_2^3 &= m\theta^1 + \bar{n}\theta^2 . \end{aligned}$$

In particular,

$$\begin{aligned} \psi_1^3 = \bar{\psi}_1^3 &\implies d\psi_1^3 = d\bar{\psi}_1^3 \\ &\implies \omega_1^2 \wedge \psi_2^3 = \omega_1^2 \wedge \bar{\psi}_2^3 \qquad \text{by Codazzi-Mainardi} \\ &\implies (n - \bar{n})\ \omega_1^2 \wedge \theta^2 = 0 . \end{aligned}$$

Applying this to X_1, X_2 yields

$$(n - \bar{n}) \cdot \omega_1^2(X_1) = 0 .$$

If $n = \bar{n}$ everywhere, then ϕ is a congruence. Assume that $n - \bar{n}$ is always $\neq 0$. Then $\omega_1^2(X_1) = 0 \implies \nabla_{X_1} X_1 = 0$, so the integral curves of X_1 are geodesics. Since they are also asymptotic curves, they must be straight lines; similarly their images, being both geodesics and asymptotic curves, are straight lines. In other words, this case involves a ruled surface being warped in such a way that the rulings remain straight.

{In general, it is easy to see that ruled surfaces can always be bent keeping their generators straight. In fact, suppose that our ruled surface is

$$f(s,t) = c(s) + t\delta(s) , \qquad |\delta| = 1 \implies \langle\delta,\delta'\rangle = 0 .$$

Then

$$E = \langle c' + t\delta', c' + t\delta' \rangle = \langle c', c' \rangle + 2t\langle c', \delta' \rangle + t^2\langle \delta', \delta' \rangle$$
$$F = \langle c', \delta \rangle$$
$$G = 1 .$$

Let $\bar{\delta}$ be any curve with

$$|\bar{\delta}| = 1 , \qquad |\bar{\delta}'| = |\delta'| .$$

In order for the surface

$$\bar{f}(s,t) = \bar{c}(s) + t\bar{\delta}(s)$$

to have the same metric as f, the curve $\bar{c}$ must satisfy

$$|\bar{c}'| = |c'|$$
$$\langle \bar{c}', \bar{\delta}' \rangle = \langle c', \delta' \rangle$$
$$\langle \bar{c}', \bar{\delta} \rangle = \langle c, \delta \rangle ,$$

which is always solvable for $\bar{c}'$. The curve $\bar{\delta}$ can have essentially any shape. For example, if $\delta' \neq 0$ everywhere (the rulings are always changing), then we could reparameterize so that $|\delta| = |\delta'| = 1$. Then all we require is $|\bar{\delta}| = |\bar{\delta}'| = 1$, so that $\bar{\delta}$ can trace out any regular curve in S^2.}

Here is one final classical problem about bendings. When does a surface M have a bending $\alpha\colon [0,1] \times M \longrightarrow \mathbb{R}^3$ such that each $\bar{\alpha}(t)(M) \subset M$? One example that immediately comes to mind is a surface of revolution. Obviously any surface isometric to a surface of revolution also has this property. (The question really has almost nothing to do with surfaces in $\mathbb{R}^3$, and is essentially intrinsic). We will show that "in general" these are the only such surfaces. What we actually assume is that the various curves $c_p(t) = \alpha(t,p)$ give a foliation of M. Since each $\bar{\alpha}(t)$ is an isometry, each curve c_p has constant geodesic curvature, for $\bar{\alpha}(t_1)\circ\bar{\alpha}(t_0)^{-1}$ is an isometry taking a neighborhood of $c(t_0)$ to a neighborhood of $c(t_1)$. Now for any curve c_p, consider its

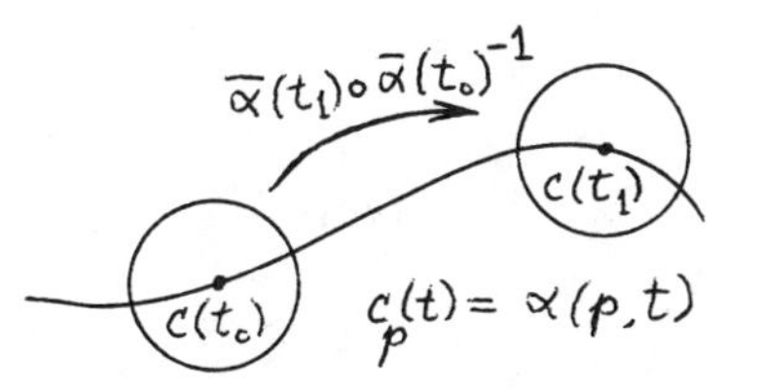

"geodesic parallels," the set of points at a fixed distance d along the geodesics perpendicular to c_p. Let q be the point on the geodesic inter-

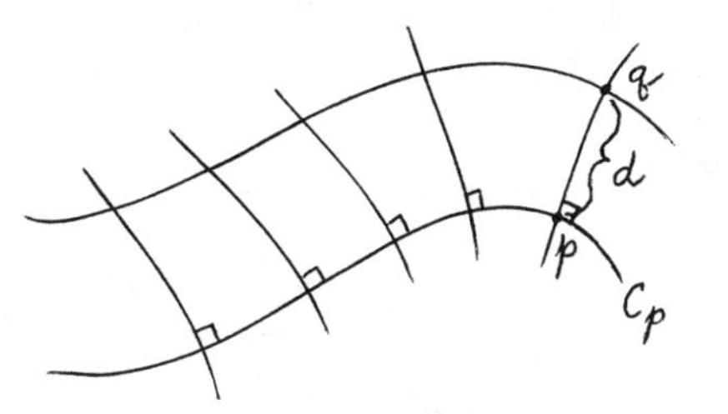

secting c_p orthogonally at p. Clearly $\alpha(t,q)$ is on the geodesic intersecting c_p orthogonally at $\alpha(t,p)$. Thus the geodesic parallels of c_p are the other curves c_q. Note that the geodesics perpendicular to c_p are also perpendicular to c_q (Problem I.9-28). Now take a coordinate system u, v such that the v-parameter curves lie along the curves c_p, while the u-parameter curves are the arclength parameterized geodesics perpendicular to all curves c_p. Then the metric has the form

$$du \otimes du + G\, dv \otimes dv \ .$$

A computation shows that the geodesic curvature of the v-parameter curve through (u,0) is

$$\kappa_u(v) = -\frac{1}{2}\frac{G_u(u,v)}{G(u,v)} = -\frac{1}{2}\frac{\partial \log G}{\partial u}(u,v) .$$

But $\kappa_u(v)$ depends only on u. So $\log G$ is of the form $a(u)+b(v)$, and thus G is of the form

$$G(u,v) = A(u)\cdot B(v) .$$

Letting v_1 be a function with

$$v_1' = \sqrt{B} ,$$

our metric takes the form

$$du\otimes du + A(u)\ dv_1\otimes dv_1 .$$

Comparing with formula (4) on p.III.231, we see that M is isometric to a surface of revolution. It doesn't seem worthwhile refining these purely local considerations by trying to analyse in detail just what happens when some of the curves c_p degenerate to points, but it certainly would be nice if one could prove that a compact surface M admitting a bending into itself is (globally) isometric to a surface of revolution.

Just for the hell of it, we will also look at a couple of classical local results about infinitesimal bendings. Let $M \subset \mathbb{R}^3$ be a surface, and let $\alpha\colon [0,1]\times M \longrightarrow \mathbb{R}^3$ be any variation of the inclusion map $i\colon M \longrightarrow \mathbb{R}^3$ whose variation vector field Z at $t = 0$ is an infinitesimal bending of M. Let X_1, X_2 be an orthonormal moving frame on M. Then

$$\left.\frac{d}{dt}\right|_{t=0} \langle \bar{\alpha}(t)_* X_i, \bar{\alpha}(t)_* X_j\rangle = \langle dZ(X_i), X_j\rangle + \langle X_i, dZ(X_j)\rangle$$

by the argument on pp.251-252; since Z is an infinitesimal bending we thus have

$$(1)\qquad \left.\frac{d}{dt}\right|_{t=0} \langle \bar{\alpha}(t)_* X_i, \bar{\alpha}(t)_* X_j \rangle = 0 \ .$$

On each surface $\bar{\alpha}(t)(M)$ we can define an orthonormal moving frame $X_1(t)$, $X_2(t)$ by applying the Gram-Schmidt orthonormalization process to $\bar{\alpha}(t)_* X_1$, $\bar{\alpha}(t)_* X_2$. Let $\theta^i(t)$ be $\bar{\alpha}(t)^*$ of the dual forms for this moving frame, so that $\theta^i(0) = \theta^i$, the dual forms for X_1, X_2. Equation (1) is easily seen to imply that

$$(2)\qquad 0 = \left.\frac{d}{dt}\right|_{t=0} \theta^i(t) = \dot{\theta}^i \ , \qquad \text{say} \ .$$

For each t we have unique forms $\omega^i_j(t)$ with

$$(3)\qquad \begin{aligned} \omega^i_j(t) &= -\,\omega^j_i(t) \\ d\theta^i(t) &= -\sum_{j=1}^{2} \omega^i_j(t) \wedge \theta^j(t) \ . \end{aligned}$$

Letting $t = 0$, we see that $\omega^i_j(t) = \omega^i_j$, the connection forms for X_1, X_2. Now differentiate (3) with respect to t. Since (Problem 2) we always have

$$(d\eta)^{\cdot} = d\dot{\eta}$$

for any 1-parameter family of forms $\eta(t)$, we obtain, using (2),

$$\begin{aligned} \dot{\omega}^i_j &= -\,\dot{\omega}^j_i \\ 0 = d\dot{\theta}^i = (d\theta^i)^{\cdot} &= -\sum_{j=1}^{2} \dot{\omega}^i_j \wedge \theta^j - \sum_{j=1}^{2} \omega^i_j \wedge \dot{\theta}^j \\ &= -\sum_{j=1}^{2} \dot{\omega}^i_j \wedge \theta^j \ . \end{aligned}$$

It follows that $\dot{\omega}^1_2 = 0$. Now we differentiate the equation

$$d\omega^1_2(t) = K(t)\theta^1(t) \wedge \theta^2(t) \ ,$$

to obtain

$$0 = d\dot{\omega}^1_2 = (d\omega^1_2)^{\boldsymbol{\cdot}} = \dot{K}\ \theta^1 \wedge \theta^2 + 0 \ .$$

Thus we see that we always have

$$\dot{K} = 0 \ .$$

It now seems a natural enough question to ask when we have $\dot{H} = 0$. The answer to this question is left to Problem 4.

Another question, especially interesting in view of Lemma 6, is to find those surfaces M which possess an infinitesimal bending Z that is everywhere tangent to M. Once again, surfaces of rotation are obvious examples. Moreover, one can easily show that if Z is an infinitesimal bending of M which is tangent to M, and $f\colon M \to \bar{M}$ is an isometry, then $\bar{Z} = f_*Z$ is an infinitesimal bending of $\bar{M}$. Again we can show that "in general" these are the only such surfaces. Given a nowhere 0 infinitesimal bending Z tangent along M, we choose an immersion $f\colon U \to M$ such that v-parameter curves lie along the integral curves of Z, and the u-parameter curves along the curves perpendicular to them. Then $I_f = f^*\langle\ ,\ \rangle$ has the form

$$I_f = E\ du \otimes du + G\ dv \otimes dv \ .$$

By assumption, Z is always proportional to $\partial f/\partial v$, and it will be convenient to write Z as

$$Z = \frac{\lambda}{\sqrt{G}} \frac{\partial f}{\partial v} = \frac{\lambda}{\sqrt{G}} f_2 .$$

Then the equations

$$0 = \langle Z_1, f_1 \rangle = \langle Z_2, f_2 \rangle , \qquad 0 = \langle Z_1, f_2 \rangle + \langle Z_2, f_1 \rangle$$

lead to

$$\frac{\partial E}{\partial v} = 0 , \qquad \frac{\partial \lambda}{\partial v} = 0 , \qquad \frac{\partial\left(\frac{\lambda}{\sqrt{G}}\right)}{\partial v} = 0 .$$

From the first we see that we can alter the immersion f so that $E = 1$ everywhere. From the third we see that we can likewise arrange that $\lambda = \sqrt{G}$ everywhere. Then our metric has the form

$$du \otimes du + \lambda(u)^2 \, dv \otimes dv ,$$

as desired.

Most of these local results are rather unsatisfying, since they usually require some subsidiary conditions of the same nature as those used in the classical classification of flat surfaces. But there are certain questions where local results are precisely what we should be interested in. We have already seen (pp. 306-308) that there are isometric compact surfaces in $\mathbb{R}^3$ which cannot be connected by a bending. But it seems likely that isometric surfaces can <u>locally</u> be connected by a bending. Actually, the argument on pp. 307-308 shows that even this is false, since a surface of positive curvature can never be bent into its mirror image. So we should instead conjecture that given any two isometric surfaces, the first can locally be connected by a bending to the second or else to its mirror image.

To investigate this question, we consider once again the Darboux equation. From the considerations on pp.208-211 we see that the immersions $f = (u,v,w)\colon U \to \mathbb{R}^3$ defined in a neighborhood of $0 \in \mathbb{R}^2$ such that

(i) $I_f = E\, dx \otimes dx + F[dx \otimes dy + dy \otimes dx] + G\, dy \otimes dy$

(ii) $w(0) = w_1(0) = w_2(0) = 0$

(iii) $u(0) = v(0) = 0$,

(iv) $u_1(0) = 0, \quad u_2(0) > 0, \quad v_1(0) > 0$

are in one-one correspondence with the solutions w of the Darboux equation which satisfy (ii). Writing the Darboux equation as on p.211 [equation (6)], we see that the following holds:

(*) Let E, F, G be the components of a metric in a neighborhood of $0 \in \mathbb{R}^2$, and let ρ, σ be two functions in a neighborhood of $0 \in \mathbb{R}$ with

$$\rho(0) = \rho'(0) = \sigma(0) = 0$$
$$\rho''(0) \neq 0 .$$

Assume E, F, G and ρ, σ are analytic, unless the curvature K satisfies $K(0) < 0$. Then there is a unique immersion $f = (u,v,w)$ defined in a neighborhood of $0 \in \mathbb{R}^2$ such that (i)-(iv) hold, and for which

$$w(x,0) = \rho(x) , \qquad w_2(x,0) = \sigma(x) .$$

From this observation it is but a short step to

<u>18. LEMMA</u>. Let $\phi\colon M \to \bar{M}$ be an isometry between two surfaces $M, \bar{M} \subset \mathbb{R}^3$ and let $X \in M_p$ be a vector such that $II(X_p,X_p)$ and $\overline{II}(\phi_*X_p,\phi_*X_p)$ are either both positive or both negative. Assume that M and $\bar{M}$ are analytic surfaces, unless $K(p) < 0$. Then there is a neighborhood U of p and a bending $\alpha\colon [0,1] \times U \to \mathbb{R}^3$ with $\bar{\alpha}(0) =$ identity and $\bar{\alpha}(1) = \phi$.

<u>Proof</u>. Choose immersions $f = (u,v,w)$ and $\bar{f} = (\bar{u},\bar{v},\bar{w})$ taking a neighborhood of $0 \in \mathbb{R}^2$ into M and $\bar{M}$, respectively, with $f(0) = p$ and $\bar{f}(0) = \phi(p)$. Without loss of generality we can assume that both f and $\bar{f}$ satisfy the conditions (ii)-(iv) above, and that $X_p = f_*((1,0))$ and $\phi_*X_p = \bar{f}_*((1,0))$. Let $I_f = I_{\bar{f}}$ have components E, F, G. For $0 \le t \le 1$, let

$$\begin{aligned}\rho_t(x) &= (1-t)w(x,0) + t\bar{w}(x,0)\\ \sigma_t(x) &= (1-t)w_2(x,0) + t\bar{w}_2(x,0)\ .\end{aligned}$$

Then (ii) gives

$$(1)\qquad \rho_t(0) = \rho_t'(0) = \sigma_t(0) = 0\ .$$

If ℓ, m, n and $\bar{\ell}$, $\bar{m}$, $\bar{n}$ are the coefficients of II_f and $II_{\bar{f}}$, then by assumption $II(X_p,X_p) = \ell(0,0)$ and $\overline{II}(\phi_*X_p,\phi_*X_p) = \ell(0,0)$ have the same sign. Now

$$\ell(0,0) = \frac{1}{\sqrt{EG-F^2}}\cdot\det\begin{pmatrix} f_{11}\\ f_1\\ f_2\end{pmatrix} \quad \text{at } (0,0) \qquad \text{[formula (A) of Chapter 3]}$$

$$= \frac{1}{\sqrt{EG - F^2}} \cdot \det \begin{pmatrix} u_{11} & v_{11} & w_{11} \\ u_1 & v_1 & w_1 \\ u_2 & v_2 & w_2 \end{pmatrix} \qquad \text{at} \quad (0,0)$$

$$= \frac{1}{\sqrt{EG - F^2}}(0,0) \cdot w_{11}(0,0) \cdot u_2(0,0) \cdot v_1(0,0) \qquad \text{by (ii) and (iv)} ,$$

and similarly for $\ell(0,0)$. Using (iv), we see that $w_{11}(0)$ has the same sign as $\ell(0,0)$, and similarly for $\bar{w}_{11}(0)$; so $w_{11}(0)$ and $\bar{w}_{11}(0)$ have the same sign. It follows that

$$\text{(2)} \qquad \rho_t''(0) \neq 0 .$$

By (*), there are unique immersions $f_t = (u_t, v_t, w_t)$ defined in a neighborhood of $0 \in \mathbb{R}^2$ such that (i)-(iv) hold, and for which

$$w_t(x,0) = \rho_t(x) , \qquad (w_t)_2(x,0) = \sigma_t(x) .$$

The uniqueness implies that $f_0 = f$ and $f_1 = \bar{f}$.

The only details which need to be checked are that all f_t can be defined in a common neighborhood of $0 \in \mathbb{R}^2$, and that the f_t vary smoothly with t. This unrewarding task is left to the reader. ■

19. THEOREM (E.E. LEVI). Let $\phi: M \to \bar{M}$ be an isometry between two surfaces $M, \bar{M} \subset \mathbb{R}^3$, and suppose that $p \in M$ and $\phi(p) \in \bar{M}$ are not planar points. Assume that M and $\bar{M}$ are analytic surfaces, unless $K(p) < 0$. Then there is a neighborhood U of p and a bending $\alpha: [0,1] \times U \to \mathbb{R}^3$ with $\bar{\alpha}(0) =$ identity and either $\bar{\alpha}(1) = \phi$ or $\bar{\alpha}(1) = R \circ \phi$, where R is a reflection.

Proof. Since p and $\phi(p)$ are not planar points, there are at most 2 asymptotic directions at these points, and thus certainly a tangent vector $X \in M_p$ such that $II(X,X)$ and $\overline{II}(\phi_*X,\phi_*X)$ are both non-zero. If they have the same sign we apply Lemma 18; if they have opposite signs we apply Lemma 18 to M and $R(\bar{M})$. ∎

We know that the reflection R has to be allowed if $K(p) > 0$. It turns out that R is unnecessary if $K(p) < 0$. First a preliminary result.

20. LEMMA. Let E, F, G be the components of a metric in a neighborhood of $0 \in \mathbb{R}^2$, and $\mathbf{v} \in \mathbb{R}^2{}_0$ a given tangent vector. Assume that E, F, G are analytic, unless $K < 0$. Then there is an immersion $f = (u,v,w)$ in a neighborhood of 0 such that I_f has components E, F, G, and $f_*(\mathbf{v})$ is not an asymptotic vector on the image of f.

Proof. Without loss of generality, we can assume that $\mathbf{v} = (1,0)$. Then choose any two functions ρ, σ satisfying

$$\rho(0) = \rho'(0) = \sigma(0) = 0$$
$$\rho''(0) \neq 0 ,$$

and consider the immersion f determined by (*), with

$$u_2(0) > 0 , \qquad v_1(0) > 0 .$$
$$w(x,0) = \rho(x) \implies w_{11}(0,0) = \rho''(0) .$$

The calculation in the proof of Lemma 18 shows that

$$\ell(0,0) = \frac{1}{\sqrt{EG - F^2}}(0,0)\cdot w_{11}(0,0)\cdot u_2(0,0)\cdot v_1(0,0) \neq 0 \ .$$

This means that $f_*(\mathbf{v})$ is not an asymptotic vector. ∎

<u>21. THEOREM (E.E. LEVI)</u>. Let $\phi\colon M \to \bar{M}$ be an isometry between two surfaces $M, \bar{M} \subset \mathbb{R}^3$, and suppose that $K(p) < 0$. Then there is a neighborhood U of p and a bending $\alpha\colon [0,1] \times U \to \mathbb{R}^3$ with $\bar{\alpha}(0) = \text{identity}$ and $\bar{\alpha}(1) = \phi$.

<u>Proof</u>. We just have to show that a neighborhood of p in M can be bent into its mirror image $R(M)$. Let $X \in M_p$ be an asymptotic vector. Then there are vectors Y arbitrarily close to X in M_p with $II(Y,Y) > 0$, as well as vectors arbitrarily close to X with $II(Y,Y) < 0$. Lemma 20 says that there is (locally) an isometry $\psi\colon M \to \tilde{M} \subset \mathbb{R}^3$ such that $\psi_*(X)$ is <u>not</u> an asymptotic vector on $\tilde{M}$. We can assume, by composing $\tilde{M}$ with a reflection if necessary, that $\widetilde{II}(\psi_*X,\psi_*X) > 0$. Then the same inequality holds for all tangent vectors of $\tilde{M}_{\psi(p)}$ in some sector containing ψ_*X. So we can choose $Y \in M_p$ with

$$II(Y,Y) > 0 \ , \qquad \widetilde{II}(\psi_*Y,\psi_*Y) > 0 \ ,$$

and it follows from Theorem 19 that there is a bending of a neighborhood of p in M onto a neighborhood of $\psi(p)$ in $\tilde{M}$. But we can also choose $Y \in M_p$ with

$$II(Y,Y) < 0 \ , \qquad \widetilde{II}(\psi_*Y,\psi_*Y) > 0 \ ,$$

and then it follows that there is a bending of a neighborhood of $R(p)$ in

$R(M)$ onto a neighborhood of $\psi(p)$ in $\tilde{M}$. Consequently, there is a bending of a neighborhood of p in M onto a neighborhood of $R(p)$ in $R(M)$. ■

This little proof, clever as it is, certainly doesn't give any idea of what is going on geometrically. E.E. Levi supplied a geometric description of the bending in the special case of a surface M of constant negative curvature whose asymptotic directions are perpendicular at $p \in M$. Rotation through

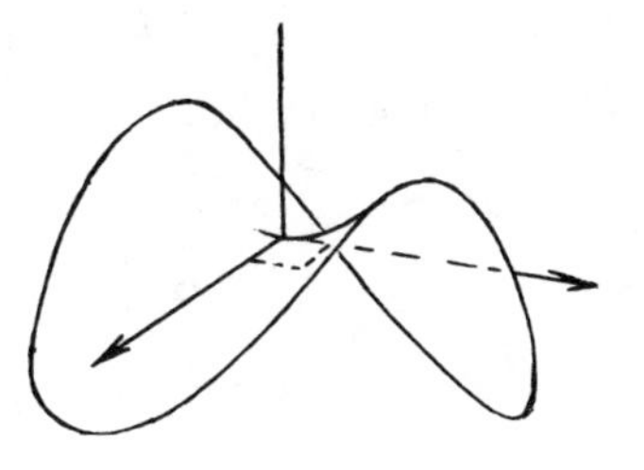

an angle of $\pi/2$ around the normal at p takes M into its reflection R through the tangent plane at p. But the isometry $M \longrightarrow R(M)$ thus obtained is not the same as $R|M$. To modify this, we consider the series of isometries obtained as follows. At time t, we first perform a rotation A_t through an angle t around the normal, and then compose $A_t|M$ with a map $B_t\colon A_t(M) \longrightarrow A_t(M)$ of the surface of constant curvature $A_t(M)$ onto itself which rotates the tangent space M_p back by an angle of $-t$. For $t = \pi/2$ we obtain the map $R|M$.

These results of E.E. Levi [1] were proved nearly 30 years after A. Voss had first explicitly pointed out that a distinction ought to be made between warpings and bendings. Levi's results were regarded as demonstrations that these distinctions really did not exist (at least locally). Of course, Theorem 19 does have the added hypothesis that p and $\phi(p)$ are not planar points;

in E.E. Levi's original theorem, there was the stronger requirement that $K(p) \neq 0$. Such requirements were regarded, if they were regarded at all, as merely technical details. Remarkably enough, H. Schilt [1] discovered that Theorem 19 is actually false if the point p is a flat point, even if all points in a neighborhood of p have $K < 0$. We will outline the arguments here, but for some of the details the reader is referred to Schilt's paper, which is very clearly written and easy to follow.

Consider a surface M which is the graph of a function $h: \mathbb{R}^2 \to \mathbb{R}$ with $0 = h(0) = h_1(0) = h_2(0)$. If the curvature $K(0) < 0$, then M looks like a "saddle," as we saw in Chapter 2. But if $K(0) = 0$, and $K < 0$ in a deleted neighborhood of 0, then it can be shown that M looks like a "generalized monkey saddle": inside a sufficiently small circle C around 0,

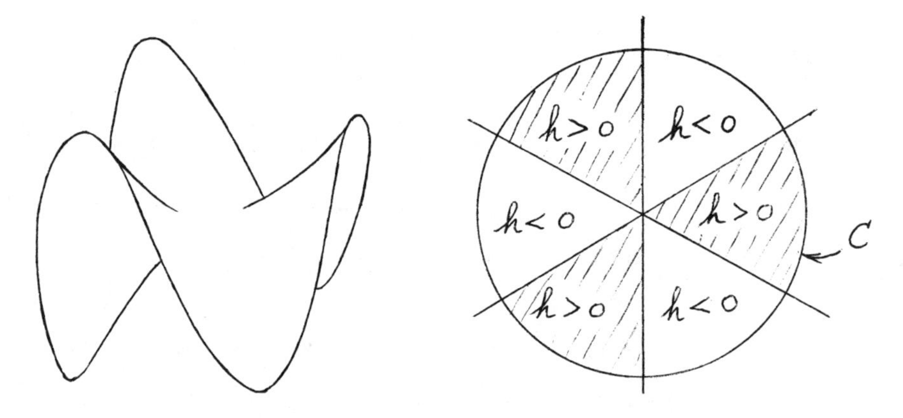

the zero set of h consists of an even number 2λ of curves starting at 0 and ending at C, with no points in common except 0; the sign of h on the sectors between these curves is constant and changes as we go from one sector to another. The number $s = \lambda - 1$ is called the <u>order</u> of the saddle point at

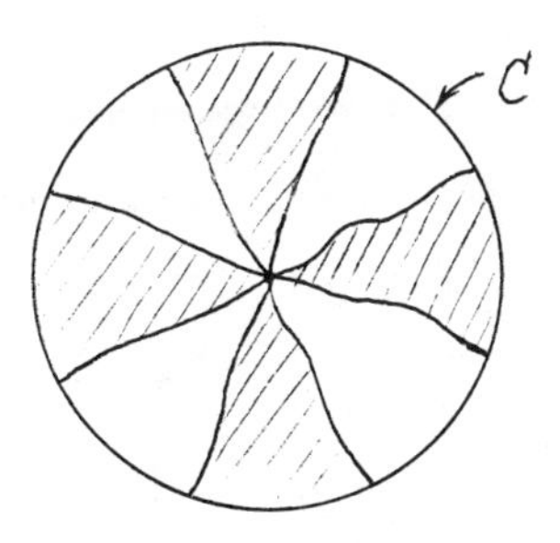

0. An ordinary saddle point, where $K(0) < 0$, has order 1, while the monkey saddle has order 2. The graph of $h(x,y) =$ Real part of $(x+iy)^{s+1}$ has order s.

The order of a saddle point can be described in another way, by considering a closed curve in M going once around p in the positively oriented sense. It turns out that the image of this curve under the normal map goes s times around the normal at p, but in the negatively oriented sense. The following picture illustrates this for the monkey saddle.

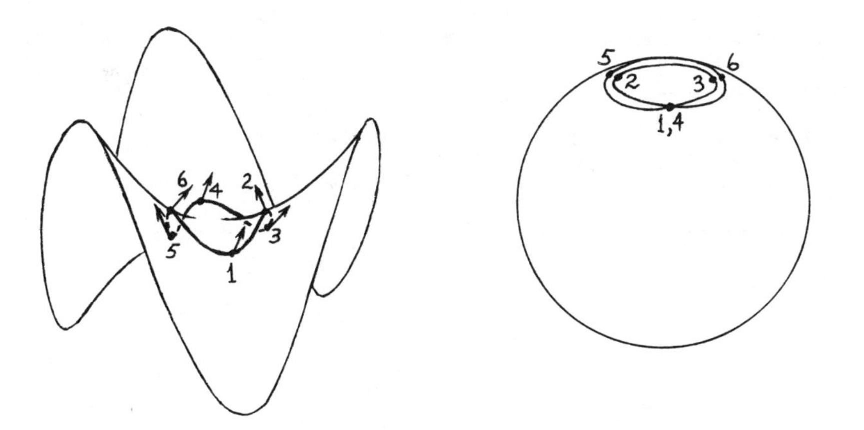

Finally, the order of a saddle point may be described in yet a third way. In the region where $K < 0$, the principal curvatures k_1, k_2 are of different signs, say $k_1 > 0 > k_2$. So we can pick out the 1-dimensional distribution of all multiples of the principal directions corresponding to the principal curvature k_1, say. Notice that if $X_p \varepsilon M_p$ is a principal direction with principal curvature k_1, and $\bar{X}_p$ is a perpendicular vector with $(X_p, \bar{X}_p)$ positively oriented, then there is just one unit asymptotic vector Y_p in the quadrant of M_p bounded by X_p and $\bar{X}_p$. Moreover, if we start with $-X_p$

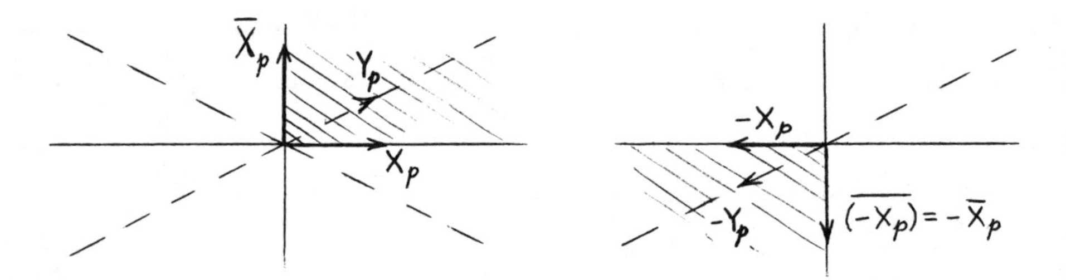

instead of X_p, then we just end up with $-Y_p$. Thus we can also pick out a 1-dimensional distribution consisting of all multiples of an asymptotic vector at each point. Another way of stating these facts is the following: the principal curves and the asymptotic curves can each be separated into two distinct families in the region where $K < 0$.* The following picture shows the projections on the (x,y)-plane of the two families of asymptotic lines (one indicated by solid lines, the other by dotted lines) for the ordinary monkey saddle. Now one can define the index of any one

*Schilt [1] tries to do this for the asymptotic curves by looking at the signs of their torsions, which have to be different by the Beltrami-Enneper theorem. But this doesn't work at points where the torsions don't exist.

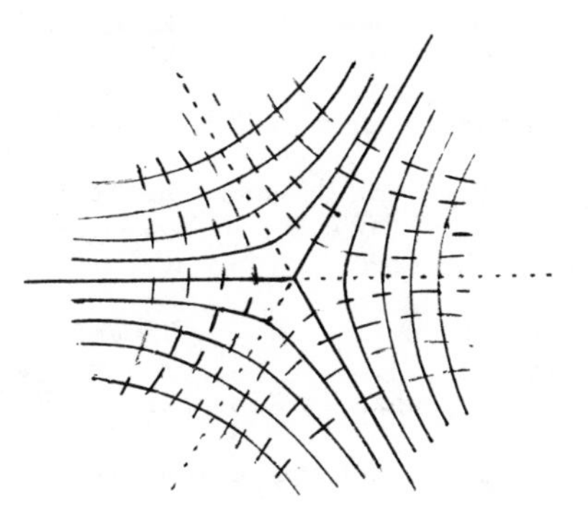

of these distributions (Addendum 2 to Chapter 4); this index i is a half-integer, and is clearly the same for both families. It is related to the order s of the saddle by

$$(*) \qquad s = 1 - 2i .$$

For example, the monkey saddle, with $s = 2$, has $i = -1/2$ (compare the above picture with the one on p. III.325). On the other hand, if 0 is a parabolic point, then the asymptotic curves have no singularity at 0, so $i = 0$ and $s = 1$, just as in the case of an ordinary saddle point.

Proving (*) is not completely straightforward, but the relation is very important, for it leads immediately to the result that s is a bending invariant: given a bending $\alpha\colon [0,1] \times M \to \mathbb{R}^3$ of the inclusion map of M into $\mathbb{R}^3$, each surface $\bar{\alpha}(t)(M)$ has a saddle at $\alpha(t,0)$, and for all t the order of this saddle is s. For the proof one merely observes that the principal curves or asymptotic curves vary continuously, and therefore always have the same index.

One might think that this result is hardly worth mentioning, on the grounds that s should actually be a warping invariant: any surface isometric to M

also ought to have a saddle of order s. But this is not true! For suppose that M is an analytic surface with a saddle of order $s > 1$. By Lemma 20, there is a surface $\bar{M}$ isometric to M such that the point corresponding to p is a parabolic point. Thus the saddle order at this point is $\bar{s} = 1$. Consequently, there is no bending from M to $\bar{M}$; indeed, no neighborhood of the saddle point on M can be bent onto its isometric image in $\bar{M}$!

These examples of isometric surfaces which are not even locally connected by a bending all have $K < 0$ in a neighborhood of the point with $K = 0$. But Hopf and Schilt [1] show that, for certain classes of surfaces, the order of contact of the graph of h with the (x,y)-plane is also a bending invariant, but not a warping invariant. This allows them to give examples of the same phenomenon, but for surfaces with $K = 0$ at one point and $K > 0$ in a neighborhood of the point. They are also able to show that Theorem 21 fails if $K(p) = 0$, even if $K < 0$ for all other points in a neighborhood of p.

It should be mentioned that the present proofs of Theorems 19 and 21 (which slightly strengthen E.E. Levi's original result) are due to Schilt, who also observed that the above analysis of the Darboux equation can be used to prove the following: If $p \in M$ is not a planar point, then [assuming M is analytic, unless $K(p) < 0$] some neighborhood of p has a non-trivial bending. Actually, this result holds without assumptions of analyticity when $K(p) > 0$, although the proof is much harder; in fact, this comes out of the proof that a convex surface with a disc deleted is bendable (see Hellwig [1]). However, the case $K(p) = 0$ is unresolved. There is a startling result of Efimov along these lines (see Efimov {1}, Chapter IX, and Hoesli [1]): There exist infinitely many examples of analytic surfaces containing a point p such that no neighborhood of p has any non-trivial analytic bendings; in fact, any smooth bending of this neighborhood into analytic surfaces is trivial. A specific example is $\{(x,y,z): z = (x^2+y^2)^5\}$ with $p = 0 \in \mathbb{R}^3$. Whether there are examples where no neighborhood has any non-trivial bendings into C^∞ surfaces is unknown, and certainly a highly intriguing question. As far as I can tell, it is not even known whether every C^∞ surface is locally warpable.

With these considerations we finally end our investigation of rigidity for surfaces in $\mathbb{R}^3$. So far we have said absolutely nothing about surfaces in S^3 or H^3. Recollection of Chapter 7F might make even the stoutest hearts quail at this prospect, but fortunately there is an incredably neat trick, due to Pogorelov { 3}, which reduces almost all such questions to the case of surfaces in $\mathbb{R}^3$. Since Pogorelov's book is written more in the style of the Russian school, and includes many results specifically tailored for such a study, we will give a treatment of the main points totally from the C^∞ point of view.

In Chapter 7A we considered the central projection $\phi\colon S^{n+} \to \mathbb{R}^n$, from the open northern hemisphere S^{n+} of S^n onto $\mathbb{R}^n$. It is easily computed that

$$\phi(x) = \left(\frac{x_1}{x_{n+1}},\ldots,\frac{x_n}{x_{n+1}}\right) \qquad \text{for } x \in S^{n+} .$$

Let $* = (0,\ldots,0,1)$ be the north pole of S^n. Then ϕ can also be described by

$$\phi(x) = \frac{x - \langle x,*\rangle\cdot *}{\langle x,*\rangle} \in \mathbb{R}^n \times \{0\} \subset \mathbb{R}^{n+1} .$$

Now let $f_1, f_2\colon M \to S^n$ be two maps of a Riemannian manifold M into S^n.

Define $\tilde{f}_1\colon M \longrightarrow \mathbb{R}^n$ (actually, into $\mathbb{R}^n \times \{0\}$) by

$$(*) \qquad \tilde{f}_1(p) = \frac{f_1(p) - \langle f_1(p), * \rangle \cdot *}{\langle f_1(p) + f_2(p), * \rangle},$$

and define $\tilde{f}_2$ similarly (this formula makes sense so long as $f_1 + f_2$ is never perpendicular to $*$, which happens, in particular, if f_1 and f_2 both go into S^{n+}).

22. <u>PROPOSITION</u>. The two maps $f_1, f_2\colon M \longrightarrow S^n$ induce the same (possibly degenerate) metric on M if and only if the two maps $\tilde{f}_1, \tilde{f}_2\colon M \longrightarrow \mathbb{R}^n$ induce the same metric on M.

<u>Proof</u>. For the $\mathbb{R}^n$-valued form $d\tilde{f}_1$ we compute from equation (*) that

$$(1) \quad \langle f_1 + f_2, * \rangle^2 \, d\tilde{f}_1 = \langle f_1 + f_2, * \rangle [df_1 - \langle df_1, * \rangle \cdot *] - \langle df_1 + df_2, * \rangle \cdot [f_1 - \langle f_1, * \rangle \cdot *] .$$

{This equation means that for $X \in M_p$ we have

$$(2) \quad \langle f_1(p) + f_2(p), * \rangle^2 \, d\tilde{f}_1(X) = \langle f_1(p) + f_2(p), * \rangle [df_1(X) - \langle df_1(X), * \rangle \cdot *] - \langle df_1(X) + df_2(X), * \rangle \cdot [f_1(p) - \langle f_1(p), * \rangle \cdot *] .\}$$

Since $\langle f_1, f_1 \rangle = 1 \implies \langle df_1, f_1 \rangle = 0$, we have

$$
\begin{aligned}
\langle f_1 + f_2, *\rangle^4 |d\tilde{f}_1|^2 &= \langle f_1 + f_2, *\rangle^2 [|df_1|^2 - \langle df_1, *\rangle^2] \\
&\quad + 2\langle f_1 + f_2, *\rangle \langle df_1 + df_2, *\rangle \langle f_1, *\rangle \langle df_1, *\rangle \\
&\quad + \langle df_1 + df_2, *\rangle^2 [1 - \langle f_1, *\rangle^2] \\
&= \langle f_1 + f_2, *\rangle^2 |df_1|^2 + \langle df_1 + df_2, *\rangle^2 \\
&\quad - [\langle f_1 + f_2, *\rangle \langle df_1, *\rangle - \langle df_1 + df_2, *\rangle \langle f_1, *\rangle]^2 \\
&= \langle f_1 + f_2, *\rangle^2 |df_1|^2 + \langle df_1 + df_2, *\rangle^2 \\
&\quad - [\langle f_2, *\rangle \langle df_1, *\rangle - \langle df_2, *\rangle \langle f_1, *\rangle]^2 ,
\end{aligned}
$$

which is symmetric in f_1 and f_2 if and only if $|df_1|^2 = |df_2|^2$. ■

Remark. Even if f_1 and f_2 are immersions, the maps $\tilde{f}_1$ and $\tilde{f}_2$ need not be. However, if $*$ is not a linear combination of $f_1(p)$ and any vector $df_1(X)$ for $X \in M_p$, then the vectors $df_1(X) - \langle df_1(X), *\rangle *$ and $f_1(p) - \langle f_1(p), *\rangle *$ in (2) are linearly independent, so $d\tilde{f}_1(X) \neq 0$ for all $X \in M_p$, and $\tilde{f}_1$ is an immersion at p.

Suppose we have two maps $f_1, f_2\colon M \to \mathbb{R}^n \subset \mathbb{R}^{n+1}$. We define $\bar{f}_1\colon M \to S^n$ by

$$
\bar{f}_1(p) = \frac{2f_1(p) + (1 - |f_1(p)|^2 + |f_2(p)|^2)\cdot *}{|2f_1(p) + (1 - |f_1(p)|^2 + |f_2(p)|^2)\cdot *|} ,
$$

and we define $\bar{f}_2\colon M \to S^n$ similarly. It is easy to compute that if we begin with two maps $f_1, f_2\colon M \to S^n$, form $\tilde{f}_1, \tilde{f}_2\colon M \to \mathbb{R}^n$ as before, and then apply the present construction to $\tilde{f}_1, \tilde{f}_2$, obtaining $\bar{\tilde{f}}_1, \bar{\tilde{f}}_2\colon M \to S^n$, then

$$\bar{\tilde{f}}_1 = f_1 \,, \qquad \bar{\tilde{f}}_2 = f_2 \,.$$

Similarly, if we start with $f_1, f_2 \colon M \to \mathbb{R}^n$, then

$$\tilde{\bar{f}}_1 = f_1 \,, \qquad \tilde{\bar{f}}_2 = f_2 \,.$$

Hence

23. COROLLARY. The two maps $f_1, f_2 \colon M \to \mathbb{R}^n$ induce the same metric on M if and only if the two maps $\bar{f}_1, \bar{f}_2 \colon M \to S^n$ induce the same metric on M.

On the other hand, our new constructions also preserve the notion of congruence.

24. PROPOSITION. Let A be an orthogonal map of $\mathbb{R}^{n+1}$, so that $A \colon S^n \to S^n$ is an isometry, and let $f_1, f_2 \colon M \to S^n$ be maps with $f_2 = A \circ f_1$. Suppose that we can define $\tilde{f}_1, \tilde{f}_2 \colon M \to \mathbb{R}^n$. Then there is a Euclidean motion $\tilde{A} \colon \mathbb{R}^n \to \mathbb{R}^n$ such that $\tilde{f}_2 = \tilde{A} \circ \tilde{f}_1$.

Proof. We just have to show that

$$|\tilde{f}_1(p) - \tilde{f}_1(q)|^2 = |\tilde{f}_2(p) - \tilde{f}_2(q)|^2$$

for all $p, q \in M$. For any x with $x + Ax$ not perpendicular to $*$, let $x^{\#} \in \mathbb{R}^{n+1}$ be a multiple of x such that

$$(1) \qquad \langle x^{\#} + Ax^{\#}, * \rangle = 1 \,.$$

Then

$$\tilde{f}_1(p) = \frac{f_1(p) - \langle f_1(p),*\rangle *}{\langle f_1(p) + A(f_1(p)),\ *\rangle} = f_1(p)^{\#} - \langle f_1(p)^{\#},*\rangle\cdot * ,$$

and similarly for $\tilde{f}_1(q)$, $\tilde{f}_2(p)$, $\tilde{f}_2(q)$. Set $z = f_1(p)^{\#} - f_1(q)^{\#}$. Then

$$|\tilde{f}_1(p) - \tilde{f}_1(q)|^2 = |z - \langle z,*\rangle *|^2 = |z|^2 - \langle z,*\rangle^2$$

$$|\tilde{f}_1(p) - \tilde{f}_2(q)|^2 = |Az - \langle Az,*\rangle *|^2 = |Az|^2 - \langle Az,*\rangle^2 .$$

But $|Az| = |z|$, while

$$\langle z + Az,\ *\rangle = \langle f_1(p)^{\#} + Af_1(p)^{\#} - [f_1(q)^{\#} + Af_1(q)^{\#}],\ *\rangle = 0 , \qquad \text{by (1)}$$

$$\Longrightarrow \langle z,*\rangle^2 = \langle Az,*\rangle^2 . \quad \blacksquare$$

<u>25. PROPOSITION</u>. Let $B\colon \mathbb{R}^n \to \mathbb{R}^n$ be an isometry, and let $f_1, f_2\colon M \to \mathbb{R}^n$ be maps with $f_2 = B\circ f_1$. Then there is an isometry $A\colon S^n \to S^n$ such that $\bar{f}_2 = A\circ\bar{f}_1$.

<u>Proof</u>. Define $\rho_1, \rho_2\colon \mathbb{R}^n \to S^n$ by

$$\rho_1(y) = \frac{2y + (1 - |y|^2 + |By|^2)\cdot *}{|\text{numerator}|}$$

$$\rho_2(y) = \frac{2y + (1 - |y|^2 + |B^{-1}y|^2)\cdot *}{|\text{numerator}|}$$

so that

$$\bar{f}_1 = \rho_1 \circ f_1 , \qquad \bar{f}_2 = \rho_2 \circ f_2 .$$

It is clear that ρ_1 and ρ_2 are continuous. We claim that ρ_1 and ρ_2 are one-one. Suppose instead that $y \neq z \in \mathbb{R}^n$, but $\rho_1(y) = \rho_1(z)$. Clearly y and z must be linearly dependent, so there is a unit vector $v \in \mathbb{R}^n$ with $y = \lambda v$ and $z = \mu v$. Then $\rho_1(y) = \rho_1(z)$ implies that

$$\frac{2\lambda}{|\text{numerator for } \rho_1(y)|} = \frac{2\mu}{|\text{numerator for } \rho_1(z)|}$$

$$\frac{1-\lambda^2+|By|^2}{|\text{numerator for } \rho_1(y)|} = \frac{1-\mu^2+|Bz|^2}{|\text{numerator for } \rho_1(z)|} ,$$

and hence

$$(1) \qquad \frac{1-\lambda^2+|By|^2}{\lambda} = \frac{1-\mu^2+|Bz|^2}{\mu} .$$

Let $B = T_w \circ C$, where C is a rotation, and T_w is translation by a vector w. Then

$$|By|^2 = |C(\lambda v)+w|^2 = \lambda^2 + 2\langle C(\lambda v),w\rangle + |w|^2$$

$$|Bz|^2 = \mu^2 + 2\langle C(\mu v),w\rangle + |w|^2 .$$

So (1) becomes

$$\frac{1+|w|^2}{\lambda} + 2\langle C(v),w\rangle = \frac{1+|w|^2}{\mu} + 2\langle C(v),w\rangle .$$

Hence $\lambda = \mu \implies y = z$. Similarly, ρ_2 is one-one.

Since ρ_1 and ρ_2 are continuous one-one maps between manifolds of the same dimension, their images are open, by Invariance of Domain

(Theorem I.1-1). So we can consider the map $\rho_2 \circ B \circ \rho_1{}^{-1}$, defined on some open set in S^n. We claim that this map preserves distances on S^n, and is thus the restriction of some isometry A. Since

$$\bar{f}_2 = \rho_2 \circ f_2 = \rho_2 \circ B \circ f_1 = \rho_2 \circ B \circ \rho_1{}^{-1} \circ \rho_1 \circ f_1 = \rho_2 \circ B \circ \rho_1{}^{-1} \circ \bar{f}_1 ,$$

this will prove the Theorem.

It suffices to show that

$$|\rho_1(y) - \rho_1(z)|^2 = |\rho_2(By) - \rho_2(Bz)|^2 ,$$

since the distance between two points in S^n is determined by their Euclidean distance. Clearly, we just have to show that for all y and z we have

$$<\rho_1(y), \rho_1(z)> = <\rho_2(By), \rho_2(Bz)> .$$

If

$$a(y) = 2y + (1 - |y|^2 + |By|^2) \cdot *$$
$$b(y) = 2By + (1 - |By|^2 + |y|^2) \cdot * ,$$

then

$$\rho_1(y) = \frac{a(y)}{|a(y)|} , \qquad \rho_2(By) = \frac{b(y)}{|b(y)|} .$$

So it suffices to show that for all $y, z \in \mathbb{R}^n$ we have

$$<a(y), a(z)> = <b(y), b(z)> . \tag{2}$$

Now

$$(3)\begin{cases} \langle a(y),a(z)\rangle = 4\langle y,z\rangle + (1-|y|^2+|By|^2)(1-|z|^2+|Bz|^2) \\ \langle b(y),b(z)\rangle = 4\langle By,Bz\rangle + (1-|By|^2+|y|^2)(1-|Bz|^2+|z|^2) \ . \end{cases}$$

Writing $B = T_w \circ C$ as before, we have

$$|By|^2 = |y|^2 + 2\langle Cy,w\rangle + |w|^2$$
$$|Bz|^2 = |z|^2 + 2\langle Cz,w\rangle + |w|^2 \ ,$$

and

$$\langle By,Bz\rangle = \langle y,z\rangle + \langle Cy+Cz, w\rangle + |w|^2 \ .$$

Substituting into (3), we obtain (2). ■

The hardest problem is to show that our construction preserves convexity. This holds only in certain circumstances, which requires a preliminary remark. We say that a hypersurface $M \subset S^{n+}$ is <u>star-shaped with respect to</u> $*$ if each geodesic ray starting from $*$, and contained in S^{n+}, intersects M exactly once. Clearly M has a natural orientation, just as in the case of hypersurfaces $M \subset \mathbb{R}^n$ star-shaped with respect to 0. Our results will hold only for imbeddings $f_1, f_2 \colon M \to S^{n+}$ or $\mathbb{R}^n$ whose images are star-shaped with respect to $*$ or 0; moreover, for some orientation on M, the induced orientations on $f_1(M)$ and $f_2(M)$ must be the natural ones.

Before giving the precise results, we consider one more preliminary. Let γ be an arc-length parameterized curve in S^n. Then the Frenet equations for S^n give

$$\kappa(s)\,\mathbf{n}(s) = \frac{D\gamma'(s)}{ds} = \mathbf{T}\gamma''(s)$$
$$= \gamma''(s) - \langle\gamma''(s),\gamma(s)\rangle\cdot\gamma(s) .$$

But

$$\langle\gamma,\gamma\rangle = 1 \implies \langle\gamma',\gamma\rangle = 0 \implies \langle\gamma'',\gamma\rangle = -\langle\gamma',\gamma'\rangle = -1 .$$

So we have

$$\gamma''(s) = \kappa(s)\,\mathbf{n}(s) - \gamma(s) .$$

Hence

$$(*) \qquad \gamma(s+h) = \gamma(s) + h\gamma'(s) + \frac{h^2}{2}\gamma''(s) + o(h^2)$$
$$= (1-\frac{h^2}{2})\gamma(s) + h\,\mathbf{t}(s) + \frac{h^2}{2}\kappa(s)\,\mathbf{n}(s) + o(h^2) .$$

26. PROPOSITION. Let M be an oriented $(n-1)$-manifold, and let $f_1, f_2\colon M \to S^{n+}$ be two imbeddings such that $f_1(M)$ and $f_2(M)$ are convex and star-shaped with respect to $*$, and such that f_1 and f_2 induce the same metric on M, and the natural orientations on $f_1(M)$ and $f_2(M)$. Suppose, moreover, that the second fundamental forms of $f_1(M)$ and $f_2(M)$ are positive semi-definite. Then the same is true for the second fundamental forms of $\tilde{f}_1(M)$ and $\tilde{f}_2(M)$ in $\mathbb{R}^n$. (Note that under the given hypotheses, $\tilde{f}_1$ and $\tilde{f}_2$ will be immersions, by the Remark after Proposition 22.)

Proof. Let c be an arc-length parameterized curve in M (with the metric induced by f_1 or f_2). Apply (*) to the arc-length parameterized curve $\gamma = f_1 \circ c$ in S^n, letting $\mathbf{t}_1$ and $\mathbf{n}_1$ be its tangent and normal, and κ_1 its curvature. We obtain

$$f_1(c(s+h)) = (1-\frac{h^2}{2})f_1(c(s)) + h\,\mathbf{t}_1(s) + \frac{h^2}{2}\kappa_1(s)\,\mathbf{n}_1(s) + o(h^2) ,$$

$$= (1-\frac{h^2}{2})x_1 + h\,\mathbf{t}_1 + \frac{h^2}{2}\kappa_1\mathbf{n}_1 + o(h^2) , \qquad \text{for short,}$$

and similarly

$$f_2(c(s+h)) = (1-\frac{h^2}{2})f_2(c(s)) + h\,\mathbf{t}_2(s) + \frac{h^2}{2}\kappa_2(s)\,\mathbf{n}_2(s) + o(h^2)$$

$$= (1-\frac{h^2}{2})x_2 + h\,\mathbf{t}_2 + \frac{h^2}{2}\kappa_2\mathbf{n}_2 + o(h^2) .$$

So

$$(1) \quad \tilde{f}_1(c(s+h)$$

$$= \frac{(1-\frac{h^2}{2})x_1 + h\,\mathbf{t}_1 + \frac{h^2}{2}\kappa_1\mathbf{n}_1 + o(h^2) - \left\langle (1-\frac{h^2}{2})x_1 + h\,\mathbf{t}_1 + \frac{h^2}{2}\kappa_1\mathbf{n}_1 + o(h^2),\ * \right\rangle *}{\left\langle (1-\frac{h^2}{2})(x_1+x_2) + h(\mathbf{t}_1 + \mathbf{t}_2) + \frac{h^2}{2}(\kappa_1\mathbf{n}_1 + \kappa_2\mathbf{n}_2) + o(h^2),\ * \right\rangle} .$$

Writing this as

$$\frac{v - \frac{h^2}{2}v + hV}{\alpha - \frac{h^2}{2}\alpha + hA} \qquad \begin{cases} v = x_1 - \langle x_1, *\rangle * \\ \alpha = \langle x_1 + x_2, *\rangle , \end{cases}$$

and noting that

$$\frac{v-\frac{h^2}{2}v+hV}{\alpha-\frac{h^2}{2}\alpha+hA} - \frac{v+hV}{\alpha+hA} = \frac{\frac{h^3}{2}(\alpha V - Av)}{(\alpha+\frac{h^2}{2}\alpha+hA)(\alpha+hA)},$$

we see that (1) can be written

$$(2)\quad \tilde{f}_1(c(s+h)) = \frac{x_1+h\,\mathbf{t}_1+\frac{h^2}{2}\kappa_1\mathbf{n}_1-\left\langle x_1+h\,\mathbf{t}_1+\frac{h^2}{2}\kappa_1\mathbf{n}_1,\ *\right\rangle *}{\left\langle x_1+x_2+h(\mathbf{t}_1+\mathbf{t}_2)+\frac{h^2}{2}(\kappa_1\mathbf{n}_1+\kappa_2\mathbf{n}_2),\ *\right\rangle} + 0(h^3),$$

where $0(h^3)$ denotes a function such that $0(h^3)/h^3$ is bounded as $h \to 0$. Expanding equation (2) out up to terms of order h^2, we find that

$$(3)\quad \tilde{f}_1(c(s+h)) - \tilde{f}_1(c(s))$$

$$= \frac{h}{\langle x_1+x_2, *\rangle}\left\{-\langle \mathbf{t}_1+\mathbf{t}_2, *\rangle\tilde{f}_1(c(s)) + \mathbf{t}_1 - \langle \mathbf{t}_1, *\rangle *\right\}$$

$$+ \frac{h^2\langle \mathbf{t}_1+\mathbf{t}_2, *\rangle}{\langle x_1+x_2, *\rangle^2}\left\{\langle \mathbf{t}_1+\mathbf{t}_2, *\rangle\tilde{f}_1(c(s)) - \mathbf{t}_1 + \langle \mathbf{t}_1, *\rangle *\right\}$$

$$+ \frac{h^2}{2\langle x_1+x_2, *\rangle}\left\{-\langle \kappa_1\mathbf{n}_1+\kappa_2\mathbf{n}_2, *\rangle\tilde{f}_1(c(s)) + \kappa_1\mathbf{n}_1 - \langle \kappa_1\mathbf{n}_1, *\rangle *\right\}$$

$$+ 0(h^3).$$

Let $\tilde{n}_1$ be the unit normal of $\tilde{f}_1(M)$ at $\tilde{f}_1(c(s))$. Clearly

$$\lim_{h\to 0}\left\langle \frac{\tilde{f}_1(c(s+h)) - \tilde{f}_1(c(s))}{h},\ \tilde{n}_1\right\rangle = 0 .$$

So equation (3) implies that

$$(4)\qquad \left\langle -\langle \mathbf{t}_1+\mathbf{t}_2, *\rangle\tilde{f}_1(c(s)) + \mathbf{t}_1 - \langle \mathbf{t}_1, *\rangle *,\ \tilde{n}_1\right\rangle = 0 .$$

Using (4), and the fact that $\langle \tilde{n}_1, * \rangle = 0$ (since $\tilde{f}_1(M)$ lies in $\mathbb{R}^n \times \{0\}$), equation (3) now gives

$$(5) \quad \langle \tilde{f}_1(c(s+h)) - \tilde{f}_1(c(s)),\ \tilde{n}_1 \rangle$$

$$= \frac{\kappa_1 h^2}{2\langle x_1 + x_2, * \rangle} \Big\langle - \langle \mathbf{n}_1, * \rangle \tilde{f}_1(c(s)) + \mathbf{n}_1,\ \tilde{n}_1 \Big\rangle$$

$$- \frac{\kappa_2 h^2}{2\langle x_1 + x_2, * \rangle} \langle \mathbf{n}_2, * \rangle \langle \tilde{f}_1(c(s)), \tilde{n}_1 \rangle \quad + \quad O(h^3) \ .$$

Since Taylor's Theorem shows that the second derivative α'' of a function α is given by

$$\alpha''(x) = \lim_{h \to 0} \frac{\alpha(x+h) + \alpha(x-h) - 2\alpha(x)}{h^2} \ ,$$

equation (5) implies that

$$\langle (\tilde{f}_1 \circ c)''(s),\ \tilde{n}_1 \rangle = \frac{\kappa_1}{\langle x_1 + x_2, * \rangle} \Big\langle - \langle \mathbf{n}_1, * \rangle \tilde{f}_1(c(s)) + \mathbf{n}_1,\ \tilde{n}_1 \Big\rangle$$

$$- \frac{\kappa_2}{\langle x_1 + x_2, * \rangle} \langle \mathbf{n}_2, * \rangle \langle \tilde{f}_1(c(s)), \tilde{n}_1 \rangle \ .$$

The term on the left is the second fundamental form of $\tilde{f}_1(M)$ applied to $((\tilde{f}_1 \circ c)'(s), (\tilde{f}_1 \circ c)'(s))$. So it suffices to show that it is always ≥ 0. Since

$$\frac{\kappa_1}{\langle x_1 + x_2, * \rangle} \geq 0 \qquad \text{and} \qquad \frac{\kappa_2}{\langle x_1 + x_2, * \rangle} \geq 0 \ ,$$

it suffices to show that

$$(6) \quad \left\langle -\langle \mathbf{n}_1, *\rangle \tilde{f}_1(c(s)) + \mathbf{n}_1, \tilde{n}_1 \right\rangle > 0 \quad \text{and} \quad -\langle \mathbf{n}_2, *\rangle \langle \tilde{f}_1(c(s)), \tilde{n}_1\rangle > 0 .$$

We can also assume that n_1 is the normal to the tangent plane of $f_1(M)$ at $f_1(c(s))$, since we can choose c so that $f_1 \circ c$ is a normal section of $f_1(M)$. Equation (6) is then proved in the following Lemma, whose statements introduces some more convenient notation.

27. LEMMA. Let P and Q be the tangent planes of $f_1(M)$ and $f_2(M)$ at the points $a_0 = f_1(p)$ and $b_0 = f_2(p)$, and let $c_0 = \tilde{f}_1(p)$. Let a_n and b_n be the unit normals to P and Q at a_0 and b_0, and let c_n be the unit normal to the tangent plane of $\tilde{f}_1(M)$ at c_0. Then

$$\left\langle -\langle a_n, *\rangle c_0 + a_n, c_n \right\rangle > 0 \quad \text{and} \quad -\langle b_n, *\rangle \cdot \langle c_0, c_n\rangle > 0 .$$

Proof. Choose positively oriented unit orthonormal vectors $a_1, \ldots, a_{n-1}$ at the point a_0 in P, let $b_1, \ldots, b_{n-1}$ be the corresponding vectors at b_0 in Q, and let $c_1, \ldots, c_{n-1}$ be the corresponding vectors at c_0 in the tangent plane of $\tilde{f}_1(M)$ at c_0. Then for some $C > 0$ we have

$$(7) \quad a_n = a_0 \times \cdots \times a_{n-1} , \qquad b_n = b_0 \times \cdots \times b_{n-1} , \qquad c_n = C \cdot * \times c_1 \times \cdots \times c_{n-1} .$$

Apply the formula for $d\tilde{f}_1$, in the proof of Proposition 22, to the tangent vector X_i in M_p such that $df(X_i) = a_i$. This gives, in the present notation,

$$\langle a_0 + b_0, *\rangle^2 c_i = \langle a_0 + b_0, *\rangle [a_i - \langle a_1, *\rangle *] - \langle a_i + b_i, *\rangle [a_0 - \langle a_0, *\rangle *] ,$$

and thus

$$(8) \qquad c_i = \frac{1}{\lambda_0^2}(\lambda_0 a_i - \lambda_i a_0) + (\cdots)* \qquad i = 1,\dots,n-1 ,$$

where

$$(9) \qquad \lambda_i = \langle a_i + b_i, *\rangle \qquad i = 0,\dots,n-1 .$$

Note also that c_0 is given by

$$(10) \qquad c_0 = \frac{a_0}{\lambda_0} + (\cdots)* .$$

From (7) and (8) we obtain

$$\begin{aligned}(11) \quad c_n &= C \cdot * \times c_1 \times \cdots \times c_{n-1} \\ &= \frac{C}{\lambda_0^{2(n-1)}} \cdot * \times (\lambda_0 a_1 - \lambda_1 a_0) \times \cdots \times (\lambda_0 a_{n-1} - \lambda_{n-1} a_0) \\ &= \frac{C}{\lambda_0^{n}} \Big\{ \lambda_0 (* \times a_1 \times \cdots \times a_{n-1}) - \lambda_1 (* \times a_0 \times a_2 \times \cdots \times a_{n-1}) - \cdots \\ &\qquad - \lambda_{n-1}(* \times a_1 \times \cdots \times a_{n-2} \times a_0) \Big\} .\end{aligned}$$

Consider first the quantity

$$- \langle b_n, *\rangle \cdot \langle c_0, c_n\rangle .$$

First of all, we have

$$(12) \qquad \langle *, b_n \rangle = \langle *, b_0 \times \cdots \times b_{n-1} \rangle \qquad \text{by (7)}$$

$$= \det \begin{pmatrix} b_0 \\ \vdots \\ b_{n-1} \\ * \end{pmatrix} .$$

Also,

$$(13) \qquad \langle c_0, c_n \rangle = \left\langle \frac{a_0}{\lambda_0} + (\cdots)*,\ c_n \right\rangle \qquad \text{by (10)}$$

$$= \frac{c}{\lambda_0^{\,n}} \langle a_0,\ * \times a_1 \times \cdots \times a_{n-1} \rangle \qquad \text{by (11)}$$

$$= -\frac{c}{\lambda_0^{\,n}} \det \begin{pmatrix} a_0 \\ \vdots \\ a_{n-1} \\ * \end{pmatrix} .$$

Since f_1 and f_2 induce the natural orientations on $f_1(M)$ and $f_2(M)$, the determinants in (12) and (13) are both positive. Hence we do indeed have

$$-\langle b_n, * \rangle \cdot \langle c_0, c_n \rangle > 0 .$$

Now consider

$$\left\langle -\langle a_n, * \rangle c_0 + a_n,\ c_n \right\rangle .$$

First of all, we have

$$(14) \quad -\langle a_n, * \rangle \cdot \langle c_0, c_n \rangle = \frac{c}{\lambda_0^{\,n}} \left[\det \begin{pmatrix} a_0 \\ \vdots \\ a_{n-1} \\ * \end{pmatrix} \right]^2 \qquad \text{by (7) and (13)} .$$

Also,

$$\langle c_n, a_n\rangle = \frac{c}{\lambda_0{}^n}\Big\{\lambda_0\langle * \times a_1 \times \cdots \times a_{n-1},\ a_0 \times \cdots \times a_{n-1}\rangle$$

$$- \lambda_1\langle * \times a_0 \times \cdots \times a_{n-1},\ a_0 \times \cdots \times a_{n-1}\rangle - \cdots$$

$$- \lambda_{n-1}\langle * \times a_1 \times \cdots \times a_0,\ a_0 \times \cdots \times a_{n-1}\rangle\Big\} \qquad \text{by (11)} .$$

Using the formula (Problem 5)

$$\langle v_1 \times \cdots \times v_n,\ w_1 \times \cdots \times w_n\rangle = \det(\langle v_i, w_j\rangle) ,$$

we obtain

$$\langle c_n, a_n\rangle = \frac{c}{\lambda_0{}^n}\Big\{\lambda_0\langle *, a_0\rangle + \lambda_1\langle *, a_1\rangle + \cdots + \lambda_{n-1}\langle *, a_{n-1}\rangle\Big\} .$$

Substituting in from (8) yields

$$\langle c_n, a_n\rangle = \frac{c}{\lambda_0{}^n} \cdot \sum_{i=0}^{n-1} \langle a_i, *\rangle^2 + \langle a_i, *\rangle\langle b_i, *\rangle$$

$$\geq \frac{c}{2\lambda_0{}^n} \cdot \sum_{i=0}^{n-1} \langle a_i, *\rangle^2 - \langle b_i, *\rangle^2 .$$

Since

$$\sum_{i=0}^{n} \langle a_i, *\rangle^2 = 1 = \sum_{i=0}^{n} \langle b_i, *\rangle^2 ,$$

we get

$$(15) \qquad \langle c_n, a_n\rangle \geq \frac{c}{2\lambda_0{}^n}(\langle *, b_n\rangle^2 - \langle *, a_n\rangle^2)$$

$$= \frac{c}{2\lambda_0{}^n}\left[\left[\det\begin{pmatrix} b_0 \\ \vdots \\ b_{n-1} \\ * \end{pmatrix}\right]^2 - \left[\det\begin{pmatrix} a_0 \\ \vdots \\ a_{n-1} \\ * \end{pmatrix}\right]^2\right].$$

From (14) and (15) we get

$$\Big\langle -\langle a_n, *\rangle c_0 + a_n,\ c_n\Big\rangle \geq \frac{c}{2\lambda_0{}^n}\left[\left[\det\begin{pmatrix} b_0 \\ \vdots \\ b_{n-1} \\ * \end{pmatrix}\right]^2 + \left[\det\begin{pmatrix} a_0 \\ \vdots \\ a_{n-1} \\ * \end{pmatrix}\right]^2\right] > 0 . \quad \blacksquare$$

There is a result analogous to Proposition 26 when we begin with imbeddings into $\mathbb{R}^n$ and construct the imbeddings into S^n, but we will not need it. It is probably already clear how the results which we have just proved can be used to transfer theorems from Euclidean space to the sphere. For example, keeping to dimension 3, which is the really interesting one, suppose we have two compact convex surfaces $M, \bar{M} \subset S^3$ (each contained in some open hemisphere), and an isometry $\alpha: M \to \bar{M}$. We claim that α is the restriction of an isometry $A: S^3 \to S^3$. Without loss of generality, we can assume that M and $\bar{M}$ are contained in S^{n+} and are star-shaped with respect to $*$. Let $f_1: M \to S^{n+}$ be the inclusion map, and let $f_2: M \to S^{n+}$ be $\alpha \circ f_1$; then f_1 and f_2 induce the same metric on M. We can also assume that M is oriented so that f_1 and f_2 induce the natural orientation on M and $\bar{M}$, by composing f_2 with a reflection if necessary. Then $\tilde{f}_1, \tilde{f}_2: M \to \mathbb{R}^3$ induce the same metric on M, by Proposition 22, and $\tilde{f}_1(M)$ and $\tilde{f}_2(M)$ have $K \geq 0$ by Proposition 26. So by Theorem 12, there is an isometry $B: \mathbb{R}^3 \to \mathbb{R}^3$ with $\tilde{f}_2 = B \circ \tilde{f}_1$. Then Proposition 25 shows that there is an isometry $A: S^3 \to S^3$ with

$$f_2 = \bar{\tilde{f}}_2 = A \circ \bar{\tilde{f}}_1 = A \circ f_1 \ ,$$

which shows that

$$\alpha \circ f_1 = A \circ f_1 \quad \Longrightarrow \quad \alpha \text{ is the restriction of } A \text{ to } f_1(M) = M \ .$$

Pogorelov states that essentially the same formulas can be used to transfer rigidity problems from hyperbolic space to Euclidean space, and he shows how problems of infinitesimal rigidity can also be transferred in this way.

We also want to add a few remarks, of a different sort, about hypersurfaces of S^{n+1} and H^{n+1}. The proof of Theorem 1 carries over almost without change to this situation, so hypersurfaces of S^{n+1} or H^{n+1} are rigid if their type number (the rank of $X_p \mapsto \nabla'_{X_p} \nu$) is ≥ 3 at each point p. The hypersurfaces of S^{n+1} and H^{n+1} with type number 2 at all points were studied by Dolbeault-Lemoine [1]. She divides them into the same three classes that E. Cartan found for hypersurfaces of $\mathbb{R}^{n+1}$, but it turns out that all hypersurfaces in one of the classes are rigid in S^{n+1} and H^{n+1} for any $n \geq 3$, while the hypersurfaces of the other two classes are rigid in S^{n+1} and H^{n+1} for any $n \geq 4$. Moreover, the hypersurfaces with type number 1 at all points are also rigid in S^{n+1} and H^{n+1} for any $n \geq 4$. This leads her to conclude that for $n \geq 4$, all hypersurfaces of S^{n+1} and H^{n+1} are rigid. Unfortunately, this does not follow directly from the preceding results, for it is conceivable that two hypersurfaces from different classes can be joined together in two isometric, but non-congruent, ways; whether this is actually possible is a question which still has to be cleared up.

The only subject left for us to consider at this point is the rigidity of submanifolds of higher codimension. There are two main results in this direction, a classical local one, and a modern global one.

The classical result involves the notion of the type number of a submanifold $M^n \subset \mathbb{R}^m$ of arbitrary codimension. First an algebraic definition. Let V be a vector space, and let $T_1,\ldots,T_k\colon V \to V$ be linearly independent linear transformations. We define the type number of $\{T_1,\ldots,T_k\}$ to be the largest integer t for which there are t vectors $v_1,\ldots,v_t \in V$ such that the kt vectors

$$T_r(v_i) \qquad 1 \le i \le t\,, \quad 1 \le r \le k$$

are linearly independent. The type number of linearly independent matrices $S_1,\ldots,S_k$ is defined as that of the corresponding linear transformations. Now for a point $p \in M^n \subset \mathbb{R}^m$ we let $k(p)$ be the rank of the map $\xi \mapsto A_\xi$ from $M_p{}^\perp$ into the space of all symmetric maps of M_p into itself (recall that $A_\xi\colon M_p \to M_p$ is defined by $\langle A_\xi(X),Y\rangle = \langle s(X,Y),\xi\rangle$). Equivalently, $k(p)$ is the dimension of the first normal space at p, which may be defined as the orthogonal complement in $M_p{}^\perp$ of $\{\xi\colon A_\xi = 0\}$ (compare Addendum 4 of Chapter 7). Set $k = k(p)$ and let $\xi_1,\ldots,\xi_k$ be a basis for the first normal space at p. If $T_r\colon M_p \to M_p$ is A_{ξ_r} for $r = 1,\ldots,k$, then $T_1,\ldots,T_k$ are linearly independent, and we can define the type number $t(p)$ of M at p to be the type number of $\{T_1,\ldots,T_k\}$; it is easily checked that this definition does not depend on the choice of $\xi_1,\ldots,\xi_k$. The following result shows that submanifolds with type number at least 2 cannot twist too much.

28. LEMMA. Let N^m be a manifold of constant curvature, and let M^n be a submanifold with normal connection D, whose first normal space $\overset{1}{\mathrm{Nor}}\, M_p$ has the same dimension k at all points, and whose type number is ≥ 2 at all points. Then $D_Z\xi \in \overset{1}{\mathrm{Nor}}\, M_p$ for any section ξ of $\overset{1}{\mathrm{Nor}}\, M_p$ and $Z \in M_p$.

Consequently, if M is connected, then it lies in some $(n+k)$-dimensional totally geodesic subspace of N.

Proof. Locally we can choose orthonormal sections $\nu_{n+1},\dots,\nu_m$ of Nor M such that $\nu_{n+1},\dots,\nu_{n+k}$ span $\overset{1}{\mathrm{Nor}}\, M$. Then $A_{\nu_r} = 0 \implies II^r = 0$ for $r > n+k$. So the Codazzi-Mainardi equations (Theorem 7-14) give

$$0 = \sum_{s=n+1}^{n+k} II^s(Y,W)\beta_s^r(X) - II^s(X,W)\beta_s^r(Y) \qquad r > n+k$$

$$\Downarrow$$

$$(1) \qquad 0 = \sum_{s=n+1}^{n+k} \beta_s^r(X)\cdot A_{\nu_s}(Y) - \beta_s^r(Y)\cdot A_{\nu_s}(X) \qquad r > n+k\,.$$

By assumption, there are $X, Y \in M_p$ such that the vectors $A_{\nu_s}(X), A_{\nu_s}(Y)$ for $s = n+1,\dots,n+k$ are linearly independent. Then (1) shows that

$$(2) \qquad \beta_s^r(X) = \beta_s^r(Y) = 0 \qquad n+1 \leq s \leq n+k\,, \quad r > n+k\,.$$

Moreover, for any $Z \in M_p$ and $r > n+k$ we have

$$0 = \sum_{s=n+1}^{n+k} \beta_s^r(X)\cdot A_{\nu_s}(Z) - \beta_s^r(Z)\cdot A_{\nu_s}(X)$$

$$= -\sum_{s=n+1}^{n+k} \beta_s^r(Z)\cdot A_{\nu_s}(X) \qquad \text{by (2)}$$

$$\Downarrow$$

(3) $\beta^r_s(Z) = 0 \qquad n+1 \le s \le n+k, \quad r > n+k.$

This shows that $D_Z \nu^s \in \overset{1}{\text{Nor}}\, M_p$ for $n+1 \le s \le n+k$, and proves the first part of the theorem.

Now consider the $(n+k)$-dimensional distribution $\Delta(p) = M_p \oplus \overset{1}{\text{Nor}}\, M_p$ along M. The first part of the theorem clearly implies that $\nabla'_Z \xi \in \Delta(p)$ for all sections ξ of Δ and $Z \in M_p$. So Δ is parallel along any curve c, by Pre-Lemma 7-7. The result then follows from Corollary 7-11. ■

Remark. A curve in $\mathbb{R}^m$ with $\kappa_1, \ldots, \kappa_m$ all non-zero represents a counter-example to Lemma 28 when the type number is < 2.

The extension of Theorem 1 of this Chapter to submanifolds of higher codimension rests on some more algebraic results.

29. LEMMETTE. Let $\phi_r, \psi_r, \bar{\phi}_r, \bar{\psi}_r \in V^*$ for $r = 1, \ldots, k$ with

$$\sum_{r=1}^{k} \phi_r \wedge \psi_r = \sum_{r=1}^{k} \bar{\phi}_r \wedge \bar{\psi}_r .$$

Suppose that $\phi_1, \ldots, \phi_k, \psi_1, \ldots, \psi_k$ are linearly independent. Then the same is true of the $\bar{\phi}_r, \bar{\psi}_r$, and the subspace $[\phi_1, \ldots, \phi_k, \psi_1, \ldots, \psi_k]$ spanned by the ϕ_r and ψ_r equals the subspace $[\bar{\phi}_1, \ldots, \bar{\phi}_k, \bar{\psi}_1, \ldots, \bar{\psi}_k]$ spanned by the $\bar{\phi}_r$ and $\bar{\psi}_r$.

Proof. Recall that for $v \in V$ and $\omega \in \Omega^k(V)$ we define $v \,\lrcorner\, \omega \in \Omega^{k-1}(V)$ by

$v \lrcorner\, \omega(v_1,\ldots,v_{k-1}) = \omega(v,v_1,\ldots,v_{k-1})$. Define a map $f\colon V \longrightarrow V^*$ by

$$f(v) = v \lrcorner \left(\sum_{r=1}^{k} \phi_r \wedge \psi_r \right) = \sum_{r=1}^{k} \phi_r(v)\cdot\psi_r - \psi_r(v)\cdot\phi_r \,.$$

Clearly

$$\text{range } f \subset [\phi_1,\ldots,\phi_k,\ \psi_1,\ldots,\psi_k] \,.$$

Moreover, by linear independence of the ϕ_r and ψ_r, there is $v \in V$ with $\phi_1(v) = 1$ and all other $\phi_r(v) = \psi_r(v) = 0$. Then $f(v) = \psi_1$, so $\psi_1 \in$ range f. Similarly, all $\phi_r, \psi_r \in$ range f, and we therefore have

$$\text{range } f = [\phi_1,\ldots,\phi_k,\ \psi_1,\ldots,\psi_k] \,.$$

Now we also have

$$f(v) = \sum_{r=1}^{k} \bar{\phi}_r(v)\cdot\bar{\psi}_r - \bar{\psi}_r(v)\cdot\bar{\phi}_r \,.$$

So

$$[\phi_1,\ldots,\phi_k,\ \psi_1,\ldots,\psi_k] = \text{range } f \subset [\bar{\phi}_1,\ldots,\bar{\phi}_k,\ \bar{\psi}_1,\ldots,\bar{\psi}_k] \,.$$

Hence the subspace on the right has dimension $\geq 2k$. So it has dimension exactly $2k$, which means that the $\bar{\phi}_r, \bar{\psi}_r$ are linearly independent and that the two subspaces are equal. ■

30. LEMMA (CHERN). Let $S_1,\ldots,S_k$, $\bar{S}_1,\ldots,\bar{S}_k$ be symmetric $n \times n$ matrices, with $S_1,\ldots,S_k$ linearly independent, of type number ≥ 3. Suppose that the

sum of the determinants of corresponding 2×2 submatrices of the S_r always equals the sum of the determinants of the corresponding 2×2 submatrices of the $\bar{S}_r$. Then we have

$$\bar{S}_r = \sum_{s=1}^{k} A_{sr} S_s$$

for some orthogonal matrix $A \in O(k)$.

Proof. Let $T_r, \bar{T}_r \colon \mathbb{R}^n \to \mathbb{R}^n$ be the linear transformations with matrices $S_r, \bar{S}_r$. The hypothesis on the determinants of $S_r, \bar{S}_r$ is equivalent to the hypothesis that the maps

$$T_r{}^*, \bar{T}_r{}^* \colon \Omega^2(\mathbb{R}^n) \to \Omega^2(\mathbb{R}^n)$$

satisfy

$$\sum_{r=1}^{k} T_r{}^* = \sum_{r=1}^{k} \bar{T}_r{}^* ;$$

in other words, for all $\phi_i, \phi_j \in V^*$ we have

$$\text{(1)} \qquad \sum_{r=1}^{k} T_r{}^*(\phi_i) \wedge T_r{}^*(\phi_j) = \sum_{r=1}^{k} \bar{T}_r{}^*(\phi_i) \wedge \bar{T}_r{}^*(\phi_j) .$$

Choose a basis $\{\phi_i\}$ for V^* such that the $3k$ vectors $\{T_r{}^*(\phi_i)\colon i = 1,2,3\}$ are linearly independent. Applying (1) with $i = 1$, $j = 2$, and using the Lemmette, we see that each $\bar{T}_r{}^*(\phi_1)$ is a linear combination of the $T_s{}^*(\phi_1)$, $T_s{}^*(\phi_2)$. Similarly, each $\bar{T}_r{}^*(\phi_1)$ is a linear combination of the $T_s{}^*(\phi_1)$,

$T_s^*(\phi_3)$. So each $\bar{T}_r^*(\phi_1)$ is a linear combination of the $T_s^*(\phi_1)$. The analogous conclusions hold for the $\bar{T}_r^*(\phi_2)$ and the $\bar{T}_r^*(\phi_3)$. Set

$$\bar{T}_r^*(\phi_1) = \sum_s B_{rs} T_s^*(\phi_1)$$
$$\bar{T}_r^*(\phi_2) = \sum_s C_{rs} T_s^*(\phi_2)$$
$$\bar{T}_r^*(\phi_3) = \sum_s D_{rs} T_s^*(\phi_3) \ .$$

Equation (1) gives us

$$\begin{cases} B\cdot C^t = I = C\cdot B^t & (i = 1,\ j = 2 \text{ and } i = 2,\ j = 1) \\ C\cdot D^t = I = D\cdot C^t & (i = 2,\ j = 3 \text{ and } i = 3,\ j = 2) \\ B\cdot D^t = I = D\cdot B^t & (i = 1,\ j = 3 \text{ and } i = 3,\ j = 1) \ . \end{cases}$$

These imply that $B = C = D$ and $B\cdot B^t = I$, so that B is orthogonal. Let $\tilde{T}_r \colon \mathbb{R}^n \longrightarrow \mathbb{R}^n$ be the linear transformation with

$$\tilde{T}_r^* = \sum_s B_{sr} T_s^* \ .$$

Then $\tilde{T}_r^*(\phi_j) = \bar{T}_r^*(\phi_j)$ for $j = 1,2,3$. We just have to show that this is also true for $j \geq 4$.

Using (1) and orthogonality of B, we see that

$$\sum_r \bar{T}_r^*(\phi_j) \wedge \bar{T}_r^*(\phi_i) = \sum_r \tilde{T}_r^*(\phi_j) \wedge \tilde{T}_r^*(\phi_i) \qquad i = 1,2,3$$

and hence

$$\sum_r \{\bar{T}_r^*(\phi_j) - \tilde{T}_r^*(\phi_j)\} \wedge \bar{T}_r^*(\phi_i) = 0 \qquad i = 1,2,3 \ .$$

The $\bar{T}_r^*(\phi_i)$ are linearly independent, so by Cartan's Lemma (Lemma 1-13 or Problem I.7-11) the $\bar{T}_r^*(\phi_j) - \tilde{T}_r^*(\phi_j)$ are a linear combination of the $\bar{T}_r^*(\phi_i)$ for each $i = 1,2$. But $\{\bar{T}_r^*(\phi_i): i = 1,2\}$ are linearly independent, since $\{T_r^*(\phi_i): i = 1,2\}$ are, so we must have $\bar{T}_r^*(\phi_j) - \tilde{T}_r^*(\phi_j) = 0$. ∎

31. THEOREM (ALLENDOERFER). Let M^n and $\bar{M}^n$ be immersed submanifolds of $\mathbb{R}^m$, and let $\phi: M \to \bar{M}$ be an isometry. Suppose that the first normal spaces of M and $\bar{M}$ have the same constant dimension k at all points. Suppose, moreover, that the type number is ≥ 3 at all points of M. Then ϕ is the restriction of a Euclidean motion.

Proof. By Lemma 28, there is no loss of generality in assuming that $m = n+k$, so that the first normal space is the whole normal space. First we will show that ϕ is locally the restriction of a Euclidean motion. Choose an orthonormal moving frame $X_1,\dots,X_n$ in a neighborhood U of p, and let $\bar{X}_i = \phi_* X_i$. Choose orthonormal sections $\nu_{n+1},\dots,\nu_m$ of the normal bundle of M, and $\bar{\nu}_{n+1},\dots,\bar{\nu}_m$ for the normal bundle of $\bar{M}$. For $q \in U$, define $n \times n$ symmetric matrices S_r, $r = n+1,\dots,m$ by

$$(S_r)_{ij} = II^r(X_i(q), X_j(q)) .$$

Define $\bar{S}_r$ similarly, for the point $\phi(q)$. Gauss' equation, in the form given on p. IV.51, shows that the S_r, $\bar{S}_r$ satisfy the hypotheses of Lemma 30. Thus we see that there is an orthogonal matrix valued function A on U with

$$\bar{S}_r = \sum_s A_{sr} S_s .$$

Using linear independence of the S_r, and smoothness of the S_r and $\bar{S}_r$, we see that the A_{sr} vary smoothly with q. Now define new sections $\nu'_{n+1},\ldots,\nu'_{n+k}$ of the normal bundle of M by

$$\nu'_r = \sum_s A_{sr}\nu_s .$$

Then for the corresponding second fundamental forms we have

$$\begin{aligned} II'^r(X,Y) &= \langle s(X,Y),\nu'_r\rangle \\ &= \langle s(X,Y), \Sigma A_{sr}\nu_s\rangle \\ &= \Sigma A_{sr} II^s . \end{aligned}$$

In particular,

$$\begin{aligned} II'^r(X_i,X_j) &= \Sigma A_{sr} II^s(X_i,X_j) \\ &= \Sigma A_{sr}(S_s)_{ij} \\ &= (\bar{S}_r)_{ij} \\ &= \overline{II}^r(\bar{X}_i,\bar{X}_j) . \end{aligned}$$

Theorem 7-19 then shows that ϕ is the restriction of a Euclidean motion on U.

We claim that this Euclidean motion is unique. This is easy to see once we note that since the first normal space is the whole normal space, every normal vector at $p \in M$ is $c''(0)$ for some arc-length parameterized curve in M. Having established uniqueness, it is clear that the local result implies the global one. ■

To be sure, the hypothesis that the type number is ≥ 3 is extremely strong for submanifolds of higher codimension, but Theorem 31 is very likely the best local result obtainable. It is therefore a pleasant surprise to find that there is a global result in this area. To end this Chapter, with its vast areas of ignorance, on a more joyful note, we quote the following beautiful recent result of J.C. Moore; the proof is somewhat lengthy, but uses only material which has already been developed here, the only somewhat non-standard result being Corollary 11-6.

THEOREM (J.C. MOORE). If $M_1,\ldots,M_k$ are compact connected Riemannian manifolds with M_i of dimension $n_i \geq 2$, then any immersion $\phi\colon M_1 \times \cdots \times M_k \to \mathbb{R}^{n_1+\cdots+n_k+k}$ is, up to a Euclidean motion, a product of immersions of the M_i as hypersurfaces.

In particular, if the M_i are compact convex surfaces in $\mathbb{R}^3$, then $M_1 \times \cdots \times M_k$ is rigid in $\mathbb{R}^{3k}$, though not locally rigid.

Addendum. Infinitesimal Bendings of Rotation Surfaces

We will be dealing with surfaces of revolution obtained by revolving a curve $c(s) = (r(s),s)$ in the (x,z)-plane around the z-axis. Define

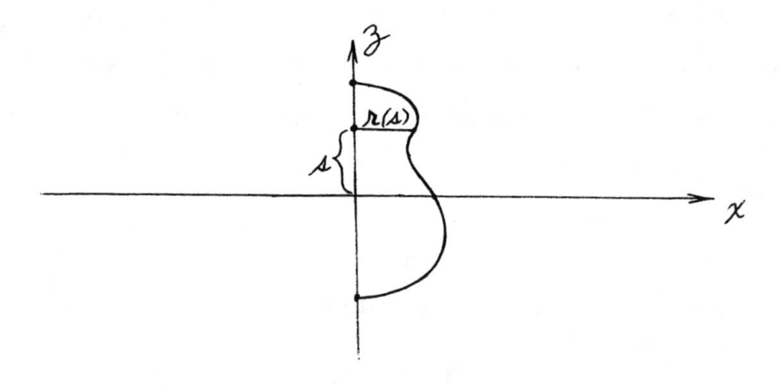

$$\gamma(t) = (\cos t, \sin t, 0) ,$$

so that in time 2π, the curve γ goes once around a unit circle in the (x,y)-plane. Then our rotation surface is given by

$$f(s,t) = r(s)\cdot\gamma(t) + s\cdot e_3 ,$$

where $e_3 = (0,0,1)$. Here $r(s)$ is smooth for $s_0 < s < s_1$, while $r'(s_0) = -\infty$ and $r'(s_1) = +\infty$. Considering for the moment only $s \in (s_0,s_1)$, any vector field Z along f can be written uniquely as a linear combination of the three vectors γ, γ', e_3, thus

$$Z(s,t) = a(s,t)\cdot\gamma(t) + b(s,t)\cdot\gamma'(t) + c(s,t)\cdot e_3 .$$

Now Z is an infinitesimal bending if and only if

$$(1) \qquad \langle f_1, Z_1 \rangle = 0 , \quad \langle f_2, Z_2 \rangle = 0 , \quad \langle f_1, Z_2 \rangle + \langle f_2, Z_1 \rangle = 0 .$$

Since

$$f_1 = r' \cdot \gamma + e_3 \qquad f_2 = r \cdot \gamma'$$

$$Z_1 = a_1 \cdot \gamma + b_1 \cdot \gamma' + c_1 e_3 \qquad Z_2 = (a_2 - b)\gamma + (b_2 + a)\gamma' + c_2 e_3 ,$$

equations (1) become

$$(2) \qquad \begin{aligned} r' a_1 + c_1 &= 0 \\ b_2 + a &= 0 \\ r'(a_2 - b) + r b_1 + c_2 &= 0 . \end{aligned}$$

Since a, b, c must be periodic in t of period 2π, it is natural to look for solutions in terms of Fourier series

$$a(s,t) = \sum_{k=-\infty}^{\infty} e^{ikt} \phi_k(s)$$

$$b(s,t) = \sum_{k=-\infty}^{\infty} e^{ikt} \psi_k(s)$$

$$c(s,t) = \sum_{k=-\infty}^{\infty} e^{ikt} \xi_k(s) ,$$

where, in order that ϕ_k, ψ_k, ξ_k should be real-valued, we must have

$$(3) \qquad \phi_{-k} = \bar{\phi}_k , \quad \psi_{-k} = \bar{\psi}_k , \quad \xi_{-k} = \bar{\xi}_k .$$

For a (complex-valued) solution involving a single k, equations (2) become

$$r'(s)\phi_k'(s) + \xi_k'(s) = 0$$

$$(4) \qquad ik\psi_k(s) + \phi_k(s) = 0$$

$$r'(s)[ik\phi_k(s) - \psi_k(s)] + r(s)\psi_k'(s) + ik\xi_k(s) = 0 \ .$$

Differentiating the third equation, and then using the first two, we obtain

$$(*) \qquad r\psi_k'' + (k^2 - 1)r''\psi_k = 0 \ .$$

Conversely, suppose we have a complex-valued function ψ_k satisfying (*). Then the first two equations of (4) can be used to determine ϕ_k and then ξ_k. If we <u>define</u> ϕ_{-k}, ψ_{-k}, ξ_{-k} by (3), then (4) also holds for $-k$. Thus we will have real-valued solutions

$$a(s,t) = e^{ikt}\phi_k(s) + e^{-ikt}\bar{\phi}_k(s)$$

$$b(s,t) = e^{ikt}\psi_k(s) + e^{-ikt}\bar{\psi}_k(s)$$

$$c(s,t) = e^{ikt}\xi_k(s) + e^{-ikt}\bar{\xi}_k(s)$$

of (2). Any finite linear combination of solutions is also a solution.

For $k = 0$, we can solve directly from (4). The first two equations give $\phi_0 = 0$, and then $\xi_0 =$ constant A. Then the third equation gives

$$0 = -r'\psi_0 + r\psi_0' = r^2\left(\frac{\psi_0}{r}\right)' \implies \psi_0 = Br \qquad \text{for some constant } B \ .$$

Thus we obtain the infinitesimal bending

$$\begin{aligned} Z_0(s,t) &= Br(s)\cdot\gamma'(t) + A\cdot e_3 \\ &= [r(s)\cdot\gamma(t) + s\cdot e_3] \times Be_3 + Ae_3 \\ &= (f(s,t) \times Be_3) + Ae_3 , \end{aligned}$$

which is trivial.

For $k = 1$, equation (*) says that ψ_1 is linear. In particular, one possible solution is

$$\psi_1(s) = Cs , \qquad C \text{ real} .$$

Then the second equation of (4) gives

$$\phi_1(s) = - iCs ,$$

and so the first equation gives $\xi_1(s) = iCr(s) + \text{constant}$; in particular, we can take

$$\xi_1(s) = iCr(s) .$$

Then we have

$$\begin{aligned} a(s,t) &= 2 \operatorname{Re} e^{it}(- iCs) = 2Cs \sin t , \\ b(s,t) &= 2 \operatorname{Re} e^{it}Cs = 2Cs \cos t , \\ c(s,t) &= 2 \operatorname{Re} e^{it}(iCr(s)) = - 2Cr(s) \sin t . \end{aligned}$$

This gives the trivial infinitesimal bending

$$\begin{aligned} \tfrac{1}{2}Z_1(s,t) &= (Cs \sin t)\gamma(t) + (Cs \cos t)\gamma'(t) - (Cr(s) \sin t)e_3 \\ &= [r(s)\cdot\gamma(t) + se_3] \times [- C \sin t\ \gamma' + C \cos t\ \gamma] \\ &= f(s,t) \times (C,0,0) . \end{aligned}$$

Similarly, we can take

$$\psi_1(s) = iCs \quad (C \text{ real}), \qquad \phi_1(s) = Cs , \qquad \xi_1(s) = - Cr(s) ,$$

obtaining the infinitesimal bending

$$\begin{aligned} \tfrac{1}{2}Z_1(s,t) &= [r(s)\cdot\gamma(t) + se_3] \times [- C \cos t\ \gamma' - C \sin t\ \gamma] \\ &= f(s,t) \times (0,-C,0) . \end{aligned}$$

Obviously <u>every</u> trivial infinitesimal bending is a linear combination of the various infinitesimal bendings Z_0, Z_1. Since the various Z_k are linearly independent, we see that any solution of (*) for $k \geq 2$ leads to an infinitesimal bending Z which is <u>not</u> trivial.

These considerations all hold only for the region $s_0 < s < s_1$, that is, for the surface of revolution minus its two "poles." Given any infinitesimal bending Z_k obtained as above, we still have to see how it behaves at the poles (one can easily see, for example, that if we had picked $\psi_1(s) = Cs + D$ with $D \neq 0$, then the corresponding solution, although not a trivial infinitesimal bending, would have a singularity at the poles). In order to do this, we consider the functions $\phi_k \circ r^{-1}$, etc., where r^{-1} really denotes two different functions, depending on which pole we are at. To be precise, we note that at either pole $r^{-1}(x)$ makes sense for x in some interval $[0,\varepsilon)$. We extend r^{-1} to a function ρ on $(-\varepsilon,\varepsilon)$ by requiring ρ to be even. We will assume that ρ is analytic at 0 (this is precisely what one needs in order for the surface f to be analytic at the poles). Setting

$$\begin{aligned} \Phi_k &= \phi_k \circ \rho \\ \Psi_k &= \psi_k \circ \rho \\ \Xi_k &= \xi_k \circ \rho , \end{aligned}$$

and noting that $\rho' = 1/r'\circ\rho$, we find that equations (4) can be written

$$
(5) \qquad
\begin{aligned}
\rho'(x)\Xi_k'(x) + \Phi_k'(x) &= 0 \\
ik\Psi_k(x) + \Phi_k(x) &= 0 \\
ik\Phi_k(x) - \Psi_k(x) + x\Psi_k'(x) + ik\rho'(x)\Xi(x) &,
\end{aligned}
$$

while equation (*) becomes

$$
(**) \qquad x\rho'(x)\Psi_k''(x) - x\rho''(x)\Psi_k'(x) - (k^2-1)\rho''(x)\Psi_k(x) = 0 .
$$

Now suppose also that $\rho''(0) \neq 0$. Then

$$
\rho'(x) = \rho''(0)x[1+\cdots] , \qquad \rho''(x) = \rho''(0)\cdot[1+***]
$$

and we can write our equation as

$$
x^2\rho''(0)[1+\cdots]\Psi_k''(x) - x\rho''(0)[1+\cdots]\Psi_k'(x) - (k^2-1)\rho''(0)[1+***]\Psi_k(x) = 0
$$

or

$$
(***) \qquad x^2\Psi_k''(x) + x\alpha(x)\Psi_k'(x) + \beta(x)\Psi_k(x) = 0 ,
$$

where α and β are analytic with

$$
\alpha(0) = -1 , \qquad \beta(0) = 1-k^2 .
$$

Now equation (***) has a "singular point" at 0; we cannot put it in the form $\Psi_k''(x) = F(x,\Psi_k(x),\Psi_k'(x))$ near 0, so we cannot apply our standard theorems. However, this singular point is a very special sort, called a "regular singular point," and there is a complete theory to cover this situation.

It is one of the standard topics in differential equations, which the reader can find in Whittaker and Watson {1; Chapter 10}. The theory shows that equation (***), or equivalently (**), has two linearly independent solutions near $x = 0$ of the form

$$\mu_1(x) = x^{1+k}\cdot(\text{analytic function})$$
$$\mu_2(x) = x^{1-k}\cdot(\text{analytic function}) + c(\log x)\cdot\mu_1(x) ,$$

where the analytic functions in question are non-zero at 0 (but c might be 0). So when $k \geq 2$, we see that (**) always has one solution which is analytic near 0, while any linearly independent solution blows up at zero. If Ψ_k is an analytic solution, so that Ψ_k vanishes up to order $k+1$ at 0, then the first and second equations of (5) determine functions Φ_k and Ξ_k which vanish up to order $k+1$ and k, respectively, at 0. So for $k \geq 2$ the infinitesimal bending Z_k then determined by $\phi_k = \Phi_k \circ r$, etc., has the form

$$Z_k(s,t) = r(s)^2\sigma(r(s),t)$$

for some analytic function σ. Now an analytic parameterization of our rotation surface near a pole is given by

$$(x,y) \mapsto (x,y,r^{-1}(\sqrt{x^2+y^2}))$$
$$= f(r^{-1}(\sqrt{x^2+y^2}), \arctan \tfrac{y}{x}) .$$

So if $\tilde{Z}_k(x,y)$ denotes the value of Z_k at this point on the rotation surface, then

$$\tilde{Z}_k(x,y) = (\sqrt{x^2+y^2})^2\sigma(\sqrt{x^2+y^2},t) = (x^2+y^2)\sigma(\sqrt{x^2+y^2},t) .$$

One can easily check that this function is* C^2 at 0.

This analysis holds at each pole, but it will usually happen that the function Ψ_k which is analytic for one choice of $\rho = r^{-1}$ will not be analytic for the other choice. So further analysis is required.

We consider a 1-parameter family of functions r_t which passes continuously from a convex function to non-convex functions. For example, we can determine r_t by the equation

$$(r_t(s)^2+s^2)^2 + 2t^2(r_t(s)^2-s^2) = 1 - 2t^2 \qquad 0 \le t^2 < \frac{1}{2} .$$

For $t^2 = 1/2$ we would have a lemniscate, and for $t = 0$ we have a circle. For $t^2 \le 1/3$ the functions are convex.

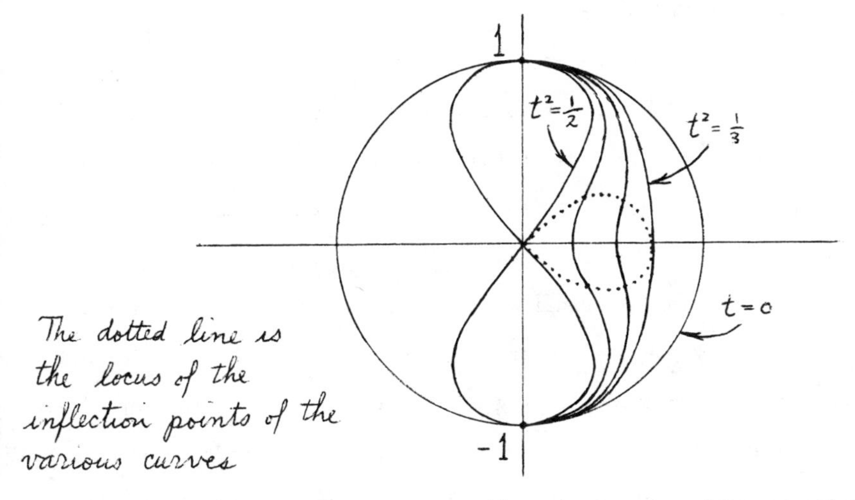

For a fixed t, which we temporarily suppress, consider equation (*)

---

*If σ involved only even powers of $\sqrt{x^2+y^2}$, then we would actually have an analytic function. Unfortunately, this can never happen, since equations (5), and the fact that σ is even, shows that Φ, Ψ, Ξ cannot all be even functions.

for $r = r_t$:

$$(*) \qquad r\psi_k'' + (k^2 - 1)r''\psi_k = 0 .$$

On any (concave) interval where $r'' > 0$ (which exist for $t^2 > 1/3$), there will be a large number of zeros of ψ_k once k is large enough. To prove this we merely choose a subinterval on which $r''/r > \varepsilon > 0$ for some ε, and apply the Sturm Comparison Theorem (9-15) to equation (*) and the equation $y'' + (k^2 - 1)\varepsilon y = 0$, which has lots of zeros for large k. Notice that on each of the outer intervals, where $r'' < 0$, the function ψ_k cannot have a positive maximum ($\psi_k > 0$, $\psi_k'' \leq 0$) or a negative minimum, so it cannot have even two zeros. Thus the total number of zeros in $(-1,1)$ is finite. For $t^2 > 1/3$ we have $r'' < 0$ everywhere on $(-1,1)$, so ψ_k has at most one zero on $(-1,1)$. Even for $t^2 = 1/3$ we easily see that ψ_k has at most 2 zeros on $(-1,1)$. Since ψ_k is a solution of a second order equation, we must have $\psi_k' \neq 0$ when $\psi_k = 0$ (assuming ψ_k is not the zero function), so ψ_k crosses the axis at each zero, rather than being tangent to it. It follows easily that any function sufficiently close to ψ_k has at least as many zeros as ψ_k.

Now pick some t_0 with $t_0^2 > 1/3$ and a k so large that any solution ψ_{k,t_0} of (*) for $r = r_{t_0}$ has more than 2 zeros on $[0,1)$. Let ψ_{k,t_0} be a solution which gives an infinitesimal bending that is C^2 at the bottom pole $s = -1$, and let $n_0 > 2$ be the number of its zeros on $[0,1)$. Now for all $t < t_0$ we will pick a continuous family of solutions $\psi_{k,t}$ of (*) for $r = r_t$, all of which also give infinitesimal bendings that are C^2 at the bottom pole. Choosing $\psi_{k,t}$ is equivalent to choosing $\Psi_{k,t}$, which is determined only up to constant factor. So we can easily arrange that each $\psi_{k,t}$ is not the zero function. For $t^2 \geq 1/3$, the function $\psi_{k,t}$ has at most two zeros in $(-1,1)$, so _a fortiori_ at most two zeros in $[0,1)$. So we can consider the greatest lower bound t_1 of all t such that $\psi_{k,t}$ has exactly n_0 zeros on $[0,1)$, and $t_1^2 > 1/3$.

We claim that ψ_{k,t_1} has a zero at 0. For suppose that all the zeros of ψ_{k,t_1} on $[0,1)$ actually occur on $(0,1)$, and let n_1 be the number of such zeros. Since any function sufficiently close to ψ_{k,t_1} has at least as many zeros on $(0,1)$ as ψ_{k,t_1} does, we easily see that $n_1 \leq n_0$. We claim that we cannot have $n_1 < n_0$. To prove this, it suffices to show that the functions $\psi_{k,t}$ for t close to t_1 have _at most_ as many zeros on $(0,1)$ as ψ_{k,t_1} does. On some interval containing the n_1 zeros of ψ_{k,t_1} we will have $r_{t_1}''/r_{t_1} < M$ for some M. The same inequality holds on this interval for t sufficiently close to t_1. Applying the (second part of) the Sturm comparison to (*) and the equation $y'' + (k^2 - 1)My = 0$, we find that the zeros of $\psi_{k,t}$ must be at least $\varepsilon = \pi/\sqrt{(k^2 - 1)M}$ apart. So if we choose a neighborhood of the graph of ψ_{k,t_1} which intersects the s axis in intervals of length $< \varepsilon$, then the $\psi_{k,t}$ whose graphs lie in this neighborhood can have at

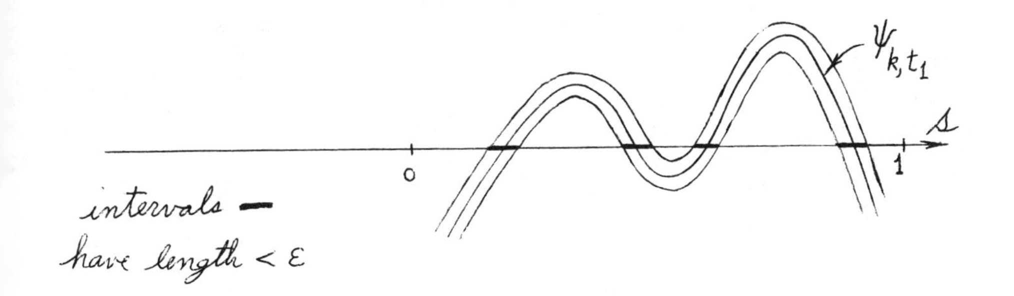

most n_1 zeros on $(0,1)$. Thus we have shown that ψ_{k,t_1} actually has exactly n_0 zeros on $(0,1)$.

But these very same arguments show that for t sufficiently close to t_1, the function $\psi_{k,t}$ has exactly n_0 zeros on $[0,1)$ <u>even for</u> $t < t_1$. This contradicts the choice of t_1, and thus shows that indeed $\psi_{k,t_1}(0) = 0$.

Now from the differential equation

$$r_{t_1}\psi_{k,t_1}'' + (k^2 - 1)r_{t_1}''\psi_{k,t_1} = 0$$

satisfied by ψ_{k,t_1}, and the fact that r_{t_1} is an even function, we easily see that $s \mapsto \psi_{k,t_1}(-s)$ also satisfies the equation. Since $\psi_{k,t_1}(0) = 0$, it follows that $\psi_{k,t_1}(-s) = c\cdot\psi_{k,t_1}(s)$ for some constant c. But this means that the infinitesimal bending determined by ψ_{k,t_1} [on the rotation surface determined by r_{t_1}] is also C^2 at the top pole.

PROBLEMS

1. Let $\alpha, \beta\colon \mathbb{R} \to \mathbb{R}^2$ be differentiable functions for which we know the functions

$$\langle\alpha,\beta\rangle, \quad \langle\alpha',\beta\rangle, \quad \langle\alpha,\alpha\rangle, \quad \langle\beta,\beta\rangle\ .$$

We want to show that we can find α and β once we know $\alpha(0)$ and $\beta(0)$.

(a) We can assume that $\langle\alpha,\alpha\rangle = \langle\beta,\beta\rangle = 1$.

(b) If we write

$$\alpha(x) = (\cos\,\theta(x),\ \sin\,\theta(x))$$

$$\beta(x) = (\cos\,\phi(y),\ \sin\,\phi(y)),$$

then $\langle\alpha,\beta\rangle$ and $\langle\alpha',\beta\rangle$ determine θ'. Hence $\alpha(0)$ determines α.

(c) Then $\beta(0)$ determines β.

2. (a) Let η be a 1-parameter family of k-forms. Use Proposition 9-10 to show that for every singular $(k + 1)$-cube c we have

$$\int_c (d\eta)^{\bullet} = \int_{\partial c} \dot{\eta} = \int_c d\dot{\eta}\ .$$

Conclude that $(d\eta)^{\bullet} = d\dot{\eta}$.

(b) Give another proof by writing $\eta(u)$ in a coordinate system $x^1,\ldots,x^n$, and noting that $\partial/\partial u$ commutes with $\partial/\partial x^j$.

3. A surface $M \subset \mathbb{R}^3$ is called isothermal if it can be covered by isothermal coordinate systems whose parameter lines lie along the lines of curvature.

(a) Surfaces of revolution are isothermal.

(b) Problem 4-8 shows that ellipsoids and hyperboloids of one or two sheets have coordinate systems (u,v) with

$$\langle\ ,\ \rangle = (u - v)\left[\frac{u}{f(u)}\, du \otimes du - \frac{v}{f(v)}\, dv \otimes dv\right],$$

where the u- and v-parameter lines are lines of curvatures. Conclude that these surfaces are isothermal.

(c) Let X_1, X_2 be a moving frame consisting of orthonormal principal vectors. Show that M is isothermal if and only if there is a nowhere zero function α such that

$$d\alpha \wedge \theta^1 + \alpha\, \omega_1^2 \wedge \theta^2 = 0 = d\alpha \wedge \theta^2 - \alpha\, \omega_1^2 \wedge \theta^1 .$$

Hint: This means that $d(u\theta^1) = d(u\theta^2) = 0$.

4. Let $M \to \mathbb{R}^3$ be a surface, and $\alpha\colon [0,1] \times M \to \mathbb{R}^3$ a variation as on p. 328.

(a) Show that

$$\dot\psi_1^3 \wedge \psi_2^3 + \psi_1^3 \wedge \dot\psi_2^3 = 0,$$

$$d\dot\psi_1^3 = \omega_1^2 \wedge \dot\psi_2^3$$

$$d\dot\psi_2^3 = -\omega_1^2 \wedge \dot\psi_1^3 .$$

(b) Show that $\dot H = 0$ if and only if

$$\dot\psi_1^3 \wedge \theta^2 = \dot\psi_2^3 \wedge \theta^1 .$$

Hint: Use the equation on p. III.98.

(c) Suppose that the moving frame X_1, X_2 on M consists of principal vectors, so that $\psi_i^3 = k_i\theta^i$, where k_1, k_2 are the principal curvatures. If p is an umbilic, then we always have $\dot H(p) = 0$. [Hint: Use the first equation of (a).] On the other hand, if $\dot H = 0$, and M has no umbilics, then

$$\dot\psi_1^3 \wedge \theta^2 = 0 = \dot\psi_2^3 \wedge \theta^1 .$$

(d) If we write

$$\psi_i^3(t) = \Sigma\, \ell_{ij}(t)\theta^j(t), \qquad \ell_{11}(0) = k_1, \quad \ell_{22}(0) = k_2 ,$$

then $\dot{H} = 0 \Rightarrow \dot{\ell}_{11} = \dot{\ell}_{22} = 0$.

(e) Conclude that if $\dot{H} = 0$, then

$$d\dot{\ell}_{12} \wedge \theta^1 - 2\dot{\ell}_{12}\omega_2^1 \wedge \theta^1 = 0$$

$$d\dot{\ell}_{12} \wedge \theta^2 - 2\dot{\ell}_{12}\omega_1^2 \wedge \theta^1 = 0 .$$

Then use Problem 3 to show that M is isothermal.

5. For vectors $v_1,\ldots,v_{n-1}$ and $w_1,\ldots,w_{n-1}$ in $\mathbb{R}^n$, show that

$$\langle v_1\times\ldots\times v_{n-1}, w_1\times\ldots\times w_{n-1}\rangle = \det(\langle v_i,w_j\rangle)$$

by noting that both sides are linear in the v_i and w_j.

Chapter 13. The Generalized Gauss-Bonnet Theorem and What It Means for Mankind

In previous Chapters we have seen that interesting and challenging questions can arise even in the lowest dimensions, and that the methods used to resolve these problems often rely more on ingenuity and hard work than on particularly sophisticated concepts -- the proofs may be involved, but they have the satisfying concreteness of geometrical arguments, and something of the charm of antique music. Nevertheless, it is futile to deny the decisive influence which has been wrought upon the shape of modern mathematics by the daemonic spirit of functorial constructions. So it is appropriate that this book end with a topic that represents one of the triumphs of machinery in mathematics. Here, at last, connections in principal bundles play their true predestined role, the invariant form of the Bianchi identities prove their superiority, and connections on arbitrary bundles are frequently invoked. As a final affirmation that we have plunged into the icy stream of modern mathematics, hardly a picture appears.

One of the star theorems of differential geometry is the Gauss-Bonnet Theorem, which for a compact oriented surface M states that

$$\int_M K\, dA = 2\pi\chi(M) .$$

Although the curvature K is defined intrinsically in terms of the metric $\langle\ ,\ \rangle$ on M, it can also be defined extrinsically when the metric $\langle\ ,\ \rangle$ on M is induced by an imbedding $M \subset \mathbb{R}^3$. In fact, if $\nu\colon M \to S^2$ is the normal map, and da is the volume element of S^2, then $K\, dA = \nu^*(da)$, so that

$$\int_M K\, dA = \int_M \nu^*(da) = (\deg \nu) \cdot \int_{S^2} da = 4\pi \cdot (\deg \nu) .$$

As we indicated in Addendum 2 to Chapter 6, one can prove, without invoking any differential geometry, that $\deg \nu = \frac{1}{2}\chi(M)$, thus proving the Gauss-Bonnet Theorem for the special case where the metric on M comes from an imbedding in $\mathbb{R}^3$. Precisely this argument was used by Heinz Hopf [2] in obtaining the first generalization of the Gauss-Bonnet Theorem. Consider a compact hypersurface $M^n \subset \mathbb{R}^{n+1}$, where n is even. If da denotes the volume element of S^n, then

$$\int_M K_n\, dV = \int_M \nu^*(da) = (\text{volume of } S^n)\cdot \deg \nu$$

$$= \frac{(\text{volume of } S^n)}{2}\cdot \chi(M)\ , \qquad \text{by Corollary 6-23.}$$

Now this result, although proved for a hypersurface of $\mathbb{R}^{n+1}$, can be formulated for any compact oriented Riemannian n-manifold $(M, \langle\ ,\ \rangle)$ of even dimension n. In fact, we have already noted (p.IV.103) that

$$(*) \qquad K_n = \frac{1}{2^{n/2} n!}\cdot \text{contraction of } (\mathcal{R}\times\cdots\times\mathcal{R}\times\mathcal{E}\times\mathcal{E})\ .$$

In a coordinate system, we have (p.IV.101)

$$K_n = \frac{1}{2^{n/2} n!}\sum_{\substack{i_1,\ldots,i_n\\ j_1,\ldots,j_n}} R_{i_1 i_2 j_1 j_2}\cdots R_{i_{n-1} i_n j_{n-1} j_n}\,\frac{\varepsilon^{i_1\cdots i_n}}{\sqrt{\det(g_{ij})}}\cdot\frac{\varepsilon^{j_1\cdots j_n}}{\sqrt{\det(g_{ij})}}\ .$$

We are thus led to conjecture that we always have

$$\int_M K_n\, dV = \frac{(\text{volume of } S^n)}{2}\cdot\chi(M)\ ,$$

whenever M is a compact oriented Riemannian manifold with n even, where K_n is defined by (*).

For the case where the metric on M comes from an imbedding $M \subset \mathbb{R}^{n+k}$ in some Euclidean space, the result was first proved by Allendoerfer [1] and Fenchel [2] . This was done by considering a closed tubular neighborhood N of M, for which $\partial N \subset \mathbb{R}^{n+k}$ is a hypersurface with a volume element $d\mathbf{V}$, say, and corresponding $\mathbf{K}_{n+k-1}$. We can assume that $n+k-1$ is even (by considering $M \subset \mathbb{R}^{n+k} \subset \mathbb{R}^{n+k+1}$ if necessary), so that the result for hypersurfaces gives

$$\int_{\partial N} \mathbf{K}_{n+k-1}\, d\mathbf{V} = \frac{(\text{volume of } S^{n+k-1})}{2} \cdot \chi(\partial N) .$$

Now it can be shown without too much difficulty that

$$\chi(\partial N) = \chi(M) \cdot \chi(S^{k-1}) = 2\chi(M) .$$

On the other hand, it also works out that

$$\int_{\partial N} \mathbf{K}_{n+k-1}\, d\mathbf{V}$$

can be computed in terms of $\int_M K_n\, dV$; when the computation is effected, we obtain the correct expression for $\int_M K_n\, dV$.

At the time of this proof, the Nash imbedding theorem was not yet known. But the Burstin-Janet-Cartan Theorem (11-9) was known. In 1943, Allendoerfer and Weil [1] proved a generalization of the Gauss-Bonnet formula for a polyhedral piece of a Riemannian manifold imbedded in Euclidean space; using this, they were able to obtain a proof of the general Gauss-Bonnet Theorem for (C^{ω})

Riemannian manifolds, by means of a triangulation. Since the Nash imbedding theorem is now available, the earlier result of Allendoerfer and Fenchel implies that the generalized Gauss-Bonnet Theorem holds for all C^∞ manifolds. But a proof of this sort is clearly unsatisfactory, not only because of the difficulty and essentially non-differential geometric nature of Nash's result, but also because an intrinsic theorem ought to have an intrinsic proof. The intrinsic proof was obtained by Chern [2] in 1944. Ensuing developments have led to a much deeper understanding of the fundamentals which are involved here, so that we can now give a completely non-computational proof of this extraordinary theorem. This proof by magic is presented in the first four sections; in the remainder of the Chapter we will contravene the rules of legerdemain, and reveal some of the mechanism behind it.

1. Operations on Bundles

In the past we have considered numerous structures on vector bundles and principal bundles, but, except in some of the problems for Chapter I.3, we have not yet examined in detail the relationships between different bundles. The simplest relation is that of equivalence $\simeq$ between two vector bundles ξ_1 and ξ_2 over the same base space X. We have also defined the notion of a bundle map from $\xi_1 = \pi_1\colon E_1 \to X_1$ to $\xi_2 = \pi_2\colon E_2 \to X_2$. This is a pair of continuous maps $(\tilde{f}, f)$, where $f\colon X_1 \to X_2$ and $\tilde{f}\colon E_1 \to E_2$; the map $\tilde{f}$ is required to satisfy $\pi_2 \circ \tilde{f} = f \circ \pi_1$, so that $\tilde{f}$ takes fibres of ξ_1 to fibres of ξ_2, and each map $\tilde{f}\colon \pi_1^{-1}(x) \to \pi_2^{-1}(f(x))$ is required to be linear. In this chapter we will re-define the notion of a <u>bundle map</u>, by adding the requirement that each $\tilde{f}\colon \pi_1^{-1}(x) \to \pi_2^{-1}(f(x))$ be an isomorphism of vector spaces (so ξ_1 and ξ_2 must have the same fibre dimension). Instead of referring to a bundle map $(\tilde{f},f)$, we will often say that $\tilde{f}$ is a <u>bundle map covering</u> f. If $\tilde{f}$ is a bundle map covering a homeomorphism $f\colon X_1 \to X_2$ from X_1 onto X_2, then we can define $\tilde{f}^{-1}\colon E_2 \to E_1$. Using the local product structure, and the fact that $A \mapsto A^{-1}$ is continuous for $A \in GL(n,\mathbb{R})$, we easily see that $\tilde{f}^{-1}$ is continuous, so that $\tilde{f}^{-1}$ is a bundle map covering f^{-1}. In particular, a bundle map covering the identity map of X is an equivalence.

Consider next two principal bundles $\xi_i = \pi_i\colon P_i \to X_i$ $(i = 1,2)$ with the same group G; we denote the action of G on the right of P_1 and P_2 by the same symbol "$\cdot$". A <u>(principal)</u> <u>bundle map</u> from ξ_1 to ξ_2 is a pair $(\tilde{f},f)$, where $f\colon X_1 \to X_2$ and $\tilde{f}\colon P_1 \to P_2$, such that $\pi_2 \circ \tilde{f} = f \circ \pi_1$, and such that

$$(*) \qquad \tilde{f}(u\cdot a) = \tilde{f}(u)\cdot a \qquad \text{for all } u \in P \text{ and } a \in G\,.$$

Notice that condition (*) already implies that $\tilde{f}$ takes fibres to fibres, and thus automatically gives us the map f. We could thus speak simply of a bundle map $\tilde{f}$. In practice, it is usually more convenient to speak of a <u>bundle map</u> $\tilde{f}$ <u>covering</u> f. Note also that $\tilde{f}\colon \pi_1^{-1}(x) \to \pi_2^{-1}(f(x))$ is clearly a homeomorphism, since the fibres of P_1 are $\{u\cdot a\colon a \in G\}$ for fixed u, and similarly for the fibres of P_2. As before, if $\tilde{f}$ is a bundle map over a homeomorphism $f\colon X_1 \to X_2$, then $\tilde{f}^{-1}$ is a bundle map over f^{-1}. When f is the identity map of X, we call $\tilde{f}$ an <u>equivalence</u>, or an isomorphism. A principal bundle $\xi = \pi\colon P \to X$ is called <u>trivial</u> if it is equivalent to the bundle $\pi'\colon X \times G \to X$, where π' is projection on the first coordinate. As we have already pointed out (p. II.353), if the principal bundle ξ has a section $s\colon X \to P$, then

ξ is trivial, for we can define a map $X \times G \longrightarrow P$ by $(x,a) \longmapsto s(x)\cdot a$.

Recall (p. II.347) that for an n-dimensional vector bundle $\xi = \pi\colon E \longrightarrow X$ we can define the principal bundle $F(\xi) = \varpi\colon F(E) \longrightarrow X$ of frames of E, with group $GL(n,\mathbb{R})$, whose fibre $\varpi^{-1}(x)$ is the set of all ordered bases $(u_1,\ldots,u_n)$ of the vector space $\pi^{-1}(x)$. If ξ_i are vector bundles over X_i and $\tilde{f}\colon E_1 \longrightarrow E_2$ is a bundle map covering $f\colon X_1 \longrightarrow X_2$, then we clearly have also a principal bundle map $\bar{f}\colon F(E_1) \longrightarrow F(E_2)$ covering f (this would not be true if we did not require a bundle map to be an isomorphism on each fibre). Conversely, a principal bundle map $\bar{f}\colon F(E_1) \longrightarrow F(E_2)$ covering f gives rise to a bundle map $\tilde{f}\colon E_1 \longrightarrow E_2$. In fact, given any frame $u = (u_1,\ldots,u_n)$ of $\pi_1^{-1}(x)$, there is a unique isomorphism $\pi_1^{-1}(x) \longrightarrow \pi_2^{-1}(f(x))$ which takes u_i to the ith member of the frame $\bar{f}(u)$. The condition (*) on $\bar{f}$ insures that this isomorphism is well-defined.

There is one operation which is special for vector bundles. Given two vector bundles $\xi_i = \pi_i\colon E_i \longrightarrow X$ over the same space X, we can form the direct sum $\pi_1^{-1}(x) \oplus \pi_2^{-1}(x)$ $[= \pi_1^{-1}(x) \times \pi_2^{-1}(x)$ as a set$]$ for each $x \in X$. Let

$$E = \bigcup_{x\in X} \pi_1^{-1}(x) \times \pi_2^{-1}(x) \subset E_1 \times E_2 ,$$

with the topology it has as a subset of $E_1 \times E_2$, and let $\pi\colon E \longrightarrow X$ be the map which takes all elements of $\pi_1^{-1}(x) \times \pi_2^{-1}(x)$ to x (thus $\pi = \pi_i | E$ for $i = 1, 2$). It is easy to check that $\pi\colon E \longrightarrow X$ is also a vector bundle, whose fibre dimension is the sum of the fibre dimensions of ξ_1 and ξ_2. This new bundle is called the Whitney sum $\xi_1 \oplus \xi_2$ of ξ_1 and ξ_2. We clearly have

$$\xi_1 \oplus \xi_2 \simeq \xi_2 \oplus \xi_1 , \qquad (\xi_1 \oplus \xi_2) \oplus \xi_3 \simeq \xi_1 \oplus (\xi_2 \oplus \xi_3) .$$

Our next construction works for either a vector bundle or a principal bundle $\xi = \pi\colon E \to Y$. Let $f\colon X \to Y$ be continuous. We can construct a

$$\begin{array}{ccc} & & E \\ & & \downarrow \pi \\ X & \xrightarrow{f} & Y \end{array}$$

[principal] bundle η over X, and a [principal] bundle map $(\tilde{f}, f)$ from η to ξ, as follows.

$$\begin{array}{ccc} X \times E \supset E' & \xrightarrow{\tilde{f}} & E \\ \downarrow \pi' & & \downarrow \pi \\ X & \xrightarrow{f} & Y \end{array}$$

Let

$$E' \subset X \times E = \{(x,e)\colon f(x) = \pi(e)\} ,$$

and let

$$\pi'\colon E' \to X \quad \text{be} \quad \pi'((x,e)) = x .$$

Thus the fibre $\pi'^{-1}(x)$ over a point $x \in X$ is just $\{x\} \times \pi^{-1}(f(x))$. In the case of a vector bundle, we use the vector space structure on $\pi^{-1}(f(x))$ to define a vector space structure on $\pi'^{-1}(x)$; in the case of a principal bundle, we use the action of G on $\pi^{-1}(f(x))$ to define the action of G on $\pi'^{-1}(x)$.

It is easy to check that $\pi' \colon E' \longrightarrow X$ is a vector bundle [principal bundle], and that $\tilde{f} \colon E' \longrightarrow E$ defined by

$$\tilde{f}((x,e)) = e$$

is a [principal] bundle map covering f. The bundle $\pi' \colon E' \longrightarrow X$ is denoted by $f^*\xi$, and is called the bundle over X induced by f and ξ. If $X \subset Y$ and $i \colon X \longrightarrow Y$ is the inclusion map, then $i^*\xi$ is equivalent to the restriction $\xi|X$ of ξ to X. If $g \colon W \longrightarrow X$ is another continuous map, then

$$g^*(f^*\xi) \simeq (f\circ g)^*(\xi) \; .$$

Finally, if $\xi = \pi \colon E \longrightarrow Y$ is a vector bundle, then

$$f^*(F(\xi)) \simeq F(f^*\xi) \; .$$

Although we used an explicit construction to define $f^*\xi$, this [principal] bundle can be characterized uniquely, up to equivalence, by the fact that there is a [principal] bundle map covering f from $f^*\xi$ to ξ. Indeed, suppose that $\eta = \pi'' \colon E'' \longrightarrow X$ is a [principal] bundle, and $\tilde{\tilde{f}} \colon E'' \longrightarrow E$ is a [principal] bundle map covering f. We define $g \colon E'' \longrightarrow E'$ by

$$g(e'') = (f(\pi''(e'')), \tilde{\tilde{f}}(e'')) \; ;$$

it is easily seen that g is an equivalence of η and $f^*\xi$.

In the case of two vector bundles $\xi_i = \pi_i \colon E_i \longrightarrow Y$ we have

$$f^*(\xi_1 \oplus \xi_2) \simeq f^*(\xi_1) \oplus f^*(\xi_2) \; .$$

The most reasonable way to prove this is to consider the explicit construction of the total space E of $f^*(\xi_1) \oplus f^*(\xi_2)$, and then define a bundle map $\tilde{f}$

covering f from E to the total space of $\xi_1 \oplus \xi_2$. On the other hand, consider two bundles $\xi_i = \pi_i\colon E_i \longrightarrow X_i$, and let $p_i\colon X_1 \times X_2 \longrightarrow X_i$ be the projections on the factors. Then we can form the bundle

$$\xi_1 \times \xi_2 = p_1{}^*(\xi_1) \oplus p_2{}^*(\xi_2)$$

over $X_1 \times X_2$; the fibre over (x_1,x_2) is essentially $\pi_1{}^{-1}(x_1) \oplus \pi_2{}^{-1}(x_2)$. When $X_1 = X_2 = X$, we have

$$\xi_1 \oplus \xi_2 \simeq \Delta^*(\xi_1 \times \xi_2) ,$$

where $\Delta\colon X \longrightarrow X \times X$ is the diagonal map, $\Delta(x) = (x,x)$.

As a somewhat more esoteric example of induced bundles, consider a vector bundle $\xi = \pi\colon E \longrightarrow X$ with a Riemannian metric $\langle\ ,\ \rangle$. Let S be the "sphere bundle"

$$S = \{e \in E\colon \langle e,e\rangle = 1\} ,$$

and denote the restriction $\pi|S$ by $\pi_0\colon S \longrightarrow X$. Then we can form the bundle $\pi_0{}^*\xi$ over S. We claim that $\pi_0{}^*\xi$ always has a nowhere-zero section. To see this, we recall the construction of $\pi_0{}^*\xi$. For each $e \in S$, the fibre of $\pi_0{}^*\xi$ over e is just

$$\{e\} \times \pi^{-1}(\pi_0(e)) = \{e\} \times \pi^{-1}(\pi(e)) .$$

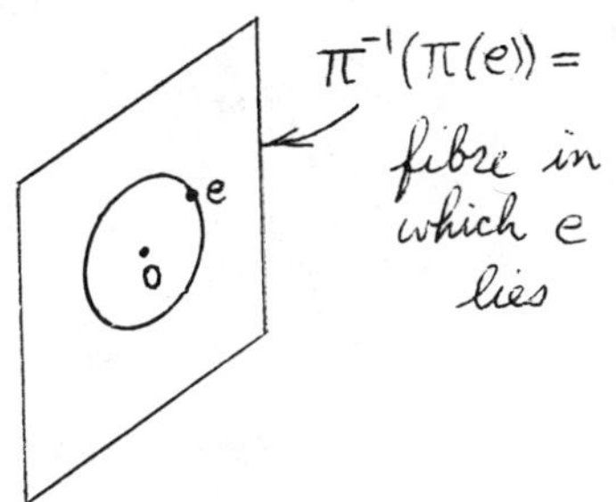

We define a section s of $\pi_0{}^*\xi$ by

$$s(e) = (e,e) \in \{e\} \times \pi^{-1}(\pi(e)) .$$

Since

$$s(e) \neq (e,0) = 0 \text{ element of fibre of } \pi_0{}^*\xi \text{ over } e ,$$

this section s is indeed nowhere zero. Similarly, if we regard X as a subspace of E (by considering X as the image of the 0 section), then the bundle $(\pi|E-X)^*\xi$ over $E-X$ has a nowhere zero section. On the other hand, the bundle $\pi^*\xi$ itself need not have such a section. Indeed, if it does, then the restriction $(\pi^*\xi)|X$ of $\pi^*\xi$ to X must have a nowhere zero section. But it is clear that $(\pi^*\xi)|X \simeq \xi$.

The most important result about induced bundles gives a condition under which $f^*\xi \simeq g^*\xi$. As a start towards this result, we note that any bundle over $[0,1]$ is trivial. The proof may be considered as an exercise for the reader; the next Lemma and Theorem establish a more general result.

1. LEMMA. Let ξ be a principal bundle over $X \times [a,b]$. Then every point $x \in X$ has a neighborhood U such that ξ is trivial over $U \times [a,b]$.

Proof. Each point $(x,t) \in \{x\} \times [a,b]$ has a neighborhood $V \times W$ such that ξ is trivial over $V \times W$. By compactness, finitely many such neighborhoods $V_1 \times W_1, \ldots, V_r \times W_r$ cover $\{x\} \times [a,b]$. We claim that the theorem holds with $U = V_1 \cap \cdots \cap V_r$. The proof will be by induction on r. For $r = 1$ it is trivial. Assume it holds for $\leq r-1$ sets. We can clearly choose a point $t_0 \in (a,b)$ such that $[a,t_0]$ and $[t_0,b]$ are each covered by $\leq r-1$ of the sets $V_i \times W_i$. Then ξ is trivial over sets $U_1 \times [a,t_0]$ and $U_2 \times [t_0,b]$. This means that there is a section s of ξ over $U_1 \times [a,t_0]$ and a section σ of ξ over $U_2 \times [t_0,b]$. On $(U_1 \cap U_2) \times \{t_0\}$ we have

$$s(x,t_0) = \sigma(x,t_0) \cdot a(x)$$

for a continuous function $x \mapsto a(x) \in G$. Then we can define a section $\bar{s}$ on $(U_1 \cap U_2) \times [a,b]$ by

$$\bar{s}(x,t) = \begin{cases} s(x,t) & a \le t \le t_0 \\ \sigma(x,t) \cdot a(x) & t_0 \le t \le b . \end{cases} \quad \blacksquare$$

Now let $j: X \times \{1\} \to X \times [0,1]$ be the inclusion, and let $p: X \times [0,1] \to X \times \{1\}$ be $p(x,t) = (x,1)$.

2. THEOREM. If $\xi = \pi: P \to X \times [0,1]$ is a principal bundle, and X is paracompact, then

$$\xi \simeq p^* j^* \xi \simeq p^*(\xi | X \times \{1\}) .$$

Proof. We want to show that there is a bundle map $\tilde{p}: P \to \pi^{-1}(X \times \{1\})$ covering p.

$$\begin{array}{ccc} P & \xrightarrow{\tilde{p}} & \pi^{-1}(X \times \{1\}) \\ \downarrow & & \downarrow \\ X \times [0,1] & \xrightarrow{p} & X \times \{1\} \end{array}$$

By Lemma 1, there is an open cover $\{U_\alpha\}$ of X such that ξ is trivial on $U_\alpha \times [0,1]$. We can assume that $\{U_\alpha\}$ is locally finite, by taking a refinement if necessary. By Theorem I.2-15 we can choose a partition of unity $\{\phi_\alpha\}$ with support $\phi_\alpha \subset U_\alpha$ (the Theorem is stated for a manifold, but holds for any normal space X if we only want the functions ϕ_α to be continuous). Let s_α be a section of ξ over $U_\alpha \times [0,1]$. Consider the map, from $\pi^{-1}(U_\alpha \times [0,1])$ to itself, defined by

$$s_\alpha(x,t) \longmapsto s_\alpha(x, \min(t + \phi_\alpha(x), 1)) .$$

This map is the identity on (boundary U_α) $\times [0,1]$. So we can extend it continuously to $X \times [0,1]$ by making it the identity on $(X - U_\alpha) \times [0,1]$. Thus we obtain a map

$$\tilde{p}_\alpha \colon P \longrightarrow P$$

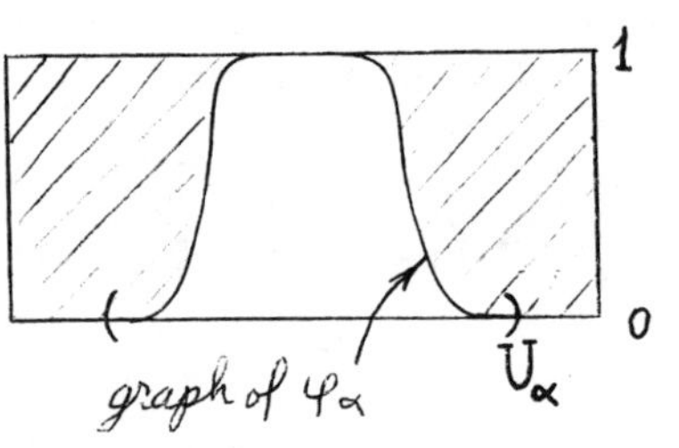

which is a bundle map from ξ to the part of ξ over the shaded set in the figure.

Suppose first that there are at most countably many such maps, $\tilde{p}_1, \tilde{p}_2, \ldots$. Define

$$\tilde{p} = \tilde{p}_1 \circ \tilde{p}_2 \circ \cdots .$$

This possibly infinite composition makes sense, since all but finitely many $\tilde{p}_i$ are the identity in a neighborhood of any point. Clearly $\tilde{p}$ is the desired bundle map.

Even if there are uncountably many maps $\{\tilde{p}_\alpha : \alpha \in A\}$, the procedure is the same. Choose any ordering on A (not necessarily a well-ordering) and define $\tilde{p}$ to be the composition of the $\tilde{p}_\alpha$ in the order given by A; in a neighborhood of any point, only finitely many $\tilde{p}_\alpha$ are not the identity map, so this makes sense. ■

In practice, it is more convenient to work with a slight restatement, and application, of Theorem 2. Let $i_t: X \to X \times [0,1]$ be $i_t(x) = (x,t)$.

3. COROLLARY. If ξ is a principal bundle over $X \times [0,1]$, and X is paracompact, then

$$i_0{}^*\xi \simeq i_1{}^*\xi .$$

Proof. Let $q: X \times [0,1] \to X$ be the projection $q(x,t) = x$. Then $j \circ p = i_1 \circ q$. So, by Theorem 2,

$$\xi \simeq p^*j^*\xi \simeq (j \circ p)^*\xi = (i_1 \circ q)^*\xi \simeq q^*i_1{}^*\xi . \tag{1}$$

On the other hand, we also have $q \circ i_0$ = identity. Consequently, equation (1) gives

$$i_0{}^*\xi \simeq i_0{}^*q^*i_1{}^*\xi \simeq [i_1 \circ (q \circ i_0)]^*\xi \simeq i_1{}^*\xi . \quad ■$$

From this we immediately obtain the result toward which we have been aiming.

4. THEOREM (THE COVERING HOMOTOPY THEOREM). If η is a principal bundle over Y and $f, g: X \to Y$ are homotopic, with X paracompact, then $f^*\eta \simeq g^*\eta$. The same result holds if η is a vector bundle.

Proof. Let $H: X \times [0,1] \to Y$ be a map with

$$H \circ i_0 = f \quad \text{and} \quad H \circ i_1 = g .$$

Applying Corollary 3 to $\xi = H^*(\eta)$ over $X \times [0,1]$, we have

$$\begin{aligned} f^*\eta = (H \circ i_0)^*\eta &\simeq i_0{}^*(H^*\eta) \\ &\simeq i_1{}^*(H^*\eta) \simeq (H \circ i_1)^*\eta = g^*\eta . \end{aligned}$$

When η is a vector bundle we have, by the remark on p. 392,

$$F(f^*\eta) \simeq f^*(F(\eta)) \simeq g^*(F(\eta)) \simeq F(g^*\eta) .$$

By the remark on p. 390, this implies that $f^*\eta \simeq g^*\eta$. ■

As a particular case of Theorem 4, note that if X is paracompact and contractible, so that the identity map $1: X \to X$ is homotopic to a constant map c, then $\xi \simeq 1^*\xi \simeq c^*\xi$, which is trivial. So any principal bundle or vector bundle over X is trivial (compare with remark 3 on p. I.644).

In applying these results, we will usually be interested only in vector bundles. But in one instance principal bundles will be used. Let $\xi = \pi: E \to X$ be a vector bundle, and let $\langle\ ,\ \rangle$ be a Riemannian metric on ξ. As in Chapter 7, we can consider the principal bundle $O(\xi) = \varpi: O(E) \to X$ with group $O(n)$, whose fibre $\varpi^{-1}(x)$ is the set of all frames of $\pi^{-1}(x)$ which are orthonormal with respect to $\langle\ ,\ \rangle$. If $\langle\ ,\ \rangle'$ is another

Riemannian metric on ξ, then we have another principal bundle $O'(\xi) = \varpi': O'(E) \to X$, consisting of frames which are orthonormal with respect to $\langle\ ,\ \rangle'$.

<u>5. COROLLARY</u>. If $\xi = \pi: E \to X$ is a vector bundle with two Riemannian metrics $\langle\ ,\ \rangle$ and $\langle\ ,\ \rangle'$, then $O(\xi) \simeq O'(\xi)$.

<u>Proof</u>. Let $q: X \times [0,1] \to X$ be the projection $q(x,t) = x$, and consider the bundle $q^*\xi$ over $X \times [0,1]$. The fibre of $q^*\xi$ over (x,t) is

$$\{(x,t)\} \times \pi^{-1}(x) .$$

The inner products $\langle\ ,\ \rangle_x$ and $\langle\ ,\ \rangle'_x$ on $\pi^{-1}(x)$ give us an inner product

$$t\langle\ ,\ \rangle_x + (1-t)\langle\ ,\ \rangle'_x$$

on $\pi^{-1}(x)$. Using this inner product on the fibre $\{(x,t)\} \times \pi^{-1}(x)$, we obtain a Riemannian metric $\langle\!\langle\ ,\ \rangle\!\rangle$ on $q^*\xi$, and we can consider the corresponding principal bundle $\mathbf{O}(q^*\xi)$. If $i_t: X \to X \times [0,1]$ is $i_t(x) = (x,t)$, then clearly

$$i_0{}^* \mathbf{O}(q^*\xi) \simeq O(\xi) \qquad \text{and} \qquad i_1{}^* \mathbf{O}(q^*\xi) \simeq O'(\xi) .$$

So the result follows from Corollary 3. ∎

2. Grassmannians and Universal Bundles

We have defined projective n-space $\mathbb{P}^n$ to be the set of all pairs $\{p,-p\}$ for $p \in S^n \subset \mathbb{R}^{n+1}$. We could also have defined $\mathbb{P}^n$ to be the set of all lines through 0 in $\mathbb{R}^{n+1}$, since each such line intersects S^n in a set $\{p,-p\}$. More generally, we define the **Grassmanian manifold** $G_n(\mathbb{R}^N)$ to be the set of all n-dimensional subspaces of $\mathbb{R}^N$ (we will always assume that $N > n$). In order to topologize $G_n(\mathbb{R}^N)$, we consider first the **Stiefel manifold** $V_n(\mathbb{R}^N)$ consisting of all n-tuples

$$(v_1,\dots,v_n) \in \mathbb{R}^N \times \cdots \times \mathbb{R}^N$$

for which $v_1,\dots,v_n$ are linearly independent. Clearly $V_n(\mathbb{R}^N)$ is an open subset of $\mathbb{R}^N \times \cdots \times \mathbb{R}^N$. We can define a map

$$\rho\colon V_n(\mathbb{R}^N) \longrightarrow G_n(\mathbb{R}^N)$$

by letting

$$\rho((v_1,\dots,v_n)) = \text{subspace of } \mathbb{R}^N \text{ spanned by } v_1,\dots,v_n .$$

We give $G_n(\mathbb{R}^N)$ the quotient topology for this map -- thus $\mathcal{U} \subset G_n(\mathbb{R}^N)$ is open if and only if $\rho^{-1}(\mathcal{U}) \subset V_n(\mathbb{R}^N)$ is open.

We can also consider the subspace $V_n^0(\mathbb{R}^N) \subset V_n(\mathbb{R}^N)$ consisting of n-tuples $(v_1,\dots,v_n) \in V_n(\mathbb{R}^N)$ which are orthonormal. If $\rho_0 = \rho|V_n^0(\mathbb{R}^N)$, then the diagram

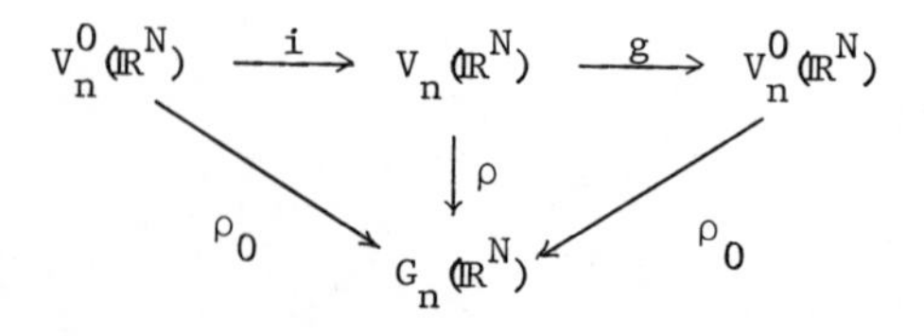

commutes, where i is the inclusion map, and $g((v_1,\ldots,v_n))$ is the n-tuple in $V_n^0(\mathbb{R}^N)$ which results by applying the Gram-Schmidt orthogonalization process to $v_1,\ldots,v_n$. From this diagram it is easy to see that the topology on $G_n(\mathbb{R}^N)$ can also be described as the quotient topology for ρ_0. Since $V_n^0(\mathbb{R}^N)$ is compact, this shows that $G_n(\mathbb{R}^N)$ is also compact.

There is yet a third description of the topology of $G_n(\mathbb{R}^N)$ which will be important later on. Consider the orthogonal group $O(N)$. If $W_0 \subset \mathbb{R}^N$ is the n-dimensional subspace spanned by $e_1,\ldots,e_n$, then we can define a map

$$O(N) \xrightarrow{\sigma} G_n(\mathbb{R}^N)$$

by

$$\sigma(A) = A(W_0) \in G_n(\mathbb{R}^N) .$$

The following diagram then commutes,

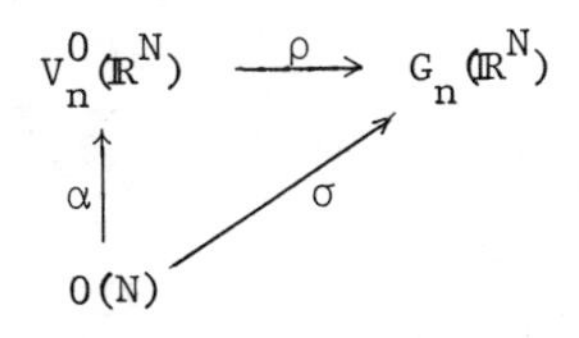

where α is the continuous map defined by

$$\alpha(A) = (A(e_1),\ldots,A(e_n)) \in V_n^0(\mathbb{R}^N) .$$

We thus see that if $\mathcal{U} \subset G_n(\mathbb{R}^N)$ is any set, then

$$(1) \qquad \rho^{-1}(\mathcal{U}) \text{ is open } \Rightarrow \alpha^{-1}(\rho^{-1}(\mathcal{U})) \text{ is open } \Rightarrow \sigma^{-1}(\mathcal{U}) \text{ is open .}$$

Notice moreover that we clearly have

$$\alpha(\sigma^{-1}(\mathcal{U})) \subset \rho^{-1}(\mathcal{U}) .$$

In addition, the map α is <u>onto</u> $V_n^0(\mathbb{R}^N)$, which easily implies that we actually have

$$\alpha(\sigma^{-1}(\mathcal{U})) = \rho^{-1}(\mathcal{U}) .$$

Finally, the map α is an <u>open</u> map, so if $\mathcal{U} \subset G_n(\mathbb{R}^N)$, then

$$(2) \qquad \begin{aligned} \sigma^{-1}(\mathcal{U}) \text{ is open } &\Rightarrow \alpha(\sigma^{-1}(\mathcal{U})) \text{ is open} \\ &\Rightarrow \rho^{-1}(\mathcal{U}) \text{ is open .} \end{aligned}$$

From (1) and (2) we see that the topology on $G_n(\mathbb{R}^N)$ can also be described as the quotient topology for σ.

This description of $G_n(\mathbb{R}^N)$ is useful for the following reason. The set

$$\sigma^{-1}(W_0) = \{A \in O(N) \colon A(W_0) = W_0\}$$

is easily seen to consist of all $N \times N$ matrices of the form

$$\begin{pmatrix} C & 0 \\ 0 & D \end{pmatrix} \qquad \text{for } C \in O(n) \text{ and } D \in O(N-n) .$$

For convenience, this group of matrices is usually denoted by $O(n) \times O(N-n)$. Now any other element of $G_n(\mathbb{R}^N)$ is of the form $B(W_0)$ for some $B \in O(N)$, and

$$\begin{aligned}\sigma^{-1}(B(W_0)) &= \{A \in O(N) : A(W_0) = B(W_0)\} \\ &= \{A \in O(N) : B^{-1}A(W_0) = W_0\} \\ &= \{A \in O(N) : B^{-1}A \in O(n) \times O(N-n)\} \\ &= \text{the left coset } B\cdot(O(n) \times O(N-n)) .\end{aligned}$$

Thus we can identify $G_n(\mathbb{R}^N)$ with the left coset space

$$O(N)/\ O(n) \times O(N-n) ,$$

where this quotient space is given the quotient topology for the natural projection map

$$O(N) \longrightarrow O(N)/\ O(n) \times O(N-n) .$$

In section 6 we will study in greater detail the quotient spaces G/H of a Lie group by a closed subgroup, and show that G/H is always a Hausdorff C^∞ manifold of dimension $\dim G - \dim H$. Thus $G_n(\mathbb{R}^N)$ will be a C^∞ manifold of dimension

$$\frac{N(N-1)}{2} - \left\{\frac{n(n-1)}{2} + \frac{(N-n)(N-n+1)}{2}\right\} = n(N-n) .$$

We can also describe this manifold structure on $G_n(\mathbb{R}^N)$ directly as follows. For any $W \in G_n(\mathbb{R}^N)$, consider the orthogonal complement $W^\perp \subset \mathbb{R}^N$. The decomposition $\mathbb{R}^N = W \oplus W^\perp$ determines an orthogonal projection

$$p\colon \mathbb{R}^N \longrightarrow W .$$

Let $\mathcal{U} \subset G_n(\mathbb{R}^N)$ be the set of all n-dimensional subspaces V with $V \cap W^\perp = \{0\}$, so that $p: V \to W$ is an isomorphism. Clearly $\rho^{-1}(\mathcal{U}) \subset V_n(\mathbb{R}^N)$ is open, so $\mathcal{U}$ is an open subset of $G_n(\mathbb{R}^N)$. Now let $w_1,\ldots,w_n$ be a fixed orthonormal basis for W, and let $w_{n+1},\ldots,w_N$ be a fixed orthonormal basis for $W^\perp$. For every $V \in \mathcal{U}$, there are unique $v_1,\ldots,v_n \in V$ with $p(v_i) = w_i$, and these v_i can be written uniquely as

$$(*) \qquad v_i = w_i + \sum_{j=n+1}^{N} a_{ij}(V)\cdot w_j \ .$$

The one-one map

$$V \mapsto (a_{ij}(V))$$

takes $\mathcal{U}$ onto the set of $n \times (N-n)$ matrices. This map is continuous, since the v_i depend continuously on V; moreover, the inverse map is

$$(a_{ij}) \mapsto \text{space spanned by the } w_i + \sum_{j=n+1}^{N} a_{ij}\cdot w_j \ ,$$

which is also continuous. Thus we have mapped $\mathcal{U}$ homeomorphically onto $\mathbb{R}^{n(N-n)}$. We leave it to the reader to check that any two such homeomorphisms are C^∞-related. Thus $G_n(\mathbb{R}^N)$ is a C^∞ manifold. The reader may also check that the map

$$G_n(\mathbb{R}^{m+n}) \to G_m(\mathbb{R}^{m+n})$$

defined by taking an n-plane W to its orthogonal m-plane $W^\perp$ is a diffeomorphism.

Over the Grassmanian manifold $G_n(\mathbb{R}^N)$ there is a natural n-dimensional bundle $\gamma^n(\mathbb{R}^N)$, constructed as follows. The total space $E(\gamma^n(\mathbb{R}^N))$ of the bundle will be the subset of $G_n(\mathbb{R}^N) \times \mathbb{R}^N$ consisting of all pairs

$$(W,w) \in G_n(\mathbb{R}^N) \times \mathbb{R}^N \qquad \text{such that} \quad w \in W ,$$

and the projection map $\pi\colon E(\gamma^n(\mathbb{R}^N)) \to G_n(\mathbb{R}^N)$ will be $\pi((W,w)) = W$. Thus the fibre $\pi^{-1}(W)$ over the point W of $G_n(\mathbb{R}^N)$ will just be W itself -- more precisely, it will be

$$\{(W,w)\colon w \in W\} .$$

The vector space structure on $\pi^{-1}(W)$ is defined by using the vector space structure on the subspace $W \subset \mathbb{R}^N$; thus

$$\begin{aligned}(W,w_1) + (W,w_2) &= (W,w_1+w_2)\\ a\cdot(W,w) &= (W,aw) .\end{aligned}$$

To show that $\gamma^n(\mathbb{R}^N)$ satisfies the local triviality condition, we consider a point $W \in G_n(\mathbb{R}^N)$, the orthogonal complement $W^\perp$, the corresponding projection $p\colon \mathbb{R}^N \to W$ and the open set $\mathcal{U} \subset G_n(\mathbb{R}^N)$ consisting of all V with $V \cap W^\perp = \{0\}$. Now we can define a map

$$\pi^{-1}(\mathcal{U}) \to \mathcal{U} \times W \approx \mathcal{U} \times \mathbb{R}^n$$

by taking

$$(V,v) \mapsto (V,p(v)) .$$

This map is easily seen to be a diffeomorphism, and is an isomorphism on each fibre, so $\gamma^n(\mathbb{R}^N)$ is a smooth vector bundle over $G_n(\mathbb{R}^N)$.

Notice that for $M > N$ there is a natural map $\alpha\colon G_n(\mathbb{R}^N) \to G_n(\mathbb{R}^M)$, since an n-dimensional subspace of $\mathbb{R}^N$ can be considered as an n-dimensional subspace of $\mathbb{R}^M$. There is also an obvious map $\tilde{\alpha}\colon E(\gamma^n(\mathbb{R}^N)) \to E(\gamma^n(\mathbb{R}^M))$ such that $(\tilde{\alpha},\alpha)$ is a bundle map from $\gamma^n(\mathbb{R}^N)$ to $\gamma^n(\mathbb{R}^M)$. Thus

$$\gamma^n(\mathbb{R}^N) \simeq \alpha^*(\gamma^n(\mathbb{R}^M)) .$$

Now consider a C^∞ manifold M^n immersed in $\mathbb{R}^{n+1}$. Since M need not be orientable, we may not be able to define the normal map $\nu\colon M^n \to S^n$. But we can certainly define a map $f\colon M \to \mathbb{P}^n = G_1(\mathbb{R}^{n+1})$, by taking $p \in M$ to the 1-dimensional subspace of $\mathbb{R}^{n+1}$ which is parallel to the line $M_p{}^\perp \subset \mathbb{R}^{n+1}{}_p$.

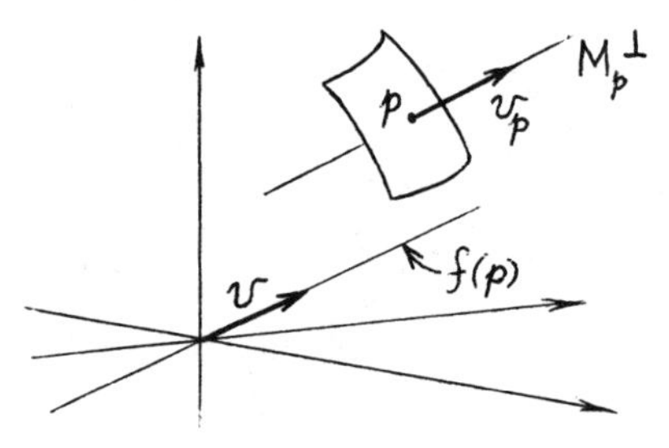

We can also define a map $\tilde{f}$ from the normal bundle Nor M of M into the total space of $\gamma^1(\mathbb{R}^{n+1})$ by sending $v_p \in M_p{}^\perp$ to $(f(p),v)$. Thus we have a bundle map $(\tilde{f},f)$ from the normal bundle Nor M to the bundle $\gamma^1(\mathbb{R}^{n+1})$; consequently, the normal bundle Nor M is equivalent to $f^*(\gamma^1(\mathbb{R}^{n+1}))$. It is even more interesting to look at the map f from M into the diffeomorphic manifold $G_n(\mathbb{R}^{n+1})$ defined by $f(p)$ = subspace of $\mathbb{R}^{n+1}$ parallel to M_p. For then we can define $\tilde{f}\colon TM \to E(\gamma^n(\mathbb{R}^{n+1}))$ by sending $v_p \in M_p$ to $(f(p),v)$. Thus we see that the tangent bundle TM is equivalent to $f^*(\gamma^n(\mathbb{R}^{n+1}))$. Moreover, this construction can be generalized. By Proposition I.2-17, any compact n-manifold M can be considered as a subspace $M^n \subset \mathbb{R}^N$ for some N. Define

$f\colon M \to G_n(\mathbb{R}^N)$ by $f(p) =$ subspace of $\mathbb{R}^N$ parallel to M_p, and define $\tilde{f}\colon TM \to E(\gamma^n(\mathbb{R}^N))$ by sending $v_p \in M_p$ to $(f(p),v)$. Then $(\tilde{f},f)$ is a bundle map from TM to the bundle $\gamma^n(\mathbb{R}^N)$. Thus the tangent bundle TM is equivalent to $f^*(\gamma^n(\mathbb{R}^N))$. Actually, the same holds for <u>all</u> bundles.

<u>6. THEOREM</u>. Let $\xi = \pi\colon E \to X$ be an n-dimensional bundle over a compact Hausdorff space X. Then for sufficiently large N there is a map $f\colon X \to G_n(\mathbb{R}^N)$ such that $\xi \simeq f^*(\gamma^n(\mathbb{R}^N))$.

If X is a smooth manifold and ξ is a smooth bundle, then f can be chosen to be a smooth map.

<u>Proof</u>. Let $U_1,\ldots,U_r$ be open sets covering X such that each $\xi|U_i$ is trivial. The Shrinking Lemma (Theorem I.2-14) holds for the cover $U_1,\ldots,U_r$ of X, for the proof merely uses the fact that X is normal. So there is an open cover $V_1,\ldots,V_r$ of X with $\bar{V}_i \subset U_i$. Similarly, there is an open cover $W_1,\ldots,W_r$ of X with $\bar{W}_i \subset V_i$. Let $\phi_i\colon X \to \mathbb{R}$ be a continuous function which is 1 on $\bar{W}_i$ and 0 outside of V_i.

By assumption on the U_i, there are equivalences

$$t_i\colon \pi^{-1}(U_i) \to U_i \times \mathbb{R}^n .$$

Composing with the projections $U_i \times \mathbb{R}^n \to \mathbb{R}^n$, we thus obtain maps

$$\tau_i\colon \pi^{-1}(U_i) \to \mathbb{R}^n$$

which are isomorphisms on each fibre. Define $\tau_i'\colon E \to \mathbb{R}^n$ by

$$\tau_i'(e) = \begin{cases} 0 & \pi(e) \notin V_i \\ \phi_i(\pi(e))\cdot\tau_i(e) & \pi(e) \in U_i \end{cases}.$$

These maps are linear on each fibre, but not one-one on all fibres. Now define

$$T\colon E \longrightarrow \mathbb{R}^n \oplus \cdots \oplus \mathbb{R}^n \approx \mathbb{R}^{rn}$$

by

$$T(e) = (\tau_1'(e),\dots,\tau_r'(e)) .$$

Then T is linear and one-one on all fibres, so each set $T(\pi^{-1}(x))$ is an n-plane in $\mathbb{R}^{rn}$. Defining $f\colon X \longrightarrow G_n(\mathbb{R}^{rn})$ and $\tilde{f}\colon E \longrightarrow E(\gamma^n(\mathbb{R}^{rn}))$ by

$$f(x) = T(\pi^{-1}(x)) = \{T(e)\colon e \in \pi^{-1}(x)\} \in G_n(\mathbb{R}^{rn})$$

$$\tilde{f}(e) = (f(\pi(e)), T(e)) \in E(\gamma^n(\mathbb{R}^{rn})) ,$$

it is easily checked that $(f,\tilde{f})$ is a bundle map from ξ to $\gamma^n(\mathbb{R}^{rn})$.

When ξ is a smooth bundle, we choose the ϕ_i to be smooth, and then f will also be smooth. ■

The map $f\colon X \longrightarrow G_n(\mathbb{R}^N)$ of Theorem 6 cannot be unique, for Theorem 4 shows that $f^*(\gamma^n(\mathbb{R}^N)) \simeq g^*(\gamma^n(\mathbb{R}^N))$ whenever f and g are homotopic. But in a certain sense this is the only extent to which the representation fails to be unique:

7. THEOREM. Let $f_0, f_1\colon X \longrightarrow G_n(\mathbb{R}^N)$ be two maps such that

$$f_0{}^*\gamma^n(\mathbb{R}^N) \simeq f_1{}^*\gamma^n(\mathbb{R}^N).$$

Consider the natural inclusion $\alpha: G_n(\mathbb{R}^N) \longrightarrow G_n(\mathbb{R}^M)$, for $M \geq 2N$. Then $\bar{f}_0 = \alpha \circ f_0$ and $\bar{f}_1 = \alpha \circ f_1$ are homotopic.

If f_0 and f_1 are smooth maps, then $\bar{f}_0$ and $\bar{f}_1$ are smoothly homotopic.

Proof. For each $x \in X$ we have two n-planes $f_0(x)$ and $f_1(x) \in G_n(\mathbb{R}^M)$. By assumption, there is a bundle map from the total space of $f_0^*(\gamma^n(\mathbb{R}^N))$ to the total space of $f_1^*(\gamma^n(\mathbb{R}^N))$. Recalling how these bundles are defined, we see that for each $x \in X$ we have an isomorphism

$$h(x): f_0(x) \longrightarrow f_1(x) ,$$

depending continuously on x. First we consider a

Special Case: For all $x \in X$ and all non-zero $v \in f_0(x)$, the vector $h(x)(v)$ is never a negative multiple of v.

In this case, for each $t \in [0,1]$ we define a map $h_t(x): f_0(x) \longrightarrow \mathbb{R}^N$ by

$$h_t(x) = (1-t)\cdot\text{identity} + t\cdot h(x) .$$

For the image $h_t(x)(f_0(x)) \subset \mathbb{R}^N$ we have

$$h_0(x)(f_0(x)) = f_0(x)$$
$$h_1(x)(f_0(x)) = f_1(x) .$$

We define a homotopy f_t between f_0 and f_1 by

$$f_t(x) = h_t(x)(f_0(x)) .$$

The assumption in our special case insures that each $h_t(x)$ is one-one on

$f_0(x)$, so that we have $f_t(x) \in G_n(\mathbb{R}^N)$. It is not hard to check that $(x,t) \mapsto f_t(x)$ is continuous on $X \times [0,1]$, and is therefore the desired homotopy.

General Case. In the general case, the above construction does not work. Moreover it may happen that the hypothesis for the special case will never occur even when we replace f_0 by some homotopic map f_0'. It is necessary to look at the compositions $\bar{f}_0, \bar{f}_1: X \longrightarrow G_n(\mathbb{R}^M)$. Note that

$$\begin{aligned} \bar{f}_i{}^*\gamma^n(\mathbb{R}^M) &= (\alpha \circ f_i)^*\gamma^n(\mathbb{R}^M) \\ &\simeq f_i{}^*\alpha^*\gamma^n(\mathbb{R}^M) \\ &\simeq f_i{}^*\gamma^n(\mathbb{R}^N) \ . \end{aligned}$$

So by hypothesis we have

$$\bar{f}_0{}^*\gamma^n(\mathbb{R}^M) \simeq \bar{f}_1{}^*\gamma^n(\mathbb{R}^M) \ .$$

Now since $M \geq 2N$, we can define a map $\mathbb{R}^M \longrightarrow \mathbb{R}^M$ by

$$(a_1,\ldots,a_N,a_{N+1},\ldots,a_{2N},\ldots) \mapsto (a_{N+1},\ldots,a_{2N},a_1,\ldots,a_N,\ldots) \ .$$

This induces a map $S: G_n(\mathbb{R}^N) \longrightarrow G_n(\mathbb{R}^n)$, which is homotopic to the identity. Thus $S \circ \bar{f}_1 \simeq \bar{f}_1$, so

$$\bar{f}_0{}^*\gamma^n(\mathbb{R}^M) \simeq \bar{f}_1{}^*\gamma^n(\mathbb{R}^M) \simeq (S \circ \bar{f}_1)^*\gamma^n(\mathbb{R}^M) \ .$$

But $\bar{f}_0$ and $S \circ \bar{f}_1$ clearly satisfy the hypotheses of the special case. So we have

$$\bar{f}_0 \simeq S \circ \bar{f}_1 \simeq \bar{f}_1 \ .$$

If f_0 and f_1 are smooth, so that $\bar{f}_0$ and $\bar{f}_1$ are smooth, then the homotopy constructed above is also smooth. ■

In algebraic topology it is customary to consider the union $G_n(\mathbb{R}^\infty)$ of the increasing sequence

$$G_n(\mathbb{R}^{n+1}) \subset G_n(\mathbb{R}^{n+2}) \subset \cdots$$

with the "weak topology": a set $\mathcal{U} \subset G_n(\mathbb{R}^\infty) = \cup\, G_n(\mathbb{R}^{n+\ell})$ is open if and only if $\mathcal{U} \cap G_n(\mathbb{R}^{n+\ell})$ is open in $G_n(\mathbb{R}^{n+\ell})$ for all ℓ. There is a natural n-dimensional bundle γ^n over $G_n(\mathbb{R}^\infty)$, defined analogously to $\gamma^n(\mathbb{R}^N)$, and this bundle has the following two properties:

(A) For every bundle ξ over a paracompact space X there is a map $f\colon X \to G_n(\mathbb{R}^\infty)$ such that $\xi \simeq f^*(\gamma^n)$,

(B) If $f_0, f_1\colon X \to G_n(\mathbb{R}^\infty)$ are maps of a paracompact space X into $G_n(\mathbb{R}^\infty)$ with $f_0{}^*\gamma^n \simeq f_1{}^*\gamma^n$, then $f_0 \simeq f_1$.

For this reason, γ^n is called the "universal n-dimensional bundle", and $G_n(\mathbb{R}^\infty)$ is called the "classifying space" for n-dimensional bundles, since equivalence classes of n-dimensional bundles over X are classified by homotopy classes of maps of X into $G_n(\mathbb{R}^\infty)$. Since $G_n(\mathbb{R}^\infty)$ is not a manifold, we do not work with these bundles. Instead we will continue to use the bundles $\gamma^n(\mathbb{R}^N)$, which we also call, somewhat sloppily, "universal bundles."

All of the preceeding discussion can be modified to deal with oriented bundles. Recall that an <u>orientation</u> μ for a vector space V is an equivalence class of ordered bases for V, where $(v_1,\ldots,v_n) \sim (w_1,\ldots,w_n)$ if and only if the matrix (a_{ij}) defined by $w_i = \Sigma\, a_{ji}v_j$ has $\det(a_{ij}) > 0$. There are only two such equivalence classes, and the one which is not μ is denoted by $-\mu$. An <u>oriented</u> vector space is a pair (V,μ), where μ is an orientation for V; the condition $(v_1,\ldots,v_n) \in \mu$ is usually expressed by saying that $v_1,\ldots,v_n$ is positively oriented (with respect to μ). Given two oriented vector spaces (V,μ) and (W,ν), we orient $V \oplus W$ by declaring $v_1,\ldots,v_n,w_1,\ldots,w_m$ to be positively oriented if $v_1,\ldots,v_n$ and $w_1,\ldots,w_m$ are positively oriented with respect to μ and ν, respectively. Thus the orientation for $W \oplus V$ is $(-1)^{mn}$ times the orientation for $V \oplus W$.

An <u>orientation</u> for a bundle $\xi = \pi\colon E \longrightarrow X$ is a collection $\mu = \{\mu_x\}$ of orientations for the fibres $\pi^{-1}(x)$, satisfying an obvious compatibility requirement, while an <u>oriented</u> bundle is a pair (ξ,μ), where μ is an orientation for ξ. An orientation μ for ξ gives us another orientation $-\mu = \{-\mu_x\}$; if X is connected, this is the only other orientation for ξ. Given two oriented bundles (ξ_1,μ_1) and (ξ_2,μ_2) over the same space X, we can define an orientation on the Whitney sum $\xi_1 \oplus \xi_2$ by using the orientation on the direct sums of fibres described in the previous paragraph; thus we can define $(\xi_1 \oplus \xi_2,\ \mu_1 \oplus \mu_2)$ to be $\xi_1 \oplus \xi_2$ with this orientation. Given an oriented bundle (ξ,μ) over Y, and a continuous map $f\colon X \longrightarrow Y$, there is an obvious way to define an orientation $f^*\mu$ for $f^*\xi$; thus we can define $f^*(\xi,\mu)$ to be the oriented bundle $(f^*\xi,f^*\mu)$.

For two bundles ξ_1, ξ_2 with orientations μ_1, μ_2, respectively, we can speak of orientation preserving bundle maps and equivalences, or we can simply speak of bundle maps and equivalences between the oriented bundles (ξ_1,μ_1) and (ξ_2,μ_2). Notice that the oriented bundles (ξ,μ) and $(\xi,-\mu)$ need not be equivalent. For example, if μ is an orientation of the tangent bundle of S^2, then (TS^2,μ) and $(TS^2,-\mu)$ are not equivalent. In fact, an orientation preserving equivalence from (TS^2,μ) to $(TS^2,-\mu)$ would give us a continuous family of isomorphisms

$$A_p\colon {S^2}_p \longrightarrow {S^2}_p , \qquad \text{with all } \det A_p < 0 .$$

Now a linear transformation $A\colon V \longrightarrow V$ from a 2-dimensional vector space V to itself has two complex eigenvalues λ_1, λ_2 and if λ_1 is not real, then $\lambda_2 = \bar{\lambda}_1$. But the condition $\det A = \lambda_1\lambda_2 < 0$ clearly implies that we do not have $\lambda_2 = \bar{\lambda}_1$, so A has two real eigenvalues of opposite signs. Thus we could use the A_p to continuously pick out a 1-dimensional subspace of ${S^2}_p$ for all $p \in S^2$, by choosing the eigenvectors with the positive eigenvalue for A_p. But such a continuous choice can not be made, by Problem I.9-7.

We define the oriented Grassmannian manifold $\tilde{G}_n(\mathbb{R}^N)$ to be the set of all oriented n-dimensional subspaces of $\mathbb{R}^N$. We have already defined the map $\rho\colon V_n(\mathbb{R}^N) \longrightarrow G_n(\mathbb{R}^N)$, where $V_n(\mathbb{R}^N)$ is the Stiefel manifold. We can define a map

$$\tilde{\rho}\colon V_n(\mathbb{R}^N) \longrightarrow \tilde{G}_n(\mathbb{R}^N)$$

by setting

$$\tilde{\rho}((v_1,\ldots,v_n)) = (\rho(v_1,\ldots,v_n),\ \mu) ,$$

where μ is the orientation of $\rho(v_1,\ldots,v_n)$ determined by the ordered basis $v_1,\ldots,v_n$. We give $\tilde{G}_n(\mathbb{R}^N)$ the quotient topology for $\tilde{\rho}$. If $\tilde{\rho}_0 = \tilde{\rho}|V_n^0(\mathbb{R}^N)$, then the diagram

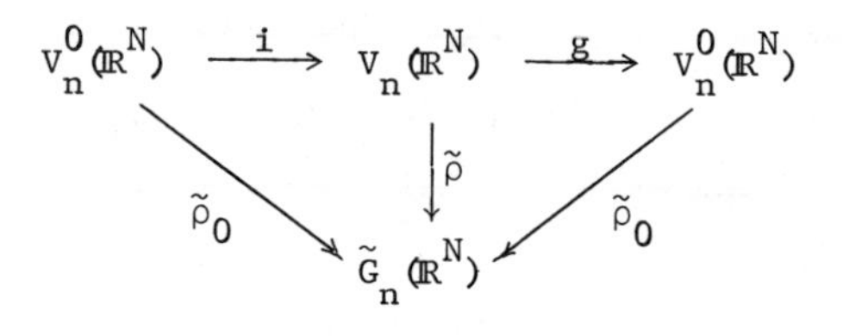

commutes; to prove this, one just has to check that the Gram-Schmidt process g preserves orientation. So the topology on $\tilde{G}_n(\mathbb{R}^N)$ could also be described as the quotient topology for $\tilde{\rho}_0$.

Similarly, we can define a map on the special orthogonal group,

$$\tilde{\sigma}: SO(N) \longrightarrow \tilde{G}_n(\mathbb{R}^N) ,$$

by

$$\tilde{\sigma}(A) = (A(W_0), \mu) ,$$

where the orientation μ on $A(W_0)$ is that determined by the ordered basis $A(e_1),\ldots,A(e_n)$. The diagram

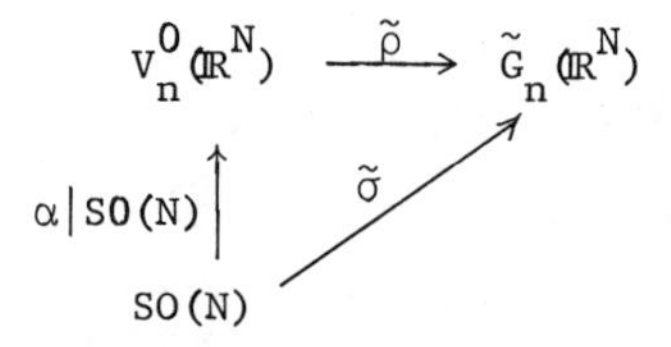

commutes, and $\alpha|SO(N)$ is onto for $N > n$. So, as before, we see that the topology on $\tilde{G}_n(\mathbb{R}^N)$ can be described as the quotient topology for $\tilde{\sigma}$. It is then easy to see that $\tilde{G}_n(\mathbb{R}^N)$ can be identified with the left coset space

$$SO(N)/\ SO(n) \times SO(N-n) .$$

There is a natural map $\tau: \tilde{G}_n(\mathbb{R}^N) \longrightarrow G_n(\mathbb{R}^N)$ defined by

$$\tau((W,\mu)) = W .$$

If $W \in G_n(\mathbb{R}^N)$ and $\mathcal{U} \subset G_n(\mathbb{R}^N)$ is the open set on p. 404, with $w_1,\ldots,w_n$ a fixed orthonormal basis for W, then every $V \in \mathcal{U}$ can be given the orientation $\mu(V)$ determined by the ordered basis $v_1,\ldots,v_n$, where v_i are the unique vectors in V with $p(v_i) = w_i$. The sets

$$\mathcal{U}^+ = \{(V,\mu(V)): V \in \mathcal{U}\}$$
$$\mathcal{U}^- = \{(V,-\mu(V)): V \in \mathcal{U}\}$$

are easily seen to be disjoint open subsets of $\tilde{G}_n(\mathbb{R}^N)$. Thus we see that $\tilde{G}_n(\mathbb{R}^N)$ is a smooth manifold, and that $\tau\colon \tilde{G}_n(\mathbb{R}^N) \to G_n(\mathbb{R}^N)$ is a two-fold covering. As a matter of fact, $\tilde{G}_n(\mathbb{R}^N)$ is clearly the oriented 2-fold covering of the non-orientable manifold $G_n(\mathbb{R}^N)$, as described in Problem 8-2 .

Over $\tilde{G}_n(\mathbb{R}^N)$ we define an oriented n-dimensional bundle $(\tilde{\gamma}^n(\mathbb{R}^N), \boldsymbol{\mu})$ as follows. The total space $E(\tilde{\gamma}^n(\mathbb{R}^N))$ consists of all pairs

$$((W,\mu),w) \in \tilde{G}_n(\mathbb{R}^N) \times \mathbb{R}^N \qquad \text{such that } w \in W ,$$

and we define $\pi((W,\mu),w) = W$. The vector space structure on $\pi^{-1}((W,\mu))$ is defined as before, and we can also define the natural orientation $\boldsymbol{\mu}$ on

$$\pi^{-1}((W,\mu)) = \{((W,\mu),w): w \in W\}$$

by using the orientation μ on W. (We ought to use a symbol like $\boldsymbol{\mu}_{n,N}$, but for simplicity we won't.) Note that $\tau^*\gamma^n(\mathbb{R}^N)$ is equivalent to the bundle $\tilde{\gamma}^n(\mathbb{R}^N)$, where we forget about the orientation $\boldsymbol{\mu}$. For $M > N$ there is a natural map $\alpha\colon \tilde{G}_n(\mathbb{R}^N) \to \tilde{G}_n(\mathbb{R}^M)$, with

$$(\tilde{\gamma}^n(\mathbb{R}^N),\mu) \simeq \alpha^*(\tilde{\gamma}^n(\mathbb{R}^M),\mu) .$$

8. THEOREM. (1) Let (ξ,μ) be an oriented n-dimensional bundle over a compact Hausdorff space X. Then for sufficiently large N there is a map $f\colon X \to \tilde{G}_n(\mathbb{R}^N)$ such that $(\xi,\mu) \simeq f^*(\tilde{\gamma}^n(\mathbb{R}^N),\mu)$. If X is a smooth manifold and ξ is a smooth bundle, then f can be chosen to be a smooth map.
(2) Let $f_0, f_1\colon X \to \tilde{G}_n(\mathbb{R}^N)$ be two maps such that $f_0{}^*(\tilde{\gamma}^n(\mathbb{R}^N),\mu) \simeq f_1{}^*(\tilde{\gamma}^n(\mathbb{R}^N),\mu)$. Then the compositions $\bar{f}_0 = \alpha\circ f_0$ and $\bar{f}_1 = \alpha\circ f_1$ are homotopic, where $\alpha\colon \tilde{G}_n(\mathbb{R}^N) \to \tilde{G}_n(\mathbb{R}^M)$ is the natural inclusion, and $M \geq 2N$. If f_0 and f_1 are smooth maps, then $\bar{f}_0$ and $\bar{f}_1$ are smoothly homotopic.

Proof. Left to the reader. ∎

3. The Pfaffian

We have already given an intrinsic expression, as well as an expression in terms of a coordinate system, for the function K_n on a compact oriented Riemannian manifold M of even dimension $n = 2m$. But the most important expression for K_n involves the curvature forms Ω^i_j for a positively oriented orthonormal moving frame $X_1,\dots,X_n$ on M. In terms of these forms, we can easily write down the n-form $K_n\,dV$ which we want to integrate over M. We will be using the symbol $\varepsilon^{j_1\cdots j_n}$ (defined on p. IV.100); notice that a sum over permutations, like

$$\sum_{\pi\in S_n} A(X_{\pi(1)},\dots,X_{\pi(n)}) ,$$

for example, can just as well be written as

$$\sum_{j_1,\ldots,j_n} \varepsilon^{j_1\cdots j_n} A(X_{j_1},\ldots,X_{j_n}) .$$

Now consider the m-fold wedge product

$$\Omega^{i_1}_{i_2} \wedge \cdots \wedge \Omega^{i_{n-1}}_{i_n} .$$

From the definition of $\wedge$ we have (remembering that the Ω^i_j are 2-forms)

$$\Omega^{i_1}_{i_2} \wedge \cdots \wedge \Omega^{i_{n-1}}_{i_n}(X_1,\ldots,X_n) =$$

$$= \frac{(2+\cdots+2)!}{2!\cdots 2!} \cdot \frac{1}{n!} \sum_{j_1,\ldots,j_n} \varepsilon^{j_1\cdots j_n} \Omega^{i_1}_{i_2}(X_{j_1},X_{j_2})\cdots\Omega^{i_{n-1}}_{i_n}(X_{j_{n-1}},X_{j_n})$$

$$= \frac{1}{2^{n/2}} \cdot \sum_{j_1,\ldots,j_n} \varepsilon^{j_1\cdots j_n} \langle R(X_{j_1},X_{j_2})X_{i_2}, X_{i_1}\rangle \cdots \langle R(X_{j_{n-1}},X_{j_n})X_{i_n}, X_{i_{n-1}}\rangle$$

$$= \frac{1}{2^{n/2}} \cdot \sum_{j_1,\ldots,j_n} \varepsilon^{j_1\cdots j_n} R_{i_1 i_2 j_1 j_2} \cdots R_{i_{n-1} i_n j_{n-1} j_n}$$ (see p. II.191).

So the formula on p. 386 gives

$$K_n = \frac{1}{2^{n/2} n!} \sum_{i_1,\ldots,i_n} \varepsilon^{i_1\cdots i_n} \cdot 2^{n/2} \cdot \Omega^{i_1}_{i_2} \wedge \cdots \wedge \Omega^{i_{n-1}}_{i_n}(X_1,\ldots,X_n) ,$$

and thus

$$\boxed{K_n \, dV = \frac{1}{n!} \sum_{i_1,\ldots,i_n} \varepsilon^{i_1\cdots i_n} \Omega^{i_1}_{i_2} \wedge \cdots \wedge \Omega^{i_{n-1}}_{i_n} .}$$

This computation shows, in particular, that the form on the right does not depend on the choice of the positively oriented orthonormal moving frame $X_1,\ldots,X_n$. It is also possible to prove this fact directly, by the following algebraic considerations.

For an $n\times n$ matrix $A = (a_{ij})$ with $n = 2m$ even, we define the Pfaffian $\mathrm{Pf}(A)$ of A by

$$\mathrm{Pf}(A) = \frac{1}{2^m m!}\sum_{i_1,\ldots,i_n} \varepsilon^{i_1\cdots i_n} a_{i_1 i_2}\cdots a_{i_{n-1} i_n} .$$

It will soon become clear why the factor $1/2^m m!$ should appear. At the moment we can account for the $1/m!$ by observing that our expression has a lot of redundancy in it. Note first that $\varepsilon^{i_1\cdots i_n}$ does not change when we interchange $i_{2\ell-1}$ and i_{2k-1} and also $i_{2\ell}$ and i_{2k}; more generally, it does not change when we perform any permutation of the pairs $(i_{2\ell-1},i_{2\ell})$. So for any set $P = \{(h_1,k_1),\ldots,(h_m,k_m)\}$ of pairs of integers between 1 and n, it makes sense to define

$$\varepsilon(P) = \varepsilon^{h_1 k_1\cdots h_m k_m} ;$$

it is not necessary to specify any ordering on the pairs (h_i,k_i) in P. Notice also that a permutation of the pairs $(i_{2\ell-1},i_{2\ell})$ does not change the factor

$$a_{i_1 i_2}\cdots a_{i_{n-1} i_n} .$$

So for each P as above we can define

$$a_P = a_{h_1 k_1}\cdots a_{h_m k_m} .$$

If $\mathcal{P}$ is the collection of all such P, we then clearly have

$$\mathrm{Pf}(A) = \frac{1}{2^m} \sum_{P \in \mathcal{P}} \varepsilon(P) a_P .$$

9. PROPOSITION. Let $n = 2m$ be even. Then for all $n \times n$ matrices A and B we have

$$\mathrm{Pf}(B^t AB) = (\det B) \cdot \mathrm{Pf}(A) .$$

In particular, if $B \in SO(n)$, then

$$\mathrm{Pf}(B^{-1}AB) = \mathrm{Pf}(A) .$$

Proof.

$$2^m m!\ \mathrm{Pf}(B^t AB)$$

$$= \sum_{i_1,\ldots,i_n} \varepsilon^{i_1\cdots i_n} \sum_{j_1,\ldots,j_n} (b_{j_1 i_1} a_{j_1 j_2} b_{j_2 i_2}) \cdots (b_{j_{n-1} i_{n-1}} a_{j_{n-1} j_n} b_{j_n i_n})$$

$$= \sum_{j_1,\ldots,j_n} \left[\sum_{i_1,\ldots,i_n} \varepsilon^{i_1\cdots i_n} b_{j_1 i_1} \cdots b_{j_n i_n} \right] \cdot a_{j_1 j_2} \cdots a_{j_{n-1} j_n}$$

$$= \sum_{j_1,\ldots,j_n} (\varepsilon^{j_1\cdots j_n} \det B) \cdot a_{j_1 j_2} \cdots a_{j_{n-1} j_n}$$

$$= 2^m m! (\det B) \mathrm{Pf}(A) . \quad \blacksquare$$

Proposition 9 was stated for matrices of real numbers, but $\mathrm{Pf}(A)$ can be defined so long as the entries of A are in some commutative algebra $\mathcal{A}$ over $\mathbb{R}$. It is easy to see that Proposition 9 still holds when A and B have entries in $\mathcal{A}$; in fact, the proof works without change. We could also deduce this extended version from the original Proposition by the "principle of extension of algebraic identities": First consider the ring $\mathbb{R}[A_{ij},B_{ij}]$ obtained by adjoining commuting indeterminants A_{ij} and B_{ij} to $\mathbb{R}$. Then the polynomials

$$\mathrm{Pf}(B^t AB) \quad \text{and} \quad (\det B)\cdot \mathrm{Pf}(A)$$

are elements of $\mathbb{R}[A_{ij},B_{ij}]$. Proposition 9 tells us that these polynomials have the same values on all (a_{ij}), (b_{ij}) for a_{ij}, $b_{ij} \in \mathbb{R}$. Therefore they are equal <u>as polynomials</u> in the indeterminants A_{ij}, B_{ij}. So these polynomials are equal when we substitute elements a_{ij}, b_{ij} of the algebra $\mathcal{A}$ for the indeterminants A_{ij}, B_{ij}. Q.E.D.

Now let us consider once again a positively oriented orthonormal moving frame $X = X_1,\ldots,X_n$ on M, with curvature forms Ω^i_j. For each $p \in M$, the direct sum

$$\mathcal{A} = \mathbb{R} \oplus \Omega^2(M_p) \oplus \Omega^4(M_p) \oplus \cdots$$

is a commutative algebra over $\mathbb{R}$, under $\wedge$. Consequently, it makes sense to write $\mathrm{Pf}(\Omega(p))$, where $\Omega(p)$ is the $n\times n$ matrix $(\Omega^i_j(p))$ of connection 2-forms at p. In fact, we clearly have

$$\mathrm{Pf}(\Omega(p)) = \frac{1}{2^m m!} \sum_{i_1,\ldots,i_n} \varepsilon^{i_1\cdots i_n}\, \Omega^{i_1}_{i_2} \wedge \cdots \wedge \Omega^{i_{n-1}}_{i_n}(p)\ .$$

Now if $X' = X\cdot a$ is another positively oriented orthonormal moving frame, then $a(p) \in O(n)$, and by Proposition II.7-15 the corresponding curvature forms (Ω'^i_j) satisfy

$$\Omega' = a^{-1}\Omega a .$$

Then Proposition 9, in its extended form, shows that

$$\mathrm{Pf}(\Omega'(p)) = \mathrm{Pf}(a^{-1}(p)\Omega(p)a(p)) = \mathrm{Pf}(\Omega(p)) ;$$

thus the form

$$\sum_{i_1,\dots,i_n} \varepsilon^{i_1\cdots i_n}\, \Omega^{i_1}_{i_2} \wedge \cdots \wedge \Omega^{i_{n-1}}_{i_n}$$

is indeed well-defined.

Later on we will need to know the important algebraic properties of $\mathrm{Pf}(A)$ which hold when A is skew-symmetric. In this case, even the expression

$$\mathrm{Pf}(A) = \frac{1}{2^m} \sum_{P\in\mathscr{P}} \varepsilon(P)a_P$$

is redundant, for the term

$$\varepsilon^{i_1\cdots i_n}\, a_{i_1 i_2} \cdots a_{i_{n-1} i_n}$$

is unchanged when we interchange $i_{2\ell-1}$ and $i_{2\ell}$. Let $\mathscr{P}' \subset \mathscr{P}$ be the collection of all $P = \{(h_1,k_1),\dots,(h_m,k_m)\} \in \mathscr{P}$ with $h_i < k_i$ for all i. Then for skew-symmetric A we clearly have

$$\mathrm{Pf}(A) = \sum_{P\in\mathscr{P}'} \varepsilon(P)a_P ,$$

which is a polynomial with integer coefficients. (It follows that Pf(A) can be defined for a skew-symmetric matrix A with entries in any commutative ring $\mathcal{a}$ with unit.)

There is an important canonical form for skew-symmetric matrices, which is merely a reformulation of the following result which has already appeared in Problem I.7-8.

10. PROPOSITION. Let V be an n-dimensional vector space, and let $\alpha \in \Omega^2(V)$. Then there is a basis $\phi_1,\ldots,\phi_n$ of V^* such that

$$\alpha = (\phi_1 \wedge \phi_2) + \cdots + (\phi_{2r-1} \wedge \phi_{2r})$$

for some r. (For $\alpha = 0$ we must allow the vacuous sum, with $r = 0$.) If $\{\phi_i\}$ is the dual basis to $\{v_i\}$, this means that

$$(1)\ \begin{cases} \alpha(v_{2i-1},v_{2i}) = -\alpha(v_{2i},v_{2i-1}) = 1 & \text{for } i \le r \\ \alpha(v_i,v_j) = 0 & \text{for all other pairs } i, j . \end{cases}$$

Proof. We use induction on n, the result being trivial for $n = 1$. Assume the result is true for dimensions $< n$, and consider a non-zero $\alpha \in \Omega^2(V)$, where V has dimension n. There exist v_1, $v_2 \in V$ with $\alpha(v_1,v_2) = 1$; let $[v_1,v_2]$ be the subspace generated by v_1 and v_2. Now consider the subspace $W \subset V$ of all $v \in V$ such that $\alpha(v_1,v) = \alpha(v_2,v) = 0$. This subspace W is $\ker f_1 \cap \ker f_2$ where $f_i \colon V \to \mathbb{R}$ is defined by $f_i(v) = \alpha(v_i,v)$. So $\dim W \ge n-2$. Moreover, we clearly have $W \cap [v_1,v_2] = \{0\}$, so $\dim W = n-2$ and $V = [v_1,v_2] \oplus W$. Since the result is assumed true for $\alpha|W \times W$, there is a basis $v_3,\ldots,v_n$ of W such that (1) holds for these v_i. Then $v_1,v_2,v_3,\ldots,v_n$ is the desired basis. ■

11. COROLLARY. Let $A = (a_{ij})$ be an $n \times n$ skew-symmetric matrix. Then there is a non-singular $n \times n$ matrix B such that

$$B^t AB = \begin{pmatrix} S & & & & & \\ & S & & & \bigcirc & \\ & & \ddots & & & \\ & & & S & & \\ & \bigcirc & & & 0 & \\ & & & & & \ddots \\ & & & & & & 0 \end{pmatrix},$$

where S is the matrix

$$S = \begin{pmatrix} 0 & 1 \\ -1 & 0 \end{pmatrix}.$$

Proof. Let $e_1,\ldots,e_n$ be the standard basis of $\mathbb{R}^n$, and define $\alpha \in \Omega^2(\mathbb{R}^n)$ by

$$\alpha(e_i,e_j) = a_{ij} .$$

Let $v_1,\ldots,v_n$ be the basis of $\mathbb{R}^n$ given by Proposition 10, and let $B = (b_{ij})$ be the matrix defined by

$$v_i = \sum_{k=1}^{n} b_{ki} e_k .$$

Then

$$\begin{aligned}\alpha(v_i,v_j) &= \alpha(\sum_k b_{ki}e_k, \sum_\ell b_{\ell j}e_\ell) \\ &= \sum_{k,\ell} b_{ki}b_{\ell j}\alpha(e_k,e_\ell) \\ &= \sum_{k,\ell} b_{ki}b_{\ell j}a_{k\ell} = (B^t AB)_{ij} ,\end{aligned}$$

which gives the desired result. ∎

Using the expression

$$\mathrm{Pf}(A) = \sum_{P\in \mathcal{P}'} \varepsilon(P)a_P$$

for skew-symmetric A, it is easy to compute that

$$\mathrm{Pf}\begin{pmatrix} S & & \bigcirc \\ & \ddots & \\ \bigcirc & & S \end{pmatrix} = 1 .$$

On the other hand, this matrix also has determinant = 1. This gives us

12. COROLLARY. For every skew-symmetric $n \times n$ matrix A with $n = 2m$ even we have

$$\{\mathrm{Pf}(A)\}^2 = \det A .$$

Proof. It suffices to prove this when $\det A \neq 0$, since both sides are continuous functions of the entries of A and the matrices with non-zero determinant are dense. By Corollary 11, there is a non-singular $n \times n$ matrix B with

$$B^t AB = \begin{pmatrix} S & & \bigcirc \\ & \ddots & \\ \bigcirc & & S \end{pmatrix} .$$

Then

$$1 = \det \begin{pmatrix} S & & \bigcirc \\ & \ddots & \\ \bigcirc & & S \end{pmatrix} = (\det B)^2 \cdot \det A ,$$

while Proposition 9 gives

$$1 = \mathrm{Pf} \begin{pmatrix} S & & \bigcirc \\ & \ddots & \\ \bigcirc & & S \end{pmatrix} = (\det B)\mathrm{Pf}(A) . \blacksquare$$

For a skew-symmetric $n \times n$ matrix A with n odd we define

$$\mathrm{Pf}(A) = 0 .$$

Note that in this case we have

$$\det A = \det A^t = \det(-A) = (-1)^n \det A \implies \det A = 0 .$$

<u>So Corollary</u> 12 <u>still holds</u>.

For even n the skew-symmetry of A is likewise crucial in Corollary 12. If we consider $\det(A_{ij})$ as a polynomial in n^2 independent commuting variables A_{ij}, then $\det(A_{ij})$ is <u>not</u> the square of another polynomial. But if we consider $\det(A_{ij})$ as a polynomial in the $n(n-1)/2$ independent commuting variables A_{ij} for $i < j$ [and define $A_{ii} = 0$, $A_{ij} = -A_{ji}$ for $i > j$], then $\det(A_{ij}) = \{\mathrm{Pf}(A)\}^2$ as polynomials in these A_{ij}, since the two polynomials give the same results when applied to all real numbers a_{ij}, $i < j$.

Since $\mathbb{R}[A_{ij}]$ is a unique factorization domain, $\mathrm{Pf}(A)$ and $-\mathrm{Pf}(A)$ are the only two polynomials with this property. The principal of extension of algebraic identities shows that $\det A = \{\mathrm{Pf}(A)\}^2$ when A is a $2m \times 2m$ skew-symmetric matrix with entries in any commutative ring $\mathcal{A}$ with unit.

{By working in the ring $\mathbb{R}[A_{ij}]$, we could have produced the Pfaffian in a neat, mysterious, way that avoids all computations. For there is a matrix X with entries in the quotient field of $\mathbb{R}[A_{ij}]$ such that

$$X^t A X = \begin{pmatrix} S & & \bigcirc \\ & \ddots & \\ \bigcirc & & S \end{pmatrix} .$$

Hence the polynomial $\det A$ in $\mathbb{R}[A_{ij}]$ is the square $(\det X)^{-2}$ in the quotient field of $\mathbb{R}[A_{ij}]$. Since $\mathbb{R}[A_{ij}]$ is a unique factorization domain, this implies that $\det A$ is already a square in $\mathbb{R}[A_{ij}]$. There are only two possible elements $\mathrm{Pf}(A)$ for $\det A$ to be the square of, and we define $\mathrm{Pf}(A)$ by requiring Pf to be $+1$ on $\begin{pmatrix} S & & \bigcirc \\ & \ddots & \\ \bigcirc & & S \end{pmatrix}$. Similarly, if we consider the ring $\mathbb{R}[A_{ij}, B_{11}, \ldots, B_{nn}]$ in the indeterminates A_{ij} for $i < j$, and B_{ij} for all i, j, then the identity $\det B^t AB = (\det B)^2 \det A$ implies that

$$\mathrm{Pf}(B^t AB) = \pm (\det B)\mathrm{Pf}A ,$$

and by choosing $B = I$ we see that the sign must be $+1$.}

As a rather trivial example of the use of polynomial rings to avoid some computations, we prove one more simple, but important property of Pf. If A and B are two square matrices, we will use $A \oplus B$ for the matrix

$$\begin{pmatrix} A & 0 \\ 0 & B \end{pmatrix} .$$

13. CORROLLARY. For all skew-symmetric matrices A and B we have

$$\mathrm{Pf}(A \oplus B) = \mathrm{Pf}(A) \cdot \mathrm{Pf}(B) .$$

Note: This result holds even when A or B, or both, is of odd order. In that case it says that $\mathrm{Pf}(A \oplus B) = 0$.

Proof. By Corollary 12 (which holds even for matrices of odd order) we have

$$(1) \qquad \{\mathrm{Pf}(A \oplus B)\}^2 = \det(A \oplus B) = \det A \cdot \det B = \{\mathrm{Pf}(A)\mathrm{Pf}(B)\}^2 ,$$

and thus

$$(2) \qquad \mathrm{Pf}(A \oplus B) = \pm\, \mathrm{Pf}(A)\mathrm{Pf}(B) .$$

If A or B is of odd order, then we already have $\mathrm{Pf}(A \oplus B) = 0$. For A and B of even order, we just need to determine the sign in (2). Now the same sign must hold in (2) for all A and B, for we may consider equation (1) as an equation in the ring $\mathbb{R}[A_{ij}, B_{ij}]$ with commuting variables A_{ij}, B_{ij} ($i < j$ in both cases). Letting A and B be of the form $S \oplus \cdots \oplus S$, we see that the $+$ sign always holds. ■

4. Defining the Euler Class in Terms of a Connection

Consider a smooth oriented n-dimensional vector bundle $\xi = \pi\colon E \longrightarrow M$, over a smooth manifold M (of any dimension). For compact orientable M we defined the Euler class $\chi(\xi) \in H^n(M)$ in Chapter I.11. To do this, we first defined the Thom class $U(\xi) \in H^n_c(E)$, and we proved (Theorem I.11-26) that

$U(\xi)$ is the unique class whose restriction to each fibre $\pi^{-1}(p)$ is the generator $\nu_p \in H_c^n(\pi^{-1}(p))$ determined by the orientation. From this result we can immediately conclude

14. LEMMA. Let $\xi = \pi\colon E \longrightarrow M$ be a smooth oriented vector bundle over a compact oriented manifold M, and let $f\colon M' \longrightarrow M$ be a smooth map, where M' is also a compact oriented manifold. If E' is the total space of $f^*\xi$, and $\tilde{f}\colon E' \longrightarrow E$ is a bundle map covering f, then

$$\tilde{f}^*(U(\xi)) = U(f^*\xi) \in H_c^n(E') .$$

Proof. Note first that $\tilde{f}$ is proper (the inverse image of a compact set is compact), so $\tilde{f}^*$ does take $H_c^n(E)$ to $H_c^n(E')$. Let $f^*\xi$ be $\pi'\colon E' \longrightarrow M'$. If $p' \in M'$ is any point, and $j_{p'}\colon \pi'^{-1}(p') \longrightarrow E'$ is the inclusion, then

$$j_{p'}{}^*\tilde{f}^*U(\xi) = (\tilde{f}\circ j_{p'})^*U(\xi) .$$

If we recall how $f^*\xi$ is defined, we see that $(\tilde{f}\circ j_{p'})^*U(\xi)$ must be the generator of $H_c^n(\pi'^{-1}(p'))$, since $j_{f(p')}{}^*U(\xi)$ is the generator of $H_c^n(\pi^{-1}(f(p')))$. This shows that $\tilde{f}^*U(\xi)$ must be $U(f^*\xi)$. ∎

We defined the Euler class $\chi(\xi)$ to be $s^*U(\xi)$, for any section s of ξ. In particular, we can choose $s = 0 =$ the zero section. Hence

15. PROPOSITION. Let $\xi = \pi\colon E \longrightarrow M$ be a smooth oriented vector bundle over a compact oriented manifold M, and let $f\colon M' \longrightarrow M$ be a smooth map, where M' is also a compact oriented manifold. Then

$$f^*\chi(\xi) = \chi(f^*\xi) \in H^n(M') .$$

<u>Proof</u>. If $0'$ denotes the zero section of $f^*\xi$, then $\tilde{f}\circ 0' = 0\circ f$. So

$$\begin{aligned}\chi(f^*\xi) &= (0')^*U(f^*\xi) \\ &= (0')^*\tilde{f}^*U(\xi) \qquad \text{by Lemma 14} \\ &= (\tilde{f}\circ 0')^*U(\xi) = (0\circ f)^*U(\xi) \\ &= f^*0^*U(\xi) = f^*\chi(\xi) . \quad \blacksquare\end{aligned}$$

As a particular consequence, we note a result which we will need later on.

<u>16. COROLLARY</u>. If n is even, then

$$\chi(\tilde{\gamma}^n(\mathbb{R}^N)) \neq 0$$

for all $N > n$.

<u>Proof</u>. Since $S^n \subset \mathbb{R}^N$ for $N > n$, we have a bundle map $(\tilde{f},f)\colon TS^n \longrightarrow E(\tilde{\gamma}^n(\mathbb{R}^N))$, as on pp. 406-407. So

$$\chi(TS^n) = f^*\chi(\tilde{\gamma}^n(\mathbb{R}^N)) .$$

But Theorem I.11-30 says that $\chi(TS^n)$ is $\chi(S^n)$ times the fundamental class of S^n, and $\chi(S^n) = 2 \neq 0$. $\blacksquare$

The Euler class has one further important property which it is not really essential to prove at this point, for it could eventually be derived from other results of this section. Nevertheless, it will motivate much of the argument to come.

17. THEOREM. Let $\xi_i = \pi_i\colon E_i \to M$ for $i = 1, 2$ be smooth oriented vector bundles over a compact oriented manifold M. Then $\chi(\xi_1 \oplus \xi_2)$ is the cup product

$$\chi(\xi_1 \oplus \xi_2) = \chi(\xi_1) \cup \chi(\xi_2) .$$

Proof. The Whitney sum $\xi_1 \oplus \xi_2$ is $\pi\colon E \to M$ where $E \subset E_1 \times E_2$ is $\{(e_1,e_2)\colon \pi_1(e_1) = \pi_2(e_2)\}$. Let $\rho_i\colon E \to E_i$ be the restriction of the projection maps $E_1 \times E_2 \to E_i$. For any $p \in M$, let

$$j\colon \pi^{-1}(p) \to E , \qquad j_i\colon \pi_i^{-1}(p) \to E_i$$

be the inclusions, and let

$$\sigma_i\colon \pi_1^{-1}(p) \times \pi_2^{-1}(p) \to \pi_i^{-1}(p)$$

be the projections. Then

$$j_i \circ \sigma_i = \rho_i \circ j .$$

So

$$\begin{aligned} j^*(\rho_1{}^*U_1(\xi_1) \cup \rho_2{}^*U_2(\xi_2)) &= (\rho_1 \circ j)^*U_1(\xi) \cup (\rho_2 \circ j)^*U_2(\xi) \\ &= \sigma_1{}^*j_1{}^*U_1(\xi) \cup \sigma_2{}^*j_2{}^*U_2(\xi) \\ &= \sigma_1{}^*\nu_1 \cup \sigma_2{}^*\nu_2 , \end{aligned}$$

where ν_i is the generator of $H_c^{n_i}(\pi_i^{-1}(p))$ determined by the orientation on $\pi_i^{-1}(p)$. But $\sigma_1{}^*\nu_1 \cup \sigma_2{}^*\nu_2$ is easily seen to be the generator in

$H_c^{n_1+n_2}(\pi_1^{-1}(p) \times \pi_2^{-1}(p))$ determined by the orientation on $\pi_1^{-1}(p) \times \pi_2^{-1}(p)$ (given on p. 412). It follows that

$$\rho_1{}^*U_1(\xi_1) \cup \rho_2{}^*U_2(\xi_2) = U(\xi) , \qquad \text{the Thom class of } \xi .$$

So if s_i are sections of ξ_i, then for the obvious section $s_1 + s_2$ of ξ we have

$$\begin{aligned}\chi(\xi) &= (s_1 + s_2)^*U(\xi) = (s_1 + s_2)^*(\rho_1{}^*U_1(\xi_1) \cup \rho_2{}^*U_2(\xi_2)) \\ &= [\rho_1 \circ (s_1 + s_2)]^*U_1(\xi_1) \cup [\rho_2 \circ (s_1 + s_2)]^*U_2(\xi_2) \\ &= s_1{}^*U_1(\xi_1) \cup s_2{}^*U_2(\xi_2) \\ &= \chi_1(\xi_1) \cup \chi_2(\xi_2) . \quad \blacksquare\end{aligned}$$

Now we are going to look at principal bundles associated with a smooth oriented n-dimensional vector bundle $\xi = \pi\colon E \to M$ over a smooth manifold M. We have already considered the principal bundle $F(\xi)$ of frames of E. If we have a Riemannian metric $\langle\ ,\ \rangle$ for ξ, then, as in Chapter 7, we can consider the bundle $O(E)$ of orthonormal frames, which is a principal bundle with group $O(n)$. Since we will be considering only paracompact manifolds M, we know (see p. II.393) that there is an Eheresmann connection ω on the bundle $O(E)$. Thus ω is a matrix of 1-forms (ω^i_j) on $O(E)$ taking values in $\mathfrak{o}(n)$; the curvature form $\Omega = D\omega$ is a matrix of 2-forms (Ω^i_j), also with values in $\mathfrak{o}(n)$. [Since we will seldom be working with TM any more, and never with moving frames, we will not resort to any special symbolism to distinguish the forms ω^i_j, Ω^i_j defined on a bundle from those defined for some moving frame.] As we pointed out in Chapter 7, a connection ω on $O(E)$ is equivalent to a covariant differentiation operator on E which is compatible with the metric $\langle\ ,\ \rangle$. In the case of a general bundle ξ over M there will be many connections compatible with the metric $\langle\ ,\ \rangle$; we cannot single one out by asking for a symmetric connection, as this concept makes sense only for the tangent bundle. Since our bundle ξ is oriented, we can also consider the bundle $SO(E)$ of positively oriented orthonormal frames; if X is connected, it is simply one of the two components of $O(E)$. The group of this bundle is $SO(n)$, whose Lie algebra is also $\mathfrak{o}(n)$. So a connection ω on $SO(n)$ again has values in $\mathfrak{o}(n)$, as does the matrix of 2-forms Ω.

Now let us specialize to the case of a smooth oriented n-dimensional vector bundle $\xi = \pi\colon E \longrightarrow M$ over M, where $n = 2m$ is <u>even</u>. If $\langle\ ,\ \rangle$ is a Riemannian metric for ξ, and ω is a connection on the corresponding principal bundle $\varpi\colon SO(E) \longrightarrow M$, then we can consider the n-form

$$2m\cdot m!\ \mathrm{Pf}(\Omega) = \sum_{i_1,\ldots,i_n} \varepsilon^{i_1\cdots i_n}\, \Omega^{i_1}_{i_2} \wedge \cdots \wedge \Omega^{i_{n-1}}_{i_n},$$

which is defined <u>on the bundle</u> $SO(E)$. The following proof is merely an invariant formulation of an argument presented in the last section.

<u>18. PROPOSITION</u>. There is a unique n-form Λ on M such that

$$\varpi^*(\Lambda) = \sum \varepsilon^{i_1\cdots i_n}\, \Omega^{i_1}_{i_2} \wedge \cdots \wedge \Omega^{i_{n-1}}_{i_n} = 2^m m!\ \mathrm{Pf}(\Omega)\ .$$

<u>Proof</u>. Given $X_1,\ldots,X_n \in M_p$, choose some $u \in \varpi^{-1}(p)$, and let $Y_1,\ldots,Y_n \in SO(E)_u$ be tangent vectors with $\varpi_* Y_i = X_i$. Clearly Λ must satisfy

$$\Lambda(X_1,\dots,X_n) = 2^m m!\ \mathrm{Pf}(\Omega)(Y_1,\dots,Y_n) ,$$

which proves uniqueness. Existence will be demonstrated once we prove that this Λ is well-defined.

Consider first what happens when we take different tangent vectors $Z_1,\dots,Z_n \in SO(E)_u$ with $\varpi_* Z_i = X_i$. Since $\varpi_*(Y_i - Z_i) = 0$, all $Y_i - Z_i$ are vertical. But $\Omega(Y,Z) = 0$ if either Y or Z is vertical. So we clearly have

$$\begin{aligned}\mathrm{Pf}(\Omega)(Y_1,\dots,Y_n) &= \mathrm{Pf}(\Omega)(Z_1,Y_2,\dots,Y_n)\\ &= \mathrm{Pf}(\Omega)(Z_1,Z_2,Y_3,\dots,Y_n) = \cdots\\ &= \mathrm{Pf}(\Omega)(Z_1,\dots,Z_n) .\end{aligned}$$

Thus the definition of Λ does not depend on the choice of the Y_i.

Now suppose we choose a different $\bar{u} \in \varpi^{-1}(p)$. Then $\bar{u} = R_A(u) = u\cdot A$ for some $A \in SO(n)$, and we can let the $\bar{Y}_i \in SO(E)_{\bar{u}}$ be $\bar{Y}_i = R_{A*}Y_i$. Then

$$\begin{aligned}\mathrm{Pf}(\Omega)(\bar{Y}_1,\dots,\bar{Y}_n) &= \mathrm{Pf}(\Omega)(R_{A*}Y_1,\dots,R_{A*}Y_n)\\ &= \mathrm{Pf}(R_A{}^*\Omega)(Y_1,\dots,Y_n)\\ &= \mathrm{Pf}(A^{-1}\Omega A)(Y_1,\dots,Y_n) &&\text{by Proposition II.8-11}\\ &= \mathrm{Pf}(\Omega)(Y_1,\dots,Y_n) &&\text{by Proposition 9.}\ \blacksquare\end{aligned}$$

We also have the following result, which is automatic when ξ is the tangent bundle of M.

19. PROPOSITION. The unique n-form Λ of Proposition 18 is closed, $d\Lambda = 0$.

Proof. Given $X_1,\ldots,X_{n+1} \in M_p$, choose $u \in \varpi^{-1}(p)$ and $Y_1,\ldots,Y_{n+1} \in SO(E)_u$ with $\varpi_* Y_i = X_i$. Let hY_i be the horizontal component of Y_i. Then

$$\begin{aligned} d\Lambda(X_1,\ldots,X_{n+1}) &= d\Lambda(\varpi_* Y_1,\ldots,\varpi_* Y_{n+1}) \\ &= d\Lambda(\varpi_* hY_1,\ldots,\varpi_* hY_{n+1}) \\ &= (\varpi^* d\Lambda)(hY_1,\ldots,hY_{n+1}) \\ &= d(\varpi^*\Lambda)(hY_1,\ldots,hY_{n+1}) \\ &= 2^m m!\ d\{Pf(\Omega)\}(hY_1,\ldots,hY_{n+1}) \\ &= 2^m m!\ D\{Pf(\Omega)\}(Y_1,\ldots,Y_{n+1}) . \end{aligned}$$

But $D\Omega = 0$ by Bianchi's identity (Theorem II.8-20), and this implies that $D\{Pf(\Omega)\} = 0$. ■

In view of Proposition 19, the n-form Λ determines a de Rham cohomology class $[\Lambda] \in H^n(M)$. The form Λ itself depends on the oriented n-dimensional bundle $\xi = \pi: E \longrightarrow M$ over M, on the choice of a metric $<\ ,\ >$ for ξ, and on the connection ω on the corresponding bundle $SO(E)$.

20. PROPOSITION. The cohomology class $[\Lambda]$ is independent of the metric $<\ ,\ >$ and of the connection ω.

Proof. Let $<\ ,\ >$ and $<\ ,\ >'$ be two metrics for ξ. By Corollary 5, the corresponding principle bundles $SO(E)$ and $SO'(E)$ are equivalent. If $\tilde{f}: SO'(E) \longrightarrow SO(E)$ is a fibre preserving diffeomorphism which commutes with the action of $SO(n)$, and ω is a connection on $SO(E)$, then $\omega' = \tilde{f}^*\omega$

is a connection on $SO'(E)$, It is easy to see that the corresponding curvature forms satisfy $\Omega' = \tilde{f}^*\Omega$, so that $\text{Pf}(\Omega') = \tilde{f}^*\text{Pf}(\Omega)$. This implies that the corresponding forms Λ' and Λ are actually equal. To complete the proof it therefore suffices to show that any two connections ω_0, ω_1 on the same principal bundle $SO(E)$ give rise to forms Λ_0 and Λ_1 whose difference is exact.

Let $q\colon M \times [0,1] \to M$ be the projection $q(p,t) = p$, and consider the bundle $q^*SO(\xi)$ over $M \times [0,1]$. There are obvious induced connections $q^*\omega_0$ and $q^*\omega_1$ on $q^*SO(\xi)$. Let $\tau\colon M \times [0,1] \to [0,1]$ be the function $\tau(p,t) = t$, and form the connection

$$\boldsymbol{\omega} = (1-\tau)(q^*\omega_0) + \tau(q^*\omega_1)$$

on $q^*SO(\xi)$, with connection form $\boldsymbol{\Omega}$, say. If $i_t\colon M \to M \times [0,1]$ is $i_t(p) = (p,t)$, then $i_0{}^*(\boldsymbol{\omega})$ can clearly be identified with ω_0, and $i_1{}^*(\boldsymbol{\omega})$ can be identified with ω_1. By Propositions 18 and 19 (which hold for manifolds-with-boundary as well as for manifolds), there is a closed n-form $\boldsymbol{\Lambda}$ on $M \times [0,1]$ which pulls back to $2^m m!\text{Pf}(\boldsymbol{\Omega})$ on the total space of $q^*SO(\xi)$. Clearly we must have

$$i_0{}^*\boldsymbol{\Lambda} = \Lambda_0 \quad \text{and} \quad i_1{}^*\boldsymbol{\Lambda} = \Lambda_1 .$$

Now Theorem I.7-17 (p. I.304) shows that $\Lambda_1 - \Lambda_0$ is exact. ■

We thus see that every oriented smooth bundle ξ over M of even fibre dimension n determines a de Rham cohomology class $C(\xi) = [\Lambda] \in H^n(M)$. Clearly $C(\xi) = C(\eta)$ if $\xi \simeq \eta$.

21. PROPOSITION. Let $\xi = \pi\colon E \to M$ be a smooth oriented bundle over M of even fibre dimension n, and let $f\colon M' \to M$ be a smooth map. Then

$$C(f^*\xi) = f^*(C(\xi)) \in H^n(M') .$$

Proof. Let E' be the total space of $f^*\xi$, and let $\tilde{f}\colon E' \to E$ be the bundle map covering f. If $\langle\ ,\ \rangle$ is a metric on E, then $\tilde{f}^*\langle\ ,\ \rangle$ is a metric on E'. Clearly there is an equivalence $\bar{f}\colon SO(E') \to SO(E)$ covering f.

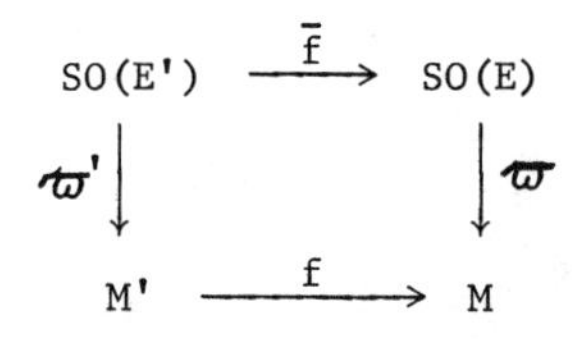

If ω is a connection on $SO(E)$, then $\bar{f}^*(\omega)$ will be a connection on $SO(E')$, and it is easy to see that the corresponding connection forms satisfy $\Omega' = \bar{f}^*\Omega$. Consequently,

$$Pf(\Omega') = Pf(\bar{f}^*\Omega) = \bar{f}^*Pf(\Omega) .$$

For the n-forms Λ on M given by Proposition 18 we then have

$$\begin{aligned} \varpi'^*(f^*\Lambda) &= \bar{f}^*\varpi^*\Lambda \\ &= 2^m m!\ \bar{f}^*Pf(\Omega) \\ &= 2^m m!\ Pf(\Omega') . \end{aligned}$$

So $f^*\Lambda$ must be the n-form Λ' on M' given by Proposition 18. ∎

We will extend the definition of C by setting $C(\xi) = 0$ when ξ is a smooth oriented bundle of odd fibre dimension. We would like to show that $C(\xi)$ is always some constant times $\chi(\xi)$. If this is the case, then we ought to have an analogue of Theorem 17 for C. And indeed we do.

22. THEOREM. Let $\xi_i = \pi_i: E_i \to M$ for $i = 1, 2$ be smooth oriented vector bundles over M, of fibre dimensions n_1 and n_2. If $n_i = 2m_i$, then

$$C(\xi_1 \oplus \xi_2) = \frac{(m_1 + m_2)!}{m_1! m_2!} C(\xi_1) \cup C(\xi_2) .$$

(For n_1 or n_2 odd, this just asserts that $C(\xi_1 \oplus \xi_2) = 0$.)

Proof. Choose Riemannian metrics $\langle\ ,\ \rangle_i$ on ξ_i, and let $\langle\ ,\ \rangle$ be the obvious metric $\langle\ ,\ \rangle_1 \oplus \langle\ ,\ \rangle_2$ on $\xi_1 \oplus \xi_2 = \pi: E \to M$. Let $\varpi_i: SO(E_i) \to M$ and $\varpi: SO(E) \to M$ be the corresponding principal bundles. Over M we consider first the principal bundle $SO(E_1)*SO(E_2)$, with group $SO(n_1) \times SO(n_2) \subset SO(n_1 + n_2)$, whose fibre over $p \in M$ is just the direct product $\varpi_1^{-1}(p) \times \varpi_2^{-1}(p)$, so that we can regard

$$SO(E_1)*SO(E_2) \subset SO(E) .$$

Let $\rho_i: SO(E_1)*SO(E_2) \to SO(E_i)$ be the obvious projection maps. If ω_i are connections on $SO(E_i)$, with curvature forms Ω_i, then

$$\rho_1{}^*\omega_1 \oplus \rho_2{}^*\omega_2 = \begin{pmatrix} \rho_1{}^*\omega_1 & 0 \\ 0 & \rho_2{}^*\omega_2 \end{pmatrix}$$

is a connection $\bar{\omega}$ on $SO(E_1)*SO(E_2)$, with curvature form

$$\bar{\Omega} = \rho_1{}^*\Omega_1 \oplus \rho_2{}^*\Omega_2 = \begin{pmatrix} \rho_1{}^*\Omega_1 & 0 \\ 0 & \rho_2{}^*\Omega_2 \end{pmatrix} .$$

The connection $\bar{\omega}$ can be extended uniquely to a connection $\tilde{\omega}$ on $SO(E)$ [the requirement $\tilde{\omega}(\sigma(M)) = M$ determines $\tilde{\omega}$ at the new vertical vectors, hence $\tilde{\omega}$ is determined at all points of $SO(E_1)*SO(E_2)$, and then at all points of $SO(E)$ by the requirement $\tilde{\omega}(R_A{}^*Y) = Ad(A^{-1})\tilde{\omega}(Y)$]. At any point $e \in SO(E_1)*SO(E_2)$ the horizontal vectors for $\tilde{\omega}$ are the same as for $\bar{\omega}$, so at e we have

$$\tilde{\Omega} = \bar{\Omega} \qquad \text{[on tangent vectors to } SO(E_1)*SO(E_2)]$$

$$\Longrightarrow Pf(\tilde{\Omega}) = Pf(\bar{\Omega}) = Pf(\rho_1{}^*\Omega_1) \wedge Pf(\rho_2{}^*\Omega_2) \qquad \text{by Corollary 13}$$

$$= \rho_1{}^*Pf(\Omega_1) \wedge \rho_2{}^*Pf(\Omega_2) .$$

So if Λ_i, Λ are the forms given by Proposition 18, then at e we must have

$$\varpi^*\Lambda = 2^{m_1+m_2}(m_1+m_2)!Pf(\tilde{\Omega}) \qquad \text{[on tangent vectors to } SO(E_1)*SO(E_2)]$$

$$= \frac{(m_1+m_2)!}{m_1!m_2!} 2^{m_1}m_1!\rho_1{}^*Pf(\Omega_1) \wedge 2^{m_2}m_2!\rho_2{}^*Pf(\Omega_2)$$

$$= \frac{(m_1+m_2)!}{m_1!m_2!} \rho_1{}^*\varpi_1{}^*\Lambda_1 \wedge \rho_2{}^*\varpi_2{}^*\Lambda_2$$

$$= \frac{(m_1+m_2)!}{m_1!m_2!} \varpi^*\Lambda_1 \wedge \varpi^*\Lambda_2 .$$

This implies that

$$\Lambda = \frac{(m_1+m_2)!}{m_1!m_2!} \Lambda_1 \wedge \Lambda_2 . \quad \blacksquare$$

Applying Theorem 22 when $n_1 = 1$, we immediately deduce that the class C has a property which we have already mentioned for χ (p. I.606):

<u>23. COROLLARY</u>. If the oriented bundle $\xi = \pi\colon E \longrightarrow M$ has a nowhere zero section s, then

$$C(\xi) = 0 .$$

<u>Proof</u>. Let $E_1 \subset E$ be

$$\bigcup_{p\in M} \mathbb{R}\cdot s(p) ,$$

and let $E_2 \subset E$ be the orthogonal complement

$$\bigcup_{p\in M} (\mathbb{R}\cdot s(p))^{\perp}$$

with respect to some Riemannian metric on E. Then $\xi_1 = \pi_1|E_1\colon E_1 \longrightarrow M$ is an oriented 1-dimensional bundle, so $\xi_2 = \pi_2|E_2\colon E_2 \longrightarrow M$ is also an oriented bundle (since ξ is oriented). Clearly $\xi \simeq \xi_1 \oplus \xi_2$. So Theorem 22 shows that $C(\xi) = 0$. ∎

But this fact practically characterizes χ:

<u>24. COROLLARY</u>. If $\xi = \pi\colon E \longrightarrow M$ is a smooth oriented vector bundle of fibre dimension n over a compact oriented manifold M, then the class $C(\xi) \in H^n(M)$ is a multiple of the Euler class $\chi(\xi)$.

Proof. Let S be the sphere bundle $S = \{e \in E: \langle e,e \rangle = 1\}$ formed with respect to some Riemannian metric on E, and let $\pi_0: S \to X$ be the restriction $\pi|S$. As we pointed out in section 2, the bundle $\pi_0{}^*\xi$ has a nowhere zero section. So Corollary 23 gives

$$\begin{aligned} 0 &= C(\pi_0{}^*\xi) \\ &= \pi_0{}^*C(\xi) \qquad \text{by Proposition 21.} \end{aligned}$$

But Theorem I.11-31 says that a class $\alpha \in H^n(M)$ satisfies $\pi_0{}^*\alpha = 0$ if and only if α is a multiple of $\chi(\xi)$. ■

If we apply this Corollary to the tangent bundle of a compact oriented manifold M of even dimension n, we find that the class $C(TM) \in H^n(M)$ is some multiple of the Euler class $\chi(TM)$. This statement is not very interesting, since $H^n(M)$ is 1-dimensional (all it tells us is that $C(TM) = 0$ if $\chi(TM) = 0$). But we obtain a statement which is interesting when we apply the Corollary to the universal bundles:

25. COROLLARY. For every even n, there is a "universal constant" A_n such that

$$C(\xi) = A_n \cdot \chi(\xi)$$

for all smooth oriented n-dimensional bundles ξ over a compact oriented manifold M.

Proof. Consider the bundles $\tilde{\gamma}^n(\mathbb{R}^N)$, for $N > n$. By Corollary 24, there are constants $A_{n,N}$ such that

$$(1) \qquad C(\tilde{\gamma}^n(\mathbb{R}^N)) = A_{n,N}\cdot\chi(\tilde{\gamma}^n(\mathbb{R}^N)) \in H^n(\tilde{G}_n(\mathbb{R}^N)) .$$

If $\alpha\colon \tilde{G}_n(\mathbb{R}^N) \longrightarrow \tilde{G}_n(\mathbb{R}^M)$ is the natural inclusion, then

$$\alpha^*(\tilde{\gamma}^n(\mathbb{R}^M)) \simeq \tilde{\gamma}^n(\mathbb{R}^N) ,$$

so Propositions 15 and 21 give

$$(2) \qquad C(\tilde{\gamma}^n(\mathbb{R}^N)) = \alpha^*C(\tilde{\gamma}^n(\mathbb{R}^M))$$

$$(3) \qquad \chi(\tilde{\gamma}^n(\mathbb{R}^N)) = \alpha^*\chi(\tilde{\gamma}^n(\mathbb{R}^M)) .$$

Equations (1)-(3) give

$$A_{n,N}\cdot\chi(\tilde{\gamma}^n(\mathbb{R}^N)) = A_{n,M}\cdot\chi(\tilde{\gamma}^n(\mathbb{R}^N)) .$$

Since $\chi(\tilde{\gamma}^n(\mathbb{R}^N)) \neq 0$ by Corollary 16, this implies that $A_{n,N} = A_{n,M}$ for all $N, M > n$. Denoting this common number by A_n, we have

$$(*) \qquad C(\tilde{\gamma}_n(\mathbb{R}^N)) = A_n\cdot\chi(\tilde{\gamma}_n(\mathbb{R}^N)) .$$

Now by Theorem 8 any smooth oriented n-dimensional handle ξ over a compact manifold M is equivalent to $f^*\tilde{\gamma}^n(\mathbb{R}^N)$ for some smooth map $f\colon M \longrightarrow \tilde{G}_n(\mathbb{R}^N)$. Thus

$$\begin{aligned} C(\xi) &= C(f^*\tilde{\gamma}^n(\mathbb{R}^N)) \\ &= f^*C(\tilde{\gamma}^n(\mathbb{R}^N)) && \text{by Proposition 21} \\ &= A_n\cdot f^*\chi(\tilde{\gamma}^n(\mathbb{R}^N)) && \text{by } (*) \\ &= A_n\cdot\chi(\xi) && \text{by Proposition 15.} \quad \blacksquare \end{aligned}$$

To see what this universal constant A_n is, we merely have to compute it in some convenient special case:

26. THE GAUSS-BONNET-CHERN THEOREM. For even $n = 2m$, the constant A_n is

$$A_n = \frac{n!}{2} \cdot \text{volume of the unit } n\text{-sphere } S^n$$
$$= \frac{n!}{2} \cdot \frac{\pi^m 2^{n+1} m!}{n!} = \pi^m m! 2^n .$$

Consequently, if $(M, \langle\ ,\ \rangle)$ is a compact oriented manifold of even dimension $n = 2m$, then

$$\int_M K_n \, dV = \frac{1}{2} \text{ volume of } S^n \cdot \chi(M)$$
$$= \frac{\pi^m 2^n m!}{n!} \chi(M) .$$

Proof. Let ξ be the tangent bundle TM of a compact oriented Riemannian n-manifold M. On p. 417 we have a formula for $K_n \, dV$ (in this formula the Ω^i_j are the curvature forms for some positively oriented orthonormal moving frame) which clearly implies that the form Λ given by Proposition 18 for the bundle $SO(\xi) = SO(TM)$ is

$$\Lambda = n! K_n \, dV .$$

So if μ is the fundamental class of M, then

$$\left(\int_M K_n \, dV\right) \cdot \mu = \frac{1}{n!}\left(\int_M \Lambda\right) \cdot \mu = \frac{1}{n!} C(\xi)$$
$$= \frac{A_n}{n!} \chi(\xi)$$

$$= \frac{A_n}{n!} \chi(M) \cdot \mu \qquad \text{by Theorem I.11-30.}$$

Hence

$$\text{(1)} \qquad \int_M K_n \, dV = \frac{A_n}{n!} \chi(M) \, .$$

Taking $M = S^n$ in (1), with $K_n = 1$, we have

$$\text{volume } S^n = \frac{A_n}{n!} \chi(S^n) = \frac{2A_n}{n!}$$

$$\Longrightarrow A_n = \frac{n!}{2} \text{ volume } S^n = \frac{n!}{2} \frac{\pi^m 2^{n+1} m!}{n!} = \pi^m m! 2^n \, , \qquad \text{by Problem I.9-14.}$$

Substituting this value of A_n back in (1) we now have, for any M,

$$\int_M K_n \, dV = \frac{\pi^m m! 2^n}{n!} \chi(M) \, . \quad \blacksquare$$

5. The Concept of Characteristic Classes

Our proof of the generalized Gauss-Bonnet theorem made essential use of the fact that both the Euler class $\chi(\xi)$ and the class $C(\xi)$ are "natural": for a bundle $\xi = \pi\colon E \to M$ and a map $f\colon M' \to M$ we have

$$\chi(f^*\xi) = f^*(\chi(\xi)) \qquad \text{and} \qquad C(f^*\xi) = f^*(C(\xi)) \, .$$

This suggests that we might obtain greater insight into the theorem by trying to find out what all such natural classes are. To be precise, we define a characteristic class of dimension k for smooth n-dimensional bundles to be a function C which associates to each smooth n-dimensional bundle $\xi = \pi: E \to M$ an element

$$C(\xi) \in H^k(M) ,$$

with the following property: if $\xi' = \pi': E' \to M'$ is another smooth n-dimensional bundle, and $(\tilde{f}, f)$ is a smooth bundle map from ξ' to ξ, then

$$C(\xi') = f^*(C(\xi)) \in H^n(M') .$$

Here is an equivalent formulation: $C(\xi) = C(\eta)$ if $\xi \simeq \eta$, and for every smooth n-dimensional bundle $\xi = \pi: E \to M$ and smooth map $f: M' \to M$ we have

$$C(f^*\xi) = f^*(C(\xi)) .$$

We can also define characteristic classes for oriented bundles; these are the characteristic classes that we will actually investigate. What we would like to do is to find out what all these characteristic classes are. This question might look hopeless, were it not for the universal bundles $\tilde{\gamma}^n(\mathbb{R}^N)$. Notice that a characteristic class C of dimension k for smooth n-dimensional bundles gives us, in particular, certain elements

$$c_N = C(\tilde{\gamma}^n(\mathbb{R}^N)) \in H^k(\tilde{G}_n(\mathbb{R}^N)) .$$

If

$$\alpha_{N,N'}: \tilde{G}_n(\mathbb{R}^N) \to \tilde{G}_n(\mathbb{R}^{N'}) \qquad N' > N$$

is the natural map, then

$$\alpha_{N,N'}{}^*(\tilde{\gamma}^n(\mathbb{R}^{N'})) \simeq \tilde{\gamma}^n(\mathbb{R}^N) \implies \alpha_{N,N'}{}^* c_{N'} = c_N .$$

Conversely, suppose we are given classes

$$c_N \in H^k(\tilde{G}_n(\mathbb{R}^N))$$

satisfying the compatibility condition

$$\text{(C)} \qquad \alpha_{N,N'}{}^* c_{N'} = c_N \qquad \text{for } N' > N .$$

Since, by Theorem 8, any oriented n-dimensional bundle ξ over a compact M is equivalent to $f^*\tilde{\gamma}^n(\mathbb{R}^N)$ for some $f\colon M \longrightarrow \tilde{G}_n(\mathbb{R}^N)$, we can <u>define</u>

$$C(\xi) = f^* c_N .$$

Then $C(\xi)$ is well-defined, for if we have

$$f\colon M \longrightarrow \tilde{G}_n(\mathbb{R}^N) \qquad \text{and} \qquad g\colon M \longrightarrow \tilde{G}_n(\mathbb{R}^{N'})$$

with

$$f^*\tilde{\gamma}^n(\mathbb{R}^N) \simeq \xi \simeq g^*\tilde{\gamma}^n(\mathbb{R}^{N'}) ,$$

then the compositions

$$\alpha_{N,N''}\circ f , \qquad \alpha_{N',N''}\circ g \qquad\qquad N'' \geq 2N,\ 2N'$$

are homotopic, so

$$\begin{aligned} f^*c_N &= f^*\alpha_{N,N''}{}^*c_{N''} && \text{by condition (C)} \\ &= (\alpha_{N,N''}\circ f)^*c_{N''} \\ &= (\alpha_{N',N''}\circ g)^*c_{N''} \\ &= g^*c_{N'} \ . \end{aligned}$$

Moreover, if $h\colon M' \longrightarrow M$ is a smooth map, then

$$\begin{aligned} C(h^*\xi) &= C(h^*f^*\tilde{\gamma}^n(\mathbb{R}^N)) \\ &= h^*f^*c_N \\ &= h^*C(\xi) \ . \end{aligned}$$

So we could just as well define a characteristic class of dimension k for smooth oriented n-dimensional bundles to be a collection of classes

$$c_N \in H^k(\tilde{G}_n(\mathbb{R}^N))$$

satisfying

$$\text{(C)} \qquad \alpha_{N,N'}{}^*c_{N'} = c_N \qquad \text{for } N' > N \ .$$

Now the problem doesn't seem quite so formidable; the main task seems to be the computation of $H^k(\tilde{G}_n(\mathbb{R}^N))$. As a matter of fact, it will turn out that the maps

$$\alpha_{N,N'}{}^*\colon H^k(\tilde{G}_n(\mathbb{R}^{N'})) \longrightarrow H^k(\tilde{G}_n(\mathbb{R}^N))$$

are isomorphisms for $N, N' > n+k$. So a characteristic class of dimension k will be just the same as an element of $H^k(\tilde{G}_n(\mathbb{R}^N))$, for any $N > n+k$.

[If we were willing to work with singular cohomology, say, on spaces which are not manifolds, then we could define a characteristic class to be simply an element of the k-dimensional cohomology of the space $\tilde{G}_n(\mathbb{R}^\infty)$, the oriented version of the space $G_n(\mathbb{R}^\infty)$ defined on p. 411 .]

Now the calculation of cohomology groups is really the business of algebraic topologists, and all sorts of machinery has been used for computing characteristic classes [for all coefficient groups]. Rather than using any of the standard methods from this field, we will compute characteristic classes by purely differential geometric methods, making essential use of the fact that the Grassmannians are coset spaces of Lie groups. Although the procedure is quite involved, along the way we will get to look at several topics which are interesting in their own right. Moreover, the analysis will motivate the definition, in section 10, of one of the famous constructions in differential geometry. Finally, the Pfaffian will arise in a completely natural way.

6. The Cohomology of Homogeneous Spaces

By a <u>homogeneous space</u> we will mean a left coset space G/H, where G is a Lie group and H is a closed subgroup. We let $\pi\colon G \longrightarrow G/H$ be the natural projection $\pi(a) = aH$, and we give G/H the quotient topology: a set $\mathcal{U} \subset G/H$ is open if and only if $\pi^{-1}(\mathcal{U}) \subset G$ is open. It is an easy exercise to show that G/H is Hausdorff. Notice also that if $V \subset G$ is any set, then

$$\begin{aligned} \pi(a) \in \pi(V) &\iff aH \in \{bH\colon b \in V\} \\ &\iff aH = bH \quad \text{for some } b \in V \\ &\iff a \in bH \quad \text{for some } b \in V\,. \end{aligned}$$

So

$$\pi^{-1}(\pi(V)) = V\cdot H = \bigcup_{h\in H} V\cdot h\ .$$

This shows that $\pi(V) \subset G/H$ is open if $V \subset G$ is open; thus π is an open map, as well as a continuous one.

For every $a \in G$ we have a map $\mathbf{L}_a\colon G/H \to G/H$ given by $\mathbf{L}_a(bH) = abH$ (the notation $L_a\colon G \to G$ will be reserved for the map given by $L_a(b) = ab$). Obviously the diagram

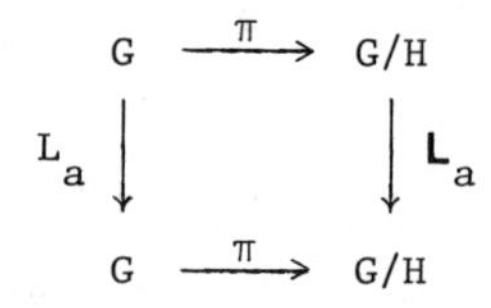

commutes, which implies that $\mathbf{L}_a$ is continuous; more generally, it is easy to see that the map

$$G \times G/H \to G \qquad \text{given by} \qquad (a,bH) \mapsto abH$$

is continuous.

In this section we will show that G/H is a manifold, and we will find a method of computing the de Rham cohomology $H^*(G/H)$ when G is compact and connected. More precisely, we will reduce the determination of the de Rham cohomology to purely algebraic calculations involving the Lie algebras of G and H. In section 9 we will carry out a sufficient portion of the algebraic calculations for the Grassmannians $\tilde{G}_n(\mathbb{R}^N) = SO(N)/\ SO(n) \times SO(N-n)$ to determine all characteristic classes for oriented bundles.

We already know, from Theorem I.10-15, that the closed subgroup H of

G is a Lie subgroup; in fact, there is a C^∞ structure on H, with the relative topology, that makes it a Lie subgroup of G.

27. PROPOSITION. Let G be a Lie group of dimension n, and H a closed subgroup of dimension d. Then G/H is a topological manifold of dimension $n-d$, and there is a unique C^∞ structure on G/H such that

(i) $\pi\colon G \longrightarrow G/H$ is C^∞

(ii) For every point of G/H there is a neighborhood $\mathcal{U}$ and a C^∞ section $s\colon \mathcal{U} \to G$ (a map $s\colon \mathcal{U} \to G$ satisfying $\pi \circ s =$ identity).

Proof. From the proof of Theorem I.10-15 we know that there is a coordinate system (x,U) around e with

$$x(e) = 0$$
$$x(U) = (-\varepsilon,\varepsilon) \times \cdots \times (-\varepsilon,\varepsilon) ,$$

such that each slice

$$x^{d+1} = \text{constant}, \ldots, x^n = \text{constant}$$

is an open subset of some left coset of H. In particular, the slice S_e through e is an open subset of H. Since H has the relative topology, this slice is of the form $V \cap H$ for some open set V. So by choosing ε smaller, if necessary, we can assume that

$$U \cap H = S_e .$$

We will now show that we can arrange for all slices to lie on different cosets of H. Choose $\varepsilon_1, \varepsilon_2 < \varepsilon$ so that the sets U_i with $x(U_i) = (-\varepsilon_i,\varepsilon_i) \times \cdots \times (-\varepsilon_i,\varepsilon_i)$ satisfy

$$U_1{}^{-1}\cdot U_1 \subset U_2\ , \qquad U_2\cdot U_2 \subset U\ .$$

If $a, b \in U_1$ satisfy $aH = bH$, then

$$b^{-1}a \in U_2 \cap H = U_2 \cap S_e \implies a \in b\cdot(U_2 \cap S_e)\ .$$

Now $b\cdot(U_2 \cap S_e)$ is connected and lies in U, so it lies in a single slice. This shows that a and b lie in the same slice. Equivalently, different slices of U_1 lie in different cosets, as desired. For convenience, we assume that U_1 is our original U.

If we now consider the "cross-section" $C \subset U$ defined by

$$C \doteq \{a \in U_1 \colon x^1(a) = \cdots = x^d(a) = 0\}\ ,$$

we see that

$$\pi|C\colon C \longrightarrow G/H$$

is one-one, with image $\pi(C) = \pi(U)$, which is open in G/H. Since π is both continuous and open, the map $\pi|C$ is a homeomorphism. The inverse homeomorphism

$$\chi = (\pi|C)^{-1}\colon \pi(U) \longrightarrow C$$

can be regarded as a map into $\mathbb{R}^{n-d}$; we will use this map as a coordinate system around the coset H in G/H. For every $a \in G$ we let χ_a be the composition

$$\pi(a\cdot U) \xrightarrow{\mathbf{L}_a{}^{-1}} \pi(U) \xrightarrow{(\pi|C)^{-1}} C \;.$$

Then for $a, b \in G$ we have

$$\chi_a \circ \chi_b{}^{-1} = (\pi|C)^{-1} \circ \mathbf{L}_{a^{-1}} \circ \mathbf{L}_b \circ \pi|C$$

$$\text{on the set } W = \chi_b(\pi(a\cdot U) \cap \pi(b\cdot U)) \;.$$

For $c \in C$ we have

$$\mathbf{L}_{a^{-1}} \mathbf{L}_b (\pi|C)(c) = a^{-1}bcH \;,$$

and it is easy to check that if $c \in W$, then

$$(\pi|C)^{-1} \mathbf{L}_{a^{-1}} \mathbf{L}_b (\pi|C)(c) = a^{-1}bc \;.$$

This shows that χ_a and χ_b are C^∞ related, so the collection $\{\chi_a\}$ determines a C^∞ structure on G/H.

To show that $\pi\colon G \to G/H$ is C^∞ at a, we have to show that the map

$$(-\varepsilon,\varepsilon) \times \cdots \times (-\varepsilon,\varepsilon) \xrightarrow{L_a \circ x^{-1}} a\cdot U \xrightarrow{\pi} \pi(a\cdot U) \xrightarrow{\chi_a} C$$

is C^∞. This map is

$$(-\varepsilon,\varepsilon) \times \cdots \times (-\varepsilon,\varepsilon) \xrightarrow{L_a \circ x^{-1}} a\cdot U \xrightarrow{\pi} \pi(a\cdot U) \xrightarrow{\mathbf{L}_a{}^{-1}} \pi(U) \xrightarrow{(\pi|C)^{-1}} C \;,$$

which equals

$$(-\varepsilon,\varepsilon)\times\cdots\times(-\varepsilon,\varepsilon)\xrightarrow{x^{-1}} U \xrightarrow{\pi} \pi(U) \xrightarrow{(\pi|C)^{-1}} C\ ;$$

the latter map is just projection on the last $n-d$ coordinates.

To prove (ii), we note that

$$s = L_a \circ \chi_a\colon \pi(a\cdot U) \longrightarrow G$$

satisfies $\pi\circ s$ = identity.

Uniqueness is left to the reader. ■

The quotient topology on G/H has the property that $f\colon G/H \to X$ is continuous if and only if $f\circ\pi\colon G\to X$ is continuous. The C^∞ structure on G/H given by Proposition 27 now has the property that $f\colon G/H\to M$ is C^∞ if and only if $f\circ\pi\colon G\to M$ is C^∞. In fact, if $f\circ\pi$ is C^∞, and s is a C^∞ section on $\mathcal{U}\subset G/H$, then $f|\mathcal{U} = f\circ\pi\circ s$ is C^∞. It is also easy to see that the map

$$G\times G/H \longrightarrow G/H\ , \qquad (a,bH)\longmapsto abH$$

is C^∞: if $s\colon\mathcal{U}\to G$ is a section, then on $G\times\mathcal{U}$ this map equals

$$G\times\mathcal{U} \xrightarrow{\text{identity}\times s} G\times G \xrightarrow{\cdot} G \xrightarrow{\pi} G/H\ .$$

In particular, each $\mathbf{L}_a\colon G/H\to G/H$ is C^∞. Finally, we recall that in section 3 we defined a C^∞ structure on $\tilde{G}_n(\mathbb{R}^N)$ geometrically. It is easily checked that this C^∞ structure satisfies (i) and (ii) when we consider

$\tilde{G}_n(\mathbb{R}^N)$ as $SO(N)/\ SO(n) \times SO(N-n)$.

For the remainder of this section we will assume that G is a compact, connected Lie group, with Lie algebra $\mathfrak{g}$, and that H is a closed subgroup with Lie algebra $\mathfrak{h} \subset \mathfrak{g}$. Before we consider the cohomology of G/H, a few preliminaries are needed. Recall (Proposition I.10-20) that any left-invariant n-form σ^n on G is also right invariant. We will choose σ^n to be the unique bi-invariant n-form with $\int_G \sigma^n = 1$; as before, for a function $f: G \to \mathbb{R}$ we often write

$$\int_G f\sigma^n \quad \text{as} \quad \int_G f(a)\, da\ .$$

For every $a \in G$ we define the map $Ad(a)\colon \mathfrak{g} \to \mathfrak{g}$ by

$$Ad(a) = (L_a \circ R_{a^{-1}})_*\colon \mathfrak{g} \to \mathfrak{g}\ .$$

When G is a subgroup of $GL(n,\mathbb{R})$, so that $\mathfrak{g}$ is a subspace of $\mathfrak{gl}(n,\mathbb{R}) = n \times n$ matrices, we have the simple formula (Problem I.10-19 or p.II.351)

$$Ad(A)M = AMA^{-1} \qquad \text{for } A \in G,\ M \in \mathfrak{g}\ .$$

Finally, recall that there is a bi-invariant Riemannian metric $\langle\ ,\ \rangle$ on the compact Lie group G. When $G = O(n)$ we can describe such a metric explicitly as follows. For $M, P \in \mathfrak{o}(n) =$ skew-symmetric $n \times n$ matrices, let

$$\langle M,P\rangle = \text{Trace } MP^t = \sum_{i,j} M_{ij}P_{ij}\ .$$

For every $A \in O(n)$ we have

$$\begin{aligned}\langle \mathrm{Ad}(A)M, \mathrm{Ad}(A)P\rangle &= \text{Trace } AMA^{-1}\cdot(APA^{-1})^t \\ &= \text{Trace } A(MP^t)A^{-1} \\ &= \text{Trace } MP^t = \langle M,P\rangle .\end{aligned}$$

So if we extend $\langle\ ,\ \rangle$ to $O(n)$ by left-invariance, then it will also be right invariant.

Now consider a k-form ω on G/H. We say that ω is invariant if $\mathbf{L}_a{}^*\omega = \omega$ for all $a \in G$. For any k-form ω on G/H we can define a new k-form

$$\omega' = \int_G (a \mapsto \mathbf{L}_a{}^*\omega)\sigma^n = \int_G \mathbf{L}_a{}^*\omega \; da \;;$$

this equation really means that

$$\omega'(X_1,\ldots,X_k) = \int_G [a \mapsto \mathbf{L}_a{}^*\omega(X_1,\ldots,X_k)]\sigma^n = \int_G \mathbf{L}_a{}^*\omega(X_1,\ldots,X_k)\; da \;.$$

It follows easily from left-invariance of σ^n that ω' is invariant. More generally, given a smooth family $a \mapsto \eta_a$ of k-forms on G/H, where η_a is defined for all a in an open set $U \subset G$, we can form

$$\int_U (a \mapsto \eta_a)\sigma^n = \int_U \eta_a \; da \;,$$

which is a k-form on all of G/H. If $X_1,\ldots,X_k$, Y are vector fields on G/H, then

$$Y\left(\int_U \eta_a(X_1,\ldots,X_k)\; da\right) = \int_U Y(\eta_a(X_1,\ldots,X_k))\; da \;;$$

this follows from Proposition 9-10 when we choose an integral curve c for Y and let

$$\Gamma(u) = [a \mapsto \eta_a(X_1(c(u)),\ldots,X_k(c(u)))]\cdot\sigma^n .$$

Using Theorem I.7-13, we then see that

$$d\left(\int_U (a \mapsto \eta_a)\sigma^n\right) = \int_U (a \mapsto d\eta_a)\sigma^n ;$$

in simpler, but rather confusing notation, we have

$$d\left(\int_U \eta_a \, da\right) = \int_U d\eta_a \, da .$$

<u>28. PROPOSITION.</u> If ω is a closed k-form on G/H, and

$$\omega' = \int_G \mathbf{L}_a{}^*\omega \, da ,$$

then $\omega - \omega'$ is <u>exact</u>.

<u>Proof</u>. For a fixed $a \in G$, the map $\mathbf{L}_a \colon G/H \to G/H$ is smoothly homotopic to the identity map $\mathbf{L}_e \colon G/H \to G/H$. In fact, we can write $a = \exp X$ for some $X \in \mathfrak{g}$ (Problem I.10-27), and consider the smooth family of maps $\mathbf{L}_{\exp tX}$ from $\mathbf{L}_e$ to $\mathbf{L}_a$. It follows from Theorem I.8-13 that there is a $(k-1)$-form η_a with

$$(*) \qquad \omega - \mathbf{L}_a{}^*\omega = d\eta_a .$$

Moreover, the proof of this Theorem gives us an explicit formula for η_a.

Now let $E \subset \mathfrak{g}$ be an open set on which exp is a diffeomorphism. From the explicit description of η_a we see that η_a varies smoothly with a for all $a \in \exp(E)$. So we can integrate equation (*) over $\exp(E)$, to obtain

$$[\text{volume} \exp(E)]\cdot\omega - \int_{\exp(E)} \mathbf{L}_a{}^*\omega \, da = d\left(\int_{\exp(E)} \eta_a \, da\right).$$

(Notice that the three forms in this equation are each defined on all of G/H.) But by Theorem 8-30, we can choose E so that $G - \exp(E)$ has measure 0. Then our equation becomes

$$\omega - \int_G \mathbf{L}_a{}^*\omega \, da = d\left(\int_{\exp(E)} \eta_a \, da\right). \quad \blacksquare$$

On the other hand, it is even easier to show

<u>29. PROPOSITION</u>. If ω is a form of G/H which is invariant and exact, then ω is actually the exterior derivative of some invariant form.

<u>Proof</u>. If $\omega = d\eta$, then

$$\omega = \int_G \mathbf{L}_a{}^*\omega \, da = \int_G \mathbf{L}_a{}^*(d\eta) \, da$$

$$= d\left(\int_G \mathbf{L}_a{}^*\eta \, da\right) = d\eta',$$

where η' is invariant. $\blacksquare$

From Propositions 28 and 29 we immediately have

30. THEOREM. The k-dimensional de Rham cohomology $H^k(G/H)$ of G/H is naturally isomorphic to

$$H^k(G/H) \approx \frac{\text{closed invariant } k\text{-forms on } G/H}{\{d\eta\colon \eta \text{ an invariant } (k-1)\text{-form on } G/H\}}\,.$$

The cup product in $H^*(G/H)$ corresponds to $\wedge$ under this isomorphism.

What really makes this result important is the fact that we can describe the invariant k-forms on G/H in terms of the left-invariant forms on G. Note that the map $\operatorname{Ad}(a)\colon \mathfrak{g} \to \mathfrak{g}$ induces maps $\operatorname{Ad}(a)^*\colon \Omega^k(\mathfrak{g}) \to \Omega^k(\mathfrak{g})$; the map $\operatorname{Ad}(a)^*$ is just $(L_a \circ R_{a^{-1}})^*$ at e. A k-form η on G will be called $\operatorname{Ad}(H)$-invariant if

$$\operatorname{Ad}(a)^*\eta(e) = \eta(e) \qquad \text{for all } a \in H\,.$$

We say that η annihilates $\mathfrak{h}$ if $\eta(e)(X_1,\ldots,X_k) = 0$ whenever some $X_i \in \mathfrak{h}$.

31. LEMMA. If $\pi\colon G \to G/H$ is the natural projection, then the map $\omega \mapsto \pi^*\omega$ is a one-one correspondence between the invariant k-forms on G/H and the left-invariant, $\operatorname{Ad}(H)$-invariant k-forms on G which annihilate $\mathfrak{h}$.

Proof. If ω is invariant, then for all $a \in G$ we have

$$L_a^*\pi^*\omega = \pi^* L_a^*\omega = \pi^*\omega\,,$$

so $\pi^*\omega$ is left-invariant. If $X_1,\ldots,X_k \in \mathfrak{g}$ and some $X_i \in \mathfrak{h}$, then $\pi_* X_i = 0$, so

$$(\pi^*\omega)(X_1,\ldots,X_k) = \omega(\pi_* X_1,\ldots,\pi_* X_i,\ldots,\pi_* X_k) = 0 ;$$

thus $\pi^*\omega$ annihilates $\mathfrak{h}$. Finally, if $a \in H$, then the map

$$\pi \circ L_a \circ R_{a^{-1}} \colon G \longrightarrow G/H$$

takes

$$b \longmapsto aba^{-1}H = abH = \mathbf{L}_a(bH) = \mathbf{L}_a(\pi(b)) ,$$

so we have

$$(1) \qquad \pi \circ L_a \circ R_{a^{-1}} = \mathbf{L}_a \circ \pi ;$$

consequently,

$$(L_a \circ R_{a^{-1}})^* \pi^*\omega = \pi^* \mathbf{L}_a{}^*\omega = \pi^*\omega ,$$

so $\pi^*\omega$ is Ad(H)-invariant.

Conversely, suppose that the k-form η on G is left-invariant, Ad(H)-invariant, and annihilates $\mathfrak{h}$. The map $\pi_*\colon \mathfrak{g} \longrightarrow (G/H)_H$ from $\mathfrak{g}$ to the tangent space of G/H at the coset H has kernel precisely $\mathfrak{h}$, and therefore induces an isomorphism

$$\pi_*\colon \mathfrak{g}/\mathfrak{h} \longrightarrow (G/H)_H .$$

We can consider $\eta(e) \in \Omega^k(\mathfrak{g}/\mathfrak{h})$, since η annihilates $\mathfrak{h}$, so there is a unique $\omega(H) \in \Omega^k((G/H)_H)$ with

$$(2) \qquad \pi^*(\omega(H)) = \eta(e) \in \Omega^k(\mathfrak{g}/\mathfrak{h}) .$$

Define ω on G/H by

$$\omega(aH) = \mathbf{L}_{a^{-1}}{}^*\omega(H) .$$

To show that ω is well-defined, consider $a, b \in G$ with $aH = bH$. Then $c = a^{-1}b \in H$. So by (1),

$$\begin{aligned} \pi^* \mathbf{L}_c{}^*\omega(H) &= (L_c \circ R_{c^{-1}})^*\pi^*\omega(H) \\ &= (L_c \circ R_{c^{-1}})^*\eta(e) \\ &= \eta(e) , \end{aligned}$$

since η is $\mathrm{Ad}(H)$-invariant. Since $\omega(H)$ is the unique element satisfying (2), we conclude that

$$\begin{aligned} \omega(H) = \mathbf{L}_c{}^*\omega(H) &= (\mathbf{L}_{a^{-1}} \circ \mathbf{L}_b)^*\omega(H) \\ &= \mathbf{L}_b{}^*(\mathbf{L}_{a^{-1}}{}^*\omega(H)) , \end{aligned}$$

and hence

$$\mathbf{L}_{b^{-1}}{}^*\omega(H) = \mathbf{L}_{a^{-1}}{}^*\omega(H) ,$$

as desired. Clearly ω is invariant. Moreover, for all $a \in G$ we have

$$\begin{aligned}
\eta(a) &= L_{a^{-1}}{}^*\eta(e) \qquad \text{since } \eta \text{ is left-invariant} \\
&= L_{a^{-1}}{}^*\pi^*\omega(H) \\
&= \pi^* L_{a^{-1}}{}^*\omega(H) \\
&= \pi^*(\omega(aH)) \\
&= (\pi^*\omega)(a) . \quad \blacksquare
\end{aligned}$$

Notice that in the first part of the proof we showed that for all $a \in H$, the given ω satisfies $(L_a \circ R_{a^{-1}})^*\pi^*\omega = \pi^*\omega$ at all points, not just at e. Since any left invariant form η on G which annihilates $\mathfrak{h}$ and satisfies $(L_a \circ R_{a^{-1}})^*\eta(e) = \eta(e)$ is $\pi^*\omega$ for an invariant form ω on G/H, it follows that such an η satisfies $(L_a \circ R_{a^{-1}})^*\eta = \eta$ at all points, for all $a \in H$; this conclusion does not follow just from the fact that η is left-invariant -- we need to know that η annihilates $\mathfrak{h}$. Similarly, but more important, since $d\omega$ is clearly invariant for all invariant ω on G/H, it follows that if η is left-invariant, $\text{Ad}(H)$-invariant, and annihilates $\mathfrak{h}$, then $d\eta$ has these same properties [but the fact that η annihilates $\mathfrak{h}$, for example, does not by itself imply that $d\eta$ annihilates $\mathfrak{h}$].

We are now ready to give a completely algebraic description of $H^k(G/H)$. Notice first that if ω is a left-invariant k-form, and $X_1,\dots,X_{k+1} \in \mathfrak{g}$, then $d\omega(X_1,\dots,X_{k+1})$ can be computed by applying Theorem I.7-13 to the left-invariant vector fields $\tilde{X}_i$ extending X_i. The terms $\omega(\tilde{X}_1,\dots,\hat{\tilde{X}}_i,\dots,\tilde{X}_{k+1})$ are all constant by left-invariance, so our formula becomes simply

$$(*) \quad d\omega(X_1,\dots,X_{k+1}) = \sum_{i<j} (-1)^{i+j}\omega([X_i,X_j],X_1,\dots,\hat{X}_i,\dots,\hat{X}_j,\dots,X_{k+1}) ,$$

which involves only the bracket operation in $\mathfrak{g}$. Now let

$$\Omega^k(\mathfrak{g}/\mathfrak{h}) = \{\omega \in \Omega^k(\mathfrak{g}) : \omega(X_1,\ldots,X_k) = 0 \text{ if some } X_i \in \mathfrak{h}\} .$$

The elements of $\Omega^k(\mathfrak{g}/\mathfrak{h})$ are clearly in one-one correspondence with the left-invariant forms on G which annihilate $\mathfrak{h}$. If $a \in H$, and $X_i \in \mathfrak{h}$, then $Ad(a)X_i \in \mathfrak{h}$, since $L_a \circ R_{a^{-1}} : H \to H$; this shows that

$$Ad(a)^* : \Omega^k(\mathfrak{g}/\mathfrak{h}) \to \Omega^k(\mathfrak{g}/\mathfrak{h}) \qquad \text{for all } a \in H .$$

Let

$$\Omega^k(\mathfrak{g}/\mathfrak{h})^H = \{\omega \in \Omega^k(\mathfrak{g}/\mathfrak{h}) : Ad(a)^*\omega = \omega \text{ for all } a \in H\} .$$

The remarks in the previous paragraph show that

$$d : \Omega^k(\mathfrak{g}/\mathfrak{h})^H \to \Omega^{k+1}(\mathfrak{g}/\mathfrak{h})^H ,$$

where d is now defined by (*). Moreover, Theorem 30 shows that

$$H^k(G/H) \approx \frac{\text{kernel } d : \Omega^k(\mathfrak{g}/\mathfrak{h})^H \to \Omega^{k+1}(\mathfrak{g}/\mathfrak{h})^H}{d(\Omega^{k-1}(\mathfrak{g}/\mathfrak{h})^H)} .$$

Even this description of $H^k(G/H)$ can be simplified. Notice that if $\mathfrak{h}' \subset \mathfrak{g}$ is a subspace with $\mathfrak{g} = \mathfrak{h} \oplus \mathfrak{h}'$, then the elements of $\Omega^k(\mathfrak{h}')$ are in one-one correspondence with the elements of $\Omega^k(\mathfrak{g}/\mathfrak{h})$: given $\omega \in \Omega^k(\mathfrak{h}')$, we define the corresponding $\bar{\omega} \in \Omega^k(\mathfrak{g}/\mathfrak{h})$ by

$$\bar{\omega}(X_1,\ldots,X_k) = \omega(\mathfrak{h}' \text{ component of } X_1,\ldots, \mathfrak{h}' \text{ component of } X_k) .$$

In particular, consider the orthogonal complement $\mathfrak{h}^{\perp} \subset \mathfrak{g}$ of $\mathfrak{h}$ with respect to $\langle\ ,\ \rangle_e$, where $\langle\ \rangle$ is a bi-invariant metric on G. Since each L_a and R_a is an isometry of $(G, \langle\ ,\ \rangle)$, the map $Ad(a)\colon \mathfrak{g} \to \mathfrak{g}$ is an isometry with respect to $\langle\ ,\ \rangle_e$. Since $Ad(a)\colon \mathfrak{h} \to \mathfrak{h}$ for $a \in H$, we also have $Ad(a)\colon \mathfrak{h}^{\perp} \to \mathfrak{h}^{\perp}$ for $a \in H$, and hence

$$Ad(a)^*\colon \Omega^k(\mathfrak{h}^{\perp}) \to \Omega^k(\mathfrak{h}^{\perp}) \qquad \text{for all } a \in H\,.$$

If we define

$$\Omega^k(\mathfrak{h}^{\perp})^H = \{\omega \in \Omega^k(\mathfrak{h}^{\perp})\colon Ad(a)^*\omega = \omega \text{ for all } a \in H\}\,,$$

then the elements of $\Omega^k(\mathfrak{h}^{\perp})^H$ are in one-one correspondence with the elements of $\Omega^k(\mathfrak{g}/\mathfrak{h})^H$. Putting all this information together, we have finally

32. THEOREM. Let G be a compact connected Lie group, and H a closed subgroup. Then the k-dimensional de Rham cohomology $H^k(G/H)$ of G/H is naturally isomorphic to

$$H^k(G/H) \approx \frac{\text{kernel } d\colon \Omega^k(\mathfrak{h}^{\perp})^H \to \Omega^{k+1}(\mathfrak{h}^{\perp})^H}{d(\Omega^{k-1}(\mathfrak{h}^{\perp})^H)}\,,$$

where

$\mathfrak{h}^{\perp} \subset \mathfrak{g}$ is the orthogonal complement of $\mathfrak{h}$ with respect to a bi-invariant metric,

and

$$\Omega^k(\mathfrak{h}^{\perp})^H = \{\omega \in \Omega^k(\mathfrak{h}^{\perp})\colon Ad(a)^*\omega = \omega \text{ for all } a \in H\}\,,$$

and d is defined by

$$d\omega(X_1,\dots,X_{k+1})$$

$$= \sum_{i<j} (-1)^{i+j}\omega(\mathfrak{h}^{\perp} \text{ component of } [X_i,X_j],\ X_1,\dots,\hat{X}_i,\dots,\hat{X}_j,\dots,X_{k+1}) ,$$

$$\text{for } X_1,\dots,X_{k+1} \in \mathfrak{h}^{\perp} .$$

The cup product in $H^*(G/H)$ corresponds to $\wedge$ under this isomorphism.

Naturally, the simplest applications of Theorem 32 will occur when H is a large subgroup, so that $\mathfrak{h}^{\perp}$ is small. As an example, we consider $G = SO(n+1)$, with $H = SO(n) \subset SO(n+1)$, so that $G/H = \tilde{G}_n(\mathbb{R}^{n+1}) \approx \tilde{G}_1(\mathbb{R}^{n+1})$ is S^n. In this case, where the geometry is so simple, it is easiest to use Theorem 30 directly. It tells us that in computing $H^k(S^n)$, it suffices to consider k-forms ω which are invariant under the action of $SO(n+1)$. In particular, at any point $p \in S^n$, the function $\omega(p) \in \Omega^k(S^n{}_p)$ must be invariant under any linear transformation $A\colon S^n{}_p \longrightarrow S^n{}_p$ which is special orthogonal with respect to the usual inner product on $S^n{}_p$. Now if $0 < k < n$ and $X_1,\dots,X_k \in S^n{}_p$ are orthonormal, then there is an A of this sort with

$$A(X_1) = X_2 , \qquad A(X_2) = X_1 , \qquad A(X_i) = X_i \qquad i = 3,\dots,k .$$

Consequently

$$\omega(p)(X_1,\dots,X_k) = \omega(p)(X_2,X_1,\dots,X_k) ,$$

so $\omega(p)(X_1,\dots,X_k) = 0$. This implies that $\omega(p) = 0$. Hence $H^k(S^n) = 0$ for $0 < k < n$. For $k = n$, we can choose $\omega(p)$ to be a multiple of the volume element $\sigma(p)$ of S^n at p. Since ω must also be invariant under special orthogonal maps taking p to any other point $q \in S^n$, we see that ω must be a constant multiple of the volume element σ. We have $d\sigma = 0$ automatically, and since there are no invariant $(k-1)$-forms, we see that $H^n(S^n) \approx \mathbb{R}$.

It will also be instructive to see what happens when we do not rely on the geometry, and use Theorem 32. The Lie algebra $\mathfrak{g} = \mathfrak{o}(n+1)$ is the set of all skew-symmetric $(n+1) \times (n+1)$ matrices, while $\mathfrak{h}$ consists of those of the form

$$\begin{pmatrix} \boxed{\;M\;} & \begin{matrix} 0 \\ \vdots \\ 0 \end{matrix} \\ 0 \cdots 0 & 0 \end{pmatrix}, \qquad M \in \mathfrak{o}(n) .$$

For the bi-invariant metric $\langle M,N\rangle = \text{Trace } MN^t = \sum_{i,j} M_{ij}N_{ij}$, the orthogonal complement $\mathfrak{h}^{\perp}$ is spanned by the n matrices

$$Y_1 = \begin{pmatrix} \boxed{\;\bigcirc\;} & \begin{matrix} 1 \\ 0 \\ \vdots \\ 0 \end{matrix} \\ -1 \; 0 \cdots & 0 \end{pmatrix}, \ldots, Y_n = \begin{pmatrix} \boxed{\;\bigcirc\;} & \begin{matrix} 0 \\ \vdots \\ 1 \end{matrix} \\ 0 \cdots -1 & 0 \end{pmatrix} .$$

For any matrix $\tilde{A} \in H$ of the form

$$\tilde{A} = \begin{pmatrix} \boxed{\;A\;} & \begin{matrix} 0 \\ \vdots \\ 0 \end{matrix} \\ 0 \cdots 0 & 1 \end{pmatrix}, \qquad A \in SO(n) ,$$

we compute that

$$\tilde{A}Y_i\tilde{A}^{-1} = \begin{pmatrix} \boxed{\;A\;} & \begin{matrix}0\\ \vdots\\ 0\end{matrix} \\ 0\ \cdots\ 0 & 1 \end{pmatrix} \begin{pmatrix} \boxed{\;\bigcirc\;} & \begin{matrix}0\\ \vdots\\ 1\\ \vdots\\ 0\end{matrix} \\ 0\ \cdots\ -1\ \cdots & 0 \end{pmatrix} \begin{pmatrix} \boxed{\;A^t\;} & \begin{matrix}0\\ \vdots\\ 0\end{matrix} \\ 0\ \cdots\ 0 & 1 \end{pmatrix}$$

$$= \begin{pmatrix} \boxed{\;A\;} & \begin{matrix}0\\ \vdots\\ 0\end{matrix} \\ 0\ \cdots\ 0 & 1 \end{pmatrix} \begin{pmatrix} \boxed{\;\bigcirc\;} & \begin{matrix}0\\ \vdots\\ 1\\ \vdots\\ 0\end{matrix} \\ -a_{1i}\ \cdots\ -a_{ni} & 0 \end{pmatrix}$$

$$= \begin{pmatrix} \boxed{\;\bigcirc\;} & \begin{matrix}a_{1i}\\ \vdots\\ a_{ni}\end{matrix} \\ -a_{1i}\ \cdots\ -a_{ni} & 0 \end{pmatrix}$$

$$= \sum_{j=1}^{n} a_{ji}Y_j \ .$$

So if we regard the Y_i simply as vectors in $\mathbb{R}^n$, then the adjoint action $\mathrm{Ad}(\tilde{A})$ on the Y_i is just the usual action of the orthogonal matrix A on the vectors Y_i. As we saw in the previous paragraph, this means that

$$\Omega^k(\mathfrak{h}^\perp)^H = \begin{cases} 0 & 0 < k < n \\ \mathbb{R} & k = n \ ; \end{cases}$$

hence $H^k(G/H) = 0$ for $0 < k < n$, and $H^n(G/H) \approx \mathbb{R}$.

7. A Smattering of Classical Invariant Theory

The simple algebraic considerations used at the conclusion of the last section won't get us very far when we replace the subgroup $H = SO(n) \subset SO(n+1)$ by a smaller subgroup. In order to analyse the more general situation in an effective way, we need to delve briefly into classical invariant theory, which was once considered the cornerstone of all mathematics, and then rapidly dwindled to a state of near extinction, although recently it has excited new interest.

As an example of the sort of question that arises in invariant theory, we consider a standard fact from algebra, to which we have already alluded on occasion. A function

$$f\colon \underbrace{\mathbb{R}\times\cdots\times\mathbb{R}}_{m} \to \mathbb{R}$$

is _symmetric_ if

$$f(x_1,\ldots,x_m) = f(x_{\pi(1)},\ldots,x_{\pi(m)})$$

for all permutations $\pi \in S_m$. Alternatively, if we define an operation of S_m on $\mathbb{R}\times\cdots\times\mathbb{R}$ by

$$\pi\cdot(x_1,\ldots,x_m) = (x_{\pi(1)},\ldots,x_{\pi(m)}) ,$$

then f is symmetric if and only if

$$f(\pi\cdot x) = f(x) \qquad \text{for all } x \in \mathbb{R}\times\cdots\times\mathbb{R} \text{ and all } \pi \in S_m .$$

As examples of symmetric functions we have the "elementary symmetric functions"

$$\sigma_1(x_1,\ldots,x_m) = \sum_{i=1}^{m} x_i\,, \qquad \sigma_2(x_1,\ldots,x_m) = \sum_{i<j} x_i x_j$$

$$\ldots$$

$$\sigma_m(x_1,\ldots,x_m) = x_1\cdots x_m\,;$$

and for all $x, y \in \mathbb{R}\times\cdots\times\mathbb{R}$ we have $y = \pi\cdot x$ for some $\pi \in S_m$ if and only if $\sigma_i(x) = \sigma_i(y)$ for all i (compare p.IV.96). From this we see immediately that any symmetric f can be written

$$f(x_1,\ldots,x_m) = F(\sigma_1(x_1,\ldots,x_m),\ldots,\sigma_m(x_1,\ldots,x_m))$$

for some function F. Indeed, we can define

$$F(s_1,\ldots,s_m) = f(x_1,\ldots,x_m) \qquad \text{for any } x_1,\ldots,x_m \text{ with } \sigma_i(x_1,\ldots,x_m) = s_i$$

(such $x_1,\ldots,x_m$ certainly exist: we can take $x_1,\ldots,x_m$ to be the roots of the polynomial

$$x^n - s_1x^{n-1} + \cdots + (-1)^n s_n = 0\ ).$$

On the other hand, if f is a polynomial, then it is by no means so evident that we can choose F to be a polynomial; the argument which establishes this fact involves a slightly delicate induction, and can be found in any standard algebra course.

Note, by the way, that the polynomials $\sigma_1,\ldots,\sigma_m$ are algebraically independent -- if p is any polynomial with

$$p(\sigma_1(x_1,\ldots,x_m),\ldots,\sigma_m(x_1,\ldots,x_m)) = 0$$

for all $x_1,\ldots,x_m$, then $p = 0$. In fact, this equation implies that

$$p(s_1,\ldots,s_m) = 0$$

for all $s_1,\ldots,s_m$, and hence that $p = 0$.

Now consider a function

$$f\colon \underbrace{\mathbb{R}^n \times \cdots \times \mathbb{R}^n}_{m \text{ times}} \to \mathbb{R} ,$$

which we will often describe as a "function of m vectors in $\mathbb{R}^n$". A typical element of $\mathbb{R}^n \times \cdots \times \mathbb{R}^n$ will be an m-tuple of vectors $(v_1,\ldots,v_m)$, and each v_r is an n-tuple $v_{r1},\ldots,v_{rn}$. We say that a function f of m vectors in $\mathbb{R}^n$ is <u>invariant under</u> $O(n)$ if

$$f(v_1,\ldots,v_m) = f(A(v_1),\ldots,A(v_m))$$

for all $v_1,\ldots,v_m \in \mathbb{R}^n$ and all $A \in O(n)$. Alternatively, if we define an action of $O(n)$ on $\mathbb{R}^n \times \cdots \times \mathbb{R}^n$ by

$$A\cdot(v_1,\ldots,v_m) = (A(v_1),\ldots,A(v_m)) ,$$

then f is invariant under $O(n)$ if and only if

$$f(A\cdot v) = f(v) \qquad \text{for all } v \in \mathbb{R}^n \times \cdots \times \mathbb{R}^n \text{ and all } A \in O(n) .$$

Similarly, we can consider functions invariant under any subgroup of $GL(n,\mathbb{R})$. In writing $A(v_r)$, we are considering an $n \times n$ matrix $A = (a_{ij})$ as a linear transformation $A\colon \mathbb{R}^n \to \mathbb{R}^n$, by the rule

$$A(e_i) = \sum_{j=1}^{n} a_{ji} e_j .$$

Since we are regarding $v_r = (v_{r1},\dots,v_{rn}) = \sum_h v_{rh}e_h$ as a row vector, this means that we have

$$A(v_r) = \sum_h v_{rh}A(e_h) = \sum_{h,j} v_{rh}a_{jh}e_j$$
$$= (\sum_h v_{rh}a_{1h},\dots,\sum_h v_{rh}a_{nh}) ,$$

and consequently

$$A(v_r) = v_r\cdot A^t$$ [the product of the $1\times n$ matrix v_r with the $n\times n$ matrix A^t],

which is slightly unpleasant, but something we can live with.

If $v, w \in \mathbb{R}^n \times \cdots \times \mathbb{R}^n$ are m-tuples of vectors with $<v_r,v_s> = <w_r,w_s>$ for all r, s, then there is $A \in O(n)$ with $w = A\cdot v$. It follows immediately that every function f of m vectors in $\mathbb{R}^n$ which is invariant under $O(n)$ can be written as

$$f(v_1,\dots,v_m) = F(<v_1,v_1>,\dots,<v_m,v_m>)$$

for some function F. For brevity, we will also write

$$f(v_1,\dots,v_m) = F(\{<v_r,v_s>\}) ,$$

and if we introduce the inner product functions

$$\iota_{rs}(v_1,\dots,v_m) = <v_r,v_s> ,$$

then we can write

$$f = F\circ(\{\iota_{rs}\}) .$$

From this general, and trivial result, however, it does not follow that every polynomial function of m vectors in $\mathbb{R}^n$ can be written as a polynomial in the ι_{rs} (a function $f\colon \mathbb{R}^n \times \cdots \times \mathbb{R}^n \to \mathbb{R}$ is a polynomial function if $f(\Sigma\, a_1^i e_i, \ldots, \Sigma\, a_m^i e_i)$ is a polynomial in the a_j^i). This deeper algebraic result is the content of the "first main theorem of invariant theory for $O(n)$". In order to prove this result, as well as the corresponding result for $SO(n)$, we will follow the classical route, which will get us to our destination in the shortest time, although it involves some unpleasant calculations, and uses some mysterious identities.

First, some preliminaries about polynomial functions f of m vectors in $\mathbb{R}^n$. We say that f is homogeneous of degree $(\alpha_1,\ldots,\alpha_m)$ if

$$f(\lambda_1 v_1,\ldots,\lambda_m v_m) = \lambda_1^{\alpha_1}\cdots\lambda_m^{\alpha_m} f(v_1,\ldots,v_m) .$$

Every polynomial function f can be written

$$(*) \qquad f = \sum_{(\alpha_1,\ldots,\alpha_m)} f_{\alpha_1,\ldots,\alpha_m} ,$$

where $f_{\alpha_1,\ldots,\alpha_m}$ is homogeneous of degree $(\alpha_1,\ldots,\alpha_m)$. For example, we might have $f\colon \mathbb{R}^2 \times \mathbb{R}^2 \to \mathbb{R}$ with

$$f(x_1,x_2,y_1,y_2) = \underbrace{3x_1{}^2x_2y_1{}^2 + 7x_1x_2{}^2y_1y_2}_{\text{degree }(3,2)} + \underbrace{8x_1{}^3}_{\text{degree }(3,0)} + \underbrace{7x_2{}^4}_{\text{degree }(4,0)} .$$

If is easy to see that the expression $(*)$ is unique. Moreover, if f is homogeneous of degree $(\alpha_1,\ldots,\alpha_m)$, then so is

$$(v_1,\ldots,v_m) \mapsto f(A\cdot v_1,\ldots,A\cdot v_m)$$

for any linear transformation A. From this we easily see that if f is invariant under any group of linear transformations, then so is $f_{\alpha_1,\ldots,\alpha_m}$. So we will henceforth consider only homogeneous polynomial functions.

Notice that if $f\colon \mathbb{R}^n \to \mathbb{R}$ satisfies $f(\lambda v) = \lambda^k f(v)$, then

$$k\lambda^{k-1}f(v) = \frac{d}{d\lambda}\lambda^k f(v) = \frac{d}{d\lambda} f(\lambda v) = \sum_{i=1}^n v_i \frac{\partial f}{\partial x_i}(\lambda v) .$$

In particular, for $\lambda = 1$ we obtain Euler's Theorem

$$kf = \sum_{i=1}^n v_i \frac{\partial f}{\partial x^i} .$$

(In the case of a polynomial function $f\colon \mathbb{R}^n \to \mathbb{R}$, this result can be verified directly.) Naturally, there is an analogous result for homogeneous functions of several vectors in $\mathbb{R}^n$.

Now let

$$e_{ri} = (0,\ldots,\underset{\substack{\uparrow \\ r^{\text{th}} \text{ place}}}{e_i},\ldots,0) \in \mathbb{R}^n\times\cdots\times\mathbb{R}^n ,$$

where 0 denotes the zero vector of $\mathbb{R}^n$, and e_i is the ith standard basis vector of $\mathbb{R}^n$. The e_{ri} form the standard basis for $\mathbb{R}^n\times\cdots\times\mathbb{R}^n$ when we identify it with $\mathbb{R}^{nm}$, so for a function

$$f\colon \underbrace{\mathbb{R}^n\times\cdots\times\mathbb{R}^n}_{m \text{ times}} \to \mathbb{R}$$

we can consider the partial derivatives

$$\frac{\partial f}{\partial e_{ri}} ;$$

these partial derivatives certainly exist if f is a polynomial function. Now for $1 \le r, s \le m$ we can consider the function

$$(D_{sr} f)(v_1, \ldots, v_m) = \sum_{i=1}^{n} v_{si} \frac{\partial f}{\partial e_{ri}}(v_1, \ldots, v_m) ;$$

in terms of the dual basis $\{\phi^{ri}\}$ to the $\{e_{ri}\}$, we can write

$$D_{sr} = \sum_{i=1}^{n} \phi^{si} \cdot \frac{\partial}{\partial e_{ri}} .$$

For example, if we denote a typical element of $\mathbb{R}^n \times \mathbb{R}^n$ by $(x_1, \ldots, x_n, y_1, \ldots, y_n)$, then for $f \colon \mathbb{R}^n \times \mathbb{R}^n \to \mathbb{R}$ we have

$$D_{21} f(x_1, \ldots, x_n, y_1, \ldots, y_n) = \sum_{i=1}^{n} y_i \frac{\partial f}{\partial x_i}(x_1, \ldots, x_n, y_1, \ldots, y_n) .$$

The operator D_{sr} is called a polarization. It is important for the following reason.

33. LEMMA. Suppose that

$$f(v_1, \ldots, v_m) = f(A(v_1), \ldots, A(v_m))$$

for all $v_1, \ldots, v_m \in \mathbb{R}^n$ and some linear transformation A. Then also

$$D_{sr} f(v_1,\ldots,v_m) = D_{sr} f(A(v_1),\ldots,A(v_m))$$

for all $v_1,\ldots,v_m \in \mathbb{R}^n$.

Proof. If (a_{ij}) is the matrix of A, then by hypothesis we have

$$f(\ldots,(v_{r1},\ldots,v_{rn}),\ldots) = f(\ldots,(\sum_{j=1}^{n} a_{1j}v_{rj},\ldots,\sum_{j=1}^{n} a_{nj}v_{rj}),\ldots) ,$$

which implies that

$$\frac{\partial f}{\partial e_{ri}}(v_1,\ldots,v_m) = \sum_{k=1}^{n} \frac{\partial f}{\partial e_{rk}}(\ldots)\, a_{ki} .$$

Therefore

$$\sum_{i=1}^{n} v_{si} \frac{\partial f}{\partial e_{ri}}(v_1,\ldots,v_m) = \sum_{k=1}^{n} \sum_{i=1}^{n} a_{ki}v_{si} \frac{\partial f}{\partial e_{rk}}(\ldots)$$

$$= \sum_{k=1}^{n} A(v_s)_k \frac{\partial f}{\partial e_{rk}}(A(v_1),\ldots,A(v_m)) . \blacksquare$$

In particular, we see that polarization takes polynomial functions invariant under a group of matrices into polynomial functions with the same property. Note that if f is homogeneous of degree $(\alpha_1,\ldots,\alpha_m)$, then Euler's theorem implies that $D_{rr}f = \alpha_r f$. Note also that polarizations take the inner product functions into sums of such:

$$D_{sr}\iota_{qp} = \delta_{rq}\iota_{sq} + \delta_{rp}\iota_{sp} .$$

Finally, consider a determinant function

$$(v_1,\ldots,v_m) \mapsto \det\begin{pmatrix} v_{r_1} \\ \vdots \\ v_{r_n} \end{pmatrix},$$

which we will denote by $\det_{r_1\cdots r_n}$. We clearly have

$$D_{sr_i} \det_{r_1\cdots r_n} = \det_{r_1\cdots r_{i-1},s,r_{i+1}\cdots r_n}$$

$$D_{sr} \det_{r_1\cdots r_n} = 0 \qquad \text{if} \quad r \neq r_1,\ldots,r_n .$$

Now we want to look at the result of composing two or more polarizations. It would be nice if

$$D_{s_2r_2} D_{s_1r_1} f \overset{(?)}{=} \sum_{i_1,i_2} \phi^{s_1i_1} \phi^{s_2i_2} \frac{\partial^2 f}{\partial e_{r_1i_1} \partial e_{r_2i_2}} .$$

But this holds only when r_2, $s_2 \neq s_1$. It will be convenient, however, to denote the right side of the above equation by a symbol that looks like a composition, even in the case where r_2 or $s_2 = s_1$. So we will use the symbol Δ_{sr} for the same operator as D_{sr}, but we will define the operator $\Delta_{s_2r_2}\Delta_{s_1r_1}$ not as a composition, but formally by

$$\Delta_{s_2r_2}\Delta_{s_1r_1} = \sum_{i_1,i_2} \phi^{s_1i_1} \phi^{s_2i_2} \frac{\partial^2}{\partial e_{r_1i_1} \partial e_{r_2i_2}} ;$$

the operators $\Delta_{s_3r_3}\Delta_{s_2r_2}\Delta_{s_1r_1}$, etc., are defined similarly. As for the actual composition $D_{s_2r_2}D_{s_1r_1}$ we have

$$D_{s_2r_2}D_{s_1r_1}f = \sum_{i_2}\phi^{s_2i_2}\frac{\partial}{\partial e_{r_2i_2}}\left(\sum_{i_1}\phi^{s_1i_1}\frac{\partial f}{\partial e_{r_1i_1}}\right)$$

$$= \sum_{i_1,i_2}\phi^{s_2i_2}\phi^{s_1i_1}\frac{\partial^2 f}{\partial e_{r_2i_2}\partial e_{r_1i_1}}$$

$$+ \delta_{s_1}^{r_2}\sum_{i_2}\phi^{s_2i_2}\frac{\partial f}{\partial e_{r_1i_2}},$$

which shows that

$$D_{s_2r_2}D_{s_1r_1} = \Delta_{s_2r_2}\Delta_{s_1r_1} + \delta_{s_1}^{r_2}\Delta_{s_2r_1}. \tag{1}$$

Now consider the operator

$$\det\begin{pmatrix} D_{11} & D_{12} \\ D_{21} & D_{22}\end{pmatrix} = \text{by definition } D_{11}D_{22} - D_{21}D_{12}.$$

From (1) we have

$$D_{11}D_{22} = \Delta_{11}\Delta_{22}, \qquad D_{21}D_{12} = \Delta_{21}\Delta_{12} + \Delta_{22},$$

so

$$D_{11}D_{22} - D_{21}D_{12} + D_{22} = \Delta_{11}\Delta_{22} - \Delta_{21}\Delta_{12}.$$

We used the indices 1 and 2 for convenience, but we clearly have the same result for any distinct indices α_1, α_2. We can write our equation as

$$\det\begin{pmatrix} D_{\alpha_1\alpha_1}+1 & D_{\alpha_1\alpha_2} \\ D_{\alpha_2\alpha_1} & D_{\alpha_2\alpha_2}\end{pmatrix} = \det\begin{pmatrix} \Delta_{\alpha_1\alpha_1} & \Delta_{\alpha_1\alpha_2} \\ \Delta_{\alpha_2\alpha_1} & \Delta_{\alpha_2\alpha_2}\end{pmatrix}. \tag{2}$$

Remarkably enough, this equation can be generalized. First we compute that

$$D_{s_3r_3}(\Delta_{s_2r_2}\Delta_{s_1r_1})f = \sum_{i_3} \phi^{s_3i_3} \frac{\partial}{\partial e_{r_3i_3}}\left(\sum_{i_1,i_2} \phi^{s_2i_2}\phi^{s_1i_1} \frac{\partial^2 f}{\partial e_{r_2i_2}\partial e_{r_1i_1}}\right)$$

$$= \sum_{i_1,i_2,i_3} \phi^{s_3i_3}\phi^{s_2i_2}\phi^{s_1i_1} \frac{\partial^3 f}{\partial e_{r_3i_3}\partial e_{r_2i_2}\partial e_{r_1i_1}}$$

$$+ \delta^{s_1}_{r_3} \sum_{i_2,i_3} \phi^{s_3i_3}\phi^{s_2i_2} \frac{\partial^2 f}{\partial e_{r_2i_2}\partial e_{r_1i_3}}$$

$$+ \delta^{s_2}_{r_3} \sum_{i_1,i_3} \phi^{s_3i_3}\phi^{s_1i_1} \frac{\partial^2 f}{\partial e_{r_2i_3}\partial e_{r_1i_1}}$$

(this formula works even if $s_1 = s_2$). Consequently,

$$(3) \quad D_{s_3r_3}\circ(\Delta_{s_2r_2}\Delta_{s_1r_1}) = \Delta_{s_3r_3}\Delta_{s_2r_2}\Delta_{s_1r_1} + \delta^{s_1}_{r_3}\Delta_{s_3r_1}\Delta_{s_2r_2} + \delta^{s_2}_{r_3}\Delta_{s_3r_2}\Delta_{s_1r_1} .$$

Now we consider the operator

$$\det\begin{pmatrix} D_{11} & \Delta_{12} & \Delta_{13} \\ D_{21} & \Delta_{22} & \Delta_{23} \\ D_{31} & \Delta_{32} & \Delta_{33} \end{pmatrix} = \text{by definition} \quad \sum_{\pi\in S_3} (\operatorname{sgn}\pi)\cdot D_{\pi(1),1}\circ(\Delta_{\pi(2),2}\Delta_{\pi(3),3}) .$$

[For each term of this sum, operators in the first column of the matrix appear on the left, followed by operators from the second column, etc.] Using (3) we have

$$\det\begin{pmatrix} D_{11} & \Delta_{12} & \Delta_{13} \\ D_{21} & \Delta_{22} & \Delta_{23} \\ D_{31} & \Delta_{32} & \Delta_{33} \end{pmatrix} = \sum_{\pi\varepsilon S_3} (\operatorname{sgn}\pi)\{\Delta_{\pi(1),1}\Delta_{\pi(2),2}\Delta_{\pi(3),3}$$

$$+ \delta^1_{\pi(3)}\Delta_{\pi(1),3}\Delta_{\pi(2),2}$$

$$+ \delta^1_{\pi(2)}\Delta_{\pi(1),2}\Delta_{\pi(3),3}\}$$

$$= \sum_{\pi\varepsilon S_3} (\operatorname{sgn}\pi)\{\Delta_{\pi(1),1}\Delta_{\pi(2),2}\Delta_{\pi(3),3}$$

[compose π with the transposition interchanging $\pi(1)$ and $\pi(3)$]

$$- \delta^1_{\pi(1)}\Delta_{\pi(3),3}\Delta_{\pi(2),2}$$

[compose π with the transposition interchanging $\pi(1)$ and $\pi(2)$]

$$- \delta^1_{\pi(1)}\Delta_{\pi(2),2}\Delta_{\pi(3),3}\}$$

$$= \sum_{\pi\varepsilon S_3} (\operatorname{sgn}\pi)\Delta_{\pi(1),1}\Delta_{\pi(2),2}\Delta_{\pi(3),3}$$

$$- 2 \sum_{\substack{\pi\,\varepsilon\, S_3 \\ \text{with} \\ \pi(1)=1}} (\operatorname{sgn}\pi)\Delta_{\pi(2),2}\Delta_{\pi(3),3}$$

$$= \det\begin{pmatrix} \Delta_{11} & \Delta_{12} & \Delta_{13} \\ \Delta_{21} & \Delta_{22} & \Delta_{23} \\ \Delta_{31} & \Delta_{32} & \Delta_{33} \end{pmatrix} - 2\det\begin{pmatrix} \Delta_{22} & \Delta_{23} \\ \Delta_{32} & \Delta_{33} \end{pmatrix} .$$

We can also write this equation as

$$\det\begin{pmatrix} D_{11}+2 & \Delta_{12} & \Delta_{13} \\ D_{21} & \Delta_{22} & \Delta_{23} \\ D_{31} & \Delta_{32} & \Delta_{33} \end{pmatrix} = \det\begin{pmatrix} \Delta_{11} & \Delta_{12} & \Delta_{13} \\ \Delta_{21} & \Delta_{22} & \Delta_{23} \\ \Delta_{31} & \Delta_{32} & \Delta_{33} \end{pmatrix} .$$

Using (2), we find that

$$\det\begin{pmatrix} D_{11}+2 & D_{12} & D_{13} \\ D_{21} & D_{22}+1 & D_{23} \\ D_{31} & D_{32} & D_{33} \end{pmatrix} = \det\begin{pmatrix} \Delta_{11} & \Delta_{12} & \Delta_{13} \\ \Delta_{21} & \Delta_{22} & \Delta_{23} \\ \Delta_{31} & \Delta_{32} & \Delta_{33} \end{pmatrix} .$$

Of course, the numbers 1, 2, 3 could be replaced by any three distinct integers $\alpha_1, \alpha_2, \alpha_3$ from 1 to m. The same general procedure yields, by induction, the result

$$(4)\quad \det\begin{pmatrix} D_{11}+(m-1) & & \cdots & D_{1m} \\ \vdots & D_{22}+(m-2) & \ddots & \vdots \\ D_{m1} & & \cdots & D_{mm} \end{pmatrix} = \det\begin{pmatrix} \Delta_{11} & \cdots & \Delta_{1m} \\ \vdots & & \vdots \\ \Delta_{m1} & \cdots & \Delta_{mm} \end{pmatrix} .$$

We introduce the Cayley Ω-process which takes a function f of n vectors in $\mathbb{R}^n$ to the function Ωf of n vectors defined by

$$\Omega f = \det\begin{pmatrix} \frac{\partial}{\partial e_{11}} & \cdots & \frac{\partial}{\partial e_{1n}} \\ \vdots & & \vdots \\ \frac{\partial}{\partial e_{n1}} & & \frac{\partial}{\partial e_{nn}} \end{pmatrix} f .$$

Notice that we could just as well write the transpose matrix here, since all partials commute. It is easily seen that

$$\Omega f(A\cdot v_1,\ldots,A\cdot v_n) = (\det A)\cdot\Omega f(v_1,\ldots,v_n) .$$

So if f is invariant under $O(n)$ or $SO(n)$, then Ωf is invariant under $SO(n)$. Using det for the function

$$(v_1,\ldots,v_n) \mapsto \det\begin{pmatrix} v_1 \\ \vdots \\ v_n \end{pmatrix} ,$$

we now have

34. THEOREM (THE CAPELLI IDENTITIES). Let f be a polynomial function of m vectors in $\mathbb{R}^n$. Then

$$\det\begin{pmatrix} D_{11}+(m-1) & \cdots & D_{1m} \\ \vdots & \ddots & \vdots \\ D_{m1} & \cdots & D_{mm} \end{pmatrix} f = \begin{cases} 0 & m > n \\ \det \Omega f & m = n \,. \end{cases}$$

Proof. Equation (4) shows that at $(v_1,\ldots,v_m)$ the left side of our equation has the value

$$\sum_{\pi\in S_m} (\operatorname{sgn}\pi)\, \Delta_{\pi(1),1}\cdots\Delta_{\pi(m),m}\, f(v_1,\ldots,v_m)$$

$$= \sum_{\pi\in S_m} (\operatorname{sgn}\pi) \sum_{i_1,\ldots,i_m=1}^{n} v_{\pi(1),i_1}\cdots v_{\pi(m),i_m} \frac{\partial^m f}{\partial e_{1,i_1}\cdots\partial e_{m,i_m}}(v_1,\ldots,v_m)$$

$$= \sum_{i_1,\ldots,i_m=1}^{n} \left[\sum_{\pi\in S_m} (\operatorname{sgn}\pi)\, v_{\pi(1),i_1}\cdots v_{\pi(m),i_m}\right] \frac{\partial^m f}{\partial e_{1,i_1}\cdots\partial e_{m,i_m}}(v_1,\ldots,v_m) \,.$$

If $i_\alpha = i_\beta$ for some $\alpha \neq \beta$, then the sum inside the parentheses is clearly 0. This always occurs if $m > n$, so we obtain 0 for the total sum in this case. When $m = n$, the sum inside the parentheses is zero unless $i_1,\ldots,i_n$ is a permutation of $1,\ldots,n$, so our total sum becomes

$$\sum_{\rho\in S_n}\left[\sum_{\pi\in S_n} (\operatorname{sgn}\pi)\, v_{\pi(1),\rho(1)}\cdots v_{\pi(n),\rho(n)}\right] \frac{\partial^n f}{\partial e_{1,\rho(1)}\cdots\partial e_{n,\rho(n)}}(v_1,\ldots,v_n)$$

$$= \sum_{\rho\in S_n} (\operatorname{sgn}\rho)\cdot\det(v_1,\ldots,v_n)\cdot \frac{\partial^n f}{\partial e_{1,\rho(1)}\cdots\partial e_{n,\rho(n)}}(v_1,\ldots,v_n)$$

$$= \det(v_1,\ldots,v_n)\cdot\Omega f(v_1,\ldots,v_n) \,. \quad \blacksquare$$

In order to make use of the Capelli identities, we introduce a partial ordering $\prec$ on the homogeneous polynomial functions of m vectors in $\mathbb{R}^n$. Let f and $\bar{f}$ be homogeneous of degree $(\alpha_1,\ldots,\alpha_m)$ and $(\bar{\alpha}_1,\ldots,\bar{\alpha}_m)$, respectively, and set $d = \alpha_1 + \cdots + \alpha_m$ and $\bar{d} = \bar{\alpha}_1 + \cdots + \bar{\alpha}_m$. Then $f \prec \bar{f}$ if and only if: $d < \bar{d}$; or $d = \bar{d}$ and $\alpha_m < \bar{\alpha}_m$; or $d = \bar{d}$ and $\alpha_m = \bar{\alpha}_m$ and $\alpha_{m-1} < \bar{\alpha}_{m-1}$; or ... . This can also be expressed a little differently. Among all homogeneous polynomial functions f of fixed total degree $d = \alpha_1 + \cdots + \alpha_m$, we can consider $(\alpha_m,\alpha_{m-1},\ldots,\alpha_1)$ as the digits of a number to the base $d+1$. We define the rank of f to be this number,

$$\operatorname{rank} f = \alpha_1 + \alpha_2(d+1) + \alpha_3(d+1)^2 + \cdots + \alpha_m(d+1)^{m-1} .$$

If f and g both have total degree d, then $f \prec g$ if and only if $\operatorname{rank} f < \operatorname{rank} g$.

Now consider the effect on f of the operator on the left side of the Capelli identities. The main term

$$(D_{11} + m - 1)\cdots D_{mm} f$$

is (by Euler's theorem) just

$$(\alpha_1 + m - 1)\cdots\alpha_m f = (\text{constant})\cdot f ,$$

and this constant is 0 only if f does not depend on v_m. All other terms will involve certain diagonal terms, which are all just multiplications by constants, and a term

$$D_{s_1r_1}\cdots D_{s_\mu r_\mu} f \quad \text{where} \quad \begin{cases} r_1 < \cdots < r_\mu \\ s_i \neq r_i \\ s_1,\ldots,s_\mu \text{ is a permutation} \\ \quad \text{of } r_1,\ldots,r_\mu . \end{cases}$$

In particular, $s_\mu < r_\mu$. But $D_{s_\mu r_\mu} f$ is homogeneous of degree

$$(\alpha_1,\ldots,\alpha_{s_\mu}+1,\ldots,\alpha_{r_\mu}-1,\ldots,\alpha_m) ,$$

which means that $D_{s_\mu r_\mu} f \prec f$. Thus

$$D_{s_1r_1}\cdots D_{s_\mu r_\mu} f = D_{s_1r_1}\cdots f^* = \mathcal{P} f^* ,$$

where $f^* \prec f$, and $\mathcal{P}$ is a composition of polarizations; since f^* is itself a polarization of f, it is invariant under $O(n)$ or $SO(n)$ if f is. So the Capelli identities show that

$$\text{(A)} \qquad (\text{constant})\cdot f = \begin{cases} \text{a sum of terms } \mathcal{P} f^* & m > n \\ \text{a sum of terms } \mathcal{P} f^* & m = n , \\ \quad + \det \cdot \Omega f \end{cases}$$

where $f^* \prec f$ is invariant under $O(n)$ or $SO(n)$ if f is, $\mathcal{P}$ is a composition of polarizations, and the constant is 0 only if f does not depend on v_m.

We are now ready to prove

<u>35. THEOREM</u>. For all m and n we have

O_n^m: Every polynomial function f of m vectors in $\mathbb{R}^n$ which is invariant under $O(n)$ can be written as a polynomial in the inner product functions ι_{rs}.

SO_n^m: Every polynomial function f of m vectors in $\mathbb{R}^n$ which is invariant under $SO(n)$ can be written as a polynomial in the functions ι_{rs} and the determinant functions $\det_{r_1 \cdots r_n}$.

Proof. Notice that a function of m vectors can always be thought of as a function of a larger number of vectors; so $O_n^m \Rightarrow O_n^{m'}$ and $SO_n^m \Rightarrow SO_n^{m'}$ automatically for $m' < m$. Note also that the determinants in SO_n^m are zero unless $m \geq n$.

The proof of O_n^m and SO_n^m proceeds in two parts.

36. LEMMA. If O_n^n [respectively SO_n^n] holds, then O_n^m [respectively SO_n^m] holds for all $m > n$.

Proof. The proof for SO will be almost exactly the same as for O, so we give only the latter. Actually, we give the proof only for O_n^{n+1}, as it will then be clear how to proceed by induction. We consider invariant homogeneous polynomial functions f of $n+1$ vectors in $\mathbb{R}^n$, of fixed total degree d. We will prove that they can be represented in the desired form by complete induction on their rank. If rank $f < (d+1)^n$, so that the degree α_n of f in v_{n+1} is 0, then f does not involve v_{n+1}, so the result follows from the hypothesis that O_n^n holds. Let $r_0 \geq (d+1)^n$. Assuming that all f of total degree d and rank $< r_0$ can be expressed in the desired form, we will show that all f of total degree d and rank r_0 can also be so expressed. The constant in equation (A) is $\neq 0$ for our f, so f is the sum of terms

$\mathcal{P} f^*$, where $f^* \prec f$ is invariant under $O(n)$ and $\mathcal{P}$ is a composition of polarizations. Since f^* is a single polarization applied to f, the total degree of f^* equals the total degree of f. Since $f^* \prec f$, the inductive assumption says that each f^* can be written

$$f^* = F^* \circ (\{\iota_{rs}\})$$

for some polynomial F^*. This implies that

$$\mathcal{P} f^* = \mathcal{F}^* \circ (\{\mathcal{P} \iota_{rs}\})$$

for some polynomial $\mathcal{F}^*$. Since each $\mathcal{P}\iota_{rs}$ is a sum of ι_{rs}'s, it follows that each $\mathcal{P} f^*$ is a polynomial in the ι_{rs}, and thus f is. Q.E.D.

We still have to show that O_n^n and SO_n^n hold. In the case of SO_n^n there is a single determinant $\det(v_{ri})$ involved. If $A = (v_{ri})$, then

$$(A \cdot A^t)_{rs} = \sum_k v_{rk} v_{sk} = \langle v_r, v_s \rangle .$$

Hence

$$[\det(v_{ri})]^2 = \det A \cdot A^t = \det(\langle v_r, v_s \rangle)$$

is a polynomial in the ι_{rs}. Thus we need only linear terms in det.

<u>37. LEMMA</u>. O_n^n holds for all n. Moreover SO_n^n holds in the strengthened form

$\mathbf{SO}^n_n$: Every polynomial function of n vectors in $\mathbb{R}^n$ which is invariant under $SO(n)$ can be written as $g + (\det)\cdot h$, where g and h are polynomials in the inner product functions ι_{rs}.

<u>Proof</u>. We use induction on n. Consider first a polynomial function $f: \mathbb{R} \to \mathbb{R}$. If f is invariant under $O(1)$, then $f(x) = f(-x)$. So $f(x)$ involves only even powers of x. Hence $f(x) = F(x^2) = F(\langle x,x\rangle)$ for some polynomial F. Moreover, <u>any</u> polynomial function $f: \mathbb{R} \to \mathbb{R}$ can be written

$$f(x) = g(x) + xh(x) = g(x) + (\det x)\cdot h(x) ,$$

where g and h involve only even powers.

To carry out the induction step,

$$(*) \qquad \{O^{n-1}_{n-1} \text{ and } \mathbf{SO}^{n-1}_{n-1}\} \Rightarrow \{O^n_n \text{ and } \mathbf{SO}^n_n\} ,$$

we first show that

$$O^{n-1}_{n-1} \Rightarrow \mathbf{SO}^{n-1}_n \quad (\Rightarrow O^{n-1}_n , \text{ since no determinants are involved}) .$$

So consider a polynomial function f of $n-1$ vectors in $\mathbb{R}^n$. Define a polynomial function $\bar{f}$ of $n-1$ vectors in $\mathbb{R}^{n-1}$ by

$$\bar{f}(w_1,\ldots,w_{n-1}) = f(\bar{w}_1,\ldots,\bar{w}_{n-1}) ,$$

where

$$\bar{w}_r = (w_{r1},\ldots,w_{r,n-1},0) .$$

If f is invariant under $SO(n)$, then $\bar{f}$ is actually invariant under $O(n-1)$. So by hypothesis, there is a polynomial $\bar{F}$ with

$$\bar{f}(w_1,\ldots,w_{n-1}) = \bar{F}(\{<w_r,w_s>\}) .$$

Now given $v_1,\ldots,v_{n-1} \in \mathbb{R}^n$, choose $A \in SO(n)$ so that all $A\cdot v_r$ lie in $\mathbb{R}^{n-1} \times \{0\}$, and hence $A\cdot v_r = \bar{w}_r$ for some $w_r \in \mathbb{R}^{n-1}$. Then

$$\begin{aligned} f(v_1,\ldots,v_{n-1}) &= f(A\cdot v_1,\ldots,A\cdot v_{n-1}) = f(\bar{w}_1,\ldots,\bar{w}_{n-1}) \\ &= \bar{f}(w_1,\ldots,w_n) = \bar{F}(\{<w_r,w_s>\}) \\ &= \bar{F}(\{<v_r,v_s>\}) . \end{aligned}$$

This completes the proof that $O_{n-1}^{n-1} \Longrightarrow \mathbf{SO}_n^{n-1}$.

Now for the proof of (*). This proof will also be by induction, using the same general scheme as in the previous Lemma, but there will be a slight complication, for we will actually be using a double induction, first on the total degree of f, and then within each total degree on the rank. In addition, the statements O_n^n and $\mathbf{SO}_n^n$ will have to be proved jointly in the induction. Thus, for a fixed total degree and rank, we will show that all f of this degree and rank which are invariant under $O(n)$ have the desired form, and also that all f of this degree and rank which are invariant under $SO(n)$ have the desired form, assuming that the same two statements hold for all f of lower degree, or of the same degree and lower rank. There is certainly no problem beginning the induction with degree 0; moreover, within any particular degree, the polynomials of sufficiently low rank will not involve v_n, so we will be back to the cases $\mathbf{SO}_n^{n-1}$ and O_n^{n-1} which we have already proved.

Now consider a particular invariant f. We use equation (A), in the case $m = n$, to see that f is a sum of terms $\mathcal{P}f^*$ plus a constant times $\det \cdot \Omega f$. The sum of the terms $\mathcal{P}f^*$ can be written as $F\circ(\{\iota_{rs}\})$, as before, and we just have to worry about $\det \cdot \Omega f$. First suppose that f is invariant under $O(n)$. Then $\Omega f \prec f$ is invariant under $SO(n)$, so by the induction hypothesis we can write

$$\Omega f = g + (\det)\cdot h ,$$

where $g, h \prec f$ are invariant under $O(n)$, and thus by the induction hypothesis expressible as polynomials in the ι_{rs}. So we have

$$f = F\circ(\{\iota_{rs}\}) + (\text{constant})\cdot\det\cdot[G\circ(\{\iota_{rs}\}) + \det\cdot H\circ(\{\iota_{rs}\})] .$$

Since f, ι_{rs}, and $\det^2$ are invariant under $O(n)$, the term $\det\cdot G\circ(\{\iota_{rs}\})$ must also be invariant under $O(n)$, which is possible only if $G = 0$. Thus f is a polynomial in the ι_{rs}.

If f is assumed invariant under $SO(n)$, then everything remains the same, except that G need not be zero. ∎

8. An Easier Invariance Problem

For an $n \times n$ matrix M, we define $f_1(M),\dots,f_n(M)$ by

$$\det(I + \lambda M) = 1 + \lambda f_1(M) + \cdots + \lambda^n f_n(M) .$$

It will also be convenient to set $f_0(M) = 1$. Then the f_i are polynomial functions of the entries of M which are invariant under the adjoint action of $GL(n,\mathbb{R})$,

$$f_i(Ad(A)M) = f_i(AMA^{-1}) = f_i(M) \qquad \text{for } A \in GL(n,\mathbb{R}) .$$

If M has eigenvalues $\lambda_1,\ldots,\lambda_n$, and σ_i denotes the $i^{\underline{th}}$ elementary symmetric polynomial, then*

$$f_i(M) = \sigma_i(\lambda_1,\ldots,\lambda_n) .$$

According to Problem I.7-15, every polynomial function on the $n \times n$ matrices $\mathfrak{gl}(n,\mathbb{R})$ which is invariant under the adjoint action of $GL(n,\mathbb{R})$ is a polynomial in the f_i (notice that we are now considering functions of a single matrix, rather than functions of many vectors). Now we want to find out which polynomial functions on $\mathfrak{o}(n)$ are invariant under the adjoint action of $O(n)$, or of $SO(n)$. The line of argument will be essentially that used in Problem I.7-15, except that in some ways it will be even easier, since we have an especially simple "canonical form" for elements of $\mathfrak{o}(n)$, which greatly strengthens Corollary 11:

38. PROPOSITION. For every $A \in \mathfrak{o}(n)$ there is a matrix $B \in O(n)$ such that

$$BAB^{-1} = \begin{pmatrix} 0 & \lambda_1 & & & \\ -\lambda_1 & 0 & & & \bigcirc \\ & & \ddots & & \\ & & & 0 & \lambda_m \\ \bigcirc & & & -\lambda_m & 0 \end{pmatrix} \quad \text{or} \quad \begin{pmatrix} 0 & \lambda_1 & & & & \\ -\lambda_1 & 0 & & & \bigcirc & \\ & & \ddots & & & \\ & & & 0 & \lambda_m & \\ \bigcirc & & & -\lambda_m & 0 & \\ & & & & & 0 \end{pmatrix}$$

(some of the λ's may also be 0).

*The λ_i are the roots of the characteristic polynomial $\chi(\lambda) = \det(\lambda I - M)$. Recall that the $\sigma_i = \sigma_i(\lambda_1,\ldots,\lambda_n)$ satisfy $\lambda^n - \sigma_1\lambda^{n-1} + \cdots = \Pi(\lambda-\lambda_i) = \chi(\lambda)$. So $\lambda^n + \sigma_1\lambda^{n-1} + \cdots = (-1)^n\chi(-\lambda) = (-1)^n\det(-\lambda I - M) = \det(\lambda I + M)$. Hence, $\det(I + \lambda M) = \lambda^n\det(I/\lambda + M) = \lambda^n[\,(1/\lambda)^n + \sigma_1(1/\lambda)^{n-1} + \cdots] = 1 + \sigma_1\lambda + \cdots$.

Proof. If $T\colon \mathbb{R}^n \longrightarrow \mathbb{R}^n$ is the linear transformation determined by A, then the skew-symmetry of A means that

$$(1) \qquad \langle Tv,w\rangle = -\langle v,Tw\rangle \qquad \text{for } v,\, w \in \mathbb{R}^n .$$

[Conversely, if this relation holds for $T\colon (V, \langle\ ,\ \rangle) \longrightarrow (V, \langle\ ,\ \rangle)$, then the matrix of T with respect to an orthonormal basis is skew-symmetric. This implies, in particular, that

$$(2) \qquad \det T = 0 \qquad \text{if } \dim V \text{ is odd .]}$$

Equation (1) implies that $(\text{image } T)^{\perp} = \ker T$. Consequently, T must be one-one on image T. Therefore $\operatorname{rank} T^2 = \operatorname{rank} T$. Moreover, this rank is even, by Corollary 11. [Alternative proof: The map $T|(\text{image } T)\colon \text{image } T \longrightarrow \text{image } T$ is one-one, so its determinant must be non-zero. Applying (2), we see that dim image T is even.]

Now A^2 is symmetric, so there is an orthonormal basis $v_1,\ldots,v_n$ of eigenvectors of T^2, with eigenvalues $\mu_1,\ldots,\mu_n$. Note that

$$\mu_j = \langle v_j,T^2v_j\rangle = -\langle Tv_j,Tv_j\rangle \le 0 .$$

By renumbering, we can assume that $\mu_1,\ldots,\mu_{2m} < 0$, and that the remaining μ's equal 0. Define a new orthonormal basis $w_1,\ldots,w_n$ by

$$\begin{aligned} w_1 &= v_1 , & w_2 &= \frac{1}{\sqrt{-\lambda_1}}\, T(v_1) \\ &\ \vdots \\ w_{2m-1} &= v_m , & w_{2m} &= \frac{1}{\sqrt{-\lambda_m}}\, T(v_m) \\ w_j &= v_j & j &> 2m . \end{aligned}$$

Then the matrix of T has the desired form with respect to the basis $w_1,\ldots,w_n$. ∎

Notice that for $M \in \mathfrak{o}(n)$ we have

$$(I+\lambda M)^t = I - \lambda M \implies \det(I+\lambda M) = \det(I-\lambda M) ,$$

so

$$1 + \lambda f_1(M) + \cdots + \lambda^n f_n(M) = 1 - \lambda f_1(M) + \cdots + (-1)^n \lambda^n f_n(M) ,$$

and hence $f_i(M) = 0$ for odd i. We will also need to use the following formula, whose verification is left to the reader:

$$\sigma_{2k}(\lambda_1,-\lambda_1,\lambda_2,-\lambda_2,\ldots,\lambda_m,-\lambda_m) = \sigma_k(-\lambda_1{}^2,\ldots,-\lambda_m{}^2) .$$

<u>39. THEOREM</u>. Let $n = 2m$ or $n = 2m+1$. Then every polynomial function f on $\mathfrak{o}(n)$ which is invariant under the adjoint action of $O(n)$ is a polynomial in $f_2,\ldots,f_{2m}$.

<u>Proof</u>. For $\lambda_1,\ldots,\lambda_m \in \mathbb{R}$, let $[\lambda_1,\ldots,\lambda_m]$ be

$$\begin{pmatrix} 0 & \lambda_1 & & & \\ -\lambda_1 & 0 & & & \\ & & \ddots & & \\ & & & 0 & \lambda_m \\ & & & -\lambda_m & 0 \end{pmatrix} \quad \text{or} \quad \begin{pmatrix} 0 & \lambda_1 & & & & \\ -\lambda_1 & 0 & & & & \\ & & \ddots & & & \\ & & & 0 & \lambda_m & \\ & & & -\lambda_m & 0 & \\ & & & & & 0 \end{pmatrix} ,$$

depending on whether $n = 2m$ or $n = 2m+1$. Notice that the eigenvalues of

$[\lambda_1,\dots,\lambda_m]$ are $i\lambda_1,-i\lambda_1,\dots,i\lambda_m,-i\lambda_m$ [and 0, if $n = 2m+1$], so

$$\begin{aligned} f_{2k}([\lambda_1,\dots,\lambda_m]) &= \sigma_{2k}(i\lambda_1,-i\lambda_1,\dots,i\lambda_m,-i\lambda_m) \\ &= \sigma_k(\lambda_1{}^2,\dots,\lambda_m{}^2) . \end{aligned}$$

Define

$$g(\lambda_1,\dots,\lambda_m) = f([\lambda_1,\dots,\lambda_m]) .$$

Then g is a polynomial function of $\lambda_1,\dots,\lambda_m$. Notice that for the matrix

$$A = \begin{pmatrix} 0 & \begin{matrix} 1 & 0 \\ 0 & 1 \end{matrix} & & & \\ \begin{matrix} 1 & 0 \\ 0 & 1 \end{matrix} & 0 & & 0 & \\ & & 1 & & \\ & 0 & & \ddots & \\ & & & & 1 \end{pmatrix} \in O(n)$$

we have

$$A\cdot[\lambda_1,\dots,\lambda_m]\cdot A^{-1} = [\lambda_2,\lambda_1,\dots,\lambda_m] .$$

Similarly, we can interchange any two λ's by some $A \in O(n)$. Thus g is symmetric in the λ's. Notice, moreover, that for the matrix

$$B = \begin{pmatrix} \begin{matrix} 0 & 1 \\ 1 & 0 \end{matrix} & & & 0 \\ & 1 & & \\ & & \ddots & \\ 0 & & & 1 \end{pmatrix} \in O(n)$$

we have

$$B\cdot[\lambda_1,\dots,\lambda_m]B^{-1} = [-\lambda_1,\lambda_2,\dots,\lambda_m] .$$

Thus

$$g(\lambda_1,\ldots,\lambda_m) = g(-\lambda_1,\lambda_2,\ldots,\lambda_m) .$$

This shows that the polynomial g does not have any terms involving λ_1 to an odd power. The same result clearly holds for all λ's, so we can write

$$g(\lambda_1,\ldots,\lambda_m) = h(\lambda_1{}^2,\ldots,\lambda_m{}^2)$$

for some polynomial h. Clearly h is symmetric in its arguments, so there is a polynomial p with

$$g(\lambda_1,\ldots,\lambda_m) = p(\sigma_1(\lambda_1{}^2,\ldots,\lambda_m{}^2),\ldots,\sigma_m(\lambda_1{}^2,\ldots,\lambda_m{}^2)) .$$

Thus we have

$$f([\lambda_1,\ldots,\lambda_m]) = p(f_2([\lambda_1,\ldots,\lambda_m]),\ldots,f_{2m}([\lambda_1,\ldots,\lambda_m])) .$$

Now for any $M \in \mathfrak{o}(n)$ there is, by Proposition 38, some $A \in O(n)$ such that $A^{-1}MA = [\lambda_1,\ldots,\lambda_m]$ for some $\lambda_1,\ldots,\lambda_m$. Since $f,f_2,\ldots,f_{2m}$ are invariant under the adjoint action of $O(n)$, the above equation yields

$$f(M) = p(f_2(M),\ldots,f_{2m}(M)) . \quad \blacksquare$$

Note that the polynomial functions $f_2,\ldots,f_{2m}$ are algebraically independent on $\mathfrak{o}(n)$ -- if

$$p(f_2(M),\ldots,f_{2m}(M)) = 0$$

for all $M \in \mathfrak{o}(n)$, then $p = 0$. Indeed, this equation implies that

$$p(\sigma_1(\lambda_1^2,\ldots,\lambda_m^2),\ldots,\sigma_m(\lambda_1^2,\ldots,\lambda_m^2)) = 0$$

for all $\lambda_1,\ldots,\lambda_m$, and thus that the polynomial

$$p(\sigma_1(x_1,\ldots,x_m),\ldots,\sigma_m(x_1,\ldots,x_m))$$

is zero whenever $x_1,\ldots,x_m$ take on positive values; but this implies that this polynomial in $x_1,\ldots,x_m$ is identically zero, and hence that $p = 0$, as we remarked on p. 467 .

With slight modifications of our previous argument we obtain

<u>40. THEOREM</u>. (1) If $n = 2m+1$, then every polynomial function f on $\mathfrak{o}(n)$ which is invariant under the adjoint action of $SO(n)$ is a polynomial in $f_2,\ldots,f_{2m}$.
(2) If $n = 2m$, then every polynomial function f on $\mathfrak{o}(n)$ which is invariant under the adjoint action of $SO(n)$ is a polynomial in $f_2,\ldots,f_{2m-2}$ and the Pfaffian Pf.

<u>Proof</u>. (1) Notice that the matrix A in the proof of Theorem 39 is actually in $SO(n)$. If n is odd, then $[\lambda_1,\ldots,\lambda_m]$ has a zero in the $(n,n)^{\text{th}}$ place, so for the matrix B in the proof of Theorem 39 we can just as well replace the 1 in the $(n,n)^{\text{th}}$ place by -1; this new B is in $SO(n)$. Now the proof of Theorem 39 goes through as before.
(2) The matrix A is still in $SO(n)$. We cannot arrange for B to be in $SO(n)$, but for the matrix

$$C = \begin{pmatrix} 0 & 1 & & & & & \\ 1 & 0 & & & & & \\ & & 0 & 1 & & & \\ & & 1 & 0 & & & \\ & & & & 1 & & \\ & & & & & \ddots & \\ & & & & & & 1 \end{pmatrix} \varepsilon \ SO(n) ,$$

we have

$$C\cdot[\lambda_1,\ldots,\lambda_m]C^{-1} = [-\lambda_1,-\lambda_2,\lambda_3,\ldots,\lambda_m] .$$

Similarly, we can send any pair of λ's to their negatives. So the symmetric function g in the proof of Theorem 39 has the property that each monomial appearing in it is either of even degree in all λ's or else of odd degree in all λ's. So g can be written

$$g(\lambda_1,\ldots,\lambda_m) = h_1(\lambda_1{}^2,\ldots,\lambda_m{}^2) + (\lambda_1\cdots\lambda_m)h_2(\lambda_1{}^2,\ldots,\lambda_m{}^2) .$$

Since g is symmetric, the term $h_1(\lambda_1{}^2,\ldots,\lambda_m{}^2)$ [= the sum of the monomials of g which are of even degree in all λ's] must be symmetric in $\lambda_1,\ldots,\lambda_m$. So h_1 is symmetric in its arguments. Thus h_2 is also symmetric in its arguments. So we can write

$$\begin{aligned} g(\lambda_1,\ldots,\lambda_m) = {} & p_1(\sigma_1(\lambda_1{}^2,\ldots,\lambda_m{}^2),\ldots,\sigma_m(\lambda_1{}^2,\ldots,\lambda_m{}^2)) \\ & + (\lambda_1\cdots\lambda_m)\cdot p_2(\sigma_1(\lambda_1{}^2,\ldots,\lambda_m{}^2),\ldots,\sigma_m(\lambda_1{}^2,\ldots,\lambda_m{}^2)) . \end{aligned}$$

Thus we have

$$\begin{aligned} f([\lambda_1,\ldots,\lambda_m]) = {} & p_1(f_2([\lambda_1,\ldots,\lambda_m],\ldots,f_{2m}([\lambda_1,\ldots,\lambda_m])) \\ & + \mathrm{Pf}([\lambda_1,\ldots,\lambda_m])\cdot p_2(f_2([\lambda_1,\ldots,\lambda_m]),\ldots,f_{2m}([\lambda_1,\ldots,\lambda_m])) . \end{aligned}$$

It follows, as before, that

$$f(M) = p_1(f_2(M),\ldots,f_{2m}(M)) + \mathrm{Pf}(M)\cdot p_2(f_2(M),\ldots,f_{2m}(M))$$

for all $M \in \mathfrak{o}(n)$. We can dispense with f_{2m}, since

$$f_{2m}(M) = \det M = \{\mathrm{Pf}(M)\}^2 . \quad \blacksquare$$

We have already observed that the polynomials $f_2,\ldots,f_{2m}$ are algebraically independent on $\mathfrak{o}(n)$. For $n = 2m$, the polynomials $f_2,\ldots,f_{2m-2}$, Pf are algebraically independent on $\mathfrak{o}(n)$. For suppose that

$$p(f_2(M),\ldots,f_{2m-2}(M),\mathrm{Pf}(M)) = 0 \qquad \text{for all } M \in \mathfrak{o}(n) .$$

Then for all $\lambda_1,\ldots,\lambda_m$ we have

$$\begin{aligned} 0 &= p(\sigma_1(\lambda_1{}^2,\ldots,\lambda_m{}^2),\ldots,\sigma_{m-1}(\lambda_1{}^2,\ldots,\lambda_m{}^2),\lambda_1\cdots\lambda_m) \\ &= p_1(\sigma_1(\lambda_1{}^2,\ldots,\lambda_m{}^2),\ldots,\sigma_{m-1}(\lambda_1{}^2,\ldots,\lambda_m{}^2),\sigma_m(\lambda_1{}^2,\ldots,\lambda_m{}^2)) \\ &\qquad + (\lambda_1\cdots\lambda_m)p_2(\sigma_1(\lambda_1{}^2,\ldots,\lambda_m{}^2),\ldots,\sigma_{m-1}(\lambda_1{}^2,\ldots,\lambda_m{}^2),\sigma_m(\lambda_1{}^2,\ldots,\lambda_m{}^2)) , \end{aligned}$$

for certain polynomials p_1 and p_2. This polynomial in $\lambda_1,\ldots,\lambda_m$ can be zero only if the two summands, representing the terms with all λ's of even degree and the terms with all λ's of odd degree, respectively, are each zero. Then as before we conclude that $p_1 = p_2 = 0$.

There is one further simple property of the functions f_k. In Corollary 13 we gave a formula for $\mathrm{Pf}(A \oplus B)$, where

$$A \oplus B = \begin{pmatrix} A & 0 \\ 0 & B \end{pmatrix} .$$

It is easy to find a formula for $f_k(A \oplus B)$, for all $A \in \mathfrak{gl}(r,\mathbb{R})$ and $B \in \mathfrak{gl}(s,\mathbb{R})$. For we have

$$\begin{aligned} \det(I_{r+s} + \lambda(A \oplus B)) &= \det((I_r + \lambda A) \oplus (I_s + \lambda B)) \\ &= \det(I_r + \lambda A) \cdot \det(I_s + \lambda B) \\ &= (1 + \lambda f_1(A) + \cdots + \lambda^r f_r(A)) \cdot (1 + \lambda f_1(B) + \cdots + \lambda^s f_s(B)) . \end{aligned}$$

So

$$f_k(A \oplus B) = \text{coefficient of } \lambda^k = \sum_{\ell=0}^{k} f_\ell(A) \cdot f_{k-\ell}(B) .$$

When A and B are skew-symmetric, we have

$$f_{2k}(A \oplus B) = \sum_{\ell=1}^{k} f_{2\ell}(A) \cdot f_{2k-2\ell}(B) .$$

9. The Cohomology of the Oriented Grassmannians

We are now ready to compute part of the cohomology of

$$\tilde{G}_n(\mathbb{R}^N) = SO(N)/\ SO(n) \times SO(N-n) = SO(N)/H .$$

The Lie algebra $\mathfrak{o}(N)$ has as a basis the matrices

$$X_\alpha^\beta = \begin{pmatrix} & 1 \\ -1 & \end{pmatrix} \qquad 1 \leq \alpha < \beta \leq N$$

which have non-zero entries only in the (α,β) and (β,α) positions. [We adopt the convention that the indices α, β range from 1 to N, while the indices i, j run from 1 to n, and r, s range from $n+1$ to N.] Let $\{\phi_\alpha^\beta\}$ be the dual basis to the $\{X_\alpha^\beta\}$. The Lie algebra $\mathfrak{h}$ consists of matrices

$$\begin{pmatrix} L_1 & 0 \\ 0 & L_2 \end{pmatrix}, \qquad \begin{array}{l} L_1 \in \mathfrak{o}(n) \\ L_2 \in \mathfrak{o}(N-n) . \end{array}$$

The orthogonal complement $\mathfrak{h}^\perp$, with respect to the bi-invariant metric on p. 453, consists of all matrices

$$\begin{pmatrix} 0 & P \\ -P^t & 0 \end{pmatrix}, \qquad P \text{ an } n\times(N-n) \text{ matrix ,}$$

and has as basis the X_i^r for $1 \le i \le n$ and $n+1 \le r \le N$; so the corresponding ϕ_i^r are a basis for $(\mathfrak{h}^\perp)^*$ [more precisely, the restrictions of the ϕ_i^r to $\mathfrak{h}^\perp$ are a basis].

The adjoint action of $SO(n)\times SO(N-n)$ on $\mathfrak{h}^\perp$ is easily computed to be

$$\begin{aligned}
\begin{pmatrix} A & 0 \\ 0 & B \end{pmatrix}\begin{pmatrix} 0 & P \\ -P^t & 0 \end{pmatrix}\begin{pmatrix} A & 0 \\ 0 & B \end{pmatrix}^{-1} &= \begin{pmatrix} A & 0 \\ 0 & B \end{pmatrix}\begin{pmatrix} 0 & P \\ -P^t & 0 \end{pmatrix}\begin{pmatrix} A^t & 0 \\ 0 & B^t \end{pmatrix} \\
&= \begin{pmatrix} A & 0 \\ 0 & B \end{pmatrix}\begin{pmatrix} 0 & PB^t \\ -P^tA^t & 0 \end{pmatrix} \\
&= \begin{pmatrix} 0 & APB^t \\ -BP^tA^t & 0 \end{pmatrix} .
\end{aligned}$$

We want to know which elements of $\Omega^k(\mathfrak{h}^\perp)$ are invariant under the induced adjoint action of $SO(n)\times SO(N-n)$. We will split this question up into two parts, by considering invariance under the adjoint action of the two subgroups

$$SO(n)\times\{I\} = \left\{\begin{pmatrix} A & 0\\ 0 & I\end{pmatrix} : A \in SO(n)\right\}$$

$$\{I\}\times SO(N-n) = \left\{\begin{pmatrix} I & 0\\ 0 & B\end{pmatrix} : B \in SO(N-n)\right\}.$$

We consider first the adjoint action of $\{I\}\times SO(N-n)$, given by

$$\begin{pmatrix} I & 0\\ 0 & B\end{pmatrix}\begin{pmatrix} 0 & P\\ -P^t & 0\end{pmatrix}\begin{pmatrix} I & 0\\ 0 & B\end{pmatrix}^{-1} = \begin{pmatrix} 0 & PB^t\\ -BP^t & 0\end{pmatrix}.$$

If we regard $\mathfrak{h}^\perp$ as the n-fold product $\mathbb{R}^{N-n}\times\cdots\times\mathbb{R}^{N-n}$ by identifying $\begin{pmatrix} 0 & P\\ -P^t & 0\end{pmatrix} \in \mathfrak{h}^\perp$ with the n-tuple of the rows of P, then this action is the usual action of $SO(N-n)$ on each factor. Since a form $\eta \in \Omega^k(\mathfrak{h}^\perp)$ can be regarded as a polynomial function on the nk-fold product $\mathbb{R}^{N-n}\times\cdots\times\mathbb{R}^{N-n}$, Theorem 35 shows that η is invariant under the adjoint action of $\{I\}\times SO(N-n)$ if and only if it is a polynomial in the inner products and determinants of the vectors involved. We have to figure out just what this means when η is a k-form, and express η in terms of the forms ϕ_i^r. *From now on we assume that* $k < N-n$.

Consider first the case where η is a 1-form, and thus a function $\eta\colon \mathfrak{h}^\perp \to \mathbb{R}$. If η is invariant under $\{I\}\times SO(N-n)$, and $M = \begin{pmatrix} 0 & P\\ -P^t & 0\end{pmatrix}$, then $\eta(M)$ can be written as a polynomial in the inner products of rows of P and in the $(N-n)\times(N-n)$ subdeterminants of P. Thus $\eta(M)$ is a polynomial in

$$\sum_{r=n+1}^{N} \phi_{i_1}^r(M)\cdot\phi_{i_2}^r(M) \qquad 1 \le i_1,\ i_2 \le n$$

and

the determinants of the matrices formed
by picking $(N-n)$ rows of P .

Multiplying M by $\alpha \in \mathbb{R}$ multiplies the first terms by α^2 and the determinants by α^{N-n}. So $\eta(M)$ cannot be linear in M unless it is zero. Thus $\Omega^1(\mathfrak{h}^\perp)^H = 0$.

Now consider a 2-form $\eta \in \Omega^2(\mathfrak{h}^\perp)$. If η is invariant under $\{I\} \times SO(N-n)$, then $\eta(M_1,M_2)$ can be written as a polynomial in

$$\sum_r \phi^r_{i_1}(M_1)\cdot\phi^r_{i_2}(M_1) , \qquad \sum_r \phi^r_{i_1}(M_2)\cdot\phi^r_{i_2}(M_2) , \qquad \sum_r \phi^r_{i_1}(M_1)\cdot\phi^r_{i_2}(M_2)$$

and

the determinants of the matrices formed by picking n_1 rows of P_1 and n_2 rows of P_2, with $n_1 + n_2 = (N-n)$.

Multiplying M_1 [or M_2] by α multiplies these determinants by α^{n_1} [or α^{n_2}]. But either $n_1 > 1$ or $n_2 > 1$, since we are assuming that $2 = k < N-n$. Consequently, since η is multi-linear, the determinants cannot be involved. Moreover, of the remaining terms, only those of the third kind can be involved. So

$$\eta = \text{a linear combination of the } \sum_r \phi^r_{i_1} \otimes \phi^r_{i_2} .$$

Since η is a 2-form, we have

$$\eta = \text{Alt}\,\eta = \text{a linear combination of the } \sum_r \phi^r_{i_1} \wedge \phi^r_{i_2} .$$

Thus $\Omega^2(\mathfrak{h}^\perp)^H$ can contain only linear combinations of the 2-forms

$$\zeta_{i_1 i_2} = \sum_r \phi^r_{i_1} \wedge \phi^r_{i_2} \qquad 1 \le i_1 < i_2 \le n .$$

For a 3-form $\eta \in \Omega^3(\mathfrak{h}^\perp)$ to be invariant under $\{I\} \times SO(N-n)$, it must be possible to write $\eta(M_1, M_2, M_3)$ as a polynomial in

$$\sum_r \phi^r_{i_1}(M_{j_1}) \cdot \phi^r_{i_2}(M_{j_2}) \qquad j_1, j_2 = 1, 2, 3 .$$

and

determinants formed from rows of P_1, P_2, P_3 .

As before, the determinants cannot be involved, since we assume $3 = k < N-n$. Then it is easy to see that no non-zero polynomial in the other terms can be multi-linear in (M_1, M_2, M_3). So $\Omega^3(\mathfrak{h}^\perp)^H = 0$.

If $\eta \in \Omega^4(\mathfrak{h}^\perp)$ is invariant under $\{I\} \times SO(N-n)$, then $\eta(M_1, \ldots, M_4)$ can be written as a polynomial in the

$$\sum_r \phi^r_{i_1}(M_{j_1}) \cdot \phi^r_{i_2}(M_{j_2}) \qquad j_1, j_2 = 1, \ldots, 4$$

(determinants are ruled out as before). Since η is multi-linear, it is easy to see that the only monomials which can appear are

$$\{\sum_r \phi^r_{i_1}(M_{j_1}) \cdot \phi^r_{i_2}(M_{j_2})\} \cdot \{\sum_r \phi^r_{i_3}(M_{j_3}) \cdot \phi^r_{i_4}(M_{j_4})\} ,$$

where $j_1, \ldots, j_4$ <u>are distinct</u>. This term can be written

$$(\sum_r \phi^r_{i_1} \otimes \phi^r_{i_2}) \otimes (\sum_r \phi^r_{i_3} \otimes \phi^r_{i_4})(M_{\pi(1)}, M_{\pi(2)}, M_{\pi(3)}, M_{\pi(4)})$$

for some permutation $\pi \in S_4$. Since η is alternating, we find that η is a linear combination of terms $\zeta_{i_1 i_2} \wedge \zeta_{i_3 i_4}$.

In general, we clearly have:

If $k < N-n$ is odd, then $\Omega^k(\mathfrak{h}^\perp)^H = 0$.

If $k < N-n$ is even, then all elements of $\Omega^k(\mathfrak{h}^\perp)^H$ can be written as linear combinations of the forms

$$\zeta_{i_1 i_2} \wedge \cdots \wedge \zeta_{i_{k-1} i_k} .$$

To determine $\Omega^k(\mathfrak{h}^\perp)^H$ completely for all even $k < N-n$, we still have to consider invariance under $SO(n) \times \{I\}$. But we already see, from Theorem 32, that

(A) If $k < N-n$ is odd, then $H^k(\tilde{G}_k(\mathbb{R}^N)) = 0$.

Moreover, the maps

$$\Omega^{k-1}(\mathfrak{h}^\perp)^H \xrightarrow{d} \Omega^k(\mathfrak{h}^\perp)^H \xrightarrow{d} \Omega^{k+1}(\mathfrak{h}^\perp)^H$$

are zero for even $k < N-n-1$, since the vector spaces on the ends are 0. Consequently,

$$H^k(\tilde{G}_n(\mathbb{R}^N)) = \frac{\ker d\colon \Omega^k(\mathfrak{h}^\perp)^H}{d(\Omega^{k+1}(\mathfrak{h}^\perp)^H)} = \Omega^k(\mathfrak{h}^\perp)^H/0 = \Omega^k(\mathfrak{h}^\perp)^H .$$

Actually, this result holds for <u>all</u> k, but we need a different argument.

Notice that something special happens when we take the bracket of two elements of $\mathfrak{h}^{\perp}$. We have

$$\begin{pmatrix} 0 & P \\ -P^t & 0 \end{pmatrix}\begin{pmatrix} 0 & Q \\ -Q^t & 0 \end{pmatrix} - \begin{pmatrix} 0 & Q \\ -Q^t & 0 \end{pmatrix}\begin{pmatrix} 0 & P \\ -P^t & 0 \end{pmatrix} = \begin{pmatrix} -PQ^t + QP^t & 0 \\ 0 & -P^tQ + QP^t \end{pmatrix},$$

which is in $\mathfrak{h}$. Consulting the statement of Theorem 32 we see that our map

$$d\colon \Omega^k(\mathfrak{h}^{\perp})^H \to \Omega^{k+1}(\mathfrak{h}^{\perp})^H$$

is always 0. Thus

(B) For all k we have $H^k(\tilde{G}_n(\mathbb{R}^N)) = \Omega^k(\mathfrak{h}^{\perp})^H$.

We now have to investigate linear combinations of the forms $\zeta_{i_1 i_2} \wedge \cdots \wedge \zeta_{i_{k-1} i_k}$. Such combinations can be described in terms of polynomial functions $f\colon \mathfrak{o}(n) \to \mathbb{R}$ which are homogeneous of degree $k/2$: if f is a sum of monomials

$$c\cdot\phi_{i_2}^{i_1}\cdot\phi_{i_4}^{i_3}\cdots\phi_{i_k}^{i_{k-1}}, \qquad 1 \le i_{2\alpha-1} < i_{2\alpha} \le n,$$

then $f(\zeta) \in \Omega^k(\mathfrak{h}^{\perp})$ will denote the k-form which is the sum of the corresponding terms

$$c\cdot\zeta_{i_1 i_2} \wedge \zeta_{i_3 i_4} \wedge \cdots \wedge \zeta_{i_{k-1} i_k}$$

(since the ζ_{ij} are 2-forms, the $\wedge$ products commute, so the order of the factors ϕ_j^i is irrelevant). Clearly every linear combination of the forms

$\zeta_{i_1 i_2} \wedge \cdots \wedge \zeta_{i_{k-1} i_k}$ is $f(\zeta)$ for some $f\colon \mathfrak{o}(n) \longrightarrow \mathbb{R}$. Since the ϕ_i^r are linearly independent, this f is unique for $k < N - n$. Moreover for homogeneous polynomials $f, g\colon \mathfrak{o}(n) \longrightarrow \mathbb{R}$ we have

$$(1) \qquad (fg)(\zeta) = f(\zeta) \wedge g(\zeta) .$$

We want to find out how $SO(n) \times I$ operates on $f(\zeta)$. Take first the special case $f = \phi_j^i$, so that $f(\zeta) = \phi_j^i(\zeta) = \zeta_{ij}$. For

$$\tilde{A} = \begin{pmatrix} A & 0 \\ 0 & I \end{pmatrix} \in SO(n) \times I$$

$$M_1 = \begin{pmatrix} 0 & P_1 \\ -P_1{}^t & 0 \end{pmatrix}, \qquad M_2 = \begin{pmatrix} 0 & P_2 \\ -P_2{}^t & 0 \end{pmatrix} \in \mathfrak{h}^{\perp}$$

we have

$$\begin{aligned}
[\mathrm{Ad}(\tilde{A})^*\zeta_{ij}](M_1,M_2) &= \sum_r \phi_i^r \wedge \phi_j^r(\mathrm{Ad}(\tilde{A})M_1, \mathrm{Ad}(\tilde{A})M_2) \\
&= \sum_r \left\{ \phi_i^r \begin{pmatrix} 0 & AP_1 \\ -P_1{}^t A^t & 0 \end{pmatrix} \cdot \phi_j^r \begin{pmatrix} 0 & AP_2 \\ -P_2{}^t A^t & 0 \end{pmatrix} - \cdots \right\} \\
&= \sum_r \{ (AP_1)_{ir}(AP_2)_{jr} - (AP_2)_{ir}(AP_1)_{jr} \} \\
&= \sum_r \sum_{\mu,\nu=1}^{n} \{ A_{i\mu}(P_1)_{\mu r} A_{j\nu}(P_2)_{\nu r} - A_{i\mu}(P_2)_{\mu r} A_{j\nu}(P_1)_{\nu r} \} \\
&= \sum_{\mu,\nu=1}^{n} A_{i\mu} A_{j\nu} [\sum_r (P_1)_{\mu r}(P_2)_{\nu r} - (P_2)_{\mu r}(P_1)_{\nu r}] ,
\end{aligned}$$

and hence

$$\mathrm{Ad}(\tilde{A})^*\zeta_{ij} = \sum_{\mu,\nu=1}^{n} A_{i\mu}A_{j\nu}\zeta_{\mu\nu} ,$$

or

$$(2) \qquad \mathrm{Ad}(\tilde{A})^*(\phi^i_j(\zeta)) = (\sum_{\mu,\nu=1}^{n} A_{i\mu}A_{j\nu}\phi^\mu_\nu)(\zeta) .$$

On the other hand, for a matrix $L \in \mathfrak{o}(n)$ we have

$$(3) \qquad \phi^i_j(\mathrm{Ad}(A)L) = \phi^i_j(ALA^t) = \sum_{\mu,\nu=1}^{n} A_{i\mu}A_{j\nu}L^\mu_\nu$$

$$\Longrightarrow \phi^i_j \circ \mathrm{Ad}(A) = \sum_{\mu,\nu=1}^{n} A_{i\mu}A_{j\nu}\phi^\mu_\nu .$$

Comparing (2) and (3), we see that

$$\mathrm{Ad}(\tilde{A})^*(\phi^i_j(\zeta)) = [\phi^i_j \circ \mathrm{Ad}(A)](\zeta) .$$

Using equation (1), we find that for all $f\colon \mathfrak{o}(n) \longrightarrow \mathbb{R}$ we have

$$(*) \qquad \mathrm{Ad}(\tilde{A})^* f(\zeta) = [f \circ \mathrm{Ad}(A)](\zeta) .$$

From equation (*) we see that a linear combination $f(\zeta)$ is invariant under all $\mathrm{Ad}(\tilde{A})^*$, and thus $f(\zeta) \in \Omega^k(\mathfrak{h}^\perp)^H \approx H^k(\tilde{G}_n(\mathbb{R}^N))$, if and only if $f\colon \mathfrak{o}(n) \longrightarrow \mathbb{R}$ is invariant under all $\mathrm{Ad}(A)$, for $A \in SO(n)$. But Theorem 40 says that all such f are polynomials in

$$f_2, \ldots, f_{2[n/2]} \qquad \text{if } n \text{ is odd}$$

$$f_2, \ldots, f_{n-2}, \mathrm{Pf} \qquad \text{if } n \text{ is even} .$$

Moreover, f is uniquely expressible as such a polynomial, since $f_2,\ldots,f_{2[n/2]}$ [or $f_2,\ldots,f_{n-2},\mathrm{Pf}$] are algebraically independent (pp.491 and 494).

Case 1. n is odd. The forms $f_2(\zeta),\ldots,f_{2[n/2]}(\zeta)$ have dimensions $4,8,\ldots,4[n/2]$. So

> if $k < N-n$ is not a multiple of 4, then $H^k(\tilde{G}_n(\mathbb{R}^N)) = 0$
>
> if $k < N-n$ is a multiple of 4, then every element of $H^k(\tilde{G}_n(\mathbb{R}^N))$ is a unique linear combination of cup products of the classes corresponding, via Theorem 32, to the forms
>
> $$f_2(\zeta),\ldots,f_{k/4}(\zeta) .$$

Case 2. n is even. The forms $f_2(\zeta),\ldots,f_{n-2}(\zeta)$, $\mathrm{Pf}(\zeta)$ have dimensions $4,8,\ldots,2n-4$, n. So

> if $k < N-n$ is odd, then $H^k(\tilde{G}_n(\mathbb{R}^N)) = 0$
>
> if $k < N-n$ is even, then every element of $H^k(\tilde{G}_n(\mathbb{R}^N))$ is a unique linear combination of cup products of the classes corresponding to the forms
>
> $$f_2(\zeta),\ldots,f_{[k/4]}(\zeta) , \quad \text{and} \quad \mathrm{Pf}(\zeta) \quad \text{if } k \geq n .$$

This can all be said more prettily if we fix n and allow N is increase:

41. PROPOSITION. If $\alpha\colon \tilde{G}_n(\mathbb{R}^N) \to \tilde{G}_n(\mathbb{R}^M)$ is the natural map, and $M > N > n+k$, then the induced map

$$\alpha^*\colon H^k(\tilde{G}_n(\mathbb{R}^M)) \to H^k(\tilde{G}_n(\mathbb{R}^N))$$

is an isomorphism.

Proof. Because of the preceeding discussion, it obviously suffices to show that the element in $H^k(\tilde{G}_n(\mathbb{R}^M))$ corresponding to f_r goes by α^* to the element in $H^k(\tilde{G}_n(\mathbb{R}^N))$ corresponding to f_r. Proving this is just a matter of unravelling definitions, and will provide a good opportunity to set straight everything done up till now. ■

Henceforth we consider only N sufficiently large so that $4[n/2] < N-n$ for odd n, and $2n-4 < N-n$ and $n < N-n$ for even n (we can take $N > 3n-2$ in both cases). Then all elements of $H^*(\tilde{G}_n(\mathbb{R}^N))$ in dimensions $< N-n$ are unique linear combinations of cup products of the classes corresponding to

$$\begin{array}{ll} f_2(\zeta),\ldots,f_{2[n/2]}(\zeta) & n \text{ odd} \\ f_2(\zeta),\ldots,f_{n-2}(\zeta),\mathrm{Pf}(\zeta) & n \text{ even}. \end{array}$$

We let

$$p_{n;k} \in H^{4k}(\tilde{G}_n(\mathbb{R}^N)) \qquad k = 1,\ldots,[n/2]$$

be the class corresponding to

$$\frac{1}{(2\pi)^{2k}} f_{2k}(\zeta),$$

and we let

$$e_n \in H^n(\tilde{G}_n(\mathbb{R}^N)) \qquad n = 2m$$

be the class corresponding to

$$\frac{1}{(2\pi)^m}\,\mathrm{Pf}(\zeta)\ .$$

We defined $p_{n;k}$ for $k = 1,\ldots,[n/2]$ for both odd and even n, just for simplicity. For even n this gives us the extra class $p_{n;n/2}$, corresponding to $f_n(\zeta)/(2\pi)^n$. It satisfies

$$p_{n;n/2} = e_n \cup e_n\ ,$$

since $e_n \cup e_n$ corresponds to

$$\frac{1}{(2\pi)^m}\,\mathrm{Pf}(\zeta) \wedge \frac{1}{(2\pi)^m}\,\mathrm{Pf}(\zeta) = \frac{1}{(2\pi)^n}\,\mathrm{Pf}^2(\zeta) \qquad \text{by equation (1) on p.502}$$

$$= \frac{1}{(2\pi)^n}\,\det(\zeta) = \frac{1}{(2\pi)^n}\,f_n(\zeta)\ .$$

In these definitions, we are taking N large, and applying Proposition 41, so that there is no need to have an extra subscript N on the symbols $p_{n;k}$ and e_n.

In accordance with our discussion in section 5, each class

$$p_{n;k} \in H^{4k}(\tilde{G}_n(\mathbb{R}^N)) \qquad N \text{ large}$$

determines a "characteristic class", that is, a function

$$\xi \longmapsto p_{n;k}(\xi) \in H^{4k}(M)$$

which assigns to a smooth oriented n-dimensional bundle $\xi = \pi\colon E \longrightarrow M$ an element of the cohomology of M. Explicitly,

$$p_{n;k}(\xi) = g^* p_{n;k}$$

where

$$g\colon M \longrightarrow \tilde{G}_n(\mathbb{R}^N) \qquad \text{satisfies} \qquad g^*\tilde{\gamma}^n(\mathbb{R}^N) \simeq \xi .$$

This characteristic class is called the kth Pontryagin class for n-dimensional bundles. For even n, we have the additional class

$$\xi \longmapsto e_n(\xi) .$$

When one is dealing with characteristic classes, the number n is usually apparent, since it is the fibre dimension of the bundle whose characteristic class is being considered. Consequently, we write simply $p_k(\xi)$ and $e(\xi)$. If ξ has fibre dimension n, then $p_k(\xi)$ is defined for $k = 1,\dots,[n/2]$; if n is even, then we also have the class $e(\xi)$, and $p_{n/2}(\xi) = e(\xi) \cup e(\xi)$.

Since all elements of $H^*(\tilde{G}_n(\mathbb{R}^N))$ in dimensions $< N-n$ are linear combinations of the $p_{n;k}$ and e_n, we see that all characteristic classes for oriented n-dimensional bundles are polynomials in the Pontryagin classes p_k, together with e if n is even. In particular, the Euler class must be representable in this way, and our notation clearly suggests that the Euler class is, in fact, just the characteristic class e. In order to prove this, we have to look a little more carefully at the universal bundles.

Consider the universal bundle $\tilde{\gamma}^n(\mathbb{R}^N) = \pi\colon E(\tilde{\gamma}^n(\mathbb{R}^N)) \longrightarrow \tilde{G}_n(\mathbb{R}^N)$. A point of $\tilde{G}_n(\mathbb{R}^N)$ is an oriented n-dimensional subspace $W \subset \mathbb{R}^N$, and the fibre $\pi^{-1}(W)$ over W is $\{(W,w)\colon w \in W\}$. So there is a natural Riemannian metric $\langle\ ,\ \rangle$ on $\tilde{\gamma}^n(\mathbb{R}^N)$: the inner product of (W,w_1), $(W,w_2) \in \pi^{-1}(W)$ is just the usual inner product of $w_1, w_2 \in \mathbb{R}^N$. For the corresponding principal bundle $\varpi\colon SO(E(\tilde{\gamma}^n(\mathbb{R}^N))) \to \tilde{G}_n(\mathbb{R}^N)$, the fibre $\varpi^{-1}(W)$ is the set of all $(W,(w_1,\ldots,w_n))$, where $(w_1,\ldots,w_n)$ is a positively oriented orthonormal n-frame in $W \subset \mathbb{R}^N$. Now we can define a map $\lambda\colon SO(N) \to SO(E(\tilde{\gamma}^n(\mathbb{R}^N)))$, from the special orthogonal group $SO(N)$ to the total space of this principal bundle, as follows: If $W_0 \subset \mathbb{R}^N$ is the subspace spanned by $e_1,\ldots,e_n$, then

$$\lambda(A) = (A(W_0),(A(e_1),\ldots,A(e_n))) \,.$$

It is easy to see that for the point $x = (W_0,(e_1,\ldots,e_n)) \in SO(E(\tilde{\gamma}^n(\mathbb{R}^N)))$, we have

$$\lambda^{-1}(x) = \{I\} \times SO(N-n) \;;$$

more generally, for any $x \in SO(E(\tilde{\gamma}^n(\mathbb{R}^N))$ the set $\lambda^{-1}(x)$ is a left coset of $\{I\} \times SO(N-n)$. So $SO(E(\tilde{\gamma}^n(\mathbb{R}^N))$ can be identified with the left coset space

$$SO(N)/\ \{I\} \times SO(N-n) \,.$$

We leave it as an exercise for the reader to show that the topology and C^∞ structure on $SO(E(\tilde{\gamma}^n(\mathbb{R}^N)))$ is the same as that on this left coset space, and that the projection

$$\varpi : SO(N)/\ \{I\} \times SO(N-n) \to SO(N)/\ SO(k) \times SO(N-n)$$

is just the natural map taking the coset $A\cdot[\{I\} \times SO(N-k)]$ to the coset $A\cdot[SO(k) \times SO(N-k)]$. Notice that the diagram

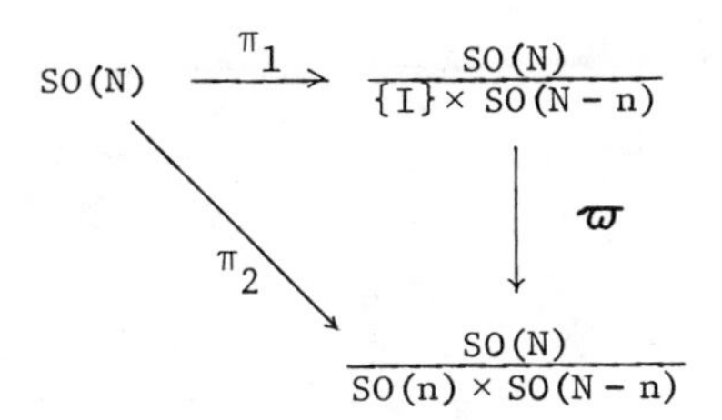

commutes, where π_1 and π_2 are the natural projections. As in section 6, we will use L_A for the left multiplication $L_A\colon SO(N) \to SO(N)$, and $\mathbf{L}_A$ for the diffeomorphism of $SO(N)/\{I\} \times SO(N-n)$ taking the coset $B\cdot[\{I\} \times SO(N-n)]$ to $AB\cdot[\{I\} \times SO(N-n)]$. We also have the map $R_A\colon SO(N) \to SO(N)$, and the map $\mathbf{R}_A$ taking the coset $B\cdot[\{I\} \times SO(N-n)]$ to $BA\cdot[\{I\} \times SO(N-n)]$. The reader should check that for $A \in SO(n)$, the map $\mathbf{R}_A$ corresponds to the right multiplication by A in the principal bundle $SO(\tilde{\gamma}^n(\mathbb{R}^N))$.

On $SO(N)$ we have the left invariant 1-forms $\tilde{\phi}_\alpha^\beta$ $(\alpha < \beta)$ whose values at $I \in SO(N)$ are the elements $\phi_\alpha^\beta \in \mathfrak{o}(N)^*$; set $\tilde{\phi}_\alpha^\beta = -\tilde{\phi}_\beta^\alpha$ for $\alpha > \beta$, and $\tilde{\phi}_\alpha^\alpha = 0$. We claim that for $i, j \le n$ there are unique 1-forms ω_j^i on $SO(N)/\{I\} \times SO(N-n)$ such that

$$(1) \qquad \pi_1{}^*\omega_j^i = \tilde{\phi}_j^i .$$

To prove this we first note that π_{1*} is always onto. Since we need to have

$$(2) \qquad \omega_j^i(\pi_{1*}X) = \tilde{\phi}_j^i(X)$$

for all tangent vectors X of $SO(N)$, this proves uniqueness. To prove existence, we need to show that definition (2) is well-defined, by showing that $\tilde{\phi}^i_j(X) = 0$ whenever $\pi_{1*}X = 0$. So suppose $X \in SO(N)_A$. Then $X = L_{A*}X_I$ for some $X_I \in \mathfrak{o}(N)$. Since $\pi_1 \circ L_A = \mathbf{L}_A \circ \pi_1$, we see that

$$\begin{aligned}
\pi_{1*}X = 0 &\implies \pi_{1*}L_{A*}X_I = 0 \\
&\implies \mathbf{L}_{A*}\pi_{1*}X_I = 0 \\
&\implies \pi_{1*}X_I = 0 \qquad \text{since } \mathbf{L}_A \text{ is a diffeomorphism} \\
&\implies X_I \text{ is of the form } \begin{pmatrix} 0 & 0 \\ 0 & * \end{pmatrix} \\
&\implies \phi^i_j(X_I) = 0 \\
&\implies \tilde{\phi}^i_j(X) = 0 \qquad \text{since } \tilde{\phi}^i_j \text{ is left-invariant.}
\end{aligned}$$

Thus the forms ω^i_j exist. Note that

$$\begin{aligned}
\mathbf{L}_A{}^*\omega^i_j(\pi_{1*}X) &= \omega^i_j(\mathbf{L}_{A*}\pi_{1*}X) \\
&= \omega^i_j(\pi_{1*}L_{A*}X) \\
&= \tilde{\phi}^i_j(L_{A*}X) \\
&= \tilde{\phi}^i_j(X) = \omega^i_j(\pi_{1*}X) .
\end{aligned}$$

So

$$\mathbf{L}_A{}^*\omega^i_j = \omega^i_j .$$

Now $\omega = (\omega^i_j)$ is an $\mathfrak{o}(n)$-valued 1-form on $SO(N)/\{I\} \times SO(N-n)$. We claim that ω is, in fact, a connection on the principal bundle $\varpi\colon SO(N)/\{I\} \times SO(N-n) \to SO(N)/SO(n) \times SO(N-n)$. We have to check that

$$\omega(\sigma(M)) = M \qquad \text{for } M \in \mathfrak{o}(n)$$

$$\omega(\mathbf{R}_{A*}Y) = A^{-1}\omega(Y)A \qquad \text{for } A \in SO(n) \text{ and } Y \text{ a tangent vector on } SO(N)/\{I\} \times SO(N-n)\ .$$

Recall that the value of $\sigma(M)$ at the coset $B\cdot[\{I\} \times SO(N-n)]$ is $c'(0)$ where

$$\begin{aligned} c(t) &= \mathbf{R}_{\exp\, tM}(B\cdot[\{I\} \times SO(N-n)]) \\ &= B(\exp\, tM)\cdot[\{I\} \times SO(N-n)] \\ &= \pi_1 L_B(\exp\, tM) \\ \Longrightarrow\ c'(0) &= \pi_{1*}L_{B*}M\ . \end{aligned}$$

Thus

$$\begin{aligned} \omega^i_j(\sigma(M) \text{ at } B\cdot[\{I\} \times SO(N-n)]) &= \omega^i_j(c'(0)) \\ &= \omega^i_j(\pi_{1*}L_{B*}M) \\ &= \tilde{\phi}^i_j(L_{B*}M) \\ &= \phi^i_j(M) = M^i_j\ , \end{aligned}$$

which proves the first condition. To prove the second, take a tangent vector Y at the coset $B\cdot[\{I\} \times SO(N-n)]$ and choose a tangent vector $X \in O(N)_B$ with $\pi_{1*}X = Y$. Then

$$\omega^i_j(Y) = \omega^i_j(\pi_{1*}X) = \tilde{\phi}^i_j(X) = \phi^i_j(L_{B^{-1}*}X)\ ,$$

while

$$
\begin{aligned}
\omega^i_j(\mathbf{R}_{A*}Y) &= \omega^i_j(\mathbf{R}_{A*}\pi_{1*}X) = \omega^i_j(\pi_{1*}R_{A*}X) \\
&= \tilde{\phi}^i_j(R_{A*}X) && R_{A*}X \text{ a tangent vector at } BA \\
&= \phi^i_j(L_{(BA)^{-1}*}R_{A*}X) \\
&= \phi^i_j(L_{A^{-1}*}R_{A*}L_{B^{-1}*}X) \\
&= \phi^i_j(\mathrm{Ad}(A^{-1})L_{B^{-1}*}X) \\
&= \phi^i_j(A^{-1}(L_{B^{-1}*}X)A) \\
&= \sum_{\mu,\nu=1}^{k}(A^{-1})^i_\mu\phi^i_j(L_{B^{-1}*}X)^\mu_\nu A^\nu_j && \text{by linearity of } \phi^i_j ,
\end{aligned}
$$

which proves the second condition.

It is easy to see which vectors $\pi_{1*}X$ are vertical or horizontal when $X \in \mathfrak{o}(N)$. First of all,

$$
\pi_{1*}X \text{ is vertical} \iff \varpi_*\pi_{1*}X = 0 \iff \pi_{2*}X = 0 \iff X \in \mathfrak{o}(n) \times \mathfrak{o}(N-n) .
$$

On the other hand,

$$
\begin{aligned}
\pi_{1*}X \text{ is horizontal} &\iff \text{all } \omega^i_j(\pi_{1*}X) = 0 \\
&\iff \text{all } \phi^i_j(X) = 0 \\
&\iff X \text{ has the form } \begin{pmatrix} 0 & * \\ * & * \end{pmatrix} .
\end{aligned}
$$

Given $X \in \mathfrak{o}(N)$, we write it as

$$
\begin{aligned}
X &= \begin{pmatrix} 0 & * \\ * & * \end{pmatrix} + \begin{pmatrix} * & 0 \\ 0 & 0 \end{pmatrix} \\
&= X_1 + X_2 .
\end{aligned}
$$

Then $\pi_{1*}X_1$ is horizontal and $\pi_{1*}X_2$ is vertical. This means that the horizontal component of $\pi_{1*}X$ is precisely

$$h(\pi_{1*}X) = \pi_{1*}X_1 .$$

To compute the curvature forms Ω^i_j for the connection ω^i_j, we use the fact (Problem 7-15) that the forms $\tilde{\phi}^\beta_\alpha$ satisfy

$$d\tilde{\phi}^\beta_\alpha = - \sum_{\gamma=1}^{N} \tilde{\phi}^\beta_\gamma \wedge \tilde{\phi}^\gamma_\alpha .$$

Then for $X, Y \in \mathcal{O}(N)$ we have

$$\begin{aligned}
\Omega^i_j(\pi_{1*}X, \pi_{1*}Y) &= d\omega^i_j(h\pi_{1*}X, h\pi_{1*}Y) \\
&= d\omega^i_j(\pi_{1*}X_1, \pi_{1*}Y_1) \\
&= d(\pi_1{}^*\omega^i_j)(X_1, Y_1) \\
&= d\tilde{\phi}^i_j(X_1, Y_1) \\
&= - \sum_{\gamma=1}^{N} \phi^i_\gamma \wedge \phi^\gamma_j(X_1, Y_1) .
\end{aligned}$$

Since

$$X_1 = X - \sum_{\alpha \text{ or } \beta > n} \phi^\beta_\alpha(X) \cdot X^\beta_\alpha , \qquad Y_1 = Y - \sum_{\alpha \text{ or } \beta > n} \phi^\beta_\alpha(X) \cdot X^\beta_\alpha ,$$

this gives

$$\begin{aligned}
\Omega^i_j(\pi_{1*}X, \pi_{1*}Y) &= \sum_r \phi^r_i \wedge \phi^r_j(X, Y) \\
&= \zeta_{ij}(X, Y) ,
\end{aligned}$$

which shows that

$$(*) \qquad \pi_1{}^*\Omega^i_j = \zeta_{ij} \qquad \text{at } SO(N)_I .$$

(In fact, we also have $\pi_1{}^*\Omega^i_j = \tilde{\zeta}_{ij}$, where $\tilde{\zeta}_{ij}$ is the left invariant form extending ζ_{ij}, since the equation $\mathsf{L}_A{}^*\omega^i_j = \omega^i_j$ implies that we also have $\mathsf{L}_A{}^*\Omega^i_j = \Omega^i_j$.)

42. THEOREM. If ξ is an oriented n-dimensional bundle, with $n = 2m$ even, then the characteristic class $e(\xi)$ is the Euler class $\chi(\xi)$.

Proof. It suffices to prove this when ξ is the universal bundle $\tilde{\gamma}^n(\mathbb{R}^N)$. We know, by Corollary 25 and Theorem 26, that the Euler class $\chi(\tilde{\gamma}^n(\mathbb{R}^N))$ is represented by the unique form Γ on $SO(N)/ SO(n) \times SO(N-n)$ such that

$$\begin{aligned} \varpi^*\Gamma &= \frac{1}{\pi^m m!\, 2^n} \sum_{i_1,\dots,i_n} \varepsilon^{i_1\cdots i_n}\, \Omega^{i_1}_{i_2} \wedge \cdots \wedge \Omega^{i_{n-1}}_{i_n} \\ &= \frac{1}{\pi^m m!\, 2^n}\, 2^m \cdot m!\ \mathrm{Pf}(\Omega) \\ &= \frac{1}{(2\pi)^m}\, \mathrm{Pf}(\Omega) . \end{aligned}$$

We want to show that Γ corresponds, via Theorem 32, to the form

$$\frac{1}{(2\pi)^m}\, \mathrm{Pf}(\zeta) .$$

Note first that if L'_A is the diffeomorphism of $SO(N)/ SO(n) \times SO(N-n)$

taking the coset $B\cdot[SO(n)\times SO(N-n)]$ to $AB\cdot[SO(n)\times SO(N-n)]$, then $L'_A\circ\varpi=\varpi\circ L_A$. So

$$\begin{aligned}\varpi^*(L'_A{}^*\Gamma) &= L_A{}^*\varpi^*\Gamma = \frac{1}{(2\pi)^m} L_A{}^*\mathrm{Pf}(\Omega)\\ &= \frac{1}{(2\pi)^m}\mathrm{Pf}(L_A{}^*\Omega) = \frac{1}{(2\pi)^m}\mathrm{Pf}(\Omega)\\ &= \varpi^*\Gamma\ .\end{aligned}$$

By uniqueness in Proposition 18 we have $L'_A{}^*\Gamma=\Gamma$ for all $A\in SO(N)$. In other words, Γ is an invariant form on $SO(N)/SO(n)\times SO(N-n)$, as defined in section 6. So the element of $\Omega^n(\mathfrak{o}(N))$ corresponding to Γ in Theorem 32 is simply $\pi_2{}^*\Gamma$ at I. But

$$\begin{aligned}\pi_2{}^*\Gamma &= \pi_1{}^*\varpi^*\Gamma\\ &= \frac{1}{(2\pi)^m}\pi_1{}^*\mathrm{Pf}(\Omega) = \frac{1}{(2\pi)^m}\mathrm{Pf}(\pi_1{}^*\Omega)\\ &= \frac{1}{(2\pi)^m}\mathrm{Pf}(\zeta) \quad \text{at } I\,, \qquad \text{by equation } (*).\ \blacksquare\end{aligned}$$

The classes $p_k(\xi)$ of an oriented bundle $\xi=\pi\colon E\to M$ may be described in exactly the same way: Choose a Riemannian metric $\langle\ ,\ \rangle$ for ξ, form the principal bundle $O(\xi)=\varpi\colon O(E)\to M$, and let ω be a connection on $O(E)$, with curvature form Ω.

<u>43. THEOREM</u>. There is a unique 4k-form Λ on M such that

$$\varpi^*\Lambda = \frac{1}{(2\pi)^{2k}} f_{2k}(\Omega)\ .$$

This form Λ is closed, and the cohomology class $[\Lambda]$ is independent of the choice of the Riemannian metric $\langle\ ,\ \rangle$ and the connection ω in terms of which Λ is defined. The cohomology class $[\Lambda]$ is precisely $p_k(\xi)$.

Proof. The proofs of Propositions 18, 19, and 20 can be adapted, essentially without modification, to prove the first two assertions. The proof of the final assertion is exactly like the proof of Theorem 42. ■

As a simple application, we consider a Riemannian manifold $(M, \langle\ ,\ \rangle)$ of constant curvature K_0. Then the curvature form Ω on $O(TM)$ satisfies

$$\Omega^i_j = K_0\ \theta^i \wedge \theta^j .$$

To calculate $f_{2k}(\Omega)$, we use the explicit formula given in Problem I.7-14, to obtain

$$f_{2k}(\Omega) = \frac{1}{(2k)!} \sum_{\substack{i_1 \cdots i_{2k} \\ j_1 \cdots j_{2k}}} \Omega^{j_1}_{i_1} \wedge \cdots \wedge \Omega^{j_{2k}}_{i_{2k}}\ \delta^{i_1 \cdots i_{2k}}_{j_1 \cdots j_{2k}}$$

$$= \frac{(K_0)^{2k}}{(2k)!} \sum_{\substack{i_1 \cdots i_{2k} \\ j_1 \cdots j_{2k}}} \theta^{j_1} \wedge \theta^{i_1} \wedge \cdots \wedge \theta^{j_{2k}} \wedge \theta^{i_{2k}}\ \delta^{i_1 \cdots i_{2k}}_{j_1 \cdots j_{2k}} .$$

In this sum, the δ factor vanishes unless $j_1, \ldots, j_{2k}$ is a permutation of $i_1, \ldots, i_{2k}$; but then $\theta^{j_1} \wedge \cdots \wedge \theta^{j_{2k}}$ has repeated factors, so it vanishes. Thus,

44. COROLLARY. If M^n is a compact manifold of constant curvature, then

$$p_k(TM) = 0 , \qquad k = 1, \ldots, [n/4] .$$

Another application of Theorem 43 gives us an analogue of Theorem 17. For a bundle ξ over M we define the <u>total Pontryagin class</u> $p(\xi)$ to be the element of $H^0(M) \oplus H^4(M) \oplus \cdots$ given by

$$p(\xi) = 1 + p_1(\xi) + \cdots + p_{[n/2]}(\xi) = p_0(\xi) + p_1(\xi) + \cdots + p_{[n/2]}(\xi) ,$$

where $1 \in H^0(M)$ is the standard element (represented by the constant function 1 on M).

<u>45. THEOREM</u>. If ξ and η are oriented bundles over the same compact manifold M, then the total Pontryagin class of $\xi \oplus \eta$ is given by the <u>Whitney product formula</u>

$$p(\xi \oplus \eta) = p(\xi) \cup p(\eta) .$$

[This means that

$$p_k(\xi \oplus \eta) = \sum_{\ell=0}^{k} p_\ell(\xi) \cup p_{k-\ell}(\eta) .]$$

<u>Proof</u>. The proof will be almost exactly like the proof of Theorem 22. For convenience we rename our bundles ξ and η as $\xi_i = \pi_i\colon E_i \to M$ for $i = 1,2$, and let $\xi_1 \oplus \xi_2 = \pi\colon E \to M$. We introduce the corresponding principal bundles $\varpi_i\colon SO(E_i) \to M$ and $\varpi\colon SO(E) \to M$, the principal bundle $SO(E_1)*SO(E_2) \to M$, and the projections $\rho_i\colon SO(E_1)*SO(E_2) \to SO(E_1)$. Choose connections ω_i on $SO(E_i)$, with curvature forms Ω_i. Then

$$\rho_1{}^*\omega_1 \oplus \rho_2{}^*\omega_2 = \begin{pmatrix} \rho_1{}^*\omega_1 & 0 \\ 0 & \rho_2{}^*\omega_2 \end{pmatrix}$$

is a connection $\bar{\omega}$ on $SO(E_1)*SO(E_2)$, with curvature form

$$\bar{\Omega} = \rho_1{}^*\Omega_1 \oplus \rho_2{}^*\Omega_2 = \begin{pmatrix} \rho_1{}^*\Omega_1 & 0 \\ 0 & \rho_2{}^*\Omega_2 \end{pmatrix},$$

and $\bar{\omega}$ can be extended uniquely to a connection $\tilde{\omega}$ on $SO(E)$. At a point $e \in SO(E_1) * SO(E_2)$ we have

$$\tilde{\Omega} = \bar{\Omega} \qquad \text{(on tangent vectors to } SO(E_1) * SO(E_2))$$

$$\Longrightarrow f_{2k}(\tilde{\Omega}) = f_{2k}(\bar{\Omega}) = \sum_{\ell=0}^{k} f_{2\ell}(\rho_1{}^*\Omega_1) \wedge f_{2k-2\ell}(\rho_2{}^*\Omega_2)$$

by the formula on p. 495

$$= \sum_{\ell=0}^{k} \rho_1{}^*f_{2\ell}(\Omega_1) \wedge \rho_2{}^*f_{2k-2\ell}(\Omega_2) .$$

So if Λ_k is the form representing $p_{2k}(\xi_1 \oplus \xi_2)$, and T^i_ℓ are the forms representing $p_{2\ell}(\xi_i)$, then at e we have [on tangent vectors to $SO(E_1)*SO(E_2)$]

$$\varpi^*\Lambda_k = \frac{1}{(2\pi)^{2k}} f_{2k}(\tilde{\Omega}) = \sum_{\ell=0}^{k} \frac{1}{(2\pi)^{\ell}} \rho_1{}^*f_{2\ell}(\Omega_1) \wedge \frac{1}{(2\pi)^{2k-\ell}} \rho_2{}^*f_{2k-2\ell}(\Omega_2)$$

$$= \sum_{\ell=0}^{k} \rho_1{}^*\varpi_1{}^*T^1_\ell \wedge \rho_2{}^*\varpi_2{}^*T^2_{k-\ell}$$

$$= \sum_{\ell=0}^{k} \varpi^*T^1_\ell \wedge \varpi^*T^2_{k-\ell} .$$

This implies that

$$\Lambda_k = \sum_{\ell=0}^{k} T^1_\ell \wedge T^2_{k-\ell} . \quad \blacksquare$$

10. The Weil Homomorphism

The invariant polynomial functions $f_{2k}\colon \mathfrak{o}(n) \to \mathbb{R}$ and $\mathrm{Pf}\colon \mathfrak{o}(n) \to \mathbb{R}$ arose naturally in our attempts to calculate the cohomology of $\tilde{G}_n(\mathbb{R}^N)$; each one gave us an element of $H^*(\tilde{G}_n(\mathbb{R}^N))$, and hence a characteristic class $\xi \mapsto C(\xi)$ for oriented bundles. On the other hand, at the end of the last section we saw how these characteristic classes $C(\xi)$ could be defined directly for the bundle ξ, by means of a connection on the associated principal bundle $SO(\xi)$. There is no reason why we cannot use exactly the same procedure for groups other than $SO(n)$.

For any Lie group G, with Lie algebra $\mathfrak{g}$, we consider the set $\mathcal{P}(\mathfrak{g})$ of functions $f\colon \mathfrak{g} \to \mathbb{R}$ which can be expressed as polynomials in $\{\phi^\alpha\}$, where $\{\phi^\alpha\}$ is a basis of $\mathfrak{g}^*$. Such functions are called <u>polynomial functions</u> on $\mathfrak{g}$ (the concept is clearly independent of the choice of basis $\{\phi_\alpha\}$), and the set of all homogeneous polynomial functions of degree k will be denoted by $\mathcal{P}^k(\mathfrak{g})$. We say that $f\colon \mathfrak{g} \to \mathbb{R}$ is $\mathrm{Ad}(G)$-<u>invariant</u> if $f \circ \mathrm{Ad}(a) = f$ for all $a \in G$. Instead of considering polynomial functions on $\mathfrak{g}$ it is often more convenient to consider the set $\mathcal{S}^k(\mathfrak{g})$ of symmetric k-linear maps $f\colon \mathfrak{g} \times \cdots \times \mathfrak{g} \to \mathbb{R}$. Given $f \in \mathcal{S}^k(\mathfrak{g})$, we define a polynomial function $\mathcal{P}f \in \mathcal{P}^k(\mathfrak{g})$ by

$$(\mathcal{P}f)(X) = f(X,\ldots,X) \qquad X \in \mathfrak{g} .$$

Conversely, given a basis $\phi^1,\ldots,\phi^r$ of $\mathfrak{g}$, and a polynomial function f of degree k on $\mathfrak{g}$, we can write it uniquely as

$$\sum_{\alpha_1,\ldots,\alpha_k=1}^{r} a_{\alpha_1\cdots\alpha_k} \phi^{\alpha_1}\cdots\phi^{\alpha_k}$$

<u>where the</u> $a_{\alpha_1\cdots\alpha_k}$ <u>are symmetric in</u> $\alpha_1,\ldots,\alpha_k$; then we define $\mathcal{S}f \in \mathcal{S}^k(\mathfrak{g})$ by

$$(\mathcal{S}f)(X_1,\ldots,X_k) = \sum a_{\alpha_1\cdots\alpha_k}\,\phi^{\alpha_1}(X_1)\cdots\phi^{\alpha_k}(X_k)\,, \qquad X_1,\ldots,X_k \in \mathfrak{g}\,.$$

It is easy to check that the maps

$$\mathcal{P}\colon \mathcal{S}^k(\mathfrak{g}) \to \mathcal{P}^k(\mathfrak{g})\,, \qquad \mathcal{S}\colon \mathcal{P}^k(\mathfrak{g}) \to \mathcal{S}^k(\mathfrak{g})$$

are inverses to each other (so $\mathcal{S}$ doesn't depend on the choice of basis). For $f \in \mathcal{S}^k(\mathfrak{g})$ and $g \in \mathcal{S}^\ell(\mathfrak{g})$ we define $fg \in \mathcal{S}^{k+\ell}(\mathfrak{g})$ by

$$fg(X_1,\ldots,X_{k+\ell}) = \frac{1}{(k+\ell)!}\sum_{\pi\in S_{k+\ell}} f(X_{\pi(1)},\ldots,X_{\pi(k)})\cdot g(X_{\pi(k+1)},\ldots,X_{\pi(k+\ell)})\,.$$

This makes $\mathcal{P}(fg) = \mathcal{P}(f)\cdot\mathcal{P}(g)$. We define $f \in \mathcal{S}^k(\mathfrak{g})$ to be $\mathrm{Ad}(G)$-<u>invariant</u> if

$$f(\mathrm{Ad}(a)X_1,\ldots,\mathrm{Ad}(a)X_k) = f(X_1,\ldots,X_k)$$
$$\text{for all } a \in G \text{ and } X_1,\ldots,X_k \in \mathfrak{g}\,;$$

then f is $\mathrm{Ad}(G)$-invariant if and only if $\mathcal{P}f$ is $\mathrm{Ad}(G)$-invariant. The set of all $f \in \mathcal{S}^k(\mathfrak{g})$ which are $\mathrm{Ad}(G)$-invariant is denoted by $I^k(G)$. Thus, $\mathcal{P}$ takes $I^k(G)$ into the set of polynomial functions on $\mathfrak{g}$ which are $\mathrm{Ad}(G)$-invariant, and $\mathcal{S}$ takes this set back to $I^k(G)$.

Now let $\pi\colon P \to M$ be a principal bundle with group G, and let ω be a connection, with curvature form Ω. Thus both ω and Ω are $\mathfrak{g}$-valued, so if $\phi^1,\ldots,\phi^r$ is a basis of $\mathfrak{g}^*$, then we can write $\omega = \sum \omega^\alpha\cdot\phi^\alpha$ and

$\Omega = \sum \Omega^{\alpha} \cdot \phi^{\alpha}$ for ordinary forms ω^{α} and Ω^{α}. Given $f \in \mathcal{P}^k(\mathfrak{g})$, we write it as a sum of terms

$$c \cdot \phi^{\alpha_1} \cdots \phi^{\alpha_k} ,$$

and then let $f(\Omega)$ be the 2k-form on P which is the corresponding sum of terms

$$c \cdot \Omega^{\alpha_1} \wedge \cdots \wedge \Omega^{\alpha_k}$$

(since the Ω^{α} are 2-forms, the order in the product $\phi^{\alpha_1} \cdots \phi^{\alpha_k}$ is irrelevant). This is the definition used previously for the case $G = SO(n)$, where the Lie algebra $\mathfrak{o}(n)$ has a natural basis $\{\phi^i_j\}_{i<j}$ (provided we use the convention that a polynomial in the ϕ^i_j $(i, j = 1,\dots,n)$ be interpreted as a polynomial in the $\{\phi^i_j\}_{i<j}$ by replacing ϕ^j_i by $-\phi^i_j$ for $j > i$). A more intrinsic description is possible when we work with $f \in \mathcal{S}^k(\mathfrak{g})$. We now define $f(\Omega)$ to be the 2k-form on P given by

$$f(\Omega)(X_1,\dots,X_{2k}) = \frac{1}{2^k} \sum_{\pi \in S_{2k}} \operatorname{sgn} \pi \; f(\Omega(X_{\pi(1)},X_{\pi(2)}),\dots,\Omega(X_{\pi(2k-1)},X_{\pi(2k)})) ,$$

where $X_1,\dots,X_{2k}$ are now tangent vectors of P. It can be checked that $(\mathcal{P}f)(\Omega) = f(\Omega)$ for $f \in \mathcal{S}^k(\mathfrak{g})$; equivalently, $(\mathcal{S}f)(\Omega) = f(\Omega)$ for $f \in \mathcal{P}^k(\mathfrak{g})$ (so the definition of $f(\Omega)$ doesn't depend on the choice of $\{\phi^{\alpha}\}$).

46. THEOREM. Let $\xi = \pi: P \to M$ be a principal bundle with group G, and let ω be a connection on P, with curvature form Ω. Then for every

$f \in I^k(G)$ there is a unique 2k-form Λ on M such that

$$\pi^*\Lambda = f(\Omega) .$$

The form Λ is closed, and its de Rham cohomology class $w_\xi(f) = [\Lambda]$ is independent of the choice of ω. For $f \in I^k(G)$ and $g \in I^\ell(G)$ we have $w_\xi(fg) = w_\xi(f) \cup w_\xi(g)$.

Proof. Exactly like the proofs of Theorems 18, 19, and 20. ■

If we set $H^*(M) = H^0(M) \oplus H^1(M) \oplus \cdots$ and $I(G) = \mathbb{R} \oplus I^1(G) \oplus \cdots$, then we have a homomorphism $w_\xi\colon I(G) \to H^*(M)$, depending only on the given principal bundle $\xi = \pi\colon P \to M$. This map is called the Weil homomorphism. It is natural, in the following sense.

47. PROPOSITION. Let $\pi\colon P \to M$ be a principal bundle with group G and let $f\colon M' \to M$ be a smooth map, inducing the map $f^*\colon H^*(M) \to H^*(M')$. Then

$$w_{f^*\xi} = f^* \circ w_\xi .$$

Proof. An elementary exercise (just like the proof of Proposition 21). ■

If we take $G = SO(n)$ in Theorem 46, and consider the functions $g_{2k} = \mathscr{S}(f_{2k}) \in I^{2k}(SO(n))$ corresponding to the polynomial functions f_{2k} on $\mathfrak{o}(n)$, then we have classes $w_\eta(g_{2k}) \in H^{4k}(M)$ for any principal $SO(n)$ bundle η over M. If $\xi = \pi\colon E \to M$ is an oriented n-dimensional vector bundle over M, then we can form the principal bundle $\eta = SO(\xi)$ by means of a Riemannian metric on ξ [all such η are equivalent by Corollary 5], and Theorem 43 amounts to the assertion that $w_\eta(g_{2k}) = p_k(\xi)$. Notice that the $SO(n)$-invariant polynomials f_{2k} on $\mathfrak{o}(n)$ are also $GL(n,\mathbb{R})$-invariant polynomials on $\mathfrak{gl}(n,\mathbb{R})$. So there are corresponding $g'_{2k} \in I^{2k}(GL(n,\mathbb{R}))$ which restrict to g_{2k} on $\mathfrak{o}(n) \times \cdots \times \mathfrak{o}(n)$. Now a connection ω for the principal bundle $\eta = SO(\xi)$ extends to a connection ω' for the principal bundle $\eta' = F(\xi)$ of frames of ξ, and Ω' is an extension of Ω. Thus the form

$$g'_{2k}(\Omega') \qquad \text{restricts to} \qquad g_{2k}(\Omega) \quad \text{on} \quad SO(E)\ .$$

This shows that $w_{\eta'}(g'_{2k}) = w_\eta(g_{2k}) = p_k(\xi)$. Since $w_{\eta'}(g'_{2k})$ doesn't depend on the particular connection Ω' for $F(\xi)$, we see that we can define $p_k(\xi)$ in terms of an arbitrary connection for $F(\xi)$; it is not necessary to use a connection which preserves inner products, and our bundle does not even have to be orientable.

On the other hand, for $n = 2m$, the Pfaffian $\mathrm{Pf}\colon \mathfrak{o}(n) \to \mathbb{R}$ is <u>not</u> $GL(n,\mathbb{R})$-invariant, nor even $GL^+(n,\mathbb{R})$-invariant, so our construction definitely requires orientability, and a connection on $SO(E)$, i.e., a connection compatible with some metric. Indeed, there are examples (see Milnor and Stasheff {1; p. 312}) of oriented bundles ξ having a connection ω with $\Omega = 0$, but with $\chi(\xi) \neq 0$; naturally such a connection cannot be compatible with any metric on ξ.

11. Complex Bundles

A <u>complex vector bundle</u> $\pi\colon E \to X$ is defined precisely like a real vector bundle, except that each fibre $\pi^{-1}(x)$ has the structure of a vector space over $\mathbb{C}$, and in all the conditions for a vector bundle, including local triviality, we replace $\mathbb{R}$ by $\mathbb{C}$ whenever it occurs; vector space isomorphisms are always understood to be isomorphisms of complex vector spaces, hence linear over $\mathbb{C}$. Linearity over $\mathbb{C}$ is also understood in the definitions of bundle maps (and equivalences) between complex bundles. The Whitney sum $\xi \oplus \eta$ of two complex bundles ξ and η is a complex bundle, and so is the induced bundle $f^*\xi$. The principal bundle $F(\xi)$ of frames is now a principal bundle with group $GL(n,\mathbb{C})$ = the set of all non-singular $n \times n$ matrices with complex entries (which may be identified in a natural way with the set of all non-singular linear transformations of $\mathbb{C}^n$). Note that the Covering Homotopy Theorem (Theorem 4) holds for complex bundles, since it holds for the corresponding principal bundles. There are two reasons for discussing complex bundles, and their characteristic classes. On the one hand, everything works out to be simpler; on the other hand, there are relations between the characteristic classes for real bundles and those for complex bundles. To discuss complex bundles, however, we need several preliminaries about complex vector spaces.

On the vector space $\mathbb{C}^n$ we could consider the bilinear function

$$(z,w) \mapsto \sum_{i=1}^{n} z^i w^i .$$

This is not an inner product, since it is not even real, and certainly not positive definite. The linear transformations $T\colon \mathbb{C}^n \to \mathbb{C}^n$ which preserve this bilinear function correspond to $n \times n$ complex matrices A such that $AA^t = I$. This group of matrices is known as the complex orthogonal group. It is of little interest to us, mainly because it is not compact. We consider instead the function $\mathbb{C}^n \times \mathbb{C}^n \longrightarrow \mathbb{C}$ given by

$$\langle z,w\rangle = \sum_{i=1}^{n} z^i \overline{w^i} .$$

More generally, for any vector space V over $\mathbb{C}$ we define an Hermitian inner product to be a map $\langle\ ,\ \rangle\colon V \times V \longrightarrow \mathbb{C}$ which is linear over $\mathbb{C}$ in the first variable, and which satisfies

$$\langle v,w\rangle = \overline{\langle w,v\rangle} \qquad (\Longrightarrow \langle v,v\rangle \text{ is real})$$

$$\langle v,v\rangle > 0 \qquad \text{for } v \neq 0 .$$

The first condition shows that $\langle\ ,\ \rangle$ is conjugate linear in the second variable ($\langle v, w_1 + w_2\rangle = \langle v,w_1\rangle + \langle v,w_2\rangle$ and $\langle v,\alpha w\rangle = \bar{\alpha}\langle v,w\rangle$). Because of the second condition, we can define $|v| = \sqrt{\langle v,v\rangle}$. We compute that

$$|v+w|^2 - |v-w|^2 = 2(\langle v,w\rangle + \overline{\langle v,w\rangle})$$

$$\Longrightarrow |v-iw|^2 - |v+iw|^2 = 2i(\langle v,w\rangle - \overline{\langle v,w\rangle}) .$$

Consequently, we can express $\langle v,w\rangle$ in terms of $|\ |$. A basis $v_1,\ldots,v_n$ of V is orthonormal with respect to an Hermitian inner product $\langle\ ,\ \rangle$ if we have, precisely as in the real case, $\langle v_i,v_j\rangle = \delta_{ij}$. We can always obtain an orthonormal basis from a given one by the Gram-Schmidt process, which works just as well for Hermitian inner products. Hence, any n-dimensional Hermitian inner product space $(V, \langle\ ,\ \rangle)$ is isomorphic to $\mathbb{C}^n$ with the standard Hermitian inner product $\langle z,w\rangle = \sum z^i \overline{w^i}$.

We define $U(n) \subset GL(n,\mathbb{C})$ to be the subgroup of all A such that $AA^* = I$, where A^* denotes the conjugate transpose of A,

$$A^* = \bar{A}^t , \qquad \text{i.e.,} \qquad A^*_{ij} = \overline{A_{ji}} .$$

We can also think of $U(n)$ as the set of all linear transformations of $\mathbb{C}^n$ which preserve the standard Hermitian inner product. It is easy to see that $U(n)$ is compact, just like $O(n)$. Thus $U(n)$ must be a Lie group (Theorem I.10-15). To see this in a more elementary way, we can consider the exponential map $\exp\colon (n \times n \text{ complex matrices}) \longrightarrow GL(n,\mathbb{C})$ defined, just as in the real case, by

$$\exp M = I + M + \frac{M^2}{2!} + \cdots .$$

Reasoning as on p. I.526, we easily see that $U(n)$ is a Lie group whose Lie algebra $\mathfrak{u}(n)$ is the set of all $n \times n$ complex matrices M with $M + M^* = 0$. Thus $M \in \mathfrak{u}(n)$ if and only if M has the form

$$\begin{pmatrix} ib_{11} & & -B^* \\ & \ddots & \\ B & & ib_{nn} \end{pmatrix} \qquad b_{ii} \text{ real} .$$

So $U(n)$ has dimension

$$n + 2(1 + \cdots + n-1) = n^2 .$$

Notice that $U(1)$ is just the set of complex numbers of absolute value 1. Hence $U(1)$ is connected. We can regard $S^{2n-1} \subset \mathbb{C}^n$ as the set of all $z \in \mathbb{C}^n$ with $|z| = 1$. So for $n \geq 2$ we can define $f\colon U(n) \longrightarrow S^{2n-1}$ by $f(A) = A(p_0)$, where p_0 is the n-tuple of complex numbers $(0,\ldots,0,1)$.

Then $f^{-1}(p_0)$ is homeomorphic to $U(n-1)$. Using induction, as in Problem I.3-30, we see that $U(n)$ is connected for all n. (Reasoning similar to that in Problem I.3-31 would show that $GL(n,\mathbb{C})$ is also connected.)

Every vector space V over $\mathbb{C}$ is also a vector space $V_{\mathbb{R}}$ over $\mathbb{R}$ [formally, $V_{\mathbb{R}}$ is V with the same addition map $V \times V \longrightarrow V$ and the multiplication $\mathbb{R} \times V \longrightarrow V$ which is the restriction of the given multiplication $\mathbb{C} \times V \longrightarrow V$]. If $v_1,\ldots,v_n$ is a basis for V over $\mathbb{C}$, then $v_1,iv_1,v_2,iv_2,\ldots,v_n,iv_n$ is a basis for $V_{\mathbb{R}}$ over $\mathbb{R}$. Let $T\colon V \longrightarrow V$ be a linear transformation (over $\mathbb{C}$) whose matrix with respect to $v_1,\ldots,v_n$ is the $n \times n$ complex matrix

$$A = (\alpha_{jk}) = (a_{jk} + ib_{jk}) ,$$

so that

$$Tv_j = \sum_{k=1}^{n} \alpha_{kj} v_k .$$

Then the matrix of $T\colon V_{\mathbb{R}} \longrightarrow V_{\mathbb{R}}$ with respect to the basis $v_1,iv_1,\ldots,v_n,iv_n$ is the $2n \times 2n$ real matrix $h(A) = (\tilde{\alpha}_{jk})$, where $\tilde{\alpha}_{jk}$ is the 2×2 block

$$\tilde{\alpha}_{jk} = \begin{pmatrix} a_{jk} & -b_{jk} \\ b_{jk} & a_{jk} \end{pmatrix} .$$

It is easy to see, using block multiplication of matrices, that

$$h\colon \{n \times n \text{ complex matrices}\} \longrightarrow \{2n \times 2n \text{ real matrices}\}$$

is a homomorphism. Hence it also gives us a homomorphism

$$h\colon GL(n,\mathbb{C}) \longrightarrow GL(2n,\mathbb{R}) ,$$

and moreover, we easily see that

$$h_* = h\colon \quad \mathfrak{gl}(n,\mathbb{C}) \to \quad \mathfrak{gl}(2n,\mathbb{R}) .$$

It is also easy to see that

$$h\colon U(n) \longrightarrow O(2n) .$$

Since $U(n)$ is connected, and h takes the identity matrix of $U(n)$ to the identity matrix of $O(2n)$, we actually have

$$h\colon U(n) \longrightarrow SO(2n) .$$

This also follows from

48. PROPOSITION. For every $n \times n$ complex matrix A we have

$$\det h(A) = |\det A|^2 .$$

Proof. The formula clearly holds for a diagonal matrix

$$A = \begin{pmatrix} a_{11} + ib_{11} & & \bigcirc \\ & \ddots & \\ \bigcirc & & a_{nn} + ib_{nn} \end{pmatrix} \implies h(A) = \begin{pmatrix} a_{11} & -b_{11} & & & \\ b_{11} & a_{11} & & \bigcirc & \\ & & \ddots & & \\ & \bigcirc & & a_{nn} & -b_{nn} \\ & & & b_{nn} & a_{nn} \end{pmatrix} .$$

So it also holds for diagonalizable matrices. But the diagonalizable matrices are dense and both sides of the equation are continuous in A. So it holds for all A. ∎

If $v_1,\ldots,v_n$ and $w_1,\ldots,w_n$ are two bases of V and A is the matrix expressing the w's in terms of the v's, then $h(A)$ is the matrix expressing the basis $w_1,iw_1,\ldots,w_n,iw_n$ in terms of $v_1,iv_1,\ldots,v_n,iv_n$. Since $\det h(A) > 0$, this shows that $V_{\mathbb{R}}$ has a natural orientation (which is but a reflection of the fact that $GL(n,\mathbb{C})$ is connected). If ξ is a complex vector bundle, then we can form a real vector bundle $\xi_{\mathbb{R}}$ by replacing each fibre by the corresponding vector space over $\mathbb{R}$; clearly $\xi_{\mathbb{R}}$ is always orientable, with a natural orientation.

For complex vector bundles it is natural to consider Hermitian metrics, which assign an Hermitian inner product to each fibre. We can prove they exist, as in the real case, by using partitions of unity (note that a positive real multiple of an Hermitian inner product is also a Hermitian inner product). Using an Hermitian inner product $\langle\ ,\ \rangle$ on the complex bundle $\xi = \pi\colon E \to X$ we can define the principal bundle $U(\xi) = \varpi\colon U(E) \to X$ with group $U(n)$, whose fibre $\varpi^{-1}(x)$ is the set of all frames of $\pi^{-1}(x)$ which are orthonormal with respect to $\langle\ ,\ \rangle$.

Corresponding to the Grassmannian $G_n(\mathbb{R}^N)$, we have the complex Grassmannian manifold $G_n(\mathbb{C}^N)$, consisting of all $W \subset \mathbb{C}^N$ which are subspaces of $\mathbb{C}^N$ (as a vector space over $\mathbb{C}$) of complex dimension n. If $V_n(\mathbb{C}^N)$ is the set of all linearly independent n-tuples $(v_1,\ldots,v_n) \in \mathbb{C}^N \times \cdots \times \mathbb{C}^N$, we define the map

$$\rho\colon V_n(\mathbb{C}^N) \to G_n(\mathbb{C}^N)$$

by letting

$$\rho((v_1,\ldots,v_n)) = \text{(complex) subspace of } \mathbb{C}^N \text{ spanned by } v_1,\ldots,v_n\ ,$$

and we give $G_n(\mathbb{C}^N)$ the quotient topology for this map. Reasoning exactly as in the real case, we see that $G_n(\mathbb{C}^N)$ can also be described as the left coset space

$$U(N)/\, U(n) \times U(N-n) \ .$$

Over $G_n(\mathbb{C}^N)$ we have a natural complex bundle $\gamma^n(\mathbb{C}^N)$ defined exactly as in the real case, and for $M > N$ there is a natural map $\alpha\colon G_n(\mathbb{C}^N) \longrightarrow G_n(\mathbb{C}^M)$ such that $\gamma^n(\mathbb{C}^N) \simeq \alpha^*\gamma^n(\mathbb{C}^M)$. The reader may easily check that Theorems 6 and 7 hold for complex bundles when we replace $\gamma^n(\mathbb{R}^N)$ by $\gamma^n(\mathbb{C}^N)$ throughout; the proofs are exactly the same.

To find the characteristic classes for complex bundles, we thus need to compute the cohomology of $U(N)/\, U(n) \times U(N-n)$. For this we need the solution to two invariance problems. First we want to consider polynomial functions on $\mathbb{C}^n \times \cdots \times \mathbb{C}^n$, by which we mean real-valued functions which are polynomials (over $\mathbb{R}$) in the real and imaginary components of the various vectors.

49. THEOREM. Every polynomial function f of m vectors in $\mathbb{C}^n$ which is invariant under $U(n)$ can be written as a polynomial in the real and imaginary parts of the Hermitian inner products.

Notice that this result is much simpler than the corresponding result for $O(n)$ and $SO(n)$, for there are no determinants involved, even though $U(n)$ is connected. The proof is also simpler, in the sense that various delicate details which arose in the proof of Theorem 35 are not needed. However, certain other considerations are required, and the proof is deferred to Addendum 1.

Another instance of the greater simplicity to be found in the complex domain is afforded by the spectral theorem, which is both more general, and easier to prove. We recall that for every linear transformation $T: \mathbb{C}^n \to \mathbb{C}^n$ there is a unique linear transformation $T^*: \mathbb{C}^n \to \mathbb{C}^n$, the "adjoint" of T, with

$$\langle Tv,w\rangle = \langle v,T^*w\rangle \qquad \text{for } v, w \in \mathbb{C}^n .$$

If T corresponds to the matrix A, then T^* corresponds to the conjugate transpose matrix A^*. We call A <u>normal</u> if $AA^* = A^*A$, and similarly for transformations. Both self-adjoint transformations $(T^* = T)$ and skew-adjoint $(T^* = -T)$ transformations are normal. If T is normal, then

$$\langle Tv,Tv\rangle = \langle v,T^*Tv\rangle = \langle v,TT^*v\rangle = \langle T^*v,T^*v\rangle .$$

Applying this to $T - \lambda I$, which is also normal, we see that

$$|(T - \lambda I)v| = |(T^* - \bar{\lambda} I)v| .$$

Now any linear transformation $T: \mathbb{C}^n \to \mathbb{C}^n$ has an eigenvector, since the equation $\det(T - \lambda I) = 0$ has a root in the field $\mathbb{C}$. The above equation shows that if T is normal, then an eigenvector v of T is also an eigenvector of T^*. Therefore the subspace $[v]$ spanned by v is invariant under T^*. Consequently, the orthogonal complement $[v]^\perp$ (under the Hermitian inner product) is invariant under $T^{**} = T$. From the invariance of both $[v]$ and $[v]^\perp$ under T, we easily see, by induction, that T has an orthonormal basis of eigenvectors. Equivalently, for every normal matrix A, there is a matrix $B \in U(n)$ such that BAB^{-1} is a diagonal matrix.

Now it is easy to give a canonical form for elements of $\mathfrak{u}(n)$.

50. PROPOSITION. For every $A \in \mathfrak{u}(n)$ there is a matrix $B \in U(n)$ such that

$$BAB^{-1} = \begin{pmatrix} i\lambda_1 & & \bigcirc \\ & \ddots & \\ \bigcirc & & i\lambda_n \end{pmatrix} \qquad \lambda_j \text{ real} .$$

Proof. Since $A = -A^*$ is normal, there is $B \in U(n)$ such that BAB^{-1} is diagonal. Moreover,

$$(BAB^{-1})^* = B^{-1*}A^*B^* = BA^*B^{-1} = -BAB^{-1} ,$$

so the diagonal entries of $B^{-1}AB$ must be pure imaginary. ■

Using this result, we can easily describe the polynomial functions on $\mathfrak{u}(n)$ which are invariant under the adjoint action of $U(n)$. Since the polynomials $f_1,\dots,f_n$ of section 8 are not real-valued on $\mathfrak{u}(n)$, it is convenient to consider instead the polynomials

$$\tilde{f}_k(M) = i^k f_k(M) = f_k(iM) ,$$

so that

$$\det(I + \lambda iM) = 1 + \lambda\tilde{f}_1(M) + \cdots + \lambda^n\tilde{f}_n(M) .$$

We also set $f_0(M) = 1$. If $M \in \mathfrak{u}(n)$, then for all real λ we have

$$(I + \lambda iM)^* = I + \lambda iM \implies \overline{\det(I + \lambda iM)} = \det(I + \lambda iM) ,$$

which shows that all $\tilde{f}_i(M)$ are real. It is easy to see, as on p. 495 , that for all $A \in \mathfrak{gl}(r,\mathbb{C})$ and $B \in \mathfrak{gl}(s,\mathbb{C})$ we have

$$\tilde{f}_k(A \oplus B) = \sum_{\ell=0}^{k} \tilde{f}_\ell(A)\cdot\tilde{f}_{k-\ell}(B) .$$

51. THEOREM. Every polynomial function f on $\mathfrak{u}(n)$ which is invariant under the adjoint action of $U(n)$ is a polynomial in $\tilde{f}_1,\ldots,\tilde{f}_n$.

Proof. For $\lambda_1,\ldots,\lambda_n \in \mathbb{C}$, let $[\lambda_1,\ldots,\lambda_n]$ be the diagonal matrix with entries $i\lambda_1,\ldots,i\lambda_n$ on the diagonal. Define

$$g(\lambda_1,\ldots,\lambda_n) = f([\lambda_1,\ldots,\lambda_n]) .$$

Then g is a polynomial in $\lambda_1,\ldots,\lambda_n$. Moreover g is symmetric, since

$$[\lambda_{\pi(1)},\ldots,\lambda_{\pi(n)}] = A\cdot[\lambda_1,\ldots,\lambda_n]A^{-1}$$

where $A \in U(n)$ is a suitable permutation matrix. So we can write

$$g(\lambda_1,\ldots,\lambda_n) = p(\sigma_1(i\lambda_1,\ldots,i\lambda_n),\ldots,\sigma_n(i\lambda_1,\ldots,i\lambda_n))$$

for some polynomial p. Then

$$f([\lambda_1,\ldots,\lambda_n]) = p(\tilde{f}_1([\lambda_1,\ldots,\lambda_n]),\ldots,\tilde{f}_n([\lambda_1,\ldots,\lambda_n])) .$$

The result follows as before, using Proposition 50. ∎

Using Theorems 49 and 51, we can carry out the whole program of section 9 for

$$G_n(\mathbb{C}^N) = U(N)/\ U(n) \times U(N-n) = U(N)/H\ .$$

A bi-invariant Riemannian metric $\langle\ ,\ \rangle$ on $U(n)$ can be defined explicitly as follows. For $M, P \in \mathfrak{u}(n)$, let

$$\langle M,P\rangle = \operatorname{Re} \operatorname{Trace} MP^* = \operatorname{Re} \sum_{i,j} M_{ij}\overline{P_{ij}}\ .$$

As in the case of $O(n)$, if we extend $\langle\ ,\ \rangle$ to $U(n)$ by left invariance, then it will also be right invariant. Now $\mathfrak{h}^\perp$ consists of all matrices

$$\begin{pmatrix} 0 & P \\ -P^* & 0 \end{pmatrix} \qquad P \text{ an } n \times (N-n) \text{ complex matrix .}$$

On $\mathfrak{h}^\perp \times \mathfrak{h}^\perp$ we have the functions

$$(P,Q) \longmapsto \sum_r P_i^r \overline{Q_j^r}$$

(which are bilinear <u>over</u> $\mathbb{R}$); their alternations ψ_{ij}, given by

$$\psi_{ij}(P,Q) = \sum_r P_i^r\overline{Q_j^r} - Q_i^r\overline{P_j^r}\ ,$$

are complex-valued alternating bilinear functions on $\mathfrak{h}^\perp$. A form $\eta \in \Omega^k(\mathfrak{h}^\perp)$ will be invariant under $\{I\} \times U(N-n)$ if and only if it is a linear combination of wedge products of the forms $\operatorname{Re} \psi_{ij}$ and $\operatorname{Im} \psi_{ij}$ (since there are no determinants to worry about in the first main theorem of invariance theory for $U(n)$, we do not need $k < N-n$). Such linear combinations can be described as $f(\psi)$ for polynomial functions $f\colon \mathfrak{u}(n) \to \mathbb{R}$ [this representation is unique in dimensions $< N-n$], and the combinations which are invariant under $U(n) \times \{I\}$

correspond to functions $f\colon \mathfrak{u}(n) \to \mathbb{R}$ which are invariant under the adjoint action of $U(n)$. Thus we have a class $\tilde{f}_i(\psi) \in H^{2i}(G_n(\mathbb{C}^N))$ for each of the polynomials $\tilde{f}_1,\ldots,\tilde{f}_n$ of Theorem 51, and every element of $H^*(U(N)/\,U(n)\times U(N-n))$ is a linear combination of cup products of these elements [it is a unique linear combination in dimensions $< N-n$].

The analogue of Proposition 41 holds, so we will consider only N with $2n < N-n$. Then all elements of $H^*(G_n(\mathbb{C}^N))$ in dimensions $< N-n$ are unique linear combinations of cup products of the classes corresponding to

$$\tilde{f}_1(\psi),\ldots,\tilde{f}_n(\psi)\ .$$

We let

$$c_{n;k} \in H^{2k}(G_n(\mathbb{C}^N)) \qquad k = 1,\ldots,n$$

be the class corresponding to

$$\frac{1}{(2\pi)^k}\,\tilde{f}_k(\psi)\ .$$

For an n-dimensional complex bundle ξ over M we define

$$c_k(\xi) = g^* c_{n;k}\ ,$$

where

$$g\colon M \to G_n(\mathbb{C}^N) \qquad \text{satisfies} \qquad g^*\gamma^n(\mathbb{C}^N) \simeq \xi\ .$$

The characteristic class $\xi \mapsto c_k(\xi)$ is called the kth Chern class for n-dimensional complex bundles; every characteristic class for n-dimensional complex bundles is a polynomial in the Chern classes $c_1,\ldots,c_n$. The class

$$c(\xi) = 1 + c_1(\xi) + \cdots + c_n(\xi) = c_0(\xi) + c_1(\xi) + \cdots + c_n(\xi)$$

is called the total Chern class of the n-dimensional complex bundle ξ.

Just as in the real case, the Chern classes of an n-dimensional complex bundle $\xi = \pi\colon E \to M$ may be described in terms of a connection. Choose any Hermitian metric for ξ, form the corresponding principal $U(n)$ bundle $U(\xi) = \varpi\colon U(E) \to M$, and let ω be a ($\mathfrak{u}(n)$-valued) connection on $U(\xi)$, with $\mathfrak{u}(n)$-valued curvature form Ω.

52. THEOREM. The kth Chern class $c_k(\xi)$ is represented by the unique form Λ on M with

$$\varpi^*\Lambda = \frac{1}{(2\pi)^k}\,\tilde{f}_k(\Omega) .$$

In other words, we have

$$c_k(\xi) = w(g_k) ,$$

where $g_k = \mathscr{S}(\tilde{f}_k) \in I^{2k}(U(n))$ corresponds to the polynomial function $\tilde{f}_k$ on $\mathfrak{u}(n)$, and w is the Weil homomorphism for $U(\xi)$.

Proof. First of all, an obvious analogue of Corollary 5 shows that the principal bundles η are all equivalent, no matter what Hermitian metric we choose. Now to prove the result, we just have to consider the universal bundle $\gamma^n(\mathbb{C}^N)$. This has a natural Hermitian metric, just like the natural Riemannian metric for $\tilde{\gamma}^n(\mathbb{R}^N)$, on pp.507-508. All the succeeding considerations also have natural analogues, and the result follows exactly as in the proof of Theorem 42 (or 43). ∎

As an immediate consequence we have

53. THEOREM. If ξ and η are complex bundles over M, then the total Chern class of $\xi \oplus \eta$ is given by the Whitney product formula

$$c(\xi \oplus \eta) = c(\xi) \cup c(\eta) .$$

Proof. Exactly like the proof of Theorem 45, using the formula on p. 533. ∎

We can also find relationships between Chern classes and Pontryagin classes. For a real vector space V, we define a complex vector space $V_{\mathbb{C}}$ by letting $V_{\mathbb{C}} = V \oplus V$, with complex multiplication determined by

$$i\cdot(v,w) = (-w,v) \qquad [\text{thus we think of } (v,w) \text{ as } v+iw] .$$

Doing this in each fibre of a real vector bundle ξ gives a complex vector bundle $\xi_{\mathbb{C}}$.

54. THEOREM. If $\xi = \pi\colon E \to M$ is an oriented n-dimensional vector bundle, then

$$c_{2k}(\xi_{\mathbb{C}}) = (-1)^k p_k(\xi) \qquad k = 1,\dots,[n/2] .$$

Proof. Choose the Hermitian metric on $\xi_{\mathbb{C}} = \pi'\colon E_{\mathbb{C}} \to M$ to be an extension of a Riemannian metric on ξ. Then $SO(E) \subset U(E_{\mathbb{C}})$, and the projection $\varpi\colon SO(E) \to M$ is the restriction of the projection $\varpi'\colon U(E_{\mathbb{C}}) \to M$. A connection ω on $SO(E)$ has a unique extension to a ($\mathfrak{u}(n)$-valued) connection ψ on $U(E_{\mathbb{C}})$ [as in the proof of Theorem 22, ψ is determined on the new vertical vectors of $SO(E)$, and hence on all of $U(E_{\mathbb{C}})$]. Let Ψ be the curvature form of ψ. At any point $e \in SO(E)$ the horizontal vectors for ψ are the same as for ω, so at e we have

$$\Psi = \Omega \qquad \text{[on tangent vectors to } SO(E)]$$
$$\Longrightarrow \tilde{f}_{2k}(\Psi) = \tilde{f}_{2k}(\Omega) = (-1)^k f_{2k}(\Omega) .$$

So if Λ represents $p_k(\xi)$ and T represents $c_{2k}(\xi_{\mathbb{C}})$, then at e we have

$$\begin{aligned} \varpi^*\Lambda &= \frac{1}{(2\pi)^k} f_{2k}(\Omega) \\ &= (-1)^k \frac{1}{(2\pi)^k} \tilde{f}_{2k}(\Psi) \\ &= (-1)^k \varpi'^* T . \end{aligned}$$

This implies that

$$\Lambda = (-1)^k T . \quad \blacksquare$$

The odd Chern classes, which are missing in Theorem 54, are easily calculated by the following considerations. Every complex vector space V gives rise to another complex vector space $\bar{V}$ in which complex multiplication $\bullet$ is defined by

$$\alpha \bullet v = \bar{\alpha} \cdot v .$$

Applying this process to each fibre of a complex bundle ξ we get the <u>conjugate bundle</u> $\bar{\xi}$. The bundles ξ and $\bar{\xi}$ are equivalent as real bundles, of course, but there may not be an equivalence which is complex linear on each fibre.

55. PROPOSITION. If ξ is a complex vector bundle, then

$$c(\bar{\xi}) = 1 - c_1(\xi) + c_2(\xi) - c_3(\xi) + \cdots .$$

Proof. If $\varpi\colon U(E) \to M$ is the associated principal bundle $U(\xi)$ for ξ, with R_A the right multiplication by $A \in U(n)$, then for the principal bundle $U(\bar{\xi})$ we may choose the same total space $U(E)$, but with the action $\bar{R}$ of $U(n)$ on the right given by

$$\bar{R}_A = R_{\bar{A}} .$$

So if ω is a connection on $U(\xi)$, with curvature form Ω, then $\bar{\omega}$ (the complex conjugate of ω) will be a connection on $U(\bar{\xi})$, with curvature form $\bar{\Omega}$. If Λ represents $c_k(\xi)$ and T represents $c_k(\bar{\xi})$, then

$$\begin{aligned}
\varpi^*T &= \frac{1}{(2\pi)^k}\,\tilde{f}_k(\bar{\Omega}) \\
&= \frac{1}{(2\pi)^k}\,\tilde{f}_k(-\Omega^t) \\
&= \frac{(-1)^k}{(2\pi)^k}\,\tilde{f}_k(\Omega) \\
&= (-1)^k \varpi^*\Lambda .
\end{aligned}$$

Hence $T = (-1)^k\Lambda$. ■

56. COROLLARY. If ξ is a real vector bundle, then

$$c_k(\xi_{\mathbb{C}}) = 0 \qquad \text{for } k \text{ odd} .$$

Proof. We just have to note that

$$\xi_{\mathbb{C}} \simeq \bar{\xi}_{\mathbb{C}} \qquad \text{(as complex bundles)} .$$

This is due to the fact that there is a natural complex isomorphism $V_{\mathbb{C}} \longrightarrow \bar{V}_{\mathbb{C}}$ for every real vector space V -- we merely take $(v,w) \longmapsto (v,-w)$. ∎

Instead of starting with a real bundle, we can instead begin with a complex bundle ξ, and regard it as an oriented real bundle $\xi_{\mathbb{R}}$ (of even dimension). In order to find its Pontryagin and Euler classes, we need a lemma concerning the homomorphism $h\colon GL(n,\mathbb{C}) \longrightarrow GL(2n,\mathbb{R})$ defined on p. 527 Since $h = h_*\colon \mathfrak{gl}(n,\mathbb{C}) \longrightarrow \mathfrak{gl}(2n,\mathbb{R})$, we have $h(\mathfrak{u}(n)) \subset \mathfrak{o}(2n)$.

<u>57. LEMMA</u>. For $M \in \mathfrak{u}(n)$ we have

$$f_{2k}(h(M)) = (-1)^k \sum_{\ell=0}^{2k} (-1)^\ell \tilde{f}_\ell(M)\tilde{f}_{2k-\ell}(M) ,$$

$$\mathrm{Pf}(h(M)) = \tilde{f}_n(M) .$$

<u>Proof</u>. For all real λ we have

$$\begin{aligned}
1 + \lambda^2 f_2(h(M)) + \cdots + \lambda^{2n} f_{2n}(h(M)) &= \det(I_{2n} + \lambda h(M)) && I_{2n} = \text{identity of } O(2n) \\
&= \det h(I_n + \lambda M) && I_n = \text{identity of } U(n) \\
&= |\det(I_n + \lambda M)|^2 && \text{by Proposition 48} \\
&= |1 - i\lambda\tilde{f}_1(M) - \lambda^2\tilde{f}_2(M) + i\lambda^3\tilde{f}_3(M) + \lambda^4\tilde{f}_4(M) - \cdots|^2 \\
&= |(1 - \lambda^2\tilde{f}_2(M) + \lambda^4\tilde{f}_4(M) - \cdots) - i(\lambda\tilde{f}_1(M) - \lambda^3\tilde{f}_3(M) + \cdots)|^2 \\
&= (1 - \lambda^2\tilde{f}_2(M) + \lambda^4\tilde{f}_4(M) - \cdots)^2 + (\lambda\tilde{f}_1(M) - \lambda^3\tilde{f}_3(M) + \cdots)^2 .
\end{aligned}$$

The coefficient of λ^{2k} on the right side is

$$\sum_{\ell \text{ even}} (-1)^{\frac{\ell}{2}}(-1)^{\frac{2k-\ell}{2}} \tilde{f}_\ell(M)\tilde{f}_{2k-\ell}(M) + \sum_{\ell \text{ odd}} (-1)^{\frac{\ell-1}{2}}(-1)^{\frac{2k-\ell-1}{2}} \tilde{f}_\ell(M)\tilde{f}_{2k-\ell}(M)$$

$$= (-1)^k\left[\sum_{\ell \text{ even}} \tilde{f}_\ell(M)\tilde{f}_{2k-\ell}(M) - \sum_{\ell \text{ odd}} \tilde{f}_\ell(M)\tilde{f}_{2k-\ell}(M)\right]$$

$$= (-1)^k \sum_\ell (-1)^\ell \tilde{f}_\ell(M)\tilde{f}_{2k-\ell}(M) .$$

For the Pfaffian we have

$$[\mathrm{Pf}(h(M))]^2 = \det h(M) = |\det M|^2 = |f_n(M)|^2$$

$$= \left|\frac{\tilde{f}_n(M)}{i^n}\right|^2 = |\tilde{f}_n(M)|^2 ,$$

and hence

$$\mathrm{Pf}(h(M)) = \pm\, \tilde{f}_n(M) .$$

To settle the sign, we consider

$$M = \begin{pmatrix} i & & \bigcirc \\ & \ddots & \\ \bigcirc & & i \end{pmatrix} \Longrightarrow h(M) = \begin{pmatrix} -S & & \bigcirc \\ & \ddots & \\ \bigcirc & & -S \end{pmatrix} \qquad S = \begin{pmatrix} 0 & 1 \\ -1 & 0 \end{pmatrix} .$$

Then

$$\mathrm{Pf}(h(M)) = (-1)^n = \det iM = \tilde{f}_n(M) ,$$

so the + sign is correct. ■

The relevance of this Lemma will become immediately apparent in the proof of our final result.

58. THEOREM. If ξ is a complex bundle of dimension n, then

$$p_k(\xi_{\mathbb{R}}) = (-1)^k \sum_{\ell=0}^{2k} (-1)^{\ell} c_{\ell}(\xi) \cup c_{2k-\ell}(\xi) \qquad k = 1,\ldots,2n ,$$

and

$$\chi(\xi_{\mathbb{R}}) = c_n(\xi) .$$

Proof. Note that a Hermitian inner product $\langle\ ,\ \rangle$ on a complex vector space V gives an ordinary inner product on $V_{\mathbb{R}}$ -- we define $v_1, iv_1, \ldots, v_n, iv_n$ to be orthonormal whenever $v_1,\ldots,v_n$ is orthonormal with respect to $\langle\ ,\ \rangle$. This inner product on $V_{\mathbb{R}}$ is well-defined, for if $w_j = \Sigma\, a_{\ell j} v_{\ell}$ is another orthonormal base, then the matrix $A = (a_{\ell j})$ is clearly in $U(n)$, so $h(A) \in SO(n)$. Choosing an Hermitian metric on $\xi = \pi\colon E \to M$, and applying this construction to each fibre, we obtain a Riemannian metric on $\xi_{\mathbb{R}}$. Moreover for the corresponding principal bundles, we have $U(E) \subset SO(E)$, and the projection $\varpi\colon U(E) \to M$ is the restriction of the projection $\varpi'\colon SO(E) \to M$. For $A \in U(n)$, the right action R_A on $U(E)$ is just the restriction of the right action $R_{h(A)}$ on $SO(E)$. A connection ω on $U(E)$ is $\mathfrak{u}(n)$-valued, as its curvature form Ω. The $h(\mathfrak{u}(n))$-valued form $h\circ\omega$ has a unique extension to a connection ψ on $SO(E)$, with curvature form Ψ. At any point $e \in U(E)$ we have

$$\Psi = h\circ\Omega \qquad \text{on tangent vectors to } U(E) .$$

So if Λ represents $p_k(\xi_{\mathbb{R}})$ and Λ_{ℓ} represents $c_{\ell}(\xi)$ for $\ell = 0,\ldots,2k$, then at e we have

$$\varpi'^*\Lambda = \frac{1}{(2\pi)^{2k}} f_{2k}(\Psi)$$

$$= \frac{1}{(2\pi)^{2k}} f_{2k}(h\circ\Omega)$$

$$= \frac{1}{(2\pi)^{2k}}(-1)^k \sum_{\ell=0}^{2k} (-1)^\ell \tilde{f}_\ell(\Omega) \wedge \tilde{f}_{2k-\ell}(\Omega) \qquad \text{by Lemma 57}$$

$$= (-1)^k \sum_{\ell=0}^{2k} (-1)^\ell \varpi^*\Lambda_\ell \wedge \varpi^*\Lambda_{2k-\ell} ,$$

which proves the first formula.

If Λ represents $\chi(\xi_{\mathbb{R}})$, then at e we have

$$\varpi'^*\Lambda = \frac{1}{(2\pi)^n} \operatorname{Pf}(\Psi)$$

$$= \frac{1}{(2\pi)^n} \operatorname{Pf}(h\circ\Omega)$$

$$= \frac{1}{(2\pi)^n} \tilde{f}_n(\Omega) \qquad \text{by Lemma 57}$$

$$= \varpi^*\Lambda_n .$$

This proves the second formula. ■

12. Valedictory

Now that we have built up so much machinery, it seems a shame not to use it. But this would really take us out of the field of differential geometry entirely. We have tried to show how the characteristic classes arise naturally, how they can be computed by differential geometric means, why they should be expressible in terms of curvature, and especially how the Euler class is expressed in terms of $K_n\, dV$, which involves $\operatorname{Pf}(\Omega)$. For further applications of these characteristic classes, the reader is urged to consult books specifically devoted to the subject, where characteristic classes are usually defined by methods of algebraic topology. One of the most famous set of notes, now finally available in book form, is Milnor and Stasheff {1}. Here the Euler class is defined essentially as we have defined it, in terms of the Thom class. But the Pontryagin and Chern classes are defined in completely different ways. For a complex n-dimensional bundle $\xi = \pi\colon E \to X$, the top Chern class $c_n(\xi)$ is <u>defined</u> by the formula of Theorem 58, as

$$c_n(\xi) = \chi(\xi_{\mathbb{R}}) \ .$$

The other Chern classes are defined inductively, as follows. Choosing an Hermitian metric for ξ, we form the associated sphere bundle $S \subset E$, and let $\pi_0\colon S \to X$ be the restriction of π. It turns out that the map

$$\pi_0{}^*\colon H^k(X) \to H^k(S)$$

is an isomorphism for $k < n$. Now $\pi_0{}^*\xi$ has a section, so it can be written as the Whitney sum

$$\pi_0{}^*\xi = \xi_1 \oplus \xi_2 \ ,$$

where ξ_2 is a trivial 1-dimensional complex bundle. The Chern class $c_{n-1}(\xi)$ is defined as

$$c_{n-1}(\xi) = (\pi_0{}^*)^{-1}(c_{n-1}(\xi_1)) \ .$$

This makes sense, since c_{n-1} is defined for the $(n-1)$-dimensional complex bundle ξ_1. Moreover, it is compatible with our definition, since it is equivalent to

$$c_{n-1}(\pi_0{}^*\xi) = c_{n-1}(\xi_1) ,$$

which is what the Whitney product formula gives when ξ_2 is trivial. Now that $c_{n-1}(\xi)$ is defined for all n-dimensional complex bundles ξ, the Chern class $c_{n-2}(\xi)$ can be defined as

$$c_{n-2}(\xi) = (\pi_0{}^*)^{-1}(c_{n-1}(\xi_1)) ,$$

and so on, by induction. After the Chern classes are defined, the Whitney product formula is proved, and the cohomology of $G_n(\mathbb{C}^\infty)$ is calculated, by means of various tricks. The Pontryagin classes are defined by the formula in Theorem 54, and all properties are derived from this definition and the properties of the Chern classes. In the whole development, there is no restriction to manifolds, and singular homology is used throughout. Moreover the coefficients are $\mathbb{Z}$, rather than $\mathbb{R}$. Integer coefficients can be used because the Thom class can be defined with integer coefficients, so the Euler class has integer coefficients, and the map $\pi_0{}^*\colon H^k(X) \to H^k(S)$ is an isomorphism with integer coefficients. This shows, by the way, that our Euler, Pontryagin, and Chern classes, defined originally with real coefficients, are actually all integral classes; that, of course, was the reason for inserting the various factors $(2\pi)^{-\alpha}$ in the definitions.

Addendum 1. Invariant Theory for the Unitary Group

At first sight, the problem of determining all invariant polynomials for $U(n)$ seems peculiarly complicated, since we are dealing with polynomial functions of the real and imaginary parts of the components of the vectors, rather than polynomial functions of the components themselves. But there is a trick which will allow us to reduce the problem to one where we study only polynomials of the latter type. The basic result which we need for such polynomials can actually be formulated for any field.

Let k be an arbitrary infinite field, and let $V = k^n$ be the standard n-dimensional vector space over k, with standard basis elements $e_1,\dots,e_n$. The group of all non-singular $n \times n$ matrices with entries in k is denoted by $GL(n,k)$. Just as in the real case, a matrix $A = (a_{ij}) \in GL(n,k)$ is also regarded as a linear transformation $A\colon V \to V$, by the rule

$$A(e_i) = \sum_{j=1}^{n} a_{ji} e_j ,$$

so that for a (row vector) $v \in k^n$ we have

$$A(v) = v \cdot A^t .$$

We also define an action of $GL(n,k)$ on the m-fold product $V \times \cdots \times V$ by

$$A\cdot(v_1,\dots,v_n) = (A(v_1),\dots,A(v_m)) .$$

A function

$$f\colon V \times \cdots \times V \to k$$

is called a polynomial function if it is a polynomial (over k) in the components of the elements of V; we also define homogeneous polynomial functions just as before. We say that f is invariant under a group $G \subset GL(n,k)$ if

$$f(A\cdot v) = f(v) \qquad \text{for all} \quad v \in V\times\cdots\times V, \quad \text{and all} \quad A \in G .$$

We still have the standard basis vectors e_{ri} for $V\times\cdots\times V$, and the partial derivatives $\partial f/\partial e_{ri}$ can be defined formally for polynomial functions f. Euler's theorem can be checked formally, polarizations are defined as before, and the Capelli identities still hold. Introducing the partial ordering on the polynomial functions as before, we still have assertion (A) on p. 481 .

Actually, we want to be even more general, and consider functions

$$f\colon \underbrace{V\times\cdots\times V}_{m} \times \underbrace{V^*\times\cdots\times V^*}_{\ell} \longrightarrow k .$$

We define an action of $GL(n,k)$ on $V\times\cdots\times V^*$ by

$$A\cdot(v_1,\ldots,v_m, \phi_1,\ldots,\phi_\ell) = (A(v_1),\ldots,A(v_m), \phi_1\circ A^{-1},\ldots,\phi_\ell\circ A^{-1}) ,$$

and we say that f is invariant under a group $G \subset GL(n,k)$ if

$$f(A\cdot(v,\phi)) = f((v,\phi)) \qquad \text{for all} \quad (v,\phi) \in V\times\cdots\times V^*, \quad \text{and} \quad A \in G .$$

For example, the "evaluations"

$$\varepsilon_{rs}(v_1,\ldots,v_m, \phi_1,\ldots,\phi_\ell) = \phi_s(v_r)$$

are invariant under all of $GL(n,k)$. It will be convenient to identify an element $\phi \in V^*$ with the column vector

$$\xi = \begin{pmatrix} \phi(e_1) \\ \vdots \\ \phi(e_n) \end{pmatrix} .$$

Then it turns out that the action of $GL(n,k)$ is

$$A\cdot(v_1,\ldots,v_m, \xi_1,\ldots,\xi_\ell) = (v_1\cdot A^t,\ldots,v_m\cdot A^t, (A^{-1})^t\cdot\xi_1,\ldots,(A^{-1})^t\cdot\xi_\ell) .$$

We call f a polynomial function if $f(v_1,\ldots,v_m, \xi_1,\ldots,\xi_\ell)$ is a polynomial in the v_{ri} and ξ_{sj} (here ξ_{sj} is the entry in the jth row of ξ_s). Notice that the evaluations are polynomial functions -- under the identification of V^* with the set of column vectors they are simply the maps

$$\varepsilon_{rs}(v_1,\ldots,v_m, \xi_1,\ldots,\xi_\ell) = v_r\cdot\xi_s .$$

Two other important types of polynomial functions are the functions

$$\det_{r_1,\ldots,r_n}(v_1,\ldots,v_m, \xi_1,\ldots,\xi_\ell) = \det\begin{pmatrix} v_{r_1} \\ \vdots \\ v_{r_n} \end{pmatrix}$$

and

$$\det^*_{s_1,\ldots,s_n}(v_1,\ldots,v_m, \xi_1,\ldots,\xi_\ell) = \det(\xi_{s_1},\ldots,\xi_{s_n}) .$$

They are invariant under the subgroup $SL(n,k) \subset GL(n,k)$ consisting of matrices of determinant 1.

We can define homogeneous polynomial functions as before, except now we must consider functions which are homogeneous of degree $(\alpha_1,\ldots,\alpha_m)$ in the V

variables, and of degree $(\beta_1,\ldots,\beta_\ell)$ in the V^* variables. For any homogeneous f, we can apply the apparatus of the Capelli identities to either the V variables or the V^* variables separately, and obtain assertion (A) on p. 481 where the partial ordering $\prec$ is applied to either the degree in V or the degree in V^*, the polarizations being applied to the variables in V or V^*, respectively.

59. THEOREM. For all m, ℓ, and n we have

> $SL_n^{m,\ell}$: Every polynomial function f on m vectors in k^n and ℓ vectors in $(k^n)^*$ which is invariant under $SL(n,k)$ can be written as a polynomial in the evaluation functions ε_{rs} and the determinant functions $\det_{r_1,\ldots,r_n}$ and $\det^*_{s_1,\ldots,s_n}$.

Proof. The proof is similar to that of Theorem 35, but the steps are easier. First we note that $SL_n^{m,\ell} \Rightarrow SL_n^{m',\ell'}$ for $m' \leq m$ and $n' \leq n$. Next we have

60. LEMMA. If $SL_n^{n-1,n-1}$ holds, then $SL_n^{m,\ell}$ holds for all $m \geq n-1$, $\ell \geq n-1$.

Proof. The argument is similar to that for Lemma 36. In the present case we can start with $SL_n^{n-1,n-1}$ rather than $SL_n^{n,n}$ because the term det appearing in assertion (A) in the case $m = n$ causes no problems -- it is one of the invariants in terms of which we are trying to express f. Q.E.D.

Now the proof of the special case to which we have reduced the problem does not even require an inductive argument:

61. LEMMA. $SL_n^{n-1,n-1}$ holds for all n.

Proof. Consider $(v_1,\ldots,v_{n-1}, \phi_1,\ldots,\phi_{n-1})$ satisfying

$$(*) \qquad \det(\phi_i(v_j)) \neq 0 .$$

Then $\bigcap_{i=1}^{n-1} \ker \phi_i$ is 1-dimensional. Let $0 \neq w \in \bigcap \ker \phi_i$. If we had

$$w = \sum_{j=1}^{n-1} a_i v_j$$

for constants a_i (necessarily not all zero), then we would have

$$0 = \phi_i(w) = \sum_{j=1}^{n-1} a_j \phi_i(v_j) \qquad j = 1,\ldots,n-1 ,$$

contradicting (*). So w is linearly independent of $v_1,\ldots,v_{n-1}$. Hence there is some $A \in SL(n,k)$ such that

$$A(e_i) = v_i \qquad i = 1,\ldots,n-1$$
$$A(e_n) = \text{a multiple of } w .$$

Then

$$(1) \quad f(v_1,\ldots,v_{n-1}, \phi_1,\ldots,\phi_{n-1}) = f(A^{-1}(v_1),\ldots,A^{-1}(v_{n-1}), \phi_1\circ A,\ldots,\phi_{n-1}\circ A)$$
$$= f(e_1,\ldots,e_{n-1}, \phi_1\circ A,\ldots,\phi_{n-1}\circ A) ;$$

note that

$$(2) \qquad (\phi_i \circ A)(e_j) = \phi_i(v_j) \qquad j = 1,\dots,n-1$$
$$(\phi_i \circ A)(e_n) = 0 .$$

Now define a polynomial function F of $(n-1)^2$ variables a_{ij}, as follows:

$$F(\{a_{ij}\}) = f(e_1,\dots,e_{n-1}, \mu_1,\dots,\mu_{n-1}) ,$$

where μ_i are the unique linear functionals with

$$\mu_i(e_j) = a_{ij} \qquad j = 1,\dots,n-1$$
$$\mu_i(e_n) = 0 .$$

Then equations (1) and (2) show that

$$(**) \qquad f(v_1,\dots,v_{n-1}, \phi_1,\dots,\phi_{n-1}) = F(\{\phi_i(v_j)\})$$

whenever (*) holds. A standard argument ("the principal of irrelevance of algebraic inequalities") shows that consequently (**) holds everywhere: for the polynomial

$$\big[f(v_1,\dots,v_{n-1}, \phi_1,\dots,\phi_{n-1}) - F(\{\phi_i(v_j)\})\big]\cdot\det(\phi_i(v_j))$$

is identically 0, hence one of the factors must be identically 0, and the second factor certainly isn't. ■

We will use Theorem 59 only for the case $SL(n,\mathbb{C})$. We easily see (compare Problem I.10-27) that the Lie algebra $\mathfrak{sl}(n,\mathbb{C})$ of $SL(n,\mathbb{C})$ consists of all $n \times n$ complex matrices with trace = 0. Similarly (compare p. 526), the group $SU(n) = U(n) \cap SL(n,\mathbb{C})$ has Lie algebra $\mathfrak{su}(n)$ consisting of all matrices

$$\begin{pmatrix} ib_{11} & & -B^* \\ & \ddots & \\ B & & ib_{nn} \end{pmatrix} \qquad \begin{array}{l} b_{ii} \text{ real} \\ \Sigma\, b_{ii} = 0 . \end{array}$$

Notice that $\mathfrak{su}(n)$ is not a complex subspace of $\mathfrak{gl}(n,\mathbb{C})$; however, it is easy to find the complex subspace $W \subset \mathfrak{gl}(n,\mathbb{C})$ spanned by $\mathfrak{su}(n)$. Note that W must contain

$$-i \cdot \begin{pmatrix} ib_{11} & & \bigcirc \\ & \ddots & \\ \bigcirc & & ib_{nn} \end{pmatrix} = \begin{pmatrix} b_{11} & & \bigcirc \\ & \ddots & \\ \bigcirc & & b_{nn} \end{pmatrix} \qquad \begin{array}{l} b_{ii} \text{ real} \\ \Sigma\, b_{ii} = 0 \end{array}$$

and

$$-i \cdot \begin{pmatrix} 0 & & -(iA)^* \\ & \ddots & \\ iA & & 0 \end{pmatrix} = -i \cdot \begin{pmatrix} 0 & & +iA^* \\ & \ddots & \\ iA & & 0 \end{pmatrix} = \begin{pmatrix} 0 & & A^* \\ & \ddots & \\ A & & 0 \end{pmatrix},$$

and thus also the matrix

$$\begin{pmatrix} 0 & & \bigcirc \\ & \ddots & \\ A & & 0 \end{pmatrix} = \begin{pmatrix} 0 & & -A^* \\ & \ddots & \\ A & & 0 \end{pmatrix} + \begin{pmatrix} 0 & & A^* \\ & \ddots & \\ A & & 0 \end{pmatrix}, \qquad \text{as well as} \qquad \begin{pmatrix} 0 & & A \\ & \ddots & \\ \bigcirc & & 0 \end{pmatrix}.$$

From this we easily see that

(*) the complex subspace of $\mathfrak{gl}(n,\mathbb{C})$ spanned by $\mathfrak{su}(n)$ is just $\mathfrak{sl}(n,\mathbb{C})$.

This simple fact leads to

62. LEMMA. Let $g\colon GL(n,\mathbb{C}) \to \mathbb{C}$ be a complex analytic function (this makes sense, since $GL(n,\mathbb{C})$ is an open subset of $\mathbb{C}^{n^2}$). If g vanishes on $SU(n)$, then g also vanishes on $SL(n,\mathbb{C})$.

Proof. Pick $Y_1, Y_2 \in \mathfrak{su}(n)$, and consider the function $h\colon \mathbb{C} \to \mathbb{C}$ defined by

$$h(z) = g(\exp(zY_1 + Y_2)) .$$

Then h is analytic and vanishes for all real z. So h vanishes for all z. Similarly, we may now prove that

$$g(\exp(z_1Y_1 + z_2Y_2)) = 0 \qquad \text{for all } z_1, z_2 \in \mathbb{C} .$$

Then (*) implies that $g(\exp(X)) = 0$ for all $X \in \mathfrak{sl}(n,\mathbb{C})$. But the image of $\exp\colon \mathfrak{sl}(n,\mathbb{C}) \to SL(n,\mathbb{C})$ is dense* in $SL(n,\mathbb{C})$, since diagonalizable matrices are certainly in the image of exp. Hence $g = 0$ on all of $SL(n,C)$. ■

63. COROLLARY. Let f be a polynomial function of m vectors of $\mathbb{C}^n$ and ℓ vectors of $(\mathbb{C}^n)^*$ [that is, f is a polynomial over $\mathbb{C}$ in the (complex) components of the vectors]. If f is invariant under $SU(n)$, then f is also invariant under $SL(n,\mathbb{C})$, and is thus a polynomial in the evaluation functions ε_{rs} and the determinant functions $\det_{r_1,\dots,r_n}$ and $\det^*_{s_1,\dots,s_n}$.

Proof. For fixed $(v,\phi) \in C^n \times \cdots \times (C^n)^*$, define $g_{(v,\phi)}\colon GL(n,\mathbb{C}) \to \mathbb{C}$ by

*Actually, the exponential map is onto $SL(n,\mathbb{C})$, but we won't prove that here.

$$g_{(v,\phi)}(A) = f(A\cdot(v,\phi)) - f((v,\phi)) .$$

Then g is complex analytic, and g vanishes on $SU(n)$ by hypothesis. So by Lemma 62, g vanishes on $SL(n,\mathbb{C})$. Since this is true for each (v,ϕ), it follows that f is invariant under $SL(n,\mathbb{C})$. ■

Now we really want to consider $\mathbb{R}$-valued functions of m vectors in $\mathbb{C}^n$ which are polynomials in the real and imaginary parts of the components of the vectors. Actually, we might as well consider complex-valued functions which are polynomials over $\mathbb{C}$ in the real and imaginary parts of the components of the vectors. Equivalently, we consider functions

$$f\colon \mathbb{C}^n \times \cdots \times \mathbb{C}^n \longrightarrow \mathbb{C}$$

which are polynomials over $\mathbb{C}$ in the components v_{ri} of the vectors and in their complex conjugates $\bar{v}_{ri}$. Given such a function f, we define a function

$$\tilde{f}\colon \underbrace{\mathbb{C}^n \times \cdots \times \mathbb{C}^n}_{m} \times \underbrace{(\mathbb{C}^n)^* \times \cdots \times (\mathbb{C}^n)^*}_{m} \longrightarrow \mathbb{C}$$

as follows:

(i) if $f(v_1,\ldots,v_n) = v_{ri}$, then $\tilde{f}(v_1,\ldots,v_n, \xi_1,\ldots,\xi_n) = v_{ri}$

(ii) if $f(v_1,\ldots,v_n) = \bar{v}_{ri}$, then $\tilde{f}(v_1,\ldots,v_n, \xi_1,\ldots,\xi_n) = \xi_{ri}$

(iii) the correspondence $f \longmapsto \tilde{f}$ is an algebra homomorphism.

Notice that $\tilde{f}$ is a polynomial function in the (complex) components of the vectors of $\mathbb{C}^n$ and $(\mathbb{C}^n)^*$. The mapping $f \longmapsto \tilde{f}$ is clearly a one-one correspondence between the polynomials in the v_{ri} and $\bar{v}_{ri}$, and the polynomials in the v_{ri} and ξ_{ri}. Note that if f is the Hermitian inner product ι_{rs} of v_r and v_s, then $\tilde{f}$ is the evaluation ε_{rs}. If f is the determinant of $v_{r_1},\ldots,v_{r_n}$, then $\tilde{f}$ is also this determinant; if f is the conjugate of this determinant, then $\tilde{f}$ is the determinant of $\xi_{r_1},\ldots,\xi_{r_n}$.

Suppose that $f(v_1,\ldots,v_n) = \bar{v}_{ri}$. Then

$$
\begin{aligned}
(1)\quad f(A\cdot(v_1,\ldots,v_n)) &= \text{the conjugate of the } i^{th} \text{ component of } v_r\cdot A^t \\
&= \text{the conjugate of } \sum_j a_{ij}v_{rj} \\
&= \sum_j \bar{a}_{ij}\bar{v}_{rj} .
\end{aligned}
$$

On the other hand, if $A \in U(n)$, so that $A^{-1} = \bar{A}^t$, then

$$
\begin{aligned}
(2)\quad \tilde{f}(A\cdot(v_1,\ldots,v_n, \xi_1,\ldots,\xi_n)) &= \tilde{f}(v_1\cdot A^t,\ldots,(A^{-1})^t\cdot\xi_1,\ldots) \\
&= i^{th} \text{ component of } (A^{-1})^t\cdot\xi_r \\
&= i^{th} \text{ component of } \bar{A}\cdot\xi_r \\
&= \sum_j \bar{a}_{ij}\xi_{rj} .
\end{aligned}
$$

Comparing (1) and (2), we see that

$$
(3)\qquad \tilde{f}\circ(A\cdot) = \widetilde{f\circ(A\cdot)} ,
$$

where $(A\cdot)$ on the left is the action of $A \in U(n)$ on $\mathbb{C}^n\times\cdots\times(\mathbb{C}^n)^*$, while $(A\cdot)$ on the right is the action of A on $\mathbb{C}^n\times\cdots\times\mathbb{C}^n$. From the way that $f\mapsto\tilde{f}$ is defined, it is clear that (3) holds for all f. Consequently,

f _is invariant under_ $U(n)$ _if and only if_ $\tilde{f}$ _is invariant under_ $U(n)$.

From this we immediately conclude

64. THEOREM. Let $f\colon \mathbb{C}^n \times \cdots \times \mathbb{C}^n \longrightarrow \mathbb{C}$ be a polynomial function in the components of the vectors of $\mathbb{C}^n$ and the conjugates of the components, which is invariant under $SU(n)$. Then f is a polynomial in the Hermitian inner products and the determinants $\det_{r_1,\ldots,r_n}$ and their conjugates $\overline{\det_{r_1,\ldots,r_n}}$.

Proof. By Corollary 63, the function $\tilde{f}$ is a polynomial in the evaluations ε_{rs} and the determinants $\det_{r_1,\ldots,r_n}$ and $\det^*_{r_1,\ldots,r_n}$. Since

$$\varepsilon_{rs} = \tilde{\iota}_{rs}, \quad \det_{r_1,\ldots,r_n} = \widetilde{\det_{r_1,\ldots,r_n}}, \quad \det^*_{r_1,\ldots,r_n} = \widetilde{\overline{\det_{r_1,\ldots,r_n}}},$$

the result follows. ∎

In order to prove Theorem 49, which gives the corresponding result for $U(n)$, we need just one more observation. Let $v_1,\ldots,v_n, w_1,\ldots,w_n \in \mathbb{C}^n$, and let $A = (v_{ij})$, $B = (w_{ij})$. Then we have

$$(A\cdot\bar{B}^t)_{ij} = \sum_k v_{ik}\bar{w}_{jk} = \langle v_i, w_j\rangle ,$$

where $\langle\ ,\ \rangle$ is the Hermitian inner product. Hence

$$(*) \qquad \det A\ \overline{\det B} = \det(\langle v_i, w_j\rangle) .$$

Proof of Theorem 49. Since f is invariant under $SU(n)$, by Theorem 64 it can be written as a polynomial in the Hermitian inner products and

$$\det_{r_1,\ldots,r_n} \quad \text{and} \quad \overline{\det_{s_1,\ldots,s_n}} .$$

The action of an element $A \in U(n)$ multiplies the latter two by

$$\det A \quad \text{and} \quad \overline{(\det A)} = \det \bar{A} = \det(A^{-1})^t = (\det A)^{-1} .$$

Hence every factor $\det_{r_1,\ldots,r_n}$ must come paired with a factor $\overline{\det_{s_1,\ldots,s_n}}$. But (*) says that the product of these two functions can be expressed in terms of the Hermitian inner product functions. ■

Addendum 2. Recovering the Differential Forms; The Gauss-Bonnet-Chern Theorem for Manifolds-with-boundary

The crucial step in our proof of the generalized Gauss-Bonnet theorem was Theorem 22, for it immediately allowed us to conclude that if $\xi = \pi\colon E \to M$ is an oriented n-dimensional vector bundle, with sphere bundle $\pi_0\colon S \to M$, then $\pi_0{}^*C(\xi) = 0$. This means that if Λ is the n-form on M representing $C(\xi)$, then the n-form $\pi_0{}^*\Lambda$ on S is exact,

$$\pi_0{}^*\Lambda = d\Phi \qquad \text{for some } (n-1)\text{-form } \Phi \text{ on } S\ .$$

In particular, suppose that $\xi = TM^n$, and let X be a unit vector field on M with a single isolated singularity, at $p \in M$ (Problem I.11-13). Let $B(\varepsilon)$ be a closed ball of radius ε around p, and set

$$M_\varepsilon = M - \text{interior } B(\varepsilon)\ .$$

Then $X(M_\varepsilon) \subset S$ is a manifold-with-boundary, the image of M_ε under the section $X\colon M - \{p\} \to S$. Now

$$\begin{aligned}\int_M \Lambda = \int_{M-\{p\}} \Lambda = \lim_{\varepsilon\to 0} \int_{M_\varepsilon} \Lambda &= \lim_{\varepsilon\to 0} \int_{M_\varepsilon} X^*\pi_0{}^*\Lambda \\ &= \lim_{\varepsilon\to 0} \int_{X(M_\varepsilon)} \pi_0{}^*\Lambda = \lim_{\varepsilon\to 0} \int_{X(M_\varepsilon)} d\Phi \\ &= \lim_{\varepsilon\to 0} \int_{\partial X(M_\varepsilon)} \Phi\ .\end{aligned}$$

Recalling the definition of the index of X at p, we easily see that

(1) $$\int_M \Lambda = (\text{index of } X \text{ at } p) \cdot \int_{\pi_0^{-1}(p)} \Phi$$

$$= \chi(M) \cdot \int_{\pi_0^{-1}(p)} \Phi \qquad \text{by Theorem I.11-30.}$$

Since for $n = 2m$ we also have

(2) $$\int_M \Lambda = \int_M n!K_n \, dV = \pi^m m! 2^n \chi(M) \qquad \text{by Theorem 26,}$$

we obtain, finally,

(3) $$\int_{\pi_0^{-1}(p)} \Phi = \pi^m m! 2^n .$$

In the original intrinsic proof of the generalized Gauss-Bonnet theorem, Chern [2] did not use Theorem 22 or Corollaries 23 and 24. Instead, by clever guess-work he explicitly constructed a form Φ with $\pi_0{}^*\Lambda = d\Phi$, and noted that it satisfied equation (3). By applying equation (1), he thus deduced equation (2), which is precisely the generalized Gauss-Bonnet theorem. As we will soon see, it is very useful to have an explicit formula for Φ when we seek a generalized Gauss-Bonnet theorem for manifolds-with-boundary.

Let $(M,\partial M)$ be a compact orientable manifold-with-boundary. The Euler characteristic $\chi(M)$ is defined, as before, by

$$\chi(M) = \dim H^0(M) - \dim H^1(M) + \cdots .$$

With some work, we could generalize Theorem I.11-5, and show that $\chi(M) = \alpha_0 - \alpha_1 + \cdots$, where α_k is the number of k-simplexes in a triangulation.

But we won't pause to prove this, because other facts about $\chi(M)$ are more important for us. First note that we can construct a compact oriented manifold DM, the double of M, by taking two disjoint copies of M, and identifying the corresponding points of ∂M. The following result is obvious in terms of

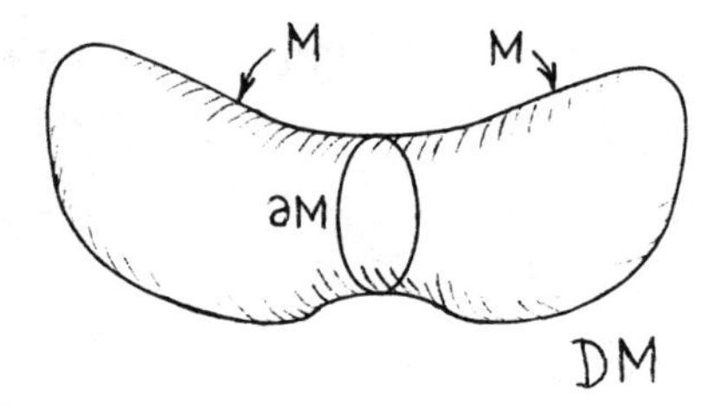

triangulations, but we will give an independent proof.

65. PROPOSITION. The Euler characteristic of DM is given by

$$\chi(DM) = 2\chi(M) - \chi(\partial M) .$$

Proof. Let U and V be open neighborhoods of the two copies of M in DM such that $H^k(U) \approx H^k(V) \approx H^k(M)$ for all k, and $H^k(U \cap V) \approx H^k(\partial M)$ for all k. Then we have the Mayer-Vietoris sequence (Theorem I.11-3)

$$0 \longrightarrow H^0(DM) \longrightarrow \cdots \longrightarrow H^k(DM) \longrightarrow H^k(U) \oplus H^k(V) \longrightarrow H^k(U \cap V) \xrightarrow{\delta} H^{k+1}(DM) \longrightarrow \cdots .$$

When we apply Proposition I.11-4, we obtain precisely the desired result. ∎

This result says quite different things when the dimension n of M is odd or even. For odd n we have $\chi(DM) = 0$ (Corollary I.11-25), so we obtain

$$\chi(M) = \frac{1}{2}\chi(\partial M) ;$$

in particular $\chi(\partial M)$ must be even. For even n, we have $\chi(\partial M) = 0$, so we have

$$(*) \qquad 2\chi(M) = \chi(DM) .$$

<u>66. COROLLARY</u>. Let M be a compact orientable manifold-with-boundary, of even dimension n. Let X be a vector field on M with only finitely many zeros, all in $M - \partial M$, such that X is outward pointing on ∂M. Then the sum of the indices of X is $\chi(M)$.

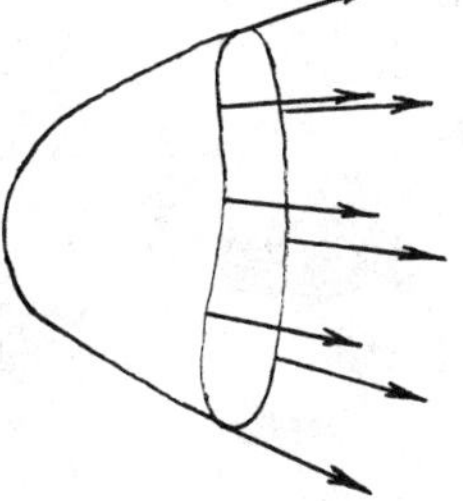

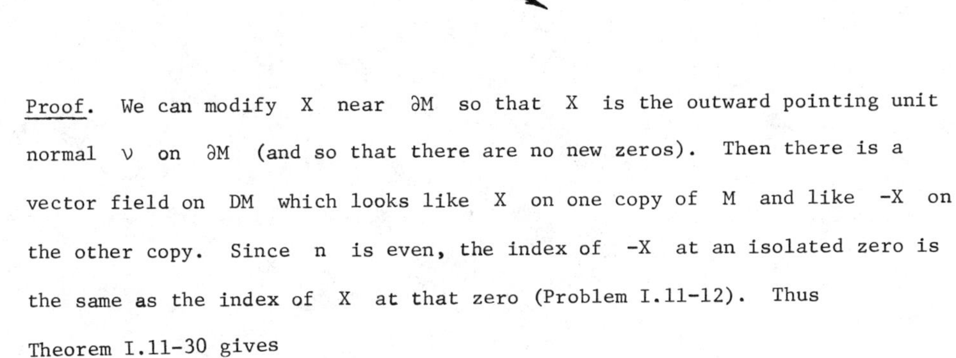

<u>Proof</u>. We can modify X near ∂M so that X is the outward pointing unit normal ν on ∂M (and so that there are no new zeros). Then there is a vector field on DM which looks like X on one copy of M and like $-X$ on the other copy. Since n is even, the index of $-X$ at an isolated zero is the same as the index of X at that zero (Problem I.11-12). Thus Theorem I.11-30 gives

$$2(\text{sum of the indices of } X) = \chi(DM) = 2\chi(M) , \qquad \text{by } (*). \quad ■$$

<u>67. COROLLARY</u>. Let M be a compact oriented Riemannian manifold-with-boundary, of even dimension $n = 2m$, with tangent bundle $\pi\colon TM \to M$, and

associated sphere bundle $\pi_0 = \pi|S\colon S \longrightarrow M$. Let ω be a connection on the principal bundle $\varpi\colon SO(TM) \longrightarrow M$, with curvature form Ω, let Λ be the unique n-form on M with $\varpi^*\Lambda = \sum \varepsilon^{i_1\cdots i_n} \Omega^{i_1}_{i_2} \wedge \cdots \wedge \Omega^{i_{n-1}}_{i_n} = 2^m m!\, \mathrm{Pf}(\Omega)$, and let Φ be an $(n-1)$-form on S with

$$\pi_0{}^*\Lambda = d\Phi\ .$$

Finally, let $\nu\colon \partial M \longrightarrow S$ be the outward pointing unit normal on ∂M. Then

$$\int_M K_n\, dV = \frac{1}{n!}\int_M \Lambda = \frac{\pi^m m!\, 2^n}{n!}\,\chi(M) \quad + \quad \frac{1}{n!}\int_{\partial M} \nu^*\Phi\ .$$

Proof. Extend ν to a vector field X on M with only finitely many zeros $p_1,\ldots,p_k \in M - \partial M$. Let $B_i(\varepsilon)$ be closed balls of radius ε around p_i which are disjoint from each other and from ∂M, and set

$$M_\varepsilon = M - \bigcup_{i=1}^{k} \text{interior } B_i(\varepsilon)\ .$$

Then, as on p. 558 , we have

$$\begin{aligned}
\int_M \Lambda &= \lim_{\varepsilon\to 0} \int_{\partial X(M_\varepsilon)} \Phi \\
&= \int_{\nu(\partial M)} \Phi \quad + \quad \sum_{i=1}^{k} \lim_{\varepsilon\to 0} \int_{\partial B_i(\varepsilon)} \Phi \\
&= \int_{\partial M} \nu^*\Phi \quad + \quad \sum_{i=1}^{k} \pi^m m!\, 2^n \cdot (\text{index of } X \text{ at } p_i) && \text{by (3), on p. 559} \\
&= \int_{\partial M} \nu^*\Phi \quad + \quad \pi^m m!\, 2^n \cdot \chi(M) && \text{by Corollary 66.} \quad \blacksquare
\end{aligned}$$

The only trouble with this result is that we can't interpret $\int_{\partial M} \nu^*\Phi$ until we have an explicit Φ with $\pi_0{}^*\Lambda = d\Phi$. Fortunately, we can pull a Φ out of the air by looking more carefully at the proofs in section 4.

First consider two connections on the principal bundle $SO(E)$, as in Proposition 20. Changing notation slightly, we denote these connections by ω and $\tilde{\omega}$, with curvature forms Ω and $\tilde{\Omega}$. Let Λ and $\tilde{\Lambda}$ be the forms with

$$\varpi^*(\Lambda) = 2^m m!\ \mathrm{Pf}(\Omega) \qquad \text{and} \qquad \varpi^*(\tilde{\Lambda}) = 2^m m!\ \mathrm{Pf}(\tilde{\Omega})\ .$$

All quantities associated with $M \times [0,1]$ will be written boldface, so the induced connections $q^*\omega$ and $q^*\tilde{\omega}$ in the proof of Proposition 20 will be denoted by

$$\boldsymbol{\omega} = q^*\omega \qquad \text{and} \qquad \tilde{\boldsymbol{\omega}} = q^*\tilde{\omega}\ ,$$

and we will set

$$\boldsymbol{\psi} = (1-\tau)\boldsymbol{\omega} + \tau\tilde{\boldsymbol{\omega}}\ .$$

Note that $\partial/\partial\tau$ is horizontal for these connections.

The proof of Proposition 20 tells us how to find a form Φ with $d\Phi = \tilde{\Lambda} - \Lambda$. To obtain Φ explicitly, we first want to describe the curvature form $\boldsymbol{\Psi}$ of $\boldsymbol{\psi}$ more explicitly. Note that tangent vectors on $M \times [0,1]$ can be considered as sums

$$X + \mu\frac{\partial}{\partial\tau} \qquad \mu \in \mathbb{R}$$

where X is a tangent vector of M, and tangent vectors on the total space of $q^*SO(\xi)$ can be considered as sums

$$Y + \mu \frac{\partial}{\partial \tau}$$

where Y is a tangent vector on $SO(E)$. We have

$$d\boldsymbol{\psi} = (1-\tau)d\boldsymbol{\omega} + \tau\, d\tilde{\boldsymbol{\omega}} + d\tau \wedge (\tilde{\boldsymbol{\omega}} - \boldsymbol{\omega}) ,$$

so for two tangent vectors $Y_1 + \mu_1\, \partial/\partial\tau$ and $Y_2 + \mu_2\, \partial/\partial\tau$ on the total space of $q^*SO(\xi)$ we have

$$\begin{aligned}\boldsymbol{\Psi}(Y_1 + \mu_1\, \partial/\partial\tau, Y_2 + \mu_2\, \partial/\partial\tau) &= d\boldsymbol{\psi}(h(Y_1 + \mu_1\, \partial/\partial\tau), h(Y_2 + \mu_2\, \partial/\partial\tau)) \\ &= [(1-\tau)d\boldsymbol{\omega} + \tau\, d\tilde{\boldsymbol{\omega}}](h(Y_1) + \mu_1\, \partial/\partial\tau, h(Y_2) + \mu_2\, \partial/\partial\tau) \\ &\quad + [d\tau \wedge (\tilde{\boldsymbol{\omega}} - \boldsymbol{\omega})](Y_1 + \mu_1\, \partial/\partial\tau, Y_2 + \mu_2\, \partial/\partial\tau) ,\end{aligned}$$

since $\tilde{\boldsymbol{\omega}} - \boldsymbol{\omega} = 0$ on vertical vectors.

Extending Y_2 to a vector field, we have

$$d\boldsymbol{\omega}(\partial/\partial\tau, h(Y_2)) = \frac{\partial}{\partial\tau}(\boldsymbol{\omega}(h(Y_2)) - h(Y_2)(\boldsymbol{\omega}(\partial/\partial\tau)) - \boldsymbol{\omega}([\partial/\partial\tau, h(Y_2)]) .$$

The first term on the right must vanish, since we can choose the vector field Y_2 to be independent of τ; the second vanishes, since $\partial/\partial\tau$ is horizontal for $\boldsymbol{\omega}$; and it is easily seen that the third term also vanishes (compare p.IV.533). So if $\boldsymbol{\Omega}$ and $\tilde{\boldsymbol{\Omega}}$ are the curvature forms for $\boldsymbol{\omega}$ and $\tilde{\boldsymbol{\omega}}$, then we have, finally,

$$\begin{aligned}\boldsymbol{\Psi} &= (1-\tau)\boldsymbol{\Omega} + \tau\tilde{\boldsymbol{\Omega}} + d\tau \wedge (\tilde{\boldsymbol{\omega}} - \boldsymbol{\omega}) \\ &= \boldsymbol{\Omega} + \tau(\tilde{\boldsymbol{\Omega}} - \boldsymbol{\Omega}) + d\tau \wedge (\tilde{\boldsymbol{\omega}} - \boldsymbol{\omega}) .\end{aligned}$$

From this expression we see that

$$2^m m!\ \mathrm{Pf}(\boldsymbol{\Psi}) = \sum_{i_1,\ldots,i_n} \varepsilon^{i_1\cdots i_n} \boldsymbol{\Psi}^{i_1}_{i_2} \wedge \cdots \wedge \boldsymbol{\Psi}^{i_{n-1}}_{i_n}$$

$$= \sum_{i_1,\ldots,i_n} \varepsilon^{i_1\cdots i_n} \boldsymbol{\Omega}^{i_1}_{i_2} \wedge \cdots \wedge \boldsymbol{\Omega}^{i_{n-1}}_{i_n}$$

$$+ d\tau \wedge \mathbf{H} ,$$

where $\mathbf{H}$ is a linear combination of terms of the form

$$(*) \quad \tau^{m-k-1} \sum_{i_1,\ldots,i_n} \varepsilon^{i_1\cdots i_n} \boldsymbol{\Omega}^{i_1}_{i_2} \wedge \cdots \wedge \boldsymbol{\Omega}^{i_{2k-1}}_{i_{2k}} \wedge (\tilde{\boldsymbol{\Omega}} - \boldsymbol{\Omega})^{i_{2k+1}}_{i_{2k+2}} \wedge \cdots \wedge$$

$$\wedge (\tilde{\boldsymbol{\Omega}} - \boldsymbol{\Omega})^{i_{n-3}}_{i_{n-2}} \wedge (\tilde{\boldsymbol{\omega}} - \boldsymbol{\omega})^{i_{n-1}}_{i_n} .$$

We won't bother keeping track of the exact coefficients involved in our calculations, since there will be a cheap way of getting them out at the end. Our expression for $2^m m!\ \mathrm{Pf}(\boldsymbol{\Psi})$ shows that the closed n-form $\boldsymbol{\Lambda}$ on $M \times [0,1]$ which pulls back to $2^m m!\ \mathrm{Pf}(\boldsymbol{\Psi})$ can be written

$$\boldsymbol{\Lambda} = \cdots + d\tau \wedge \boldsymbol{\eta} ,$$

where $\boldsymbol{\eta}$ is a linear combination of forms which pull back* to the forms (*). Now Theorem I.7-17 (p. I.304) says that

*It is easy to see that there <u>are</u> forms with this property, since $\boldsymbol{\Omega}$, $\tilde{\boldsymbol{\Omega}}$, and $\tilde{\boldsymbol{\omega}} - \boldsymbol{\omega}$ vanish on vertical vectors. The proof is similar to that of Proposition 18, except that in the second part we explicitly write out the value of the form (*) on $R_{A*}Y_1,\ldots,R_{A*}Y_{n-1}$, noting that $R_A{}^*\boldsymbol{\omega} = A^{-1}\boldsymbol{\omega} A$ and $R_A{}^*\tilde{\boldsymbol{\omega}} = A^{-1}\tilde{\boldsymbol{\omega}} A$, by the definition of a connection. Then we check that this is $\det A = 1$ times the value of the form on $Y_1,\ldots,Y_{n-1}$, the computation being similar to that in the proof of Proposition 9.

$$\tilde{\Lambda} - \Lambda = i_1{}^*\boldsymbol{\Lambda} - i_0{}^*\boldsymbol{\Lambda} = d(I\boldsymbol{\Lambda}) \,,$$

where

$$I\boldsymbol{\Lambda}(p)(X_1,\ldots,X_{n-1}) = \int_0^1 \boldsymbol{\eta}(p,t)(i_{t*}X_1,\ldots,i_{t*}X_{n-1})\,dt \,.$$

In this integral, t will enter only in the factors t^{m-k-1}. All other terms are independent of t, since the connections $\boldsymbol{\omega}$ and $\tilde{\boldsymbol{\omega}}$ are independent of t -- for a tangent vector Y on $SO(E)$ we have $\boldsymbol{\omega}(i_{t*}Y) = \omega(Y)$, and similarly for $\tilde{\boldsymbol{\omega}}, \boldsymbol{\Omega}, \tilde{\boldsymbol{\Omega}}$. So we see, finally, that

(A) $\tilde{\Lambda} - \Lambda = d\Phi$, where the $(n-1)$-form Φ on M is a linear combination of $(n-1)$-forms which pull back to the forms

$$(**)\quad \sum_{i_1,\ldots,i_n} \varepsilon^{i_1\cdots i_n}\, \Omega^{i_1}_{i_2} \wedge \cdots \wedge \Omega^{i_{2k-1}}_{i_{2k}} \wedge (\tilde{\Omega}-\Omega)^{i_{2k+1}}_{i_{2k+2}} \wedge \cdots \wedge (\tilde{\Omega}-\Omega)^{i_{n-3}}_{i_{n-2}} \wedge (\tilde{\omega}-\omega)^{i_{n-1}}_{i_n}$$

on $SO(E)$.

Now we will apply this to a special, complicated, case. Let $\xi = \pi\colon E \to M$ be an n-dimensional vector bundle with a Riemannian metric $\langle\ ,\ \rangle$, let $SO(\xi) = \varpi\colon SO(E) \to M$ be the corresponding principal bundle, and let $\pi_0\colon S \to M$ be the corresponding sphere bundle. Set

$$\zeta = \pi_0{}^*\xi = p\colon \mathcal{E} \to M \;;$$

the total space E consists of all pairs (e,v) where $e \in S$ and $v \in \pi(e)$. For the corresponding principal bundle

$$SO(\zeta) = \mathcal{P}: SO(E) \longrightarrow S \ ,$$

the fibre over e is the set of all (e,u) where $u \in \varpi^{-1}(\pi(e))$ is an orthonormal frame at $\pi(e)$. The principal bundle map

$$\tilde{\pi}_0: SO(E) \longrightarrow SO(E) \qquad \begin{array}{ccc} SO(E) & \xrightarrow{\tilde{\pi}_0} & SO(E) \\ \downarrow \mathcal{P} & & \downarrow \varpi \\ S & \xrightarrow{\pi_0} & M \end{array}$$

which covers π_0 takes (e,u) to u.

Recall that we can write

$$\zeta = \zeta_1 \oplus \zeta_2 \ ,$$

where

$\zeta_1 = p_1: E_1 \longrightarrow M$ is an $(n-1)$-dimensional bundle,

$\zeta_2 = p_2: E_2 \longrightarrow M$ is a trivial 1-dimensional bundle;

the total space $E_2 \subset E$ consists of all pairs $(e,\mu e)$ for $e \in S$ and $\mu \in \mathbb{R}$, while $E_1 \subset E$ consists of all pairs (e,v) where $v \in \pi(e)$ is orthogonal to e. For the corresponding principal bundles

$$SO(\zeta_i) = \mathcal{P}_i: SO(E_i) \longrightarrow S \ ,$$

the fibre $\mathcal{P}_2^{-1}(e)$ contains just the two pairs (e,e) and $(e,-e)$, while the fibre $\mathscr{P}_1^{-1}(e)$ consists of all pairs (e,υ) where υ is an ordered $(n-1)$-tuple of orthonormal vectors at $\pi(e)$ all of which are perpendicular to e. We have a natural inclusion

$$\iota\colon SO(E_1) \longrightarrow SO(E) ,$$

which takes (e,υ) to (e,u) where u is the frame whose first $n-1$ members come from υ, while $u_n = e$. Clearly,

$$\mathscr{P}\circ\iota = \mathscr{P}_1 .$$

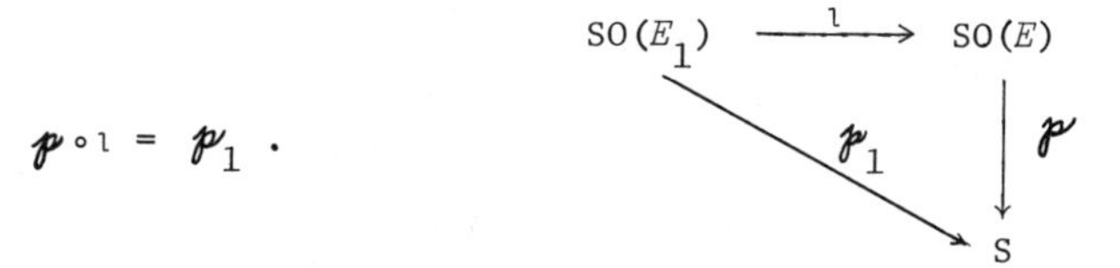

Now let ω be a connection on $SO(E)$, with curvature form Ω. Then

$$\gamma = \tilde{\pi}_0{}^*\omega$$

is a connection on $SO(E)$, whose curvature form Γ is $\tilde{\pi}_0{}^*\Omega$. Hence if Λ is the unique form on M with

$$\varpi^*\Lambda = 2^m m!\ \mathrm{Pf}(\Omega) ,$$

then

$$(1) \qquad \mathscr{P}^*\pi_0{}^*\Lambda = \tilde{\pi}_0{}^*\varpi^*\Lambda = 2^m m!\ \mathrm{Pf}(\tilde{\pi}_0{}^*\Omega) = 2^m m!\ \mathrm{Pf}(\Gamma) .$$

We can also use γ to determine a connection γ_1 on $SO(E_1)$ by

$$(\gamma_1)^i_j = \iota^*\gamma^i_j \qquad i,\ j = 1,\dots,n-1 .$$

The two conditions

$$\gamma_1(\sigma(M)) = M \qquad \text{for } M \in \mathfrak{o}(n-1)$$

$$R_A{}^*\gamma_1 = \mathrm{Ad}(A^{-1})\gamma_1 \qquad \text{for } A \in SO(n-1)$$

follow from the corresponding conditions for γ -- if we regard $SO(E_1) \subset SO(E)$ via the map ι, then $\sigma(M)$ in $SO(E_1)$ is the restriction of $\sigma(M)$ in $SO(E)$ for $M \in \mathfrak{o}(n-1)$, and R_A in $SO(E_1)$ is the restriction of $R_{\tilde{A}}$ in $SO(E)$, where $\tilde{A} = \begin{pmatrix} A & 0 \\ 0 & 1 \end{pmatrix}$.

Although $(\gamma_1)^i_j = \iota^*\gamma^i_j$ for $i, j = 1,\ldots,n-1$, it does not follow for these same values of i and j that the curvature forms Γ_1 and Γ satisfy $(\Gamma_1)^i_j = \iota^*\Gamma^i_j$; for the horizontal component in $SO(E_1)$ is different than in $SO(E)$. To find the correct relationship, we use the second structural equation (Theorem II.8-16),

$$d\gamma(Y_1,Y_2) = -[\gamma(Y_1),\gamma(Y_2)] + \Gamma(Y_1,Y_2)$$

for tangent vectors Y_1, Y_2 on $SO(E)$, which means that

$$(2) \qquad d\gamma^i_j(Y_1,Y_2) = -\sum_{k=1}^{n} \gamma^i_k(Y_1)\gamma^k_j(Y_2) - \gamma^i_k(Y_2)\gamma^k_j(Y_1) \ + \ \Gamma^i_j(Y_1,Y_2)\ .$$

Similarly, for tangent vectors Z_1, Z_2 on $SO(E_1)$, we have

$$d(\gamma_1)^i_j(Z_1,Z_2) = -\sum_{k=1}^{n-1} (\gamma_1)^i_k(Z_1)(\gamma_1)^k_j(Z_2) - (\gamma_1)^i_k(Z_2)(\gamma_1)^k_j(Z_1)$$
$$+ (\Gamma_1)^i_j(Z_1,Z_2)$$

$$\Longrightarrow \quad (3) \qquad d\gamma^i_j(\iota_* Z_1, \iota_* Z_2) = -\sum_{k=1}^{n-1} \gamma^i_k(\iota_* Z_1)\gamma^k_j(\iota_* Z_2) - \gamma^i_k(\iota_* Z_2)\gamma^k_j(\iota_* Z_1)$$
$$+ (\Gamma_1)^i_j(Z_1, Z_2) \ .$$

Comparing (2) and (3), we see that

$$(4) \qquad (\Gamma_1)^i_j = \iota^*\Gamma^i_j + \iota^*(\gamma^i_n \wedge \gamma^j_n) \qquad i, j = 1, \ldots, n-1 \ .$$

Now we are going to apply the construction in the proof of Theorem 22. The principal bundle $SO(E_2)$ for the 1-dimensional bundle ζ_2 is just 2 copies of M; the only connection on $SO(E_2)$ is $\gamma_2 = 0$. The bundle $SO(E_1)*SO(E_2) \subset SO(E)$ in the proof of Theorem 22 is just 2 copies of $SO(E_1)$; the first copy may be identified with $\iota(SO(E_1))$. The connection

$$\bar{\gamma} = \rho_1{}^*\gamma_1 \oplus \gamma_2{}^*\gamma_2$$

on $SO(E_1)*SO(E_2)$ which is constructed in the proof of Theorem 22 is just $\iota^*\gamma_1$ on the first copy, and similarly the curvature form $\bar{\Gamma}$ is just $\iota^*\Gamma_1$ on the first copy. As before, we extend the connection $\bar{\gamma}$ to a connection $\tilde{\gamma}$ on $SO(E)$, with curvature form $\tilde{\Gamma}$. We have unique forms $T, \tilde{T}$ on S with

$$\mathfrak{p}^*T = 2^m m! \ \mathrm{Pf}(\Gamma)$$
$$\mathfrak{p}^*\tilde{T} = 2^m m! \ \mathrm{Pf}(\tilde{\Gamma}) \ .$$

But equation (1) says that $T = \pi_0{}^*\Lambda$, while, as in the proof of Theorem 22, we have $\mathrm{Pf}(\tilde{\Gamma}) = 0$ at points of $SO(E_1)*SO(E_2)$, which implies that $\tilde{T} = 0$. Assertion (A) on p. 566 thus shows that

$\pi_0{}^*\Lambda = d\Phi$, where the $(n-1)$-form Φ on S is a linear combination of $(n-1)$-forms which pull back, via $\mathscr{p}^*$, to the forms

$$\sum_{i_1,\ldots,i_n} \varepsilon^{i_1\cdots i_n} \Gamma^{i_1}_{i_2} \wedge \cdots \wedge \Gamma^{i_{2k-1}}_{i_{2k}} \wedge (\tilde{\Gamma}-\Gamma)^{i_{2k+1}}_{i_{2k}} \wedge \cdots \wedge (\tilde{\Gamma}-\Gamma)^{i_{n-3}}_{i_{n-2}} \wedge (\tilde{\gamma}-\gamma)^{i_{n-1}}_{i_n} .$$

But on tangent vectors to $SO(E_1)*SO(E_2)$ = 2 copies of $SO(E_1)$, we clearly have

$$(\tilde{\gamma}-\gamma)^i_j = \begin{cases} 0 & i,\ j \neq n \\ \iota^*\gamma^i_j & i \text{ or } j = n \end{cases} \qquad \text{on the first copy,}$$

while equation (4) shows that we also have

$$(\tilde{\Gamma}-\Gamma)^i_j = \begin{cases} \iota^*(\gamma^i_n \wedge \gamma^j_n) & i,\ j \neq n \\ -\ \iota^*\Gamma^i_j & i \text{ or } j = n \end{cases} \qquad \text{on the first copy.}$$

Hence, remembering that $\gamma = \tilde{\pi}_0{}^*\omega$ and $\Gamma = \tilde{\pi}_0{}^*\Omega$, we can conclude that

$\pi_0{}^*\Lambda = d\Phi$, where the $(n-1)$-form Φ on S is a linear combination of $(n-1)$-forms Φ_k, $0 \leq k \leq m-1$, such that

$$\mathscr{p}_1{}^*\Phi_k = \iota^*\mathscr{p}^*\Phi_k = \iota^*\tilde{\pi}_0{}^*\left(\sum_{i_1,\ldots,i_{n-1}} \varepsilon^{i_1\cdots i_{n-1}} \Omega^{i_1}_{i_2} \wedge \cdots \wedge \Omega^{i_{2k-1}}_{i_{2k}} \wedge \omega^{i_{2k+1}}_n \wedge \cdots \wedge \omega^{i_{n-1}}_n\right);$$

in this sum, the indices i_α run from 1 to $n-1$.

This can be put in a more useful form by introducing the "last vector" map $\ell\colon SO(E) \to S$ defined by

$$\ell(u) = u_n .$$

We clearly have

$$\ell \circ \tilde{\pi}_0 \circ \iota = \mathscr{P}_1\colon SO(E_1) \to S .$$

It is easy to check that there are unique forms Θ_k on S such that

$$\ell^*\Theta_k = \sum_{i_1,\dots,i_{n-1}} \varepsilon^{i_1 \cdots i_{n-1}} \Omega_{i_2}^{i_1} \wedge \dots \wedge \Omega_{i_{2k}}^{i_{2k-1}} \wedge \omega_n^{i_{2k+1}} \wedge \dots \wedge \omega_n^{i_{n-1}} ;$$

we use the procedure given in the footnote on p. 565, noting that in the second part we only want to consider $A \in SO(n)$ with $\ell(u\cdot A) = \ell(u)$, so that $A = \begin{pmatrix} B & 0 \\ 0 & 1 \end{pmatrix}$ for $B \in SO(n-1)$. But now we have

$$\mathscr{P}_1{}^*\Phi_k = \iota^*\tilde{\pi}_0{}^*\ell^*\Theta_k = \mathscr{P}_1{}^*\Theta_k$$

$$\Longrightarrow \quad \Phi_k = \Theta_k .$$

So we can state, finally,

(B) There are constants $a_0,\dots,a_{m-1}$ such that

$$\pi_0{}^*\Lambda = \sum_{k=0}^{m-1} a_k\Phi_k ,$$

where the Φ_k are the unique forms on S such that

$$\ell^*\Phi_k = \sum_{i_1,\ldots,i_{n-1}} \varepsilon^{i_1\cdots i_{n-1}} \Omega_{i_2}^{i_1} \wedge \cdots \wedge \Omega_{i_{2k}}^{i_{2k-1}} \wedge \omega_n^{i_{2k+1}} \wedge \cdots \wedge \omega_n^{i_{n-1}} .$$

Note that the constants $a_0,\ldots,a_{m-1}$ do not depend on the bundle, or anything else; they are certain combinatorial terms in a formal calculation which is exactly the same for all bundles. If we apply Corollary 67, we obtain

$$(*) \qquad \int_M K_n \, dV = \frac{\pi^m m! 2^n}{n!} \chi(M) + \frac{1}{n!} \sum_{k=0}^{m-1} a_k \int_{\partial M} \nu^*\Phi_k .$$

If we choose a positively oriented orthonormal moving frame $\mathbf{X} = (X_1,\ldots,X_n)$ on ∂M with $X_n = \nu$, then

$$\nu = \ell \circ \mathbf{X} , \qquad \mathbf{X} : \partial M \to SO(E) ,$$

so

$$\begin{aligned} \nu^*\Phi_k &= \mathbf{X}^*\ell^*\Phi_k \\ &= \sum_{i_1,\ldots,i_{n-1}} \varepsilon^{i_1\cdots i_{n-1}} (\mathbf{X}^*\Omega_{i_2}^{i_1}) \wedge \cdots \wedge (\mathbf{X}^*\Omega_{i_{2k}}^{i_{2k-1}}) \\ &\qquad \wedge (\mathbf{X}^*\omega_n^{i_{2k+1}}) \wedge \cdots \wedge (\mathbf{X}^*\omega_n^{i_{n-1}}) . \end{aligned}$$

Recall that ω and Ω are forms on $SO(E)$; the terms $\mathbf{X}^*\omega_n^{i_{2k+1}}$ and $\mathbf{X}^*\Omega_{i_2}^{i_1}$ are just the corresponding connection and curvature forms for the moving frame $\mathbf{X}$. This allows us to give an invariant definition of $\nu^*\Phi_k$ similar to the invariant definition of $\Lambda/n! = K_n \, dV$: if dV_{n-1} denotes the volume element on ∂M, then

$$\nu^*\Phi_k = K_k \, dV_{n-1} ,$$

where K_k can be written as a contraction of tensor products of the tensor $\mathcal{E}$ on ∂M (contravariant of order $n-1$), the curvature tensor $\mathcal{R}$ for ∂M, and the tensor

$$(X,Y) \mapsto \langle \nabla_X \nu, Y \rangle ,$$

which is just the second fundamental form of ∂M in M.

To calculate the constants $a_0, \ldots, a_{m-1}$, we just have to apply equation (*) to products $M = D^{k+1} \times S^{n-k-1}$, with $\partial M = S^k \times S^{n-k-1}$; then the only non-zero boundary integral is the one involving $\nu^*\Phi_k$. The explicit calculations are left to the reader -- after doing them, it should be fun to compare with Chern's paper [2].

Bibliography

Part A of the bibliography describes the main topics of differential geometry not included in these volumes, as well as many subsidiary topics. Part B lists books, monographs, etc., referred to by curly brackets { }, while Part C lists individual journal articles, referred to by square brackets [ ]. Part B has a supplement (p. 632), mainly for books published since 1975. The list of journal articles is rudimentary, containing only a few articles not specifically referred to in this volume, or in the mini-bibliographies of the preceeding two; no serious attempt has been made to provide historical background or to attribute theorems justly. A more than adequate supply of references can be obtained from the extensive bibliographies in some of the books listed in Part B, and of course there's always *Mathematical Reviews*.

A. Other Topics in Differential Geometry

I. Major topics everyone should know something about.*

(a) Complex manifolds. This is the main topic which was hardly touched upon in these notes. Some people heartily dislike the subject, with its penetrating algebraic odor, but for others it has a seductive appeal. A differential geometer whose work often uses the simplifications effected by considering the complex domain explained to me that the additional structure of complex manifolds makes them more interesting, just as two sexes are more interesting than one. A basic treatment of complex manifolds is given in Kobayashi and Nomizu {1; v.2, ch.9}. See also Chern {1}, Weil {1}, Wu {1}, and Yano {2}. For more emphasis on the analysis aspects, see Griffiths {1}, Morrow and Kodaira {1}, and Wells {1}.

(b) Homogeneous spaces. In Chapter 13 we defined a "homogeneous space" to be a quotient space G/H. The terminology comes from the fact that these are precisely the manifolds M on which G acts transitively (see Warner {1; 120ff.} or Wolf {1; 11-13}); one should also be aware of the usage in the case of Riemannian manifolds (cf. Kobayashi and Nomizu {1; v.1, 176}). Once the identification with G/H is made, further study of these spaces becomes rather algebraic. See Kobayashi and Nomizu {1; v.2, ch.10}.

(c) Symmetric Spaces. A very nice brief introduction to (Riemannian) symmetric spaces can be found in Milnor {2; §20, 21}. Increasing detail, and algebra, can be found in Wolf {1}, Kobayashi and Nomizu {1; v.2, ch.11}, Boothby and Weiss {1}, Loos {1}, and the standard treatise Helgason {1}.

* As evidenced by the fact that they get a chapter apiece in Kobayashi and Nomizu {1}.

(d) Mappings. Numerous types of maps between Riemannian manifolds are of importance -- isometries, similarities (which multiply the metrics by a constant), conformal maps (which multiply the metrics by a function), affine maps (which take geodesics into geodesics), and projective maps (which take geodesics into reparameterized geodesics); the latter two can be defined for arbitrary connections. We might also mention [essential] volume preserving maps, which preserve volumes of open subsets [up to a constant factor]. Certain classical relations between such maps, not mentioned here, may be found, for example, in Laugwitz {1; 147-161}. In preference to Theorem 13.4.6, one may consult Lemma 1 on p.242 of Kobayashi and Nomizu {1; v.1} (Laugwitz defines "irreducible" incorrectly); their proof of Lemma 2 on the same page is perhaps also somewhat preferable to Laugwitz's proof of the corresponding Theorem 13.6.2. One other classical result may be found in Haack {1; 130-133} or Kreysig {1; 267-269}. Naturally, the study of maps from S^2 to $\mathbb{R}^2$ has received special attention. Although cartography is an independent subject, the reader will probably find more than enough information about it in classical differential geometry books, for example Kreysig {1} and Laugwitz {1}, or Scheffers {1; v.2, 36-53} and Strubecker {1; v.2, 170-201} for more examples. In addition to the maps themselves, one can study vector fields which represent "infinitesimal" versions of them. In particular, the infinitesimal versions of isometries are the "Killing vector fields". For basic information on mappings and infinitesimal mappings see Kobayashi and Nomizu {1; v.1, ch.6}, Lichnerowicz {2}, and Yano {3}.

II. Other topics of substantial interest.

(a) Classical curve and surface theory. There is still a lot of information to be mined here, though the ore naturally tends to decrease in quality, rapidly passing the point of diminishing returns.

In our discussions of curve theory, we initially sought the limiting circle through 3 points on a plane curve. This osculating circle is often said to have second order contact with the curve. The notion of contact of curves and surfaces is described in Struik {1; 23}, somewhat more carefully in Goetz {1; 37, 44} and Kreysig {1; 47-51, see especially Theorem 14.3}, and in detail in Favard {1; Part 1, ch.2}. Innocuous as the concept may seem, it is sometimes useful to have precise information about it (cf. do Carmo and Warner [1; p.136]).

In the theory of space curves it is natural to seek an "osculating sphere" having third order contact with the curve -- see Blaschke {1; v.1, 33}, Eisenhart {2, 37}, Goetz {1; 77}, Kreysig {1; 51}, Struik {1; 25}, or Gerretsen {1; 91}, which gives the analogous considerations in $\mathbb{R}^n$. The condition that a curve lie on some sphere is usually determined by setting equal the radii of all osculating spheres. Problem III.4-2 gives a different approach (generalized in Gerretsen {1; 78}); some calculation is required to establish the equivalence of the two answers.

A standard topic in curve theory is the study of involutes and evolutes, which was originated (cf. Coolidge {1; 319}) by Huygens in order to construct a pendulum whose period is independent of its amplitude (for ordinary pendulums this is only approximately so, for small amplitudes). This property is possessed by a pendulum whose weighted end describes a cycloid, hence the problem of finding a curve whose "involute", traced out by unwinding a thread from it, will be a cycloid. The desired curve (the "evolute" of the cycloid) turns out to be another cycloid. [Unfortunately, this ingeniously designed Huygens pendulum was superceded by a pendulum suspended from a spring, which turns out to work just as well.] Two other familiar curves might be mentioned here: the evolute of the tractrix is a catenary. The standard material, none too interesting, on involutes and evolutes (which are also defined for space curves) can be found in Eisenhart {2; 43-45}, Gerretsen {1; 83-87}, Goetz {1; 65-70}, Kreysig {1; 52-54}, Struik {1; 39-41}, and Strubecker {1; v.1, 222-226}. Guggenheimer {1; 35-47} gives a treatment requiring less differentiability, by means of the Riemann-Stieltjes integral (or see Ostrowski [1]), and finds all plane curves similar to their evolutes (pp. 59-61). It is strange that the theory of envelopes is so seldom mentioned in connection with involutes and evolutes, for the evolute of a plane curve is the envelope of its normals, and thus the locus of the centers of its osculating circles. Even the latter fact is seldom mentioned (cf. Guggenheimer {1; 43-44}). As the figure on p. II.7 seems to indicate, an osculating circle of a curve separates the parts of the curve with smaller curvature from those with larger curvature. A direct verification may be given (cf. Goetz {1; 84}), but it is even easier to prove a much stronger result, due to Kneser. If c is a curve, parameterized by arclength, with κ nowhere 0, then the curve of centers of the osculating circles is $\gamma(s) = c(s) + \mathbf{n}(s)/\kappa(s)$. The parameter s is not arclength for γ. Instead we have

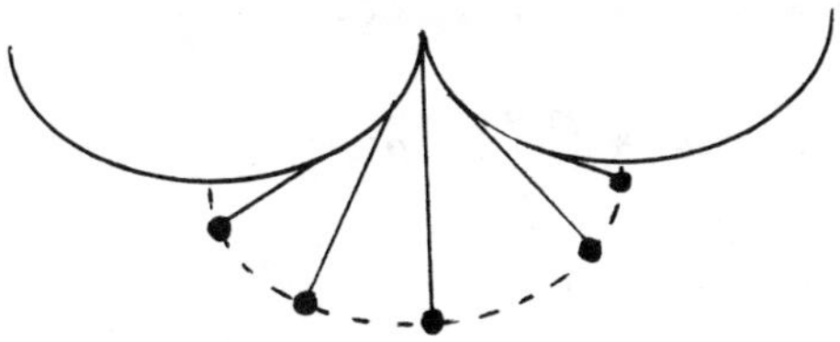

$$\begin{aligned}
\gamma'(s) &= c'(s) + \left(\frac{1}{\kappa}\right)'(s)\,\mathbf{n}(s) + \frac{1}{\kappa(s)}\,\mathbf{n}'(s) \\
&= c'(s) + \left(\frac{1}{\kappa}\right)'(s)\,\mathbf{n}(s) + \frac{1}{\kappa(s)}\cdot(-\kappa(s)c'(s)) \quad \text{by Serret-Frenet} \\
&= \left(\frac{1}{\kappa}\right)'(s)\,\mathbf{n}(s) \; .
\end{aligned}$$

Now $\mathbf{n}$ is not constant on any interval, so if $\kappa' \neq 0$ everywhere, then γ' is not constant on any interval, and hence no portion of γ is a straight line. Therefore

$$\begin{aligned}
|\gamma(s_1) - \gamma(s_2)| &< \text{length of } \gamma \text{ from } \gamma(s_1) \text{ to } \gamma(s_2) \\
&= \int_{s_1}^{s_2} |\gamma'(t)|\,dt = \left|\frac{1}{\kappa}(s_2) - \frac{1}{\kappa}(s_1)\right| \; ,
\end{aligned}$$

where the left side is the distance between the centers of the osculating circles at s_1 and s_2, while the right side is the difference in the radii of these osculating circles. Hence one must lie inside the other. Thus all the osculating circles are nested (and the smaller ones, containing points of the curve with larger curvature, must lie inside the larger ones). This gives a striking example of a family of curves (the osculating circles) none of which intersect, but which nevertheless have an envelope (the original curve). There is a beautiful little book of Boltyanskii {1}, unfortunately out of print, which makes the study of envelopes seem very pretty. In differential geometry books, emphasis is usually placed on envelopes of families of surfaces; see Eisenhart {2; 59-65} and Kreysig {1; 253-263}. There is a fairly thorough treatment of envelopes in Favard {1; Part 1, ch.3}. We should not fail to mention that the conjugate locus of a point is just the envelope of the geodesics through the point (refer again to the picture on p. IV.325). An argument of Carathéodory shows that the conjugate locus always has at least 4 cusps (cf. Blaschke {1; v.1, 231}). Similar arguments show that the evolute (= envelope of the normals) of a closed plane curve must have at least 4 cusps. This actually follows immediately from the 4 vertex theorem, but it can also be used to prove the 4 vertex theorem, as well as a 4 vertex theorem for the hyperbolic plane (oral communication by A. Weinstein).

Naturally, many special sorts of space curves are investigated in the classical books. Special mention may be made of Bertrand curves -- see Blaschke {1; v.1, 35}, Eisenhart {2; 39-41}, Gerretsen {1; 83}, Goetz {1; 74-76}, Haack {1; 29}, or Strubecker {1; v.1, 228-238}. They are of some interest, as they may be used to prove (Catalan's theorem) that the only ruled minimal surface is the helicoid -- see do Carmo {1}. (This also follows from a classical result of Schwarz that a minimal surface containing a straight line is taken into itself by a rotation of π around the line. See, for example, Blaschke {1}.) Perhaps this is also a suitable place to mention an elementary theorem of Beltrami: the tangent developable of c intersects the osculating plane of c at $c(t)$ in a curve γ whose curvature at $c(t)$ is 3/4 the curvature of c at this point.

Moving on to global theorems, we first mention that a simple proof of Theorem II.1-8 can be given when κ is nowhere 0 by noting that the curve is locally convex, and then using a theorem of Schmidt (cf. Stoker {1; 46-47}) that local convexity implies convexity. (But for this proof we need to know that the curve bounds a region, to which Schmidt's theorem is applied. The proof given previously, using the "Hopf Umlaufsatz" (Theorem II.1-7), not only works for $\kappa \geq 0$ or $\kappa \leq 0$, but also proves directly that the curve lies on one side of each of its tangent lines, a criterion for convexity which does not use the fact that the curve bounds a region.) Schmidt's result holds in all dimensions, and could also be used to prove Hadamard's theorem (III.2-11) for imbedded surfaces. Some theorems of Schwarz, Schur, and Schmidt are especially interesting because they are global theorems about non-closed curves -- see Blaschke {1; v.1, 61-64}, as well as Chern {3; 35-38}. Guggenheimer {1; 31} proves one of these theorems in the special case where both curves are planar; Hilbert and Cohn-Vossen {1; 211-212} show how to obtain the general case from this, and then give some further discussion. Compare also Blaschke {1; v.1, §39}. Our old friend, the four vertex theorem, can be proved from the planar case (Guggenheimer {1; 30-32} or Fog [1]). By the way, in our proof of the 4 vertex theorem we only obtained 4 points where $\kappa' = 0$, but it is easy to actually obtain 4 relative maxima or minima of κ. Other interesting global theorems about non-closed curves are due to Vogt and Ostrowski (see Guggenheimer {1; 49-53} or Ostrowski [2]). Laugwitz {1; 198-202} has some results on curves of constant width, which may also be proved by more elementary means -- see, for example, the beautiful book of Yaglom and Boltyanskiĭ {1; ch.7}. Notice that the last formula on p. III.334 can be written $\tau = (\arctan \kappa_g)'$. It follows from this that $\int \tau \, ds = 0$ for a closed curve lying on a sphere. Conversely, if this holds for all closed curves on a surface, then it is part of a plane or sphere (Scherrer [1]). The same results hold for $\int \kappa^n \tau \, ds$ (Saban [1]). Finally, we mention that any curve in S^2 is the unit tangent $\mathbf{t}$ of some curve c of constant torsion. Such curves were studied classically (see, for example, Blaschke {1; v.1, 47}, and Darboux {1; §36, 39 and v.4, Note 7, §7}), but only recently have closed curves of constant torsion been discovered (Weiner [1]; compare p. IV.163).

Classical surface theory is of course much more extensive, and there is so much material contained in the standard classical books that there is no point trying to list the main topics. In Part B mention is sometimes made of specific information contained in particular books. For serious digging be sure not to forget the Encyklopädia der Mathematischen Wissenschaften.

See also the references under III.(c).

(b) Extremal and isoperimetric problems. Various extremal and isoperimetric problems for curves are treated in Blaschke {1; v.1, ch.2}. See Chapter 8 of the same volume for surfaces, and Chapters 2 and 6 of the second volume for similar problems in special affine geometry. For various sorts of solutions to the isoperimetric problem see also Blaschke and Reichardt {1; §28}, Chern {3; 25-29}, and Guggenheimer {1}. The last has a solution in the plane (pp. 79-84) which generalizes (p.289) to convex surfaces, by means of Steiner's formulas relating the area A and enclosed volume V of a compact convex surface $M \subset \mathbb{R}^3$ to the area $A(\varepsilon)$ and enclosed volume $V(\varepsilon)$ of its parallel surface $\{p + \varepsilon\nu(p) : p \in M\}$:

$$A(\varepsilon) = A + 2\varepsilon \int_M H\, dA + 4\pi\varepsilon^2$$

$$V(\varepsilon) = V + \varepsilon A + \varepsilon^2 \int_M H\, dA + \frac{4}{3}\pi\varepsilon^3 .$$

(The first formula follows from Problem III.3-12, and the second by integrating with respect to ε. It can also be proved by approximating the surface by convex polyhedra. In this case H measures dihedral angles and K measures vertex angles; for more details see pp. 168-170 of the article by Santalo in Chern {3}.) See also Blaschke {3} or Santalo {1} for a treatment of the isoperimetric problem by integral geometry (III.(a)). There is a detailed discussion of the isoperimetric problem in Blaschke {2}, but for the final word (including the isoperimetric problem in the spaces of constant curvature) see Hadwiger {1} and references therein.

(c) Closed geodesics. For brief remarks on the existence of closed geodesics see Blaschke {1; v.1, 211-212}. See also p. 233 for surfaces on which all geodesics are closed; for more details consult Berger {1}, which also proves the theorem of L.W. Green [1] that the sphere is the only surface on which every point has a unique conjugate point. For modern treatments of closed geodesics see Schwartz {1} and especially Flaschel and Klingenberg {1}. Recent books are Besse {1} and Klingenberg {3} (see the Supplement to Part B).

(d) Holonomy. The holonomy group of a connection on a principal bundle P with group G is the subgroup of all $a \in G$ such that a fixed $u \in P$ can be joined to $u \cdot a$ by a horizontal curve. Thus, the holonomy group measures the extent to which the distribution of horizontal subspaces is not integrable. {In classical mechanics (V.(a)) a system of "constraints" which can be described by a suitable distribution is called "holonomic" if the distribution is integrable, and "non-holonomic" otherwise. Thus, the "holonomy group" really should be called the "non-holonomy group", since it measures the extent to which a distribution is non-integrable.} Holonomy groups are studied in great detail in Kobayashi and Nomizu {1}, with basic properties treated in chapter 2 of volume 1, and applications throughout. In particular, we have the "holonomy theorem" of Ambrose and Singer (first proved, or at least stated, by E. Cartan), which describes the Lie algebra of the holonomy group in terms of the curvature form of the connection. It should perhaps be pointed out that this is in some sense a global version of a classical description of the curvature tensor in terms of parallel translation around an "infinitesimal parallelogram" -- see

Eisenhart {1; 65}, Kreysig {1; 295}, or Laugwitz {1; 108}, or slightly different versions in Bishop and Crittendon {1; 97}, Nelson {1; 77}, or Singer and Thorpe {1; 170-174}. It should also be noted that the holonomy theorem gives an immediate proof of the Test Case. The most important application of holonomy groups for Riemannian manifolds is the de Rham decomposition theorem (Kobayashi and Nomizu {1; v.1, 187ff.}).

(e) Reducing the group of a bundle; G-structures. The proof of the holonomy theorem uses the concept of a reduction of the group G of a principal bundle P to a subgroup H. This is, by definition, a subset P' of P such that $u \cdot a \in P'$ for all $u \in P'$ and $a \in H$, so that P' is a principal bundle with group H. The prime example is a reduction of the group $GL(n,\mathbb{R})$ of the bundle of frames of M to the subgroup $O(n)$. Any Riemannian metric $\langle\ ,\ \rangle$ gives such a reduction -- we define P' to be the set of all frames which are orthonormal with respect to $\langle\ ,\ \rangle$. Conversely, given any such reduction, we can define $\langle\ ,\ \rangle$ by declaring the frames in P' to be orthonormal. Similarly, an orientation on M is equivalent to a reduction of the group $GL(n,\mathbb{R})$ to the group $GL^+(n,\mathbb{R})$ of matrices of positive determinant. It could hardly be supposed that mathematicians would not get around to generalizing these examples: a reduction of the bundle of frames on M to a subgroup $G \subset GL(n,\mathbb{R})$ is called a G-structure. For the theory of G-structures see the last chapter of Sternberg {1}, and Kobayashi {1}.

(f) Contact transformations and contact structures. At each point p of M we can consider the set of $(n-1)$-dimensional subspaces of the tangent space M_p. With the notation of Chapter V.13, this set would be denoted by $G_{n-1}(M_p)$; in the terminology of topic III.(g) it is the set of 1st order (n - 1)-dimensional contact elements at p. We can form the manifold $C^1_{n-1}M = \bigcup_{p \in M} G_{n-1}(M_p)$ of all these contact elements, and any immersion $f\colon M \longrightarrow N$ gives rise to a map $f_*\colon C^1_{n-1}M \longrightarrow C^1_{n-1}N$. An arbitrary smooth map $g\colon C^1_{n-1}M \longrightarrow C^1_{n-1}N$ was classically called a contact transformation if it satisfies the following condition, which is automatic for f_*: for every hypersurface $P \subset M$, there is a hypersurface $Q \subset N$ such that the set of tangent spaces of Q is just the image under g of the set of tangent spaces of P. This geometric definition is unfortunately rather vague, since we want to allow the possibility, for example, that Q is a single point $q \in N$ and g takes all tangent spaces of P into $G_{n-1}(N_q)$. But it is not hard to derive a precise analytic condition which captures the geometric content. The manifold $C^1_1\mathbb{R}^2$, for example, has a covering by two coordinate systems, one of which is defined on the set U of all directions not parallel to the y-axis -- we use the coordinates a, b of the point $p \in \mathbb{R}^2$ as two coordinates on U and the slope m of the line in $\mathbb{R}^2_p$ as the third coordinate. For a curve c in $\mathbb{R}^2$ we have

$$m(\text{subspace spanned by } c'(t)) = \frac{db(c'(t))}{da(c'(t))} .$$

From this it is not hard to see that a map $g\colon U \longrightarrow U$ should be called a contact transformation if and only if $g^*(db - m\, da) = \alpha(db - m\, da)$ for a nowhere 0 function α. For $C^1_n\mathbb{R}^{n+1}$, with $x^1,\ldots,x^n,z$ as coordinates for the point, and y^i $(i = 1,\ldots,n)$ as the slope of the intersection of the n-dimensional subspace with the (x^i,z)-plane, we have the analogous criterion, in terms of the form $dz - \Sigma\, y^i dx^i$. For arbitrary manifolds these conditions can be formulated on coordinate neighborhoods. Although the classical reference Lie and Scheffers {2} will present problems, it is delightfully concrete and filled with examples; see also Eisenhart {4; ch.6} and Favard {1; part 1, ch.4}.

Nowadays, these motivating geometric considerations are almost never mentioned (an exception is Hermann {2; ch.3}). The modern approach to the subject may be found in Kobayashi and Nomizu {1; v.2, 381-382} and Kobayashi {1; 28ff.}; it involves the notion of a contact structure, which also plays an important role in classical mechanics (V(a)). An important tool in the study of contact structures is a theorem of Darboux, which is proved in Kobayashi {1; Appendix 1}; a very different proof is given in Lang {1; ch.5, §7}. The proof in Godbillion {1; 115-121} or Sternberg {1; 135-141} is of interest, as it uses the "characteristic system" of an ideal of differential forms. This "characteristic system" is related to the characteristics of a PDE; to see this made more explicit one may consult Dieudonné {1; v.4, 92-118}. The "Legendre transformation" is a contact transformation which is often used in PDEs (see, for example, Courant and Hilbert {1; v.2, 32-39}). Legendre transformations are also used in the calculus of variations and classical mechanics (cf. Abraham {1}, Godbillion {1}, or Sternberg {1}). For the connection between the two, try Hermann {2; ch.6, §9}.

We could also consider maps $g\colon C^k_r M \longrightarrow C^k_r$, defined on $k^{\underline{th}}$ order r-dimensional contact elements [cf. III(g)], satisfying an analogous geometric condition. But these are essentially of the form f_* for $r < n-1$, or an extension of a contact transformation $g\colon C^1_{n-1}M \longrightarrow C^1_{n-1}$ for $r = n-1$ (Knebelman [1]). Similarly, one can define "infinitesimal contact transformations", but they always come from a map $f\colon M \longrightarrow N$. See Eisenhart {4; 252} or Lie and Scheffers {2; ch.4, §2} for a classical statement of this fact, and

Kobayashi {1; 30} for a modern version.

(g) The Laplacian and Hodge theory. Berger, Gauduchon, and Mazet {1} is an excellent introduction to the significance of the Laplacian, though somewhat out of date because of the recent rapid progress in this field. For many applications of Hodge theory similar to Bochner's Theorem (IV.7-63), see Yano and Bochner {1}, and Yano {1} (which also has applications of integral formulas similar to those in Chapter V.12). See also Ruse, Walker, and Willmore {1}. A simple application to prove Poincarè duality is given in Warner {1}; for applications to complex manifolds see, for example, Weil {1}.

III. Other Geometries

(a) Finsler geometry. For a brief introduction, see Laugwitz {1; §15}. For an extended treatment see Rund {1} and Matsumoto {1}. The reprinting of the thesis of Finsler {1}, with a bibliography up to that time, may be of interest.

(b) Integral geometry. Blaschke {3} and Santalo {1} are very nice introductions to this subject. The article by Santalo in Chern {3} gives references to later work. It is of interest to compare the arguments on p. 167 with the proofs of Fenchel's theorem and the Fary-Milnor theorem on pp. 33-35.

(c) Line geometry. Here one studies the manifold consisting of all lines in $\mathbb{R}^3$. I haven't the slightest idea what is done, but there are supposed to be some nice things. See Blaschke {1; v.1, ch.9}, Eisenhart {2; ch.12}, Favard {1; pt. 2, sec. 1, ch. 5}, Forsyth {1; ch.12} and Hlavatý {2}. See also (e). The manifold of all circles in $\mathbb{R}^3$ has also been studied. See Eisenhart {2; ch.13} and Forsyth {1; ch.12}.

(d) Affine geometry. For special affine curve theory see Blaschke {1; v.2} and Favard {1; pt.2, sec.2, ch.1}, as well as Guggenheimer {1; §8-3}. For special affine surface theory see Blaschke {1; v.2} and Favard {1; pt.2, sec.2, ch.2}. See also Flanders [1] and the references listed under it. Not much seems to have been written about general affine invariants. See Guggenheimer {1; ch.7-3, probs. 10-12} and Dieudonné {1; v.4, ch.20, §14, probs. 11, 12}, which first describes an affine normal ("pseudo-normale") for a hypersurface of any manifold with a torsion-free connection on its bundle of frames [compare with the second part of (e)] and then specializes to special affine geometry of $\mathbb{R}^n$ (it will probably make things much easier to re-write the problem so that it deals with moving frames, rather than with the bundle of frames itself). On the other hand, there is a considerable literature on what might be called "general linear geometry" -- properties of submanifolds of $\mathbb{R}^n$ invariant under all linear maps (but not translations!), as well as properties invariant under all elements of $SL(n,\mathbb{R})$, again excluding translations -- see Salkowski {1}, and especially Schirokow and Schirokow {1}. This certainly seems like a strange geometry to study, but see Laugwitz {2}.

(e) Projective differential geometry. This is the study of properties of submanifolds of projective space $\mathbb{P}^n$ which are invariant under the projective group (cf. Hartshorne {1} for basic terminology). An eminent differential geometer, who perhaps prefers to remain anonymous, has said that the problem with projective differential geometry is that the projective group is too large to allow any interesting local results, while no one has ever discovered any interesting global ones. The best introduction is probably Fubini and Čech {1}, which also introduces E. Cartan's methods (using moving frames -- compare (g)) as worked out in Cartan {5}. Other texts are Bol {1},

Favard {1; pt.2, sec.3}, Lane {1}, and Wilczynski {1}, one of the earliest works in the field [cf. (g)]. For "line geometry" in the projective case see Švec {1}.

Just as Riemannian geometry generalizes the differential geometry of $\mathbb{R}^n$, so one might expect to generalize projective differential geometry to an arbitrary manifold M by forming the union of the projective spaces obtained from each tangent space M_p, constructing a corresponding principal bundle P, and considering some canonical connection ω on P. All these steps can be carried out, but ω is not characterized so simply as in the Riemannian case by having vanishing torsion (which is defined only for connections on the bundle of frames); instead, there is a unique ω which satisfies certain identities like the Bianchi identities. This is described in Cartan {5}, but it will probably be much easier to read Kobayashi {1; 127-138}.

(f) Other esoteric geometries. See Blaschke {1; v.3} for the interesting, but complicated, geometries of Möbius, Laguerre, and Lie. In all three geometries, the points are basically the circles in the plane, or the spheres in space, etc., and properties are sought which are invariant under various groups of maps on these circles or spheres. Möbius geometry involves those maps which take the set of all circles or spheres through a fixed point into another set of the same sort. Such maps are always induced by similarities and inversions in the plane, space, etc., so that Möbius geometry reduces to the study of this group of maps (the Möbius group) on these spaces. For dimensions $n > 2$ it is thus "conformal geometry". In the case of the plane one might also study properties of curves invariant under all analytic maps. I know of only one strange result in this direction -- see Theorem 1-4 of Ahlfors {1}. The more complicated geometries of Laguerre and Lie are allied to the notion of contact transformations (II.(f)). Laguerre geometry involves the maps which take a set of circles tangent to a line (or of spheres tangent to a plane) into another set of this sort, while the geometry of Lie involves the group of maps which simply take circles or spheres which are tangent to each other to pairs of the same sort.

Another weird topic is the theory of webs -- see Blaschke {4} and Blaschke and Bol {1}.

All sorts of other oddities may be found by consulting Mathematical Reviews and the journal Tensor.

(g) Lie's theory of differential invariants, and E. Cartan's general method of moving frames. Lie's theory is the one topic which I greatly regret not having written up, for it is used extensively in certain early work which is far more impenetrable than other classical material. In particular, Lie's theory was used in the first investigations of special affine surface theory by Pick [1] and in early work in projective differential geometry (cf. Wilczynski {1}). Unfortunately, a reasonable exposition would probably require close to a hundred pages, which wouldn't fit in anywhere, for the material on first order linear PDEs from Chapter V.10 is needed, while the theory relates most directly to Chapter III.2 (which is already too long). Matters were in no way helped by my lack of understanding, nor, since this delayed its treatment until last, by my lack of endurance.

The most interesting part of Lie's theory applies to situations like Euclidean, special affine, or projective differential geometry, where we seek properties of submanifolds of M which are invariant under a group G of diffeomorphisms [i.e., submanifolds of homogeneous spaces (I.(b))]. It thus connects directly with the famous definition proposed by Felix Klein [1] of geometry as the theory of geometric invariants of a transitive transformation group.* The idea is to find "(geometric) differential invariants of order k for r-dimensional submanifolds of M", a simple example of which is the curvature κ of a curve c in $\mathbb{R}^n$. This is an "invariant" (it is the same for a curve and its composition with a Euclidean motion) "of order 2" (one can compute it at any point knowing only the first two derivatives of the curve at that point) which is independent of the parameterization ("geometric"). Such invariants can be thought of as functions on a suitable space. First we construct the "k-jets" of maps of $(\mathbb{R}^r,0)$ into (M,p); these are equivalence classes $j_0^k(f)$ of maps $f\colon (\mathbb{R}^r,0) \longrightarrow (M,p)$, where $f \sim g$ if all mixed partials of order $\leq k$ of all component functions of f and g are equal at 0 (cf. topic VI(c)). On the set of k-jets represented by immersions $f\colon (\mathbb{R}^r,0) \longrightarrow (M,p)$ we introduce a further equivalence relation by declaring $j_0^k(f_1)$ and $j_0^k(f_2)$ equivalent if in a neighborhood of $0 \in \mathbb{R}^r$ we have $f_2 = f_1 \circ \alpha$ for some diffeomorphism α. These new equivalence classes are the kth order r-dimensional <u>contact elements</u> of M at p (for $k=1$ they may be identified with the r-dimensional subspaces of M_p). The set $C_r^k M$ of all such contact elements at all $p \in M$ is a manifold, and any diffeomorphism $\phi\colon M \longrightarrow M$ gives rise to a diffeomorphism $\phi_*\colon C_r^k M \longrightarrow C_r^k M$. A geometric differential invariant of order k for r-dimensional submanifolds of M under a group of diffeomorphisms G is just a function $F\colon C_r^k M \longrightarrow \mathbb{R}$ invariant under G (i.e. satisfying $F \circ \phi_* = F$ for all $\phi \in G$). This set-up is briefly discussed in Hermann {2; ch.3, §14}.

---

*This definition is of course woefully inadequate for describing such subjects as Riemannian geometry, though some attempts have been made to bring even these within its compass -- see the references at the end of Chapter 2 of Veblen and Whitehead {1}.

We can also consider "differential invariant tensors". If γ is a $k^{\underline{th}}$ order r-dimensional contact element of M at p, represented by a k-jet $j_0^k(f)$ for $f\colon (\mathbb{R}^r,0) \longrightarrow (M,p)$, then the $1^{\underline{st}}$ order r-dimensional contact element represented by the 1-jet $j_0^1(f)$ depends only on γ -- it may be thought of as the "tangent space" $T\gamma$ of γ. A function F on C_r^kM such that each $F(\gamma)$ is a bilinear function $F(\gamma)\colon T\gamma \times T\gamma \longrightarrow \mathbb{R}$, for example, may be thought of as a $k^{\underline{th}}$ order covariant tensor of order 2 for r-dimensional submanifolds of M; it is easy to formulate the invariance conditions for such F.

There is actually a reasonable way to compute such differential invariants, or at least to formulate the computations (in practice they become hopeless quite quickly unless one introduces some extraneous geometric insight). Any $X \in \mathfrak{g}$ = Lie algebra of G induces a 1-parameter family of diffeomorphisms $\{\exp tX\}$ of M, hence a family $\{(\exp tX)_*\}$ on C_r^kM, and hence a vector field $X^{(k)}$ on C_r^kM. If G is connected, and thus generated by the elements $\exp tX_i$ for a basis $\{X_i\}$ of $\mathfrak{g}$, then F is invariant if and only if $X_i^{(k)}(F) = 0$ for all i; each of these equations is simply a linear first order PDE in terms of coordinates on C_r^kM. This allows one to compute the invariants F once one picks a natural coordinate system on C_r^kM and figures out appropriate methods for evaluating $X_i^{(k)}(\rho)$ for each coordinate function ρ. There is no difficulty solving each particular equation $X_i^{(k)}F = 0$, by the methods of Chapter V.10, §1. We find that F must be constant along certain curves in C_r^kM, or equivalently, that F must be expressible as a function of certain combinations of the coordinate functions. The problems arise when we seek a solution F of the equations $X_i^{(k)}F = 0$ for all i; we then have to guess a single function which can be expressed as a function of each of the different combinations of coordinate functions which arise for each i. Even without performing the calculations, however, one can decide how many invariants there should be. We seek a submanifold $\mathcal{O} \subset C_r^k M$ which intersects each "orbit" $\{\phi_* \gamma: \phi \in G\}$, $\gamma \in C_r^k M$ just once. The invariant functions on $C_r^k M$ are in one-one correspondence with the functions on $\mathcal{O}$, so the number of "independent" ones is the dimension of $\mathcal{O}$, thus the dimension of $C_r^k M$ minus the dimension of an orbit. Differential invariant tensors can be treated similarly.

Unfortunately, there is no really good reference for this topic. One can try Lie and Scheffers {1; ch.22}, but it will be much easier to read Scheffers {1}. Plane curves are treated in volume 1, part 1, §8, 9, space curves in volume 1, part 2, §12, and surfaces in volume 2, part 3, §10. All the information here will be quite new, because curves and surfaces are determined up to congruence by certain <u>functions</u>, not by tensors; for curve theory this means that no use is made of the parameterization by arclength (which is really equivalent to using the first fundamental form of the image curve). There are also some (rather misleading) calculations in volume 2, part 3, §6 which essentially determine all invariants on the space of jets (i.e., the functions on surfaces which are invariant under Euclidean motions but <u>not</u> under change of parameter). The main problem with this reference is that it doesn't illustrate the general methods of computation outlined above. The most modern reference for these methods, Guggenheimer {1; ch.7-2}, is frustratingly old-fashioned in its language. It might help to mention that the formally introduced new "variables" $x_i' = dx_i/dx_1$, x_i'',... are merely certain natural coordinates for 1-dimensional contact elements in $\mathbb{R}^m$: on the set of 1-dimensional contact elements γ represented by curves whose images can be parameterized as $x_1 \longmapsto (x_1, f_2(x_1),\ldots,f_m(x_1))$ [for unique $f_2,\ldots,f_m$] we let $x_i'(\gamma) = f_i'$, etc. Similarly, on the set of 2-dimensional contact elements γ in $\mathbb{R}^3$ which are represented by immersions of $\mathbb{R}^2$ whose images can be parameterized as $(x_1,x_2) \longmapsto (x_1,x_2,f(x_1,x_2))$ we have the coordinates $p(\gamma) = \partial f/\partial x$, $q(\gamma) = \partial f/\partial y$, $r(\gamma) = \partial^2 f/\partial x^2$, etc. Beware of misprints in some of the computations of the "prolongations". There are extensive calculations in Forsythe {1; §132-146}, but they are not carried out directly on the contact elements. Instead, "relative invariants of order w" are computed first; these are functions F on the jets such that

$$F(j_0^k(f\circ\alpha)) = (\det \alpha)^w F(j_0^k(f))$$

for any diffeomorphism α. Clearly the product of relative invariants whose weights add up to 0 will give invariants on the contact elements. (Similar tricks are used in the works of Pick and Wilcyzniski quoted above, as well as in the much more straightforward calculations in Blaschke {1; v.3}.) One may also try the introductory chapter of Schirokow and Schirokow {1}.

Lie's methods are supposedly applicable, in ways I don't understand at all, to such problems as the equivalence of Riemannian manifolds, and even more general questions. See Lie and Scheffers {1; ch.23}, Veblen {1; ch.3, §20-22, ch.5, 6}, and Wright {1}. It might also be mentioned that there is a theory involving even more general notions than jets and contact elements, the "geometric objects" -- see Yano {3; ch.2} or Aczél and Gołab {1}.

E. Cartan's general method of moving frames is a sort of dual to Lie's method which allows computations to be made more easily. The general features of the theory are well illustrated by the development of special affine surface theory in Chapter III.2 -- see especially p. III.152. We consider moving frames X_1, X_2, X_3 along $M^2 \subset \mathbb{R}^3$ with $\det(X_1,X_2,X_3) = 1$. An arbitrary moving frame of this sort is what Cartan calls a "zeroth order frame". A "first order frame" is one which is adapted to M (i.e., X_1, X_2 are tangent to M). To define "second order frames" we now try to specialize the first order frames as much as possible by seeking an appropriate condition on the dual and connection forms. As we observed on p. III.122, the condition $\psi_i^3 = \theta^i$ has just the "invariant" property we need -- it depends only on the value of the frame at p [i.e., if X_1, X_2, X_3 and $\bar{X}_1$, $\bar{X}_2$, $\bar{X}_3$ are adapted moving frames with dual and connection forms θ^i, ψ_j^i and $\bar{\theta}^i$, $\bar{\psi}_j^i$, respectively, and the two frames agree at p, then $\psi_i^3(p) = \theta^i(p)$ if and only if $\bar{\psi}_i^3(p) = \bar{\theta}^i(p)$]. This definition of second order frames already gives an invariant tensor, the special affine first fundamental form. To obtain this specialization we used the calculations on pp. III.120-122. To be sure, these calculations were not very difficult, but that is because we already knew what we were looking for; we didn't even bother to compute the ψ'^i_j in general, since they would not be involved in our invariant condition. It would have been much harder to simply guess an invariant condition without the previous geometric motivation (indeed, the difficulties which arise here are exactly dual to the problem in Lie's method of finding a common solution to the equations $X_i^{(k)}F = 0$); this is an aspect of the theory which Cartan always carefully suppressed in his expositions of it. Now we can seek "third order frames" by specializing the second order frames. An appropriate condition is $\psi_3^3 = 0$; it gives us the special affine normal. The verification that this condition is invariant is given on pp. III.152-153. It would clearly be a lot easier to discover *ab initio* than the corresponding invariant condition (p. III.148) for first order frames; in general, one always works with the highest order frames already successfully discovered. Specializing the third order frames would lead us to the special affine second fundamental form (X_1 and X_2 would be the eigenvectors of $\mathcal{I}\mathcal{I}$ with respect to $\mathcal{I}$). For more examples and details, see Cartan {6} or Favard {1}.

IV. The Russian School; Synthetic Differential Geometry

A thorough treatment of the foundations of surface theory without differentiability hypotheses is given in Alexandrov {1}. As an introduction, the reader may prefer to consult Busemann {1} or relevant portions of Efimov {1} and Pogorelov {1}, {2}, {3}. For an extensive connected account of further developments in the theory see Pogorelov {5}. We might also mention Pogorelov {6}, where the geometric results are used to obtain stronger-than-usual results about PDEs.

Although most of the material in these references pertains to convex surfaces (for which one may also consult Bonneson and Fenchel {1}, Blaschke {2}, Hadwiger {1}, and Yaglom and Boltyanskiĭ {1}), there is also an elaborate theory which investigates the most general sorts of surfaces, or even arbitrary metric spaces. One treatment can be found in Alexandrov and Zalgaller {1}, while a somewhat different direction is taken in Blumenthal {1}, Blumenthal and Menger {1}, and Rinow {1}. In yet another vein, we have the work of Busemann, which represents the very antithesis of Riemann's approach (whereby a mechanism for measuring lengths of arbitrary curves is postulated, and geodesics are defined as curves of minimal lengths). In Busemann {3} postulates are instead given for the geodesics, and many relations of classical differential geometry are derived from them; although the arguments are often involved, just as a rigorous axiomatic development of Euclidean geometry would be, the strength of the results is often startling. For related results see Busemann {2}.

V. Applications to physics.

(a) Classical mechanics. It turns out that differential geometry provides the natural language for classical mechanics, for the two equivalent basic formulations of the subject, via Lagrange's equations and Hamilton's equations, take place on the tangent bundle and cotangent bundle, respectively, of a suitable manifold. A discussion of mechanics which manages to make some interesting points in a short space, but which doesn't make very clear which manifold one is working on, may be found in Laugwitz {1; §14}. Another brief discussion, without this shortcoming, is contained in Bishop and Goldberg {1; ch.6}. One can consult pp. 141-147 of Sternberg {1}, which also mentions other aspects of the subject as part of an extensive discussion of the calculus of variations, in the succeeding chapter. Similarly, see Hermann {1; ch.16}. The short book by Godbillion {1} reaches its climax in the final, 9 page, chapter on mechanics, which begins by defining a "mechanical system" as a triple (M,T,π) where M is a manifold, T a differentiable function on TM, and π a semi-basic form on TM. A serious study of mechanics will be found in Abraham {1}, beginning in the third chapter, where it is admitted however, that the treatment "possibly...will seem severely unmotivated without some background in classical mechanics..." Aside from this difficulty common to all these books, the most heartbreaking omission is any adequate discussion of the fictional "forces of constraint" which are involved in such useful abstractions as "rigid" bodies. These can usually be formulated as particular subspaces that the velocity vector in $\mathbb{R}^{3N}$ of the system of N particles is constrained to lie in. When these subspaces form an integrable distribution (the constraints are "holonomic") the integral submanifolds form a lower dimensional "configuration space" and the basic principle is that the motion of the system with these forces of constraint is determined by restricting the original equations on $T\mathbb{R}^{3N}$ or $T^*\mathbb{R}^{3N}$ to equations on the tangent or cotangent bundle of the configuration space. (Some details are given in Hermann {2; ch.2}.) For physics books on classical mechanics the following references may be of use: Corbin and Stehle {1} (the most modern in spirit of the elementary books), Pars {1}, and Whittaker {1}.

(b) General Relativity. Fortunately, I was ultimately not foolish enough to attempt writing anything on this vast subject which I do not understand. My conscience is set at ease by recommending the monumental book of Misner, Thorne, and Wheeler {1}, which is probably owned by every relativist in the world. For more mathematical treatments you may prefer Sacks and Wu {1}, which places great emphasis on foundational points, or Weinberg {1} (see the Supplement to Part B), or Hawking and Ellis {1}, which is quite advanced.

VI. Miscellaneous

(a) Calculus of variations; Hamilton-Jacobi theory. A subject closely linked with mechanics (V.(a)). The two great classical works usually referred to are Carathéodory {1} and E. Cartan {4}. See also Abraham {1}, H. Cartan {1}, Godbillion {1}, Hermann {1; pt.2}, Rund {2}, and Sternberg {1; ch.4}.

(b) Sprays. This is a topic which I have assiduously avoided learning, convinced that one can get by without it, and suspicious that it's just a complicated new way of saying something old. Less obstinate readers may wish to consult Gromoll, Klingenberg, and Meyer {1; 60}, Lang {1; ch.4, §3-5, ch.6, §6}, or Sternberg {1; 199, 361}.

(c) Jets. These, and the contact elements [cf. III.(g)], are natural structures to consider in differential geometry, but they are only just beginning to be used in any serious way. For basic definitions, see Bourbaki {1; §12} or Dieudonné {1; v.3 (ch.16.5 - Problem 9, ch.16.9 - Problem 1), v.4 (ch.20.1 - Problem 3)}. For some applications, see Kobayashi {1; 139ff.}.

(d) Other definitions of connections. Gromoll, Klingenberg, and Meyer {1; 43} gives a definition of a connection in terms of a map $K: TTM \longrightarrow TM$. This is useful in dealing with infinite dimensional manifolds; see Flaschel and Klingenberg {1; ch.1}. Connections have also been defined as sprays with certain properties, and as a splitting of the "jet exact sequence". I personally feel that the next person to propose a new definition of a connection should be summarily executed.

(e) Reducing differentiability hypotheses. It is of some interest to some people (analysts) to establish results with the minimum differentiability assumptions. Although the work of the Russian school sometimes eclipses such efforts, this is not always true, and in any case few mathematicians seem to find it a sufficiently compelling argument to go that route. In any proof of classical geometry one can always carefully count how many times one differentiates, but this usually turns out to be one or two more times than one really has to, if one is sufficiently clever. So the problem of finding minimal differentiability hypotheses (and examples to prove they are minimal) is not easy. A long series of papers on this subject was published by Hartmann and Wintner, mainly in the American Journal of Mathematics, beginning in 1947.

(f) Transversality. Not a part of differential geometry, really, but of differential topology. Nevertheless, it is probably a wise move to learn the basic ideas. See Sternberg {1; 64ff.} or Guillemin and Pollack {1}, with its many beautiful applications.

(g) Polyhedral geometry, models, constructive aspects, etc. Perhaps the oldest contribution in this direction was the argument of Hilbert and Cohn-Vossen {1; pp. 194-195} proving Gauss' Theorema Egregium for polyhedral surfaces. Recent work of T. Banchoff and others has carried this approach much further. One may also consult Sauer {1} and Kruppa {1}.

B. Books

During the compilation of this bibliography, certain instincts urged me to seek encyclopaedic completeness, while healthier ones advised selectivity and utility. From this conflict resulted the usual unsatisfactory compromise, wherein the advantages of neither course of action is retained. I have tried to single out sources which might be particularly valuable, but this applies mainly to books concerned with the topics covered in these five volumes; many others will have already been mentioned in Part A.

★ *Encyklopädia der Mathematischen Wissenschaften*, Volume III, Part 3D, B.G. Teubner, Leipzig, 1902-1927.

Abraham, R.
{1} *Foundations of Mechanics*, W.A. Benjamin, New York, 1967. (MR 36 #3527.) [see the Supplement]

Aczél, J. and Gołab, S.
{1} *Funktionalgleichungen der Theorie der Geometrischen Objekte*, Pánstwowe Wydawnictwo Naukowe, Warsaw, 1960. (MR 24 #A3588.)

Ahlfors, L.V.
{1} *Conformal Invariants*, McGraw-Hill, New York, 1973.

Alexandrov, A.D.
{1} *Die Innere Geometrie der Konvexen Flächen*, Akademie-Verlag, Berlin, 1955. (MR 17, 74.)
★ {2} *Kurven und Flächen*, VEB Deutscher Verlag der Wissenschaften, Berlin, 1959.

A very nice elementary introduction to curves and surfaces which mentions some things you might not see otherwise (e.g. why a pail with a curved rim is stronger than one with a plain rim). For an English translation see Chapter 7 of

Alexandrov, A.D., Kolmogorov, A.N., and Lavrent'ev, M.A. (eds.)
{1} *Mathematics. Its Contents, Methods and Meaning*, 2nd ed., M.I.T. Press, Cambridge (Mass.), 1969. (MR 39 #1258a-c.)

Alexandrov, A.D. and Zalgaller, V.A.
{1} *Intrinsic Geometry of Surfaces*, American Mathematical Society, Providence, R.I., 1967. (MR 35 #7267.)

Auslander, L.
{1} *Differential Geometry*, Harper & Row, New York, 1967. (MR 35 #2208.)

This is an attempt to construct an introductory course in differential geometry from the point of view of Lie groups, with the fundamental equations of surface theory arising from the equations of structure of SO(3). Later chapters cover Riemannian geometry. The treatment of geodesic completeness (pp. 203-214) may be of interest, and the Poincaré upper half space is discussed in some detail (pp. 223-236). In particular, there is a description of the various one-parameter subgroups of the group of isometries. The orbits of these subgroups are the geodesics, geodesic circles, horocycles, and equidistant curves (for these are the curves of constant curvature).

Auslander, L. and MacKenzie, R.E.
{1} *Introduction to Differentiable Manifolds*, McGraw-Hill, New York, 1963. (MR 28 #4462.)

Berger, M.
★{1} *Lectures on Geodesics in Riemannian Geometry*, Tata Institute of Fundamental Research, Bombay, 1965. (MR 35 #6100.)

These notes cover many topics, frequently with details not to be found elsewhere.

see also Lascoux, A.

Berger, M., Gauduchon, P., and Mazet, E.
★{1} *Le Spectre d'une Variété Riemannienne*, Springer-Verlag, Berlin, 1971. (MR 43 #8025.)

Berger, M. and Gostiaux, B.
★{1} *Geometrie Differentielle*, Armand Colin, Paris, 1972.

A very beautiful recent text on differentiable manifolds, with some differential geometry included. §6.7-6.9 and 7.5 prove the Gauss-Bonnet theorem for submanifolds of Euclidean space by the method of Allendoerfer and Fenchel mentioned at the beginning of Chapter 13 (p. V.387). Chapter 9 treats global properties of curves, including an elementary proof of the Jordan curve theorem for smooth curves, Whitney's theorem on smooth homotopy of closed curves, and the formula of Fabricius-Bjerre-Halpern, which relates the number of double points and inflection points of a closed curve to the number of double tangents (lines tangent to the curve at two different points).

Bianchi, L.
{1} *Vorlesungen über Differentialgeometrie*, B.G. Teubner, Leipzig, 1899.

This is a translation, with some additions, of the first edition of the original Italian work. Naturally it contains a considerable number of results from classical surface theory, but it differs from many classical books by also treating surfaces in the spaces of constant curvature. See in particular §348, which considers a surface M in the upper half space model of H^3.

For $p \in (x,y)$-plane, let γ be the geodesic intersecting M orthogonally and approaching p, and let $f(p)$ be the other point in the (x,y)-plane which γ approaches. Then f is holomorphic if and only if M has (intrinsic) curvature 0. Also note that §110 gives a nicer treatment of the problem considered on pp. V.317-319; Bianchi shows directly that the curve $t \mapsto (u(t,0),v(t,0),w(t,0))$ has the same curvature and torsion as c. (The whole problem is simply ignored by Darboux {1}. By the way, there is no adequate treatment anywhere in the classical literature of the case where $\kappa = \tilde{\kappa}_g$.) There is a later, 4 volume, Italian edition, not translated, which treats special questions of surface theory in great detail.

Bishop, R.L. and Crittenden, R.J.
{1} *Geometry of Manifolds*, Academic Press, New York, 1964. (MR 29 #6401.)

Very compactly written, with many results merely quoted or left as exercises. Particular attention may be called to some of the problems on pp. 106-107, 110, 114, 134. There is a study of complete simply connected manifolds of constant curvature (§9.5) rather different from the elementary one outlined in Problem III.1-5, a more elaborate study of convex neighborhoods, using the second variation (§11.8), and some applications of the second variation to theorems on the volumes of balls (pp. 256-257). A bibliography of 95 items.

Bishop, R.L. and Goldberg, S.I.
{1} *Tensor Analysis on Manifolds*, Macmillan, New York, 1968. (MR 36 #7057.)

Blaschke, W.
★ {1} *Differential Geometrie*, Volumes 1, 2, Chelsea, New York, 1967; Volume 3, Springer, Berlin, 1929.

Although this book is quite old-fashioned, I nevertheless find it very stimulating, perhaps because the author is more interested in genuine geometric questions, especially global ones, than in the formalities of calculations. More topics are covered here than in almost any other classical book, and there is an extraordinary number of interesting exercises, remarks, and sidelights.

Volume 1, §72 shows that if the geodesic circles are the same as the curves of constant κ_g, then K is constant, while §84 proves the more difficult result (mentioned on p. IV.449) that K is constant if all curves of constant κ_g are closed. The manipulations of §94 (used in the next section for a proof of Christoffel's theorem) are mysterious; Problem III.3-8 may be used instead. See also the funny result on p. 121. It is interesting to find that the general formula for variation of area was already given by Gauss (§109). Classical results of Schwarz (one mentioned under topic II.(a)) are given in §110, 111, while §116 gives the second variation of area (in a special case), and mentions a classical condition of Schwarz for a minimal surface to be a local minimum for area. For a modern presentation, see Barbosa and do Carmo [1]. The proof of the related Theorem IV.9-39 is from Rado {1}. §117 gives the first variation of H and K (we essentially found the first variation of H in order to find the second variation of area). Problem 2 in §118 mentions interesting properties of associated minimal surfaces, for example, the tangent planes at corresponding points are parallel. Conversely, if there is an isometry between two surfaces such that tangent planes are parallel, then they are either congruent surfaces or associated minimal surfaces (to be taken with a grain of salt -- one of the surfaces could be the union, along a common curve, of a piece congruent to part of the other surface and another piece associated to a part of the other surface which is a minimal surface).

Volume 2 covers affine differential geometry. There are very many nice geometric interpretations of the invariants arising here, as well as many global results.

For Volume 3 see topic III.(f).

{2} *Kreis und Kugel*, de Gruyter & Co., Berlin, 1956. (MR 17, 1123.)
{3} *Vorlesungen über Integralgeometrie*, 2nd ed., Chelsea, New York, 1949.
{4} *Einführung in die Geometrie der Waben*, Birkhäuser, Basel, 1955. (MR 17, 780.)

Blaschke, W. and Bol, G.
{1} *Geometrie der Gewebe*, Springer, Berlin, 1938.

Blaschke, W. and Leichtweiss, K.
{1} *Elementare Differentialgeometrie*, 5th ed., Springer-Verlag, Berlin, 1973.

This is a modernization of Blaschke's book which preserves the style of the original. Numerous new problems and references of interest.

Blaschke, W. and Reichardt, H.
{1} *Einführung in die Differentialgeometrie*, 2nd ed., Springer-Verlag, Berlin, 1960. (MR 22 #7062.)

This is an attempt to modernize Blaschke's book by writing everything in terms of moving frames. It may be consulted for a few interesting points hard to find elsewhere, especially §56, 57, 77, and 69, Problem 19.

Blumenthal, L.M.
{1} *Theory and Applications of Distance Geometry*, 2nd ed., Chelsea, New York, 1970. (MR 42 #3678.)

Blumenthal, L.M. and Menger, K.
{1} *Studies in Geometry*, W.H. Freeman, San Francisco, 1970. (MR 42 #8370.)

Bochner, S.
see Yano, K.

Bol, G.
{1} *Projektive Differentialgeometrie*, 3 vols., Vandenhoeck & Ruprecht, Göttingen, 1950. (MR 16, 1150.)

Extensive bibliography, extending that of Fubini and Čech {1}.

see also Blaschke, W.

Boltyanskii, V.G.
★{1} *Envelopes*, Pergamon Press, Macmillan, New York, 1964. (MR 31 #2438.)

see also Yaglom, I.M.

Bonneson, T. and Fenchel, W.
{1} *Theorie der Konvexen Körper*, Springer, Berlin, 1934.

Boothby, W. and Weiss, G.L. (eds.)
{1} *Symmetric Spaces: Short Courses Presented at Washington University*, Dekker, New York, 1972.

Bourbaki, N.
{1} *Variétés Différentielles et Analytiques. Fascicule de Résultats/ Paragraphes* 1 à 7 and /*Paragraphes* 8 à 15, Hermann, Paris, 1971. (MR 43 #6834.)

Bourbaki is the originator of that famous pedagogical method whereby one begins with the general and proceeds to the particular only after the student is too confused to understand even that any more. His influence is to be seen everywhere, probably in these volumes too. Bourbaki has apparently decided that the theory of manifolds has now entered that domain of "dead" mathematics to which he hopes to give definitive form. In this summary of results the corpse is laid out to public view; the complete autopsy is eagerly awaited.

Boys, C.V.
★{1} *Soap Bubbles, Their Colors and the Forces Which Mold Them*, 3rd ed., Dover, New York, 1959.

Brickell, F. and Clark, R.S.
{1} *Differentiable Manifolds*, Van Nostrand Reinhold, London, 1970.

Busemann, H.
{1} *Convex Surfaces*, Interscience, New York, 1958. (MR 21 #3900.)
{2} *Recent Synthetic Differential Geometry*, Springer-Verlag, Berlin, 1970. (MR 45 #5936.)
{3} *The Geometry of Geodesics*, Academic Press, New York, 1955. (MR 17, 779.)

Campbell, J.E.
{1} *A Course of Differential Geometry*, Oxford University Press, Oxford, 1926.

§149-154 prove that any n-dimensional Riemannian manifold can be imbedded in an (n+1)-dimensional Einstein space of vanishing scalar curvature. I know of no other reference for this result.

Carathéodory, C.
{1} *Calculus of Variations and Partial Differential Equations of the First Order*, Holden-Day, San Francisco, 1965. (MR 33 #597.)

Carmo, M. do
{1} *Differential Geometry of Curves and Surfaces*, Prentice Hall, Englewood Cliffs, N.J., 1976
{2} *Notas de Geometria Riemanniana*, Instituto de Matematica Pura e Aplicada, Rio de Janeiro, 1972.

Carrell, J.B.
see Dieudonné, J.A.

Cartan, É.J.
{1} *Oeuvres Complètes*, 3 vols. in 6 parts, Gauthier-Villars, Paris, 1952-1955. (MR 14, 343; 15, 383; 16, 697.)

The greatest differential geometer of the previous generation. Few have read his works, many pretend to have read them, and every one agrees that every one should read them. I get shell-shock every time I try. Fortunately, most of his books have by now been re-worked in modern presentations, and it's still worth looking at the originals after you know more or less what they're about.

{2} *Les Systèmes Différentiels Extérieurs et leurs Applications Géométriques*, Hermann, Paris, 1971.

As an introduction, try Godbillion {1}.

{3} *Lecons sur la Géométrie des Espaces de Riemann*, 2nd ed., Gauthier-Villars, Paris, 1963.

This is probably the easiest and most important of Cartan's books. Most of the material has been covered somewhere in these five volumes. There are few other classical books with a thorough description of the spaces of constant curvature -- unfortunately, here it is approached *via* projective geometry.

{4} *Lecons sur les Invariants Intégraux*, Hermann, Paris, 1971.

The hardest part of this book is the Cartan-Kähler theorem. For further reading in modern sources, see Dieudonné {1; v.4, ch.18, §8-18} or the very useful discussion in Jacobowitz and

Moore [1]. The second half of the book gives applications to all sorts of questions in differential geometry, mainly from surface theory. Here you are on your own.

{5} *Lecons sur la Théorie des Espaces à Connexion Projective*, Gauthier-Villars, Paris, 1937.
{6} *La Théorie des Groupes Finis et Continus et la Géométrie Différentielle traitées par la Méthode du Repère Mobile*, Gauthier-Villar, Paris, 1937.

Cartan, H.
{1} *Formes Différentielles*, Hermann, Paris, 1967. (MR 37 #6858.)

A neat little presentation of the elements of manifold theory, and the calculus of variations. In this midst of all this elegance the proof on p. 162 is truly startling.

Chavel, I.
★ {1} *Riemannian Symmetric Spaces of Rank One*, M. Dekker, New York, 1972.

Discusses a class of spaces connected with the sphere theorem, when the curvature is allowed to take on values K with $1/4 \leq K \leq 1$.

Cech, E.
see Fubini, G.

Chen, B.-Y.
★ {1} *Geometry of Submanifolds*, Dekker, New York, 1973.

Many results, using various techniques, about submanifolds (especially in the spaces of constant curvature), involving many of the topics in Volumes 3-5. Extensive bibliography.

Chern, S.-S.
{1} *Complex Manifolds without Potential Theory*, Van Nostrand, Princeton, N.J., 1967. (MR 37 #940.)
{2} *Topics in Differential Geometry* (mimeographed notes), The Institute for Advanced Study, Princeton, N.J., 1951. (MR 19, 764.)
{3} (ed.) *Studies in Global Geometry and Analysis*, Prentice-Hall, Englewood Cliffs, N.J., 1967. (MR 35 #1429.)

Chernov, I.
{1} *1,000 Best Short Games of Chess*, Simon and Schuster, New York, 1955.

Choquet-Bruhat, Y.
{1} *Géométrie Différentielle et Systèmes Extérieurs*, Dunod, Paris, 1968. (MR 38 #5118.)

Clark, R.S.
see Brickell, F.

Cohn-Vossen, S.
see Hilbert, D.

Coolidge, J.L.
{1} *A History of Geometrical Methods*, Dover, New York, 1963. (MR 28 #3357.)

I can't understand any of the mathematics in this book, but there must be something to be learned from a man who can say "as Darboux once remarked to me..."

Corbin, H. and Stehle, P.
{1} *Classical Mechanics*, 2nd ed., Wiley, New York, 1960. (MR 22 #3131.)

Courant, R.
{1} *Dirichlet's Principle, Conformal Mapping and Minimal Surfaces*, Interscience, New York, 1950. (MR 12, 90.)

Courant, R. and Hilbert, D.
{1} *Methods of Mathematical Physics*, vol. 2, Interscience, New York, 1962. (MR 25 #4216.)

Crittenden, R.J.
see Bishop, R.L.

Darboux, G.
★ {1} *Lecons sur la Théorie Générale des Surfaces*, 4 vols., 3rd ed., Chelsea, New York, 1972.

The great classic. Always referred to in hushed tones of awe. Unreadable, but in this new printing by Chelsea, at just $60 for 2200 pages, how can you resist?

{2} *Lecons sur les Systèmes Orthogonaux et les Coordonnées Curvilignes*, Gauthier-Villars, Paris, 1898.

de Rham, G.
see Rham, G. de

Dieudonné, J.A.
{1} *Éléments d'Analyse*, volumes 3, 4, Gauthier-Villars, Paris, 1971. (MR 42 #5266.)

Dieudonné, J.A. and Carrell, J.B.
{1} *Invariant Theory, Old and New*, Academic Press, New York, 1971. (MR 42 #4828.)

do Carmo, M.
see Carmo, M. do

Efimov, N.W.

★ {1} *Flächenverbiegung im Grossen*, mit einem Nachtrag von E. Rembs und K.P. Grotemeyer, Akademie-Verlag, Berlin, 1957. (MR 19, 59.)

A wonderful review article, covering many different aspects of rigidity, with additions by E. Rembs and K.P. Grotemeyer on more recent (but usually rather specialized) results which just double the size of the original. In addition to providing an introduction to the methods of A.D. Alexandrov, there are descriptions of the work of H. Lewy, Liebmann, Cohn-Vossen, etc. which cannot be found collected together anywhere else. I'm afraid, however, that the purported proof of Satz X on p. 177 is invalid -- the mistake appears at the very last step; only Pogorelov's difficult proof (for the more general situation) is available for this result.

The original article also appears in the A.M.S. Translations, series 1, volume 6.

Eisenhart, L.P.

★ {1} *Riemannian Geometry*, Princeton University Press, Princeton, N.J., 1966.

As good a compendium as any of the basic material of Riemannian geometry. In addition to isometric correspondence, conformal correspondence is also allotted an important role. (However, the discussion of the conformal tensor (§28) is treated better in Gerretsen {1}.) The important classical notion of "scalar curvature" is defined on p. 83. Appendix 3 gives the form of the metric in normal coordinates, and Appendix 22 gives the surprising result that the Codazzi-Mainardi equations follow from the Gauss equations for "general" hypersurfaces of $\mathbb{R}^n$, $n \geq 4$. Apart from this, the reader might like to look at the theorems on pp. 124, 155, 179, 182, 183, 184, 199.

 {2} *A Treatise on the Differential Geometry of Curves and Surfaces*, Dover, New York, 1960. (MR 22 #5936.)

This is a subset of Darboux, though it uses classical calculations rather than the moving frames which Darboux introduced for surface theory. A good way to get into more esoteric aspects of surface theory. Many results given in the problems.

{3} *Non-Riemannian Geometry*, American Mathematical Society, New York, 1927.

I.e., the study of general connections.

{4} *Continuous Groups of Transformations*, Princeton University Press, Princeton, N.J., 1933.

{5} *Transformations of Surfaces*, 2nd ed., Chelsea, New York, 1962.

A complement to Eisenhart {2}, going into greater detail on various classical transformations of surfaces. It doesn't look very interesting, but parts may turn out to be important, as they involve contact transformations (topic II.(f)).

{6} *An Introduction to Differential Geometry, with use of the tensor calculus*, Princeton University Press, Princeton, N.J., 1940. (MR 2, 154.)

Ellis, G.F.R.
see Hawking, S.W.

Favard, J.
★{1} *Cours de Géomêtrie Différentielle Locale*, Gauthier-Villars, Paris, 1957. (MR 18, 668.)

A complete development of differential geometry, covering many topics in considerable detail, using E. Cartan's general method of moving frames (topic III.(g)), which is expounded at the beginning, before any geometry has been done. Shades of Bourbaki!

Fenchel, W.
see Bonnesen, T.

Ferus, D.
{1} *Totale Absolutkrümmung in Differentialgeometrie und -topologie*, Springer-Verlag, Berlin, 1968. (MR 40 #3468.)

Finsler, P.
{1} *Über Kurven und Flächen in Allgemeinen Räumen*, Birkhäuser, Basel, 1951. (MR 13, 74.)

Flanders, H.
{1} *Differential Forms*, Academic Press, New York, 1963. (MR 28 #5397.)

Flaschel, P. and Klingenberg, W.
{1} *Riemannsche Hilbertmannigfaltigkeiten. Periodische Geodätsche*, Springer-Verlag, Berlin, 1972.

Forsyth, A.R.
★{1} *Lectures on the Differential Geometry of Curves and Surfaces*, Cambridge University Press, Cambridge, 1912.

Like Eisenhart {2}, a subset of Darboux, and a good way to get into surface theory. §132-146 will probably be incomprehensible (cf. topic III.(g)).

Fubini, G. and Čech, E.
{1} *Introduction a la Géométrie Projective Différentielle des Surfaces*, Gauthier-Villars, Paris, 1931.

Large bibliography.

Gauduchon, P.
see Berger, M.

Gauss, C.F.
★ {1} *General Investigations of Curved Surfaces*, Raven Press, Hewlett, New York, 1965. (MR 32 #20.)

Gerretsen, J.C.H.
{1} *Tensor Calculus and Differential Geometry*, P. Noordhoff N.V., Groningen, 1962. (MR 25 #1494.)

An introduction to Riemannian geometry about midway between Eisenhart's presentation and a completely modern one. Considerably less material than in Eisenhart, but there is a discussion of integrability conditions, which Eisenhart always treats rather shabbily, so it is a preferable source for certain results, notably the Weyl-Schouten conditions for a conformally flat space (p. 188).

Gilkey, P.B.
★ {1} *The Index Theorem and the Heat Equation*, Publish or Perish, Boston, 1974.

A brief introduction to pseudo-differential operators, which are used to obtain both Hodge's theorem and the Gauss-Bonnet-Chern theorem.

Godbillon, C.
{1} *Géométrie Différentielle et Mécanique Analytique*, Hermann, Paris, 1969. (MR 39 #3416.)

Goetz, A.
{1} *Introduction to Differential Geometry*, Addison Wesley, Reading, Mass., 1968. (MR 42 #2370.)

Gołab, S.
see Aczél, J.

Goldberg, S.I.
{1} *Curvature and Homology*, Academic Press, New York, 1962. (MR 25 #2537.)

Basically a compendium of results.

see also Bishop, R.L. *and* Weber, W.C.

Gostiaux, B.
see Berger, M.

Goursat, É.
{1} *Cours D'Analyse Mathématique*, 5th ed., vol. 1, Gauthier-Villars, Paris, 1933.

The exercises for chapter 12 give a classical style proof of Laguerre's theorem (p. III.282), as well as the result of Beltrami given in Problem III.4-4.

Greub, W., Halperin, S., and Vanstone, R.
{1} *Connections, Curvature, and Cohomology*, 3 vols., Academic Press, New York, 1972, 1973, ____

A very thorough treatise on the subjects indicated by the title, with information on an extremely large number of topics. Rather heavy on the symbolism, which is frequently neither standard nor felicitous. Large bibliography, especially for volume 2.

Griffiths, P.
{1} *Topics in Algebraic and Analytic Geometry*, Princeton University Press, Princeton, N.J., 1974.

Gromoll, D., Klingenberg, W., and Meyer, W.

{1} *Riemannsche Geometrie im Grossen*, Springer-Verlag, Berlin, 1968. (MR 37 #4751.)

These lecture notes, by three important differential geometers, give a modern presentation of intrinsic Riemannian geometry that starts right at the beginning and gets into recent deep global results. In addition, there are hundreds of examples and exercises that significantly extend the theory. By now many of the proofs have undoubtedly been simplified, and eventually the notes may become outdated, especially in light of Cheeger and Ebin {1} (see the Supplement), but they are definitely worth owning. It would be impossible to mention here all the topics covered, but the Sphere Theorem is certainly one of the most significant. There is a version of Klingenberg's Theorem (IV.8-35) for arbitrary dimensions (p. 254), and attention might also be called to the Lemma on p. 198 -- it gives an elementary version of the Morse theory proof of the Cartan-Hadamard Theorem (IV.8-13) in Milnor {2}.

Guggenheimer, H.W.
{1} *Differential Geometry*, McGraw-Hill, New York, 1963. (MR 27 #6194.)

The one-parameter subgroups of the group of isometries of the hyperbolic plane are discussed in detail (pp. 276-278). The discussion of space curves in affine geometry (pp. 170-172) is especially interesting; it turns out that the analysis given in Chapter III.2 is not really adequate. Attention might also be called to exercise 12 on p. 288.

Guillemin, V. and Pollack, A.

★ {1} *Differential Topology*, Prentice Hall, Englewood Cliffs, N.J., 1974.

A beautiful extension, and complement, of Milnor {1}.

Haack, W.

{1} *Elementare Differentialgeometrie*, Birkhäuser Verlag, Basel, 1955. (MR 18, 596.)

This seems to be the only book that treats infinitesimal bending by means of moving frames, and is the source for the material on pp. V.328-330. The author's definition seems to imply that the infinitesimal bending is non-trivial if and only if $\dot{\psi}_1^3$ or $\dot{\psi}_2^3$ is not identicaly zero, but I don't see offhand how to prove it. This fact is implicitly used in Problem V.12-4, which is a theorem of Weingarten (p. 217); an analogous theorem (p. 218) tells when "the lines of curvature remain lines of curvature in an infinitesimal bending" (i.e., when $\dot{\ell}_{12} = 0$). Eisenhart {2; 387} may be consulted for another proof, as well as for many examples of the "isothermal" surfaces arising in Problem V.12-4 (in particular, minimal surfaces are isothermal). For some modern theorems on infinitesimal bending that are basically translations of results about PDEs, see pp. 219-226. Finally, there is a proof (pp. 131-133) that only surfaces of constant curvature can be mapped conformally on the plane in such a way that the geodesics go to circles or straight lines. The proof is really unsatisfactory, for it uses certain lines that only exist on complex surfaces, and thus requires that the original surface be analytic. (I also know of no example to show that conformality of the mapping is a necessary hypothesis.) Bibliography of about 60 items.

Hadwiger, H.

{1} *Vorlesungen über Inhalt, Oberfläche und Isoperimetrie*, Springer-Verlag, Berlin, 1957. (MR 21 #1561.)

Halperin, S.

see Greub, W.

Hartshorne, R.

{1} *Foundations of Projective Geometry*, W.A. Benjamin, New York, 1967. (MR 36 #5801.)

Hawking, S.W. and Ellis, G.F.R.

{1} *The Large Scale Structure of Space-time*, Cambridge University Press, Cambridge, 1973.

Helgason, S.

{1} *Differential Geometry and Symmetric Spaces*, Academic Press, New York, 1962. (MR 26 #2986.)

See Chapter I, §12 for a completely invariant definition of sectional curvature which does not involve the curvature tensor.

Hermann, R.
{1} *Differential Geometry and the Calculus of Variations*, Academic Press, New York, 1968. (MR 38 #1635.)
{2} *Geometry, Physics, and Systems*, Dekker, New York, 1973.

Hicks, N.J.
{1} *Notes on Differential Geometry*, Van Nostrand, Princeton, N.J., 1965. (MR 31 #3936.)

On pp. 122-123 there are some rigidity results of a simple nature. On p. 154 there is a direct proof that the conjugate values of a geodesic are isolated, and pp. 168-169 classify the constant curvature simply connected manifolds as is done in Bishop and Crittenden {1}.

Hilbert, D.
see Courant, R.

Hilbert, D. and Cohn-Vossen, S.
★ {1} *Geometry and the Imagination*, Chelsea, New York, 1952. (MR 13, 766.)

An almost universally admired book, that discusses the visual and intuitive aspects of geometry rather than developing machinery and proofs. A refreshing view of the geometry of curves and surfaces in the differential geometry section, with references to all sorts of material that you won't find any where else.

Hlavatý, V.
{1} *Differentialgeometrie der Kurven und Flächen und Tensorrechnung*, P. Noordhoff, Groningen, 1939.
{2} *Differential Line Geometry*, P. Noordhoff, Groningen, 1953. (MR 15, 252.)

Hu, S.T.
{1} *Differentiable Manifolds*, Holt, Rinehart and Winston, New York, 1969. (MR 39 #6343.)

Huck, H., Roitzsch, R., Simon, U., Vortisch, W., Walden, R., Wegner, B., and Wendland, W.
★ {1} *Beweismethoden der Differentialgeometrie im Grossen*, Springer-Verlag, Berlin, 1973.

Rigidity theorems by means of integral formulas and the index method.

Ishihara, S.
see Yano, K.

John, F.
★ {1} *Partial Differential Equations*, Springer-Verlag, Berlin, 1971. (MR 46 #3960.)

The source for most of the material in the beginning of Chapter V.10. Probably the best introduction to PDEs, with a good bibliography for more advanced study.

Kähler, E.
{1} *Einführung in die Theorie der Systeme von Differentialgleichungen*, Chelsea, New York, 1949.

Kamber, F. and Tondeur, P.
{1} *Flat Manifolds*, Springer-Verlag, Berlin, 1968. (MR 38 #6618.)

Killing, W.
{1} *Die Nicht-euklidischen Raumformen in Analytischer Behandlung*, G.B. Teubner, Leipzig, 1885.

Klingenberg, W.
{1} *Eine Vorlesung über Differentialgeometrie*, Springer-Verlag, Berlin, 1973. [see the Supplement]
see also Flaschel, P. *and* Gromoll, D.

Kobayashi, S.
{1} *Transformation Groups in Differential Geometry*, Springer-Verlag, Berlin, 1972.
{2} *Hyperbolic Manifolds and Holomorphic Mappings*, Dekker, New York, 1970. (MR 43 #3503.)

Kobayashi, S. and Nomizu, K.

{1} *Foundations of Differential Geometry*, 2 vols., Interscience, New York, 1963, 1969. (MR 27 #2945; 38 #6501.)

This will probably become the standard reference for this generation. A complete treatment of the foundations, and the definitive exposition of the principal bundle point of view. Not exactly the sort of book to read like a novel, but one you should certainly have. There is a very large bibliography.

Kobayashi, S., Obata, M., and Takahashi, T. (eds.)
{1} *Differential Geometry*, Kinokuniya, Tokyo, 1972.

Kodaira, K.
see Morrow, J.

Kolmogorov, A.N.
see Alexandrov, A.D.

Kreyszig, E.
{1} *Introduction to Differential Geometry and Riemannian Geometry*, University of Toronto Press, Toronto, 1968. (MR 37 #2096.)

See p. 267 for the theorem on mapping of surfaces mentioned under Haack {1}. There are many interesting references to classical papers.

Kruppa, E.
{1} *Analytische und Konstruktive Differentialgeometrie*, Springer, Wien, 1957. (MR 19, 165.)

Kulk, W.v.d.
see Schouten, J.A.

Lanczos, C.
{1} *Space Through the Ages*, Academic Press, New York, 1970. (MR 42 #5747.)

Historical, somewhat popularized, account.

Lane, E.P.
{1} *A Treatise on Projective Differential Geometry*, University of Chicago Press, Chicago, 1942. (MR 4, 114.)

Large bibliography.

Lang, S.

{1} *Introduction to Differentiable Manifolds*, 2nd ed., Interscience, New York, 1972. (MR 27 #5192.)

The standard reference, with everything done neatly and cleanly, for how things go in the infinite dimensional case.

Lascoux, A. and Berger, M.
{1} *Variétés Kähleriennes Compactes*, Springer-Verlag, Berlin, 1970. (MR 43 #3979.)

Laugwitz, D.
{1} *Differential and Riemannian Geometry*, Academic Press, New York, 1965. (MR 30 #2406.)

Despite the usual problems with differentials, this book is a rather nice introduction to classical differential geometry, as well as modern material. On p. 131 there is an interesting formula for Δn, where n is the normal map of an immersion $f: M^2 \to \mathbb{R}^3$ (compare with the formula $\Delta f = 2Hn$, p. IV.198); note that the first term in the formula is just $-2\langle dH, df\rangle$. The references on this page may also be of interest. §15 treats not only Finsler metrics, but also "systems of paths" (vector fields on the tangent bundle).

{2} *Differentialgeometrie in Vektorräumen*, Friedr. Vieweg & Sohn, Braunschweig, 1965. (MR 32 #406.)

Lavrent'ev, M.A.
see Alexandrov, A.D.

Lawson, H.B. Jr.
{1} *Lectures on Minimal Submanifolds*, Volume 1, Publish or Perish, Inc., Berkeley, 1979.

A good review source for minimal surfaces and higher dimensional analogues, with a discussion of the Plateau problem, and bibliography.

{2} *Minimal Varieties in Real and Complex Geometry*, University of Montréal Press, Montréal, 1972.

Leichtweiss, K.
see Blaschke, W.

Lelong-Ferrand, J.
{1} *Géométrie Différentielle*, Masson, Paris, 1963. (MR 27 #648.)

Levi-Civita, T.
{1} *The Absolute Differential Calculus*, Blackie & Son, London, 1961.

Of historical interest only.

Lichnerowicz, A.
{1} *Theorie Globale des Connexions et des Groupes d'Holonomie*, Edizioni cremonese, Rome, 1955. (MR 19, 453.)
{2} *Géométrie des Groupes de Transformations*, Dunod, Paris, 1958. (MR 23 #A1329.)

Lie, S. and Scheffers, G.W.
{1} *Vorlesungen über Continuierliche Gruppen*, B.G. Teubner, Leipzig, 1893.
{2} *Geometrie der Berührungstransformationen*, B.G. Teubner, Leipzig, 1896.

Loos, O.
{1} *Symmetric Spaces*, 2 vols., W.A. Benjamin, New York, 1969. (MR 39 #365a,b.

Lyusternik, L.A.
★{1} *Shortest Paths. Variational Problems*, Pergamon Press, Macmillan, New York, 1964. (MR 31 #2644.)

A beautiful little book that gives elementary treatments of variational problems. There is a geometric proof of Clairaut's theorem in which the surface of revolution is approximated by a union of frustra of cones, for which the theorem is checked directly.

MacKenzie, R.E.
see Auslander, L.

Malliavin, P.
{1} *Géométrie Différentielle Intrinsèque*, Hermann, Paris, 1972.

Matsumoto, M.
{1} *The Theory of Finsler Connections*, Okayama University, Okayama, 1970. (MR 42 #2409.)

Matsushima, Y.
{1} *Differentiable Manifolds*, Marcel Dekker, New York, 1972.

Mazet, E.
see Berger, M.

Menger, K.
see Blumenthal, L.M.

Meyer, W.
see Gromoll, D.

Milnor, J.W.

★ {1} *Topology from the Differentiable Viewpoint*, University Press of Virginia, Charlottesville, Virginia, 1965. (MR 37 #2239.)

This book is not really about differential geometry at all, but about differential topology. But this is a field of related interest, and, besides, anything written by Milnor is beautiful. See, in particular, §2 for the hard version of Sard's theorem.

★ {2} *Morse Theory*, Princeton University Press, Princeton, N.J., 1963. (MR 31 #6249.)

Although this book is really devoted to more advanced material from differential topology, a substantial part contains material from differential geometry. See, in particular, §6 and §19, where the Cartan-Hadamard theorem is proved using Morse theory (compare the remarks under Gromoll, Klingenberg, and Meyer {1}).

Milnor, J.W. and Stasheff, J.D.

★ {1} *Characteristic Classes*, Princeton University Press, Princeton, N.J., 1974.

The first characteristic classes considered here are the Stiefel-Whitney classes, with $\mathbb{Z}_2$ coefficients. Since the actual construction of these classes, in §8, involves the Steenrod squares, it might be advisable to simply skip this section. §§9-15 proceed to define the Euler, Pontryagin, and Chern classes in the way mentioned in the "Valedictory" to Chapter V.13. The remaining §§ give applications beyond those interspersed in the preceding ones. We should explicitly mention that §11 identifies the Euler class of TM without invoking triangulations. §6 describes the cell structure for the Grassmannians by means of Schubert varieties. This cell structure was originally used by Ehresmann [1], [2] to compute the (integral) cohomology of the Grassmannians. From this one immediately obtains the real cohomology, which of course agrees with our calculations in Chapter V.13. Naturally it would be of interest to connect the two calculations directly, by determining the integrals over the Schubert varieties of the forms giving the Chern, Pontryagin, and Euler classes. I know of no reference to such calculations, except for the brief note of Chern [4] and the mimeographed notes of Chern {2}.

Misner, C.W., Thorne, K.S., and Wheeler, J.A.
{1} *Gravitation*, Freeman, San Francisco, 1973.

Morrow, J. and Kodaira, K.
{1} *Complex Manifolds*, Holt, Rinehart and Winston, New York, 1971. (MR 46 #2080.)

Munkres, M.R.
★ {1} *Elementary Differential Topology*, Princeton University Press, Princeton, N.J., 1963. (MR 29 #623.)

This little book proves almost all the facts about differentiable manifolds which arise in the description of C^k-manifolds, and which a differential geometer might want to know, even though they are not part of differential geometry *per se*.

Narasimhan, R.
{1} *Analysis on Real and Complex Manifolds*, Masson, Paris, 1968. (MR 40 #4972

Nelson, E.
{1} *Tensor Analysis*, Princeton University Press, Princeton, N.J., 1967.

An unorthodox approach to tensors, which has some neat things. See, in particular, the discussion of the bracket (pp. 30-36), the interpretation of curvature (p. 77), and the approach to the operator δ which uses proposition IV.7-62 as the definition, instead of invoking the $*$ operator (pp. 96-100).

Nomizu, K.
see Kobayashi, S.

Obata, M.
see Kobayashi, S.

O'Neill, B.
{1} *Elementary Differential Geometry*, Academic Press, New York, 1966. (MR 34 #3444.)

This book expounds moving frames for undergraduates. Mention might be made of the material on pp. 330-333 and exercises 10-13 on pp. 337-338, generalizing our discussion of geodesics on surfaces of revolution (pp. III.314-319). Also exercise 8 on p. 387 gives the Gauss-Bonnet theorem for surfaces-with-boundary.

Osserman, R.

{1} *Survey of Minimal Surfaces*, Van Nostrand Reinhold, New York, 1969. (MR 41 #934.)

A good place to learn more about minimal surfaces, and minimal submanifolds, with a large bibliography. I would probably have gone out of my mind trying to write Chapter IV.9 if it hadn't been for this book, which clears up things that standard texts have contentedly left in a muddle for years. For example, nowhere else are the poles of g considered in the Enneper-Weierstrass representation of minimal surfaces. Also, almost all classical books consider all minimal surfaces to be complex surfaces. This is valid, since minimal surfaces are analytic, and can therefore be complexified, but it usually leads to results that have no real geometric significance. For example, a theorem of Lie (cf. Blaschke {1; v.1, 236}, Haack {1; 140}, or Kreysig {1; 244}) says that every minimal surface is a "translation surface", of the form $(s,t) \mapsto f(s) + g(t)$. But this is only true if f and g are complex-valued; it is easy to see that the only minimal surfaces of this form for $\mathbb{R}^3$-valued functions f and g is Scherk's minimal surface (in fact, that's how the surface was discovered).

Pars, L.
{1} *A Treatise on Analytical Dynamics*, Wiley, New York, 1965.

Petrov, A.Z.
{1} *Einstein Spaces*, Pergamon Press, Oxford, 1969. (MR 39 #6225.)

Pogorelov, A.V.
{1} *Die Eindeutige Bestimmung Allgemeiner Konvexer Flächen*, Akademie-Verlag, Berlin, 1956. (MR 18, 330.)
{2} *Die Verbiegung Konvexer Flächen*, Akademie-Verlag, Berlin, 1957. (MR 19, 305.)
{3} *Topics in the Theory of Surfaces in Elliptic Space*, Gordon and Breach, New York, 1961. (MR 26 #6908.)
{4} *Einige Untersuchungen zur Riemannschen Geometrie im Grossen*, VEB Deutscher Verlag der Wissenschaften, Berlin, 1960. (MR 22 #5946.)

Concerning the imbedding of 2-dimensional Riemannian manifolds in a given 3-dimensional Riemannian manifold.

{5} *Extrinsic Geometry of Convex Surfaces*, American Mathematical Society, Providence, R.I., 1973.
{6} *Monge-Ampere Equations of Elliptic Type*, P. Noordhoff, Groningen, 1964. (MR 31 #4993.)
{7} *Differential Geometry*, P. Noordhoff, Groningen.
[see the Supplement]

Pollack, A.
see Guillemin, V.

Quan, P.M.
{1} *Introduction a la Geométrie des Variétés Différentiables*, Dunod, Paris, 1969.

Rado, T.
{1} *On the Problem of Plateau*, Springer, Berlin, 1933.

Raschewski, P.K.
{1} *Riemannsche Geometrie und Tensoranalysis*, Deutscher Verlag der Wissenschaften, Berlin, 1959. (MR 21 #2258.)

Reichardt, H.
see Blaschke, W.

Rham, G. de
{1} *Variétés Différentiables. Formes, Courants, Formes Harmoniques*, Hermann, Paris, 1955. (MR 16, 957.)

Contains, among other things, a proof of Hodge's theorem by means of "currents" -- these are essentially differential forms whose coefficients are distributions (in the analysts' sense).

★ Riemann, B.
{1} *Collected Works*, Dover, New York, 1953. (MR 14, 610.)

Rinow, W.
{1} *Die Innere Geometrie der Metrischen Räume*, Springer, Berlin, 1961. (MR 23 #A1290.)

Roitzsch, R.
see Huck, H.

Rund, H.
{1} *The Differential Geometry of Finsler Spaces*, Springer, Berlin, 1959. (MR 21 #4462.)
{2} *The Hamilton-Jacobi Theory in the Calculus of Variations; Its Role in Mathematics and Physics*, Van Nostrand, London, 1966. (MR 37 #5752.) [see the Supplement]

Ruse, H.S., Walker, A.G., and Willmore, T.J.
{1} *Harmonic Spaces*, Edizioni Cremonese, Rome, 1961. (MR 25 #5456.)

Salkowski, E.
{1} *Affine Differentialgeometrie*, de Gruyter, Berlin, 1934.

Santaló Sors, L.A.
{1} *Introduction to Integral Geometry*, Hermann, Paris, 1953. (MR 15, 736.)

Sauer, R.
{1} *Differenzengeometrie* [sic], Springer, Berlin, 1970. (MR 41 #7544.)

Scheffers, G.W.
{1} *Anwendung der Differential- und Integral-Rechnung auf Geometrie*, 2 vols., Veit & Co., Leipzig, 1901-1902.

A classical book with quite extended discussions of various topics, too numerous to be listed here. Special mention may be made of the analytic determination of all geodesic maps between open subsets of the plane (v.2, 429-432). The unique features of the book have been mentioned under topic III.(g).

see also Lie, S.

Schild, A.
see Synge, J.L.

Schirokow, P.A. and Schirokow, A.P.
{1} *Affine Differentialgeometrie*, Teubner, Leipzig, 1962. (MR 27 #660.)

Contains an extensive bibliography, and historical remarks in the forward.

Schouten, J.A.
{1} *Ricci-Calculus*, 2nd ed., Springer-Verlag, Berlin, 1954. (MR 16, 521.)

This book was written to show the great superiority of the classical notation. Reading it, one can see why differential geometry was once given up for dead. There are super-, sub-, pre- and post-scripts, including dots, brackets, etc. It is certainly a triumph of the printer's art, but is also supposed to be important for its considerations of non-symmetric connections (sometimes considered in General Relativity). There is an enormous bibliography. Amusing footnotes on pp. 118, 160, 172.

Schouten, J.A. and Kulk, W. v.d.
{1} *Pfaff's Problems and its Generalization*, Clarendon Press, Oxford, 1949. (MR 11, 179.)

Schouten, J.A. and Struik, D.J.
{1} *Einführung in die Neueren Methoden der Differentialgeometrie*, 2nd ed., 2 vols., P. Noordhoff, n.v., Groningen-Batavia, 1935-

In volume 2 see p. 78 for references to generalizations of the Beltrami-Enneper theorem (another reference is Hayden [1]), pp. 94-116 for a detailed classification of points in a submanifold, and p. 136 for results related to the invariance of K_r when r is even (p. IV.103). The notation, though classical, is quite manageable.

Schwartz, J.T.
{1} *Nonlinear Functional Analysis*, Gordon and Breach, New York, 1969.
{2} *Differential Geometry and Topology*, Gordon and Breach, New York, 1968.

See p. 87 for a direct verification of the fact that all partial derivatives of g_{ij} at the center of a normal coordinate system are expressible in terms of the covariant derivatives of the curvature tensor (compare pp. IV.224-225). The original proof of this fact was by Vermeil [1]. By the way, "normal coordinates" have been used for various tensors other than a Riemannian metric (cf. references at the end of Veblen {1; ch.6}, and also Weitzenböck {1; §19-22}).

Segre, B.
{1} *Some Properties of Differentiable Varieties and Transformations; With Special Reference to the Analytic and Algebraic Cases*, Springer, Berlin, 1957. (MR 16, 679.)

Simon, U.
see Huck, H.

Singer, I.M. and Thorpe, J.A.
{1} *Lecture Notes on Elementary Topology and Geometry*, Scott, Foresman, Glenview, Ill., 1967. (MR 35 #4834.)

Another non-classical introduction to surface theory for undergraduates, this time in terms of principal bundles. The trick is that for surface theory the principal bundle degenerates to the sphere bundle and everything is much easier. Nevertheless, the book is not easy going. Chapter 6 contains a weird proof of the de Rham theorem.

Ślebodziński, W.
{1} *Exterior Forms and Their Application*, Państwowe Wydawnictwo Naukowe, Warsaw, 1970. (MR 42 #672.)

Sommerville, D.M.Y.
★ {1} *Bibliography of Non-euclidean Geometry including the theory of parallels, the foundations of geometry, and space of n dimensions*, Harrison & Sons, London, 1911.

Sorani, G.
{1} *An Introduction to Real and Complex Manifolds*, Gordon and Breach, New York, 1969. (MR 41 #6220.)

Stasheff, J.D.
see Milnor, J.W.

Stehle, P.
see Corbin, H.

Sternberg, S.
{1} *Lectures on Differential Geometry*, Prentice-Hall, Englewood Cliffs, N.J., 1964. (MR 33 #1797.)

The author's heart was really in the last chapter, on G-structures, and perhaps in the fourth chapter on the calculus of variations. But see also p. 45 for the hard version of Sard's theorem, p. 63 for the Whitney imbedding theorem, and p. 256 for Whitney's theorem on smooth homotopy of plane curves.

Stoker, J.J.
{1} *Differential Geometry*, Wiley-Interscience, New York, 1969. (MR 39 #2072.)

Strubecker, K.
{1} *Differentialgeometrie*, 3 vols., Walter de Gruyter & Co., Berlin, 1964, 1969, 1969. (MR 16, 954; 20 #4273; 21 #878.)

Three pretty little paperback volumes with lots of details about elementary classical topics, and dozens of excellent pictures.

Struik, D.J.
{1} *Lectures on Classical Differential Geometry*, Addison-Wesley, Reading, Mass., 1961. (MR 12, 227.)

See pp. 153-156 for a simple classical proof of the Gauss-Bonnet theorem.

see also Schouten, J.A.

Sulanke, R. and Wintgen, P.
{1} *Differentialgeometrie und Faserbündel*, Birkhäuser Verlag, Basel, 1972.

Švec, A.
{1} *Projective Differential Geometry of Line Congruences*, Czechoslovak Academy of Sciences, Prague, 1965. (MR 33 #7949.)

Synge, J.L. and Schild, A.
{1} *Tensor Calculus*, University of Toronto Press, Toronto, 1962. (MR 11, 400.)

Takahashi, T.
see Kobayashi, S.

Thomas, T.Y.
{1} *The Differential Invariants of Generalized Space*, Cambridge University Press, Cambridge, 1937.
{2} *Concepts from Tensor Analysis and Differential Geometry*, 2nd ed., Academic Press, New York, 1965. (MR 32 #4623.)

Thorne, K.S.
see Misner, C.W.

Thorpe, J.A.
see Singer, I.M.

Tondeur, P.
see Kamber, F.

Vaisman, I.
{1} *Cohomology and Differential Forms*, Dekker, New York, 1973.

The sheaf theory here is more advanced than in Warner {1}. There is a brief account of the theory of Allendoerfer and Eels [1], which describes the *integral* cohomology of M in terms of forms with singularities.

Vanstone, R.
see Greub, W.

Veblen, O.
{1} *Invariants of Quadratic Differential Forms*, Cambridge University Press, Cambridge, 1927.

Veblen, O. and Whitehead, J.H.C.
{1} *The Foundations of Differential Geometry*, Cambridge University Press, Cambridge, 1932.

Vortisch, W.
see Huck, H.

Vranceanu, G.
{1} *Lectii de Geometrie Diferentiala*, 4 vols., Editura Academiei Republicii Socialiste România, Bucharest, 1968. (MR 39 #6181.)

The first volume was originally written in French, and then translated into Rumanian, the language in which the last three volumes were originally written. The second and third volumes have been translated into French, and the first two volumes have been translated into German. Various changes have been made in some of the translations. The review cited above is for Volume 4 only, but it contains a complete list of all other reviews of previous volumes.

Walden, R.
see Huck, H.

Walker, A.G.
see Ruse, H.S.

Warner, F.W.
★ {1} *Foundations of Differentiable Manifolds and Lie Groups*, Scott, Foresman and Co., Glenview, Ill., 1971. (MR 45 #4312.)

The first part of this book treats manifold theory and Lie groups concisely, but with all the necessary details. The relationship between the functors $\mathcal{J}^k(V)$ and $\Omega^k(V)$ and the algebraists' $\otimes^k V$ and $\Lambda^k V$ are spelled out in detail in Chapter 2. The last theorem of that chapter formally states a fact which we have frequently used (e.g. in the proof of Theorem IV.7-20). The third chapter gives a little more detail on Lie groups. The adjoint representation is treated explicitly, as is the universal covering group of a Lie group. See also exercise 15 on p. 134 for a proof that the exponential map of $GL(n,\mathbb{C})$ is onto and exercise 18 on p. 135 for the classification of Abelian Lie groups.

The second and third parts of the book give connected presentations of material available nowhere else. The sheaf-theoretic proof of de Rham's theorem is given in the second part: Alexander-Spanier, Čech, and singular cohomologies are all mentioned, the isomorphism between de Rham and singular cohomology is shown to be given by integration, and the wedge product is shown to correspond to cup product.* (Although a simple proof of the de Rham theorem was outlined in Problem I.11-14, the sheaf-theoretic proof explains, in some sense, "why" the theorem is true, for it shows that only certain basic facts ($d^2 = 0$, the local converse, and existence of partitions of unity) matter, while the details of how they are obtained is irrelevant.) The third part of the book gives a completely self-contained elementary treatment of the Hodge theorem.

Watson, G.N.
see Whittaker, E.T.

Weatherburn, C.E.
{1} *An Introduction to Riemannian Geometry and the Tensor Calculus*, Cambridge University Press, Cambridge, 1957.

The historical note at the end may be of interest -- it contains some names that you have probably never even seen before.

Weber, W.C. and Goldberg, S.I.
{1} *Conformal Deformations of Riemannian Manifolds*, Queen's University, Kingston, Ontario, 1969. (MR 40 #1938.)

Wegner, B.
see Huck, H.

Weil, A.
{1} *Introduction à l'Étude des Variétés Kählériennes*, Hermann, Paris, 1958. (MR 22 #1921.)

---

*The reader might be concerned, as I was, whether this holds for our wedge product, with the factor $(p+q)!/p!q!$, or for the wedge product without this factor. The answer is that it holds for either! One way to see this is the following. In $\Lambda^k(V)$, regarded as a quotient vector space of $\otimes^k V$, there is a natural $\wedge$ product ($v_1 \wedge \cdots \wedge v_k$ is the residue class of $v_1 \otimes \cdots \otimes v_k$), and this wedge product on each $\Lambda^k(M_p{}^*)$ gives a wedge product on differential forms if we regard a k-form as a function $p \mapsto \omega(p) \in \Lambda^k(M_p{}^*)$. It is then this wedge product that corresponds to cup product under the de Rham isomorphism, so long as we define integration so that the integral of $f\,dx^1 \wedge \cdots \wedge dx^n$ on $\mathbb{R}^n$ is just the ordinary integral of f. Now we can also get a wedge product on $\Omega^k(V)$ by means of an isomorphism of $\Omega^k(V)$ with $\Lambda^k(V^*)$. One isomorphism will give our wedge product, while another will give the product without the factor $(p+q)!/p!q!$, but in either case the wedge product will correspond to cup product. If this explanation seems too paradoxical, the following may help. For a k-form ω on M and a singular k-cube $c\colon [0,1]^k \to M$, we define $\int_c \omega$ as the ordinary integral of f over $[0,1]^k$ where

$$c^*\omega = f\,dx^1 \wedge \cdots \wedge dx^k .$$

In this definition, we naturally use the same $\wedge$ in $dx^1 \wedge \cdots \wedge dx^k$ as we use in taking the product $\omega \wedge \eta$ on our manifold M; if we change the definition of $\wedge$, then we also end up changing the definition of $\int_c \omega$, and the change involves just the right factor so that $\wedge$ still corresponds to cup product.

Weiss, G.L.
see Boothby, W.

Weitzenböck, R.
{1} *Invariantentheorie*, P. Noordhoff, Groningen, 1923.

Famous for the message contained in the initial letters of the sentences in the forward. In addition to the material of a differential geometric nature, some of the classical invariant theory might be of interest, for example, apolarity conditions. One might also consult Dieudonné and Carrell [1].

Wells, R.O. Jr.
{1} *Differential Analysis on Complex Manifolds*, Prentice-Hall, Englewood Cliffs, N.J., 1973.

Wendland, W.
see Huck, H.

Weyl, H.
{1} *The Classical Groups, their Invariants and Representations*, 2nd ed., Princeton University Press, Princeton, N.J., 1953. (MR 1, 42.)

This book certainly qualifies as a classic, if unreadability is one of the criteria. It was the source for the proof of the first main theorem of invariance theory for $O(n)$ and $SL(n,k)$. One might also consult Dieudonné and Carrell [1].

Wheeler, J.A.
see Misner, C.W.

Whitehead, J.H.C.
see Veblen, O.

Whittaker, E.T.
{1} *A Treatise on the Analytical Dynamics of Particles and Rigid Bodies*, 4th ed., Cambridge University Press, Cambridge, 1959. (MR 21 #2381.)

Whittaker, E.T. and Watson, G.N.
{1} *A Course of Modern Analysis*, 4th ed., Cambridge University Press, Cambridge, 1962.

Wilczynski, E.J.
{1} *Projective Differential Geometry of Curves and Ruled Surfaces*, B.G. Teubner, Leipzig, 1906.

Willmore, T.J.
{1} *An Introduction to Differential Geometry*, Oxford University Press, London, 1959. (MR 28 #2482.)

see also Ruse, H.S.

Wintgen, P.
see Sulanke, R.

Wolf, J.A.
★ {1} *Spaces of Constant Curvature*, 3rd ed., Publish or Perish, Boston, 1974.

After a rapid run through Riemannian geometry, this book gets down to the main task of classifying complete manifolds of constant curvature. This is basically an algebraic problem, since all such manifolds have the standard simply connected examples as their universal covering spaces. The case of n-dimensional manifolds M of constant positive curvature is easy for n even: M is S^n or $\mathbb{P}^n$ (p. 74 or Kobayashi and Nomizu {1; v.1, Note 4}). Other cases are not so easy, and all sorts of interesting material gets woven together in the search for the final classification.

Wright, J.E.
{1} *Invariants of Quadratic Differential Forms*, Cambridge University Press, Cambridge, 1908.

Wu, H.-H.
{1} *The Equidistribution Theory of Holomorphic Curves*, Princeton University Press, Princeton, N.J., 1970.

Yaglom, I.M. and Boltyanskiǐ, V.G.
{1} *Convex Figures*, Holt, Rinehart and Winston, New York, 1961. (MR 23 #A1283.)

Yano, K.
★ {1} *Integral Formulas in Riemannian Geometry*, Dekker, New York, 1970. (MR 44 #2174.)

Large bibliography.

{2} *Differential Geometry on Complex and Almost Complex Spaces*, Macmillan, New York, 1965. (MR 32 #4635.)
{3} *The Theory of Lie Derivatives and its Applications*, North-Holland, New York, 1957. (MR 19, 576.)

Large bibliography.

Yano, K. and Bochner, S.
★ {1} *Curvature and Betti Numbers*, Princeton University Press, Princeton, N.J., 1953. (MR 15, 989.)

Yano, K. and Ishihara, S.
{1} *Tangent and Cotangent Bundles: Differential Geometry*, Dekker, New York, 1973.

Zalgaller, V.A.
see Alexandrov, A.D.

Supplement

Abraham, R. and Marsden, J.
{1} *Foundations of Mechanics*, 2nd ed., Addison-Wesley, Reading, Mass., 1978.

Arnold, V.I.
{1} *Mathematical Methods in Classical Mechanics*, Springer-Verlag, New York, 1978.

Besse, A.L.
{1} *Manifolds All of Whose Geodesics are Closed*, Springer-Verlag, New York, 1978.

Boothby, W.
{1} *An Introduction to Differentiable Manifolds and Riemannian Geometry*, Academic Press, New York, 1975.

Brakke, K.A.
{1} *The Motion of a Surface by its Mean Curvature*, Princeton University Press, Princeton, N.J., 1978.

Cheeger, J. and Ebin, D.
{1} *Comparison Theorems in Riemannian Geometry*, North-Holland, Amsterdam, 1974.

Dodson, C.T.J. and Poston, T.
{1} *Tensor Geometry. The Geometric Viewpoint and its Uses*, Pitman, London, 1977.

Particular attention is given to indefinite Riemannian metrics.

Ebin, D.
see Cheeger, J.

Forbes, W.F.
see Rund, H.

Klingenberg, W.
{2} *A Course in Differential Geometry*, Springer-Verlag, New York, 1978.
{3} *Lectures on Closed Geodesics*, Springer-Verlag, New York, 1978.

Lovelock, D.
see Rund. H.

Marsden, J.
see Abraham, R.

Millman, R.S. and Parker, G.D.
{1} *Elements of Differential Geometry*, Prentice Hall, Englewood Cliffs, N.J., 1977.

Parker, G.D.
see Millman, R.S.

Pogorelov, A.V.
{8} *The Minkowski Multidimensional Problem*, V.H. Winston & Sons, Washington, D.C., John Wiley & Sons, New York, 1978.

Poston, T.
see Dodson, C.T.J.

Rund, H. and Lovelock, D.
{1} *Tensors, Differential Forms, and Variational Principles*, John Wiley & Sons, New York, 1975.

Rund, H. and Forbes, W.F. (eds.)
{1} *Topics in Differential Geometry*, Academic Press, New York, 1976.

Sachs, R.K. and Wu, H.
{1} *General Relativity for Mathematicians*, Springer-Verlag, New York, 1977.

Weinberg, S.
{1} *Gravitation and Cosmology*, Wiley, New York, 1972.

Wu, H.
see Sachs, R.K.

C. Journal Articles

Alexandrov, A.D.
[1] *Uniqueness theorems for surfaces in the large. I*, Vestnik Leningrad. Univ. 11 (1956), no. 19, 5-17. (Russian.) (MR 19, 167.); Amer. Math. Soc. Transl. (2) 21 (1962), 341-354. (MR 27 #698a.)
[2] *A characteristic property of spheres*, Ann. Mat. Pura Appl. (4) 58 (1962), 303-315. (MR 26 #722.)
[3] *Zur Theorie der gemischten Volumina von konvexen Körpern*, Mat. Sb. 2 (44) (1937), 947-972, 1205-1238, 3 (45) (1938), 27-46, 227-251. (Russian. Germany summary.)
[4] *On a class of closed surfaces*, Mat. Sb. 4 (1938), 69-77. (Russian.)

Allendoerfer, C.B.
[1] *The Euler number of a Riemannian manifold*, Amer. J. Math. 62 (1940), 243-248. (MR 2, 20.)
[2] *Rigidity for spaces of class greater than one*, Amer. J. Math. 61 (1939), 633-644. (MR 1, 28.)
[3] *The imbedding of Riemann spaces in the large*, Duke Math. J. 3 (1937), 317-333.

This paper is the source for Chapter IV.7, Addendum 4.

Allendoerfer, C.B. and Eels, J. Jr.
[1] *On the cohomology of smooth manifolds*, Comment. Math. Helv. 32 (1958), 165-179. (MR 21 #868.)

Allendoerfer, C.B. and Weil, A.
[1] *The Gauss-Bonnet theorem for Riemannian polyhedra*, Trans. Amer. Math. Soc. 53 (1943), 101-129. (MR 4, 169.)

Barbosa, J.L. and do Carmo, M.
[1] *On the size of a stable minimal surface in* $\mathbb{R}^3$, (Proc. Symp. Pure Math. 27(1974), A.M.S., to appear).

Beez, R.
[1] *Zur Theorie des Krümmungsmasses von Mannigfaltigkeiten höhere Ordnung*, Zeitschrift für Mathematik und Physik 21 (1876), 373-401.

Bianchi, L.
[1] *Sulle superficie a curvatura nulla in geometria ellittica*, Ann. Mat. Pura Appl. 24 (1896), 93-129.

Bieberbach, L.
[1] *Hilberts Satz über Flächen konstanter negativer Krümmung*, Acta Math. 48 (1926), 319-327.

This paper gives a rigorous proof of Hilbert's theorem which is closer to the original argument of Hilbert [1] than the one we have given (it also contains an unfounded criticism of Holmgren's proof).

Blaschke, W.
[1] *Kreis und Kugel*, Jber. Deutsch. Math.-Verein. 24 (1915-1916), 195-207.

Bol, G.
[1] *Über Nabelpunkte auf einer Eifläche*, Math. Z. 49 (1944), 389-410. (MR 7, 29.)

Carmo, M. do and Lima, E.
[1] *Isometric immersions with semi-definite second quadratic forms*, Arch. Math. (Basel) 20 (1969), 173-175. (MR 39 #6214.)
[2] *Immersions of manifolds with non-negative sectional curvatures*, Bol. Soc. Brasil. Mat. 2 (1972), 9-22.

Carmo, M. do and Warner, F.W.
[1] *Rigidity and convexity of hypersurfaces in spheres*, J. Differential Geometry 4 (1970), 133-144. (MR 42 #1014.)

Note that the assertion, on p. 140, that $\tilde{x}$ and $\tilde{y}$ are imbeddings needs some argument. (See the last part of the statement of Proposition V.12-26; as far as I can tell, Theorem 1 of Chapter IV of Pogorelov {3} does not claim that $\tilde{x}$ and $\tilde{y}$ are always immersions, as asserted in the editor's footnote, but only that their images have a tangent space at each point.)

see also Barbosa, J.L.

Carrell, J.B.
see Dieudonné, J.A.

Cartan, É.
[1] *La déformation des hypersurfaces dans l'espace euclidien réel à n dimensions*, Bull. Soc. Math. France 44 (1916), 65-99.
[2] *Les surfaces qui admettent une seconde forme fondamentale donneé*, Bull. Sci. Math. (2) 67 (1943), 8-32. (MR 7, 30.)
[3] *Sur les variétés de courbure constante d'un espace euclidien ou non euclidien*, Bull. Soc. Math. France 47 (1919), 125-160, 48 (1920), 132-208.

Chern, S.-S.
[1] *Integral formulas for hypersurfaces in Euclidean space and their applications to uniqueness theorems*, J. Math. Mech. 8 (1959), 947-956. (MR 22 #4997.)
[2] *A simple intrinsic proof of the Gauss-Bonnet formula for closed Riemannian manifolds*, Ann. of Math. (2) 45 (1944), 747-752. (MR 6, 106.)

See also the next paper, especially for the formula for manifolds-with-boundary.

[3] *On the curvatura integra in a Riemannian manifold*, Ann. of Math. (2) 46 (1945), 674-684. (MR 7, 328.)

[4] *On the characteristic classes of Riemannian manifolds*, Proc. Nat. Acad. Sci. U.S.A. 33 (1947), 78-82. (MR 8, 490.)
[5] *On a theorem of algebra and its geometrical application*, J. Indian Math. Soc. 8 (1944), 29-36. (MR 6, 216.)
[6] *An elementary proof of the existence of isothermal parameters on a surface*, Proc. Amer. Math. Soc. 6 (1955), 771-782. (MR 16, 856.)

This is the standard reference, but for more details I used Bers' NYU notes on Riemann surfaces (1957-1958). See also the papers cited in the review, and in the immediately preceeding review on the same page.

Chern, S.-S. and Lashof, R.K.
[1] *On the total curvature of immersed manifolds*, Amer. J. Math. 79 (1957), 306-318. (MR 18, 927.); II, Michigan Math. J. 5 (1958), 5-12. (MR 20 #4301.)

Christoffel, E.B.
[1] *Ueber die Transformation der homogenen Differentialausdrücke zweiten Grades*, J. Reine Angew. Math. 70 (1869), 46-70.

I could not resist mentioning this extraordinarily impressive paper, which appeared just one year after the publication of Riemann's inaugural lecture. Christoffel defines covariant differentiation, uses it to define the curvature tensor, and solves the problem of determining when two Riemannian manifolds are locally isometric. This is done essentially as in Chapter V.7, Addendum 3, although the argument is not phrased in terms of the bundle of frames -- the latter formulation comes from Cartan {3}.

Cohn-Vossen, S.
[1] *Unstarre geschlossene Flächen*, Math. Ann. 102 (1929), 10-29.
[2] *Zwei Sätze über die Starrheit der Eiflächen*, Nachr. Ges. Wiss. Göttingen, Math.-Phys. Kl. 1927, 125-134.

For another proof of the infinitesimal bendability of convex surfaces with a disc deleted, see Süss [1].

Courant, R.
[1] *Soap film experiments with minimal surfaces*, Amer. Math. Monthly 47 (1940), 167-174. (MR 1, 270.)

Courant, R. and Lax, P.
[1] *On nonlinear partial differential equations with two independent variables*, Comm. Pure Appl. Math. 2 (1949), 255-273. (MR 11, 441.)

Note that the region R_δ on p. 263 is defined incorrectly -- it has very steep sides, rather than the opposite. In the footnote on p. 260 the authors make a more serious error; it amounts to asserting that the matrix A on p. V.110 need not be assumed invertible, because this can always be achieved by an affine transformation of the plane. But, in fact, this works only if the dimension of ker A is only 1. Finally, the process by which the authors reach their equation (4.6) on p. 261 is to me mysteriously complicated. For the arguments on pp. V.112-113 I used Lax's NYU notes on PDEs (1949-1950).

Cutler, E.H.
[1] *On the curvatures of a curve in Riemann space*, Trans. Amer. Math. Soc. 33(1931), 832-838.

This paper contains a generalization of Theorem IV.7-9: Let N be an arbitrary Riemannian manifold, let $M \subset N$ be a j-dimensional totally geodesic submanifold, and let $c: [a,b] \to N$ be an arclength parameterized curve with $\kappa_1, \ldots, \kappa_{j-1}$ nowhere zero, and κ_j everywhere zero. Suppose that $c(a) \in M$ and $\mathbf{v}_i(a) \in M_{c(a)}$ for $i = 1, \ldots, j$. Then $c([a,b]) \subset M$.

Dieudonné, J.A. and Carrell, J.B.
[1] *Invariant theory, old and new*, Advances in Math. 4 (1970), 1-80.

Also available in book form, see Dieudonné, J.A. and Carrell, J.B. {1}.

do Carmo, M.
see Carmo, M. do

Dolbeault-Lemoine, S.
[1] *Sur la déformabilité des variétés plongées dans espace de Riemann*, Ann. Sci. Ecole Norm. Sup. (3) 73 (1956), 357-438. (MR 18, 819.)

Eels, J. Jr.
see Allendoerfer, C.B.

Efimov, N.V.
[1] *Generation of singularities on surfaces of negative curvature*, Mat. Sb. 64 (106) (1964), 286-320. (Russian.) (MR 29 #5203.)

Ehresmann, C.
[1] *Sur la topologie de certains espaces homogènes*, Annals of Math. 35 (1934), 396-443.
[2] *Sur la topologie de certaines variétés algébriques réeles*, J. Math. Pures Appl. (9) 16 (1937), 69-100.

Fenchel, W.
[1] *Über Krummung und Windung geschlossener Raumkurven*, Math. Ann. 101 (1929), 238-252.

[2] *On total curvatures of Riemannian manifolds: I*, J. London Math. Soc. 15 (1940), 15-22. (MR 2, 20.)

Fenchel, W. and Jessen, B.

[1] *Mengenfunktionen und konvexe Körper*, Danske Videns. Selskab. Math-Fysiske Medd. 16 (1938), 1-31.

Fialkow, A.

[1] *Hypersurfaces of a space of constant curvature*, Ann. of Math. 39 (1938), 762-785.

Firey, W.J.

[1] *The determination of convex bodies from their mean radius of curvature functions*, Mathematika 14 (1967), 1-13. (MR 36 #788.)

The next paper treats Christoffel's problem for non-differentiable convex surfaces.

[2] *Christoffel's problem for general convex bodies*, Mathematika 15 (1968), 7-21. (MR 37 #5822.)

[3] *Intermediate Christoffel-Minkowski problems for figures of revolution*, Israel J. Math. 8 (1970), 384-390. (MR 42 #6719.)

This paper considers the functions P_i for $1 < i < n$.

Flanders, H.

[1] *Local theory of affine hypersurfaces*, J. Analyse Math. 15 (1965), 353-387. (MR 32 #403.)

This paper uses moving frames to describe the (local) theory of hypersurfaces M^n of $\mathbb{R}^{n+1}$ in special affine geometry. Global questions of special affine geometry for surfaces in $\mathbb{R}^3$ were first studied by Blaschke (see Blaschke {1; v.2}). Santaló [1] used moving frames, and derived integral formulas, for the same purpose, and Hsiung and Shahin [1] do the same thing for hypersurfaces in higher dimension. Section 10 of Flanders' paper gives a formula for the special affine second fundamental form $\mathfrak{U}$ -- more precisely, it gives a formula for $\mathfrak{U}(c',c',c')$ for a curve c in M (compare Blaschke {1; v.2, §46} and Hsiung and Shahin [1, §4]). Section 2 indicates one crucial difference between the cases $n = 2$ and $n \geq 3$: in the latter case, the special affine Codazzi-Mainardi equations are a consequence of the apolarity conditions; this essentially depends on the Bianchi identities (compare Blaschke {1; v.2; §65,66}). In ordinary geometry there is a similar situation for $n \geq 4$ (see Eisenhart {1; Appendix 22}).

Gardner, R.B.

[1] *Subscalar pairs of metrics and hypersurfaces with a non-degenerate second fundamental form*, J. Differential Geometry 6 (1972), 437-458.

Green, L.W.

[1] *Auf Wiedersehensflächen*, Ann. of Math. (2) 78 (1963), 289-299. (MR 27 #5206.)

Greene, R.E. and Wu, H.
[1] *On the rigidity of punctured ovaloids*, Ann. of Math. (2) 94 (1971), 1-20. (MR 44 #7490.); II, J. Differential Geometry 6 (1972), 459-472.

Gromoll, D. and Meyer, W.
[1] *On complete open manifolds of positive curvature*, Ann. of Math. (2) 90 (1969), 75-90. (MR 40 #854.)

Gromov, M.L. and Rokhlin, V.A.
[1] *Embeddings and immersions in Riemannian geometry*, Russian Math. Surveys 25 (1970), no. 5, 1-57. (MR 44 #7571.)

Grove, V.G.
[1] *On closed convex surfaces*, Proc. Amer. Math. Soc. 8 (1957), 777-786. (MR 19, 167.)

Hamburger, H.
[1] *Beweis einer Carathéodoryschen Vermutung. Teil I*, Ann. of Math. 41 (1940), 63-86. (MR 1, 172.); *II*, Acta Math. 73 (1941), 175-228; *III*, Acta Math. 73 (1941), 229-332. (MR 3, 310.)

Hartman, P. and Nirenberg, L.
[1] *On spherical image maps whose Jacobians do not change sign*, Amer. J. Math. 81 (1959), 901-920. (MR 23 #A4106.)

Hayden, H.A.
[1] *Asymptotic lines in a* V_m *in a* V_n, Proc. London Math. Soc. (2) 33 (1931-32), 22-31.

Heinz, E.
[1] *On Weyl's embedding problem*, J. Math. Mech. 11 (1962), 421-454. (MR 25 #2565.)

Hellwig, G.
[1] *Über die Verbiegbarkeit von Flächenstücken mit positiver Gauszscher Krümmung*, Arch. Math. (Basel) 6 (1955), 243-249. (MR 16, 1148.)

Hermann, R.
[1] *The second variation for minimal submanifolds*, J. Math. Mech. 16 (1966), 473-491. (MR 34 #8348.)

Chapter IV.9, Addendum 4 was based on the presentation in this paper.

Hilbert, D.
[1] *Ueber Flächen von constanter Gausscher Krümmung*, Trans. Amer. Math. Soc. 2 (1901), 87-99.

Hoesli, R.J.
[1] *Spezielle Flächen mit Flachpunkten und ihre lokale Verbiegbarkeit*, Composito Math. 8 (1950), 113-141. (MR 12, 357.)

Holmgren, E.
[1] *Sur les surfaces à courbure constante négative*, C.R. Acad. Sci. Paris Ser. A-B 134 (1902), 740-743.

Hopf, H.
[1] *Über Flächen mit einer Relation Zwischen den Hauptkrümmungen*, Math. Nachr. 4 (1951), 232-249. (MR 12, 634.)
[2] *Über die Curvatura integra geschlossener Hyperflächen*, Math. Ann. 95 (1925), 340-367.

Hopf, H. and Schilt, H.
[1] *Über Isometrie und stetige Verbiegung von Flächen*, Math. Ann. 116 (1939), 58-75.

Hopf, H. and Voss, K.
[1] *Ein Satz aus der Flächentheorie im Grossen*, Arch. Math. (Basel) 3 (1952), 187-192. (MR 14, 583.)

This is the source for Theorem V.12-17; it relates this result to another, apparently quite different one.

Horn, R.A.
[1] *On Fenchel's theorem*, Amer. Math. Monthly 78 (1971), 380-381. (MR 44 #2142.)

Hsiung, C.-C. and Shahin, J.K.
[1] *Affine differential geometry of closed hypersurfaces*, Proc. London Math. Soc. (3) 17 (1967), 715-735. (MR 36 #2069.)

Jacobowitz, H.
[1] *Local isometric embeddings of surfaces into Euclidean four space*, Indiana Univ. Math. J. 21 (1971/72), 249-254. (MR 46 #6247.)
[2] *Extending isometric embeddings*, (Proc. Symp. Pure Math. 27 (1974), A.M.S., to appear).

This paper, together with helpful hints from the author, was the source for the proof of Theorem V.11-9.

Jacobowitz, H. and Moore, J.D.
[1] *The Cartan-Janet theorem for conformal embeddings*, Indiana Univ. Math. J. 23 (1973), 187-203.

This paper considers the problem of conformal embedding from two different approaches -- the first using the methods to be found in the proof of Theorem V.11-9, and the second using the methods of Chapter V.11, Addendum. There is a long discussion of differential systems which should be useful for reading Cartan {4}.

Jessen, B.
see Fenchel, W.

Klein, F.

[1] *Vergleichende Betrachtungen über neuere geometrische Forschungen,* Math. Ann. 43 (1893), 63-100.

Also in volume 1, pp. 460-497 of Klein's Gesammelte Mathematische Abhandlungen, 3 vols., Springer, Berlin, 1921-23, and in English translation in Bull. N.Y. Math. Soc. 2 (1892), p. 215.

Klotz, T.

[1] *On G. Bol's proof of Carathéodory's conjecture*, Comm. Pure Appl. Math. 12 (1959), 277-311. (MR 22 #11352.)

Klotz Milnor, T.

[2] *Efimov's theorem about complete immersed surfaces of negative curvature*, Advances in Math. 8 (1972), 474-543.

Knebelman, M.S.

[1] *Contact transformations*, Ann. of Math. 2 (39) (1938), 507-515.

Kuiper, N.H.

[1] *On C^1-isometric imbeddings. I, II*, Nederl. Akad. Wetensch. Proc. Ser. A. 58 = Indag. Math. 17 (1955), 545-556, 683-689. (MR 17, 782.)

The results of this paper are strengthened somewhat in the next.

[2] *Isometric and short imbeddings*, Nederl. Akad. Wetensch. Proc. Ser. A. 62 = Indag. Math. 21 (1959), 11-25. (MR 20 #7316.)

[3] *On surfaces in euclidean three-space*, Bull. Soc. Math. Belg. 12 (1960), 5-22. (MR 23 #A609.)

This paper is the source for the arguments on pp. III.413-421.

Lashof, R.K.

see Chern, S.-S.

Lawson, H.B.

[1] *Complete minimal surfaces in S^3*, Ann. of Math. (2) 92 (1970), 335-374. (MR 42 #5170.)

Lax, P.

see Courant, R.

Levi, E.E.

[1] *Sulla deformazione delle superficie flessibili ed inestendibili*, Atti Accad. Torino 43 (1907-1908), 292-302.

Lewy, H.

[1] *An example of a smooth linear partial differential equation without solution*, Ann. of Math. (2) 66 (1957), 155-158. (MR 19, 551.)

[2] *On the existence of a closed surface realising a given Riemannian metric*, Proc. Nat. Acad. Sci. USA 24 (1938), 104-106.

[3] *On differential geometry in the large I (Minkowski's problem)*, Trans. Amer. Math. Soc. 43 (1938), 258-270.

[4] *Über das Anfanswertproblem einer hyperbolischen nichtlinearen partiellen Differentialgleichung zweiter Ordnung mit zwei unabhängigen Veränderlichen*, Math. Ann. 98 (1927), 179-191.

[5] *Neuer Beweis des analytischen Charakters der Lösungen elliptscher Differentialgleichungen*, Math. Ann. 101 (1929), 609-619.

The last two papers are the source for Chapter V.10, sections 8 and 9.

Liebmann, H.

[1] *Ein Satz über endliche einfach zusammenhängende Flächenstücke negativer Krümmung*, Berichte über die Verhandlungen der Könglich Sächsischen Gesellschaft der Wissenschaften zu Leipzig, Mathematisch-physische classe 52 (1900), 28-36.

[2] *Die Verbiegung von geschlossenen und offenen Flächen positiver Krümmung*, Bayer. Akad. Wiss. Math.-Physik Kl. S.-B. (Munich) 1919, 267-291.

[3] *Ueber die Verbiegung der geschlossenen Flächen positiver Krümmung*, Math. Ann. 53 (1900), 81-112.

This paper gives a geometric proof of the infinitesimal rigidity of convex surfaces, as well as an early proof that the sphere is unwarpable (Theorem III.5-2).

Lima, E.

see Carmo, M. do

Massey, W.S.

[1] *Surfaces of Gaussian curvature zero in Euclidean 3 space*, Tôhoku Math. J. (2) 14 (1962), 73-79. (MR 25 #2527.)

This paper gives the first of our three proofs of Theorem III.5-9. For other formulations of the theorem, see Hartman and Nirenberg [1] and Pogorelov [3], [4].

Meyer, W.

see Gromoll, D.

Moore, J.D.

[1] *Isometric immersions of Riemannian products*, J. Differential Geometry 5 (1971), 159-168. (MR 46 #6249.)

[2] *Isometric immersions of space forms in space forms*, Pacific J. Math. 40 (1972), 157-166. (MR 46 #4442.)

This is the source for Problem V.11-2.

see also Jacobowitz, H.

Nash, J.

[1] *C^1 isometric imbeddings*, Ann. of Math. (2) 60 (1954), 383-396. (MR 16, 515.)

[2] *The imbedding problem for Riemannian manifolds*, Ann. of Math. (2) 63 (1956), 20-63. (MR 17, 782.)

Nirenberg, L.

[1] *The Weyl and Minkowski problems in differential geometry in the large*, Comm. Pure Appl. Math. 6 (1953), 337-394. (MR 15, 347.)

[2] *Rigidity of a class of closed surfaces*, Nonlinear Problems (Proc. Sympos., Madison, Wis., 1962), pp. 177-193. Univ. of Wisconsin Press, Madison, Wis., 1963. (MR 27 #697.)

see also Hartman, P.

Nitsche, J.C.C.

[1] *Elementary proof of Bernstein's theorem on minimal surfaces*, Ann. of Math. (2) 66,(1957), 543-544. (MR 19, 878.)

O'Neil, B.

[1] *Isometric immersions which preserve curvature operators*, Proc. Amer. Math. Soc. 13 (1962), 759-763. (MR 26 #721.)

Olowjanischnikow, S.P.

[1] *On the bending of infinite convex surfaces*, Mat. Sb. 18 (60) (1946), 429-440. (Russian. English summary.) (MR 8, 169.)

Ostrowski, A.

[1] *Un'applicazione dell'integrale di Stieltjes alla teoria elementare delle curve piane*, Atti Accad. Naz. Lincei. Rend. Cl. Sci. Fis. Mat. Nat. (8) 18 (1955), 373-375. (MR 17, 780.)

[2] *Über die Verbindbarkeit von Linien- und Krümmungselementen durch monoton gekrümmte Kurvenbögen*, Enseignement Math. (2) 2 (1956), 277-292. (MR 19, 58.)

Perron, O.

[1] *Über Existenz und Nichtexistenz von Integralen partieller Differentialgleichungssysteme im reelen Gebiet*, Math. Z. 27 (1928), 549-564.

Pick, G.

[1] *Über affine Geometrie IV: Differentialinvarianten der Flächen gegenüber affinen Transformationen*, Berichte über die Verhandlungen der Königlich Sächsischen Gesellschaft der Wissenschaften zu Leipzig, Mathematisch-physische Klasse 69 (1917), 107-136.

Pogorelov, A.W.

[1] *Isometric transformation of punctured convex surfaces*, Dokl. Akad. Nauk SSSR 137 (1961), 1307-1308. = Soviet Math. 2, 475-476. (MR 25 #1520.)

[2] *On the rigidity of general infinite convex surfaces with integral curvature* 2π, Dokl. Akad. Nauk SSSR 106 (1956), 19-20. (Russian.) (MR 17, 888.)

[3] *Continuous maps of bounded variations*, Dokl. Akad. Nauk SSSR 111 (1956), 757-759. (Russian.) (MR 19, 309.)

[4] *Extensions of the theorem of Gauss on spherical representation to the case of surfaces of bounded extrinsic curvature*, Dokl. Akad. Nauk SSSR 111 (1956), 945-947. (Russian.) (MR 19, 309.)

Rembs, E.
[1] *Zur Verbiegung von Flächen im Grossen*, Math. Z. 56 (1952), 271-279. (MR 14, 901.)

Rokhlin, V.A.
see Gromov, M.L.

Ryan, P.J.
[1] *Homogeneity and some curvature conditions for hypersurfaces*, Tôhoku Math. J. (2) 21 (1969), 363-388. (MR 40 #6458.)

Saban, G.
[1] *Nuove caratterizzazioni della sfera*, Atti Accad. Naz. Lincei. Rend. Cl. Sci. Fis. Mat. Nat. 25 (1958), 457-464. (MR 21 #5967.)

Sacksteder, R.
[1] *On hypersurfaces with no negative sectional curvatures*, Amer. J. Math. 82 (1960), 609-630. (MR 22 #7087.)

Santaló, L.A.
[1] *Geometria diferencial afin y cuerpos convexos*, Math. Notae 16 (1958), 20-42.

Scherrer, W.
[1] *Eine Kennzeichnung der Kugel*, Vierteljschr. Naturforsch. Ges. Zürich 85 Beiblatt (Festschrift Rudolf Feuter) 1940, 40-46. (MR 3, 89.)

Schilt, H.
[1] *Über die isolierten Nullstellen der Flächenkrümmung und einige Verbiegbarkeitssätze*, Composito Math. 5 (1937), 239-283.
see also Hopf, H.

Shahin, J.K.
see Hsiung, C.C.

Stoker, J.J.
[1] *Über die Gestalt der positiv gekrümmten offenen Flächen im dreidimensionalen Raume*, Composito Math. 3 (1936), 55-89.

Süss, W.
[1] *Zur relativen Differentialgeometrie III: Über Relativ-Minimalflächen und Verbiegung*, Japan J. Math. 4 (1928), 203-207.

Vermeil, H.
[1] *Bestimmung einer quadratischen Differentialform aus der Riemannschen und den Christoffelschen Differentialinvarianten mit Hilfe von Normalkoordinaten*, Math. Ann. 79 (1918), 289-312.

Volkov, Ju. A. and Vladimirova, S.M.
[1] *Isometric immersions of the Euclidean plane in Lobačevskiĭ space*, Mat. Zametki 10 (1971), 327-332. = Math. Notes 10 (1971), 619-622. (MR 45 #2624.)

Voss, K.
[1] *Eine Bemerkung über die Totalkrümmung geschlossener Raumkurven*, Arch. Math. (Basel) 6 (1955), 259-263. (MR 17, 75.)
see also Hopf, H.

Walden, R.
[1] *Eindeutigkeitssätze für II-isometrische Eiflächen*, Math. Z. 120 (1971), 143-147. (MR 44 #2182.)

Warner, F.W.
see Carmo, M. do

Weil, A.
see Allendoerfer, C.B.

Weiner, J.L.
[1] *Closed curves of constant torsion*, (Proc. Symp. Pure Math. 27 (1974), A.M.S., to appear).

Wu, H.
see Greene, R.E.

Wunderlich, W.
[1] *Über ein abwickelbares Möbiusband*, Monatsch. Math. 66 (1962), 276-289. (MR 26 #680.)

NOTATION INDEX

Chapter 13 (cont.)

INDEX

(The index does not include entries for the Bibliography.)

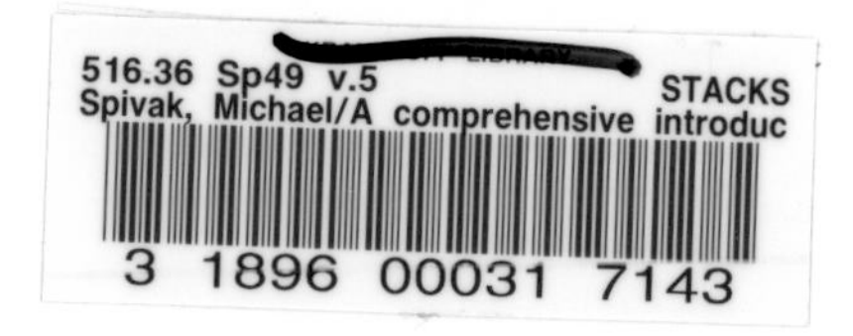